Ceramic nanocomposites

Related titles:

Advances in ceramic matrix composites
(ISBN 978-0-85709-120-8)

Sintering of advanced materials
(ISBN 978-1-84569-562-0)

High-energy ball milling
(ISBN 978-1-84569-531-6)

Details of these books and a complete list of titles from Woodhead Publishing can be obtained by:

- visiting our web site at www.woodheadpublishing.com
- contacting Customer Services (e-mail: sales@woodheadpublishing.com; fax: +44 (0) 1223 832819; tel.: +44 (0) 1223 499140 ext. 130; address: Woodhead Publishing Limited, 80 High Street, Sawston, Cambridge CB22 3HJ, UK)
- in North America, contacting our US office (e-mail: usmarketing@woodheadpublishing.com; tel.: (215) 928 9112; address: Woodhead Publishing, 1518 Walnut Street, Suite 1100, Philadelphia, PA 19102-3406, USA)

If you would like e-versions of our content, please visit our online platform: www.woodheadpublishingonline.com. Please recommend it to your librarian so that everyone in your institution can benefit from the wealth of content on the site.

We are always happy to receive suggestions for new books from potential editors. To enquire about contributing to our Composites Science and Engineering series, please send your name, contact address and details of the topic/s you are interested in to francis. dodds@woodheadpublishing.com. We look forward to hearing from you.

The team responsible for publishing this book:

Commissioning Editor: Jess Rowley
Publications Coordinator: Lucy Beg
Project Editor: Diana Paulding
Editorial and Production Manager: Mary Campbell
Production Editor: Adam Hooper
Project Manager: Sheril Leich
Copyeditor: Fiona Chapman
Proofreader: Lou Attwood
Cover Designer: Terry Callanan

Woodhead Publishing Series in Composites Science and Engineering:
Number 46

Ceramic nanocomposites

Edited by
Rajat Banerjee and Indranil Manna

WOODHEAD
PUBLISHING

Oxford   Cambridge   Philadelphia   New Delhi

Published by Woodhead Publishing Limited,
80 High Street, Sawston, Cambridge CB22 3HJ, UK
www.woodheadpublishing.com
www.woodheadpublishingonline.com

Woodhead Publishing, 1518 Walnut Street, Suite 1100, Philadelphia, PA 19102-3406, USA

Woodhead Publishing India Private Limited, 303 Vardaan House, 7/28 Ansari Road, Daryaganj, New Delhi – 110002, India
www.woodheadpublishingindia.com

First published 2013, Woodhead Publishing Limited
© Woodhead Publishing Limited, 2013. The publisher has made every effort to ensure that permission for copyright material has been obtained by authors wishing to use such material. The authors and the publisher will be glad to hear from any copyright holder it has not been possible to contact.
The authors have asserted their moral rights.

British Library Cataloguing in Publication Data
A catalogue record for this book is available from the British Library.

Library of Congress Control Number: 2013938683

ISBN 978-0-85709-338-7 (print)
ISBN 978-0-85709-349-3 (online)

The publisher's policy is to use permanent paper from mills that operate a sustainable forestry policy, and which has been manufactured from pulp which is processed using acid-free and elemental chlorine-free practices. Furthermore, the publisher ensures that the text paper and cover board used have met acceptable environmental accreditation standards.

Typeset by Data Standards Ltd, Frome, Somerset, UK
Printed by Lightning Source

Contents

Contributor contact details

(* = main contact)

Editors

Dr Rajat Banerjee
Central Glass and Ceramic
  Research Institute
Jadavpur
Kolkata 700032
India
E-mail: rajatbanerjee@hotmail.com

Dr Indranil Manna
Indian Institute of Technology
Kharagpur
India
E-mail: imanna@cgcri.res.in

Chapter 1

Dr Nripati Ranjan Bose
Consultant
Sols 4 All Consultants
UTSA(Luxury)-The Condoville
Flat No. LX-30204, P.O. New
  Town
Rajarhat, Kolkata 700156
India
E-mail: dr.nrbose@gmail.com

Chapter 2

Professor D. Dutta Majumder*
Director–Secretary
Institute of Cybernetics Systems and
  Information Technology;
Professor Emeritus, ECSU
Indian Statistical Institute
155 Ahokegarh
Kolkata-108
India
E-mail: ddmdr@hotmail.com

Debosmita Dutta Majumder and
  Sankar Karan
Institute of Cybernetics Systems and
  Information Technology
Kolkata
India
E-mail: smita80@gmail.com;
  sankar.karan@g-mail.com

Chapter 3

Dr Rajat Banerjee
Central Glass and Ceramic
  Research Institute
Jadavpur
Kolkata 700032
India
E-mail: rajatbanerjee@hotmail.com

Chapter 4

Dr Pavol Hvizdoš
Institute of Materials Research
Slovak Academy of Sciences
Watsonova 47
04353 Kosice
Slovakia
E-mail: hvizdosp@yahoo.com

Chapter 5

Professor Vikas Tomar
Associate Professor
School of Aeronautics and
    Astronautics
Purdue University
West Lafayette, IN
47907
USA
E-mail: tomar@purdue.edu

Chapter 6

Ms Fei He
Mechanical Engineering
    Technology
Purdue University
P.O. Box 481160
Niles, IL
60714
USA
E-mail: fayeengineer@gmail.com

Chapter 7

Professor Aldo R. Boccaccini*
Institute of Biomaterials
University of Erlangen-Nuremberg
Cauerstrasse 6
91058
Erlangen
Germany

E-mail: aldo.boccaccini@ww.
    uni-erlangen.de

Dr Tayyab Subhani
Center for Advanced Structural
    Ceramics
Department of Materials
Imperial College London
South Kensington Campus
London
SW7 2AZ
UK
E-mail: t.subhani08@imperial.ac.uk

Professor Milo S. P. Shaffer
Department of Chemistry
Imperial College London
South Kensington Campus
London
SW7 2AZ
UK
E-mail: m.shaffer@imperial.ac.uk

Chapter 8

Dr Xinhua Liang*
Department of Chemical and
    Biological Engineering
University of Colorado
Boulder, CO
80309-0424
USA
E-mail: xinhua.liang@colorado.edu

Current mailing address:
Department of Chemical and
    Biochemical Engineering
Missouri University of Science and
    Technology
Rolla, MO
65409-1230
USA
E-mail: liangxin@mst.edu

Dr David M. King and Professor
    Alan W. Weimer
Department of Chemical and
    Biological Engineering
University of Colorado
Boulder, CO
80309-0424
USA
E-mail: david.m.king@colorado.
    edu; alan.weimer@colorado.edu

Chapter 9

Professor Anahit O. Tonoyan* and
    Professor S. P. Davtyan
Department of Chemical
    Technology and Ecological
    Engineering
State Engineering University of
    Armenia
105 Teryan Str
0009 Yerevan
Armenia
E-mail: atonoyan@seua.am;
    atonoyan@mail.ru

Chapter 10

Dr Indranil Manna*
Indian Institute of Technology
Kharagpur
India
E-mail: imanna@cgcri.res.in

Dr Gayatri Paul
CSIR-Central Glass and Ceramic
    Research Institute
Jadavpur
Kolkata-32
India

Chapter 11

Dr Indranil Manna*
Indian Institute of Technology
Kharagpur
India
E-mail: imanna@cgcri.res.in

Dr Gayatri Paul
CSIR-Central Glass and Ceramic
    Research Institute
Jadavpur
Kolkata-32
India

Chapter 12

Professor Krystyna Wieczorek-
    Ciurowa
Faculty of Chemical Engineering
    and Technology
Cracow University of Technology
Warszawska 24
31-155 Kraków
Poland
E-mail: kwc@pk.edu.pl

Chapter 13

Dr Zak Fang
University of Utah
USA
E-mail: zak.fang@utah.edu

Chapter 14

Professor Philippe Dubois
University of Mons and Materia
    Nova Research Centre
Belgium
E-mail: philippe.dubois@
    umons.ac.be

Chapter 15

Professor Binod Kumar
University of Dayton Research
  Institute
KL 502C
300 College Park
Dayton, OH
45469-0170
USA
E-mail: Binod.Kumar@udri.
  udayton.edu

Chapter 16

Dr Nere Garmendia
FideNa
Centro de I + D "Jerónimo de
  Ayanz"
Campus de Arrosadía
C/Tajonar s/n
31006 Pamplona
Navarra
Spain
E-mail: nere.garmendia@fidena.es

Chapter 17

Professor Masami Okamoto
Advanced Polymeric
  Nanostructured Materials
  Engineering
Graduate School of Engineering
Toyota Technological Institute
2-12-1 Hisakata
Tempaku
Nagoya 468-8511
Japan
E-mail: okamoto@toyota-ti.ac.jp

Woodhead Publishing Series in Composites Science and Engineering

Part I

Properties

Thermal shock resistant and flame retardant ceramic nanocomposites

N . R . B O S E , Sols 4 All Consultants, India

DOI: 10.1533/9780857093493.1.3

Abstract: This chapter discusses the performance behaviour of ceramic nanocomposites under conditions of thermal shock, i.e. when they are subjected to sudden changes in temperature during either heating or cooling or may be in flame propagating zones. For example, during emergency shut-downs of gas turbines, cool air is drawn from the still spinning compressor and driven through the hot sections: the temperature at the turbine outlet decreases by more than 800°C within one second and ceramic nanocomposite materials are an appropriate choice for such application. Furthermore, such a situation may arise about 100 times during the lifetime of a modern gas turbine engine. Similarly, in the nuclear industries, apart from the moderate shocks inflicted during start-up and shut-down of the system, the plasma-facing material can suffer rapid heating due to plasma discharge. Thus, when a body is subjected to a rapid temperature change such that non-linear temperature gradients appear, stresses arise due to the differential expansion of each volume element at a different temperature. The design principles for the fabrication of high-performance thermal shock resistant ceramic nanocomposites with improved mechanical properties are highlighted in this chapter. Moreover, the pertinent factors such as interface characteristics, densification methods, superplasticity and the role of nano-size particulate dispersion, which are responsible for the development of thermal shock resistant and flame retardant nanoceramic materials, are addressed and reviewed. Various test methods for the characterisation and evaluation of ceramic nanocomposites are described. Finally, the new concept of materials design for future structural ceramic nanocomposites is discussed for safe applications in high-temperature thermal shock zones.

Key words: thermal shocks, densification, superplasticity, interface, ceramic nanocomposites.

1.1   Introduction

Dramatic improvements in toughness, strength, creep strength and thermoresistance of ceramic-matrix composites have been achieved by incorporating either nanocrystalline oxide/non-oxide ceramic particles or their hybrid combination in a microcrystalline matrix. The other group of nanophase ceramic composites is nanocrystalline matrix composites, also called nanoceramics, in which the matrix grain size is below 100 nm. The nano–nano type microstructure can also be formed when the second phase is also nanoscaled. The main objective of this chapter is to present the performance behaviour of ceramic nanocomposites under conditions of thermal shock, i.e. when they are subjected to sudden changes in temperature during either heating or cooling or may be in flame propagating zones. Such conditions are possible in the high-temperature applications for which ceramic nanocomposite materials can be selected (Ohnabe *et al.* 1991). The importance of the use of thermal shock resistant ceramic nanocomposite materials was reported by Baste in 1993 during steady-state operation of a gas turbine (Baste 1993). While thermal shock is not a concern during steady-state operation of a gas turbine, it becomes of great importance during emergency shut-downs, when cool air drawn from the still spinning compressor is driven through the hot sections and can result in a temperature decrease of more than 800°C at the turbine inlet within one second. An additional factor is that such a situation may arise about 100 times during the lifetime of a modern gas turbine engine.

The use of thermal shock resistant ceramic nanocomposite materials was also illustrated by Jones *et al.* (2002) for fusion energy applications. In the nuclear industries, SiC reinforced with SiC fibres has been proposed as a structural material for the first wall and blanket in several conceptual design studies of future fusion power reactors. In this case, apart from the moderate shocks inflicted during start-up and shut-down of the system, the plasma-facing material can suffer rapid heating due to plasma discharge. When a body is subjected to a rapid temperature change such that a non-linear temperature gradient appears, stresses arise due to the differential expansion of each volume element at a different temperature. The temperature at each point changes with time at a rate that depends on the surface heat transfer coefficient (HTC) between the media at different temperature and the body, the shape of the body and its thermal conductivity. High HTCs, large dimensions and low thermal conductivities result in large temperature gradients and, thus, large stresses. The dimensionless parameter, the Biot modulus (Bi), can be used to describe the heat transfer condition (Kastritseas *et al.* 2006)

$$Bi = l(h/k) \qquad\qquad [1.1]$$

where l is a characteristic material dimension (e.g. the half-thickness of a plate), h is the HTC between the body and the medium and k is the thermal conductivity of the body. The larger the value of Bi, the larger is the rate of heat transfer between a medium of different temperature and the body. The sudden temperature change (ΔT) that generates non-linear temperature gradients in a body and, as a consequence, thermal stresses is termed 'thermal shock'. If ΔT is positive (i.e. the temperature reduces) the material is subjected to a cold shock, whereas if ΔT is negative the material is subjected to a hot shock. The term refers to a single thermal cycle ($N = 1$) in contrast to terms such as thermal cycling, cyclic thermal shock and thermal fatigue, which apply to multiple thermal cycles ($N > 1$) (Kastritseas et al. 2006).

Although ceramic materials based on oxides, non-oxides, nitrides, carbides of silicon, aluminium, titanium and zirconium, alumina, etc. possess some very desirable characteristics (e.g. high strength and hardness, excellent high-temperature structural applications, chemical inertness, wear resistance and low density), they are not very robust under tensile and impact loading and, unlike metals, they do not show any plasticity and are prone to catastrophic failure under mechanical or thermal loading (thermal shock) (Banerjee and Bose 2006). Many researchers have focused attention on both oxide and non-oxide ceramic materials for the improvement of microstructure. However, many problems are not yet solved. For instance, non-oxide ceramic materials such as Si_3N_4 and SiC suffer from degradation of mechanical properties at high temperature due to slow crack growth caused by the softening of grain boundary impurity phases associated with sintering additives. Oxide ceramic materials such as Al_2O_3, MgO and ZrO_2 suffer from relatively low fracture toughness and strength, significant strength degradation at high temperatures and poor creep, fatigue and thermal shock resistance.

Attempts have been made by many researchers to solve these problems, as well as brittleness problems, by incorporating secondary phases such as particulates, platelets, whiskers and fibres in the ceramic materials (Becher and Wei 1984, Buljun et al. 1987, Claussen 1985, Greskovich and Palm 1980, Homeney et al. 1987, Izaki et al. 1980, Lange 1973, 1974, 1982, Nakahira et al. 1987, Niihara et al. 1986, Nishida et al. 1987, Shalek et al. 1986, Tiegs and Bechere 1987, Wahi and Ilschner 1980). For non-oxide ceramic materials such as Si_3N_4, SiC has been incorporated for a second-phase dispersion to take advantage of the characteristics of both phases. (Becher and Wei 1984, Homeney et al. 1987, Nakahira et al. 1987, Tiegs and Bechere 1987). Recently, the Si_3N_4/SiC whisker composite system has demonstrated significant improvements in fracture toughness and high-temperature fracture behaviour, but with a decrease in fracture strength (Becher and Wei 1984, Nakahira et al. 1987) For the SiC particulate dispersion system, Lange (1973) investigated the mechanical properties of the system by dispersing SiC particles (average size: 5,9,32 µm) to Si_3N_4 matrix, which showed that the fracture energy increased in the largest particle size dispersion series whereas the system was inferior in room-temperature fracture strength. Greskovich and Palm (1980) reported that the fracture toughness of the Si_3N_4/SiC composite system, upon dispersing submicron particles of SiC, was independent of the volume fraction of SiC. There seem to be no reported investigations that have succeeded in improving toughness, strength and high-temperature fracture behaviour at the same time. For oxide ceramic materials such as Al_2O_3, MgO and ZrO_2, second-phase dispersions have also been added (Banerjee and Bose 2006, Claussen 1985, Kastritseas et al. 2006, Lange 1974, 1982, Niihara et al. 1986, Uchiyama et al. 1989, Wahi and Ilschner 1980). ZrO_2 dispersion has been observed to dramatically improve the fracture toughness and strength of Al_2O_3 due to its stress-induced transformation (Lange 1982, Wahi and Ilschner 1980). The significant increase in both fracture toughness and strength of Al_2O_3 and MgO has also been achieved by incorporating SiC whiskers (Banerjee and Bose 2006, Claussen 1985, Kastritseas et al. 2006, Lange 1974, 1982, Niihara et al. 1986, Uchiyama et al. 1989, Wahi and Ilschner 1980). However, ZrO_2 reinforced composites show strength degradation at low temperatures, while the SiC whisker reinforced composites are difficult to prepare by pressureless sintering.

To overcome these problems, Niihara and colleagues have investigated fabrication processes and developed ceramic nanocomposites (Izaki et al. 1991, Niihara 1990, 1989, Niihara and Nakahira 1988, 1990a, 1990b, Niihara et al. 1988a, 1989a, 1990a,1990b, 1999, Suganuma et al. 1991, Wakai et al. 1990). They significantly improved mechanical properties even at high temperatures by incorporating nano-size dispersoids within the matrix grains and at the grain boundaries. There are also several materials scientists all over the world investigating various novel fabrication processes and who have developed ceramic nanocomposites with high-temperature strength, wear resistance, toughness, hardness, chemical inertness, flame retardancy and thermal shock resistance. In this chapter, the design principles for the fabrication of high-performance thermal shock resistant ceramic nanocomposites with improved mechanical properties will be highlighted. The pertinent factors such as interface characteristics, densification methods and the role of nano-size particulate dispersions, which are responsible for the development of thermal shock resistant and flame retardant nanoceramic materials, will also be addressed and reviewed. Various test methods for the characterization and evaluation of ceramic nanocomposites will be described. Finally, the new concept of materials design for future structural ceramic nanocomposites will be discussed for safe applications in high temperature induced thermal shock zones.

1.2   Design of thermal shock resistant and flame retardant ceramic nanocomposites

Various design methodologies have been used by materials scientists for the development of thermal shock resistant and flame retardant ceramic nanocomposites. In such design it is necessary to consider the cause and effect of various parameters such as processing, the functions of selective ingredients, interface conditions, grain sizes, grain boundary sliding, creep resistance, oxidation resistance, densification of ingredients, dispersions, strengthening and toughening mechanisms, thermal expansion mismatch at the grain boundaries, etc. Materials scientists are considering all such viable parameters in the design of various types of ceramic nanocomposite materials that will be particularly effective in thermal shock and flame retardant applications.

1.2.1   Design of sintering techniques

Ceramic nanocomposites were first successfully designed by the chemical vapour deposition (CVD) (Niihara and Hirai 1986). TiN particles or whiskers of diameter approximately 5 nm were dispersed within Si_3N_4 matrix grains. The CVD process is a very suitable method to disperse nano-size second phases into matrix grains or at grain boundaries. However, this process is not applicable for the fabrication of large and complex-shaped components for mass production and it is also very expensive. There are some low-cost but effective processes to obtain nano-sized ceramic powders and nanocomposite powders, such as sol–gel, micro emulsion, autoignition, co-deposition and high-energy ball milling (HEBM). One of the principal problems is the ability to consolidate nanopowders to high relative density without grain growth. To obtain dense bulk nanoceramics, it is essential to decrease either the sintering temperature or retaining time at the highest point, or to employ hot-pressing (HP), gas pressure sintering (GPS) or fast consolidation techniques such as microwave sintering and spark plasma sintering (SPS). Pressureless sintering can be employed to fabricate complex shapes, and significant stabilization can be achieved by using a protective powder bed while the products generally indicate low density and the process requires large amounts of additives for densification (Munakata *et al.* 1986). In hot-pressing, powder mixtures with additives are heated to high temperatures under an applied uniaxial pressure. Traditional hot-pressing uses 20–30 MPa pressure, which enhances both the rearrangement of particles and grain boundary diffusion. Hot-pressing offers the ability to fabricate dense products, but also limits the products to simple shapes (Hwang and Chen 1994). Other promising established techniques are isostatic pressure and hot isostatic pressing (HIP). These techniques are very attractive for making dense ceramic products with a negligible residual glassy grain boundary phase and hence better high-temperature properties. However, HP and HIP techniques are very costly. They become economic only when a large number of samples are involved (Niihara and Hirai 1986). Composite formation with nanostructured coatings based on ceramic powders is very attractive for cutting tools and wear-resistant tools. Pulsed electric current sintering (PECS) technique can be used for the fabrication of such nanocomposites.

1.2.2 Thermal shock and flame retardant behaviour of ceramic nanocomposites

Ceramic materials have a greater thermal shock sensitivity than metals and can suffer catastrophic failure due to thermal shock because of their unfavourable ratio of stiffness and thermal expansion to strength and thermal diffusivity, and their limited plastic deformation. The stress field that develops in a thermally shocked component can be explained by calculating the thermal shock induced stresses along the x- and y-axes of a plate. Considering a plate with Young's modulus E, Poisson's ratio v and coefficient of thermal expansion (CTE) α, initially held at temperature T_i: if the top and bottom surfaces of the plate come into sudden contact with a medium of lower temperature $T\infty$ they will cool and try to contract. However, the inner part of the plate initially remains at a higher temperature, which hinders contraction of the outer surfaces, giving rise to tensile surface stresses balanced by a distribution of compressive stresses at the interior. By contrast, if the surfaces come into contact with a medium of higher temperature $T\infty$ they will try to expand. As the interior will be at a lower temperature, it will constrain expansion of the surfaces, thus giving rise to compressive surface stresses balanced by a distribution of tensile stresses at the interior (Kastritseas et al. 2006).

If perfect heat transfer between the surfaces and the medium is assumed (i.e. if $B_i \to \infty$) the surface immediately adopts the new temperature while the interior of the plate remains at T_i. Following Munz and Fett (1999), this case corresponds to having a plate that can expand freely in the z-direction with suppressed expansion in the x- and y-directions. In the absence of displacement restrictions, the plate would expand along the x- and y-directions by thermal strains of:

$$\varepsilon_x = \alpha(T\infty - T_i) \qquad\qquad [1.2]$$

$$\varepsilon_y = \alpha(T\infty - T_i) \qquad\qquad [1.3]$$

Since thermal expansion in both directions is completely suppressed, elastic strains are created that compensate the thermal strains, i.e.

$$\varepsilon_{el,x} + \varepsilon_{th,x} \qquad\qquad [1.4]$$

$$\varepsilon_{el,y} + \varepsilon_{th,y} \qquad\qquad [1.5]$$

Equations 1.1–1.5 yield

$$\varepsilon_{el,x} = -\varepsilon_{th,x} = -\alpha(T_\infty - T_i) = \alpha(T_i - T\infty) = \alpha\Delta T \qquad [1.6]$$

$$\varepsilon_{el,y} = -\varepsilon_{th,y} = -\alpha(T_\infty - T_i) = \alpha(T_i - T\infty) = \alpha\Delta T \qquad [1.7]$$

The elastic strains cause 'thermal stresses' along the x- and y- axes and can be written as:

$$\varepsilon_{el,x} = (\sigma_x^{TS}/E) - (v\sigma_y^{TS})/E \qquad\qquad [1.8]$$

$$\varepsilon_{el,y} = (\sigma_y^{TS}/E) - (v\sigma_x^{TS})/E \qquad\qquad [1.9]$$

By substituting equations 1.6 and 1.7 into equations 1.8 and 1.9 respectively and solving first for σ_x^{TS} and then for σ_y^{TS}, we can obtain the thermal shock induced stresses along the x- and y-axes as:

$$\sigma_x^{TS} = \sigma_y^{TS} = E\alpha\Delta T/(1 - v) \qquad\qquad [1.10]$$

Equation 1.10 shows that thermal shock induces a biaxial stress field whose maximum value depends on the elastic properties of the material and the imposed temperature differential. However, if the rate of heat transfer is not infinite, the thermal shock induced stresses will gradually build up and after some time will reach a peak value that will be a fraction of the value given by equation 1.10. The solution requires detailed transient stress analysis as reported elsewhere (Becher 1981, Becher and Warwick 1993, Becher *et al.* 1980, Cheng 1951, Lu and Fleck 1998, Manson 1966, Wang and Singh 1994).

1.3 Types and processing of thermally stable ceramic nanocomposites

The concept of thermally stable structural ceramic nanocomposites was first proposed by Niihara in 1991. Niihara (1991) divided the nanocomposites into three types – intragranular, intergranular and nano–nano composites. In the intra- and intergranular nanocomposites, nano-sized particles were dispersed mainly within the matrix grains or at the grain boundaries of the matrix, respectively. The main purpose of these composites was to improve not only mechanical properties such as hardness, fracture strength, toughness and reliability at room temperature but also high-temperature mechanical properties such as hardness, strength, and creep and fatigue fracture resistances. The nano–nano composites were composed of dispersoids and matrix grains of nanometre-size. The primary purpose of this composite was to add new functions such as machinability and superplasticity (as in metals) to ceramics.

1.3.1 Processing

The processing of thermally stable ceramic nanocomposites is based mainly on results obtained on Al_2O_3/SiC and Si_3N_4/SiC systems. The Al_2O_3/SiC nanocomposite system has been researched intensively because it has been reported to have significantly improved mechanical properties over basic Al_2O_3 (Davidge et al. 1997, Jhao et al. 1993, Niihara, 1991). In Al_2O_3/SiC nanocomposites, SiC 'nano' phases of order 100–200 nm in size were added into the Al_2O_3 matrix (typically 1–15 vol% SiC) and dispersed during sintering or hot-pressing within the matrix grains or at the grain boundaries. Although hot-pressed Al_2O_3/SiC materials have been reported to have the most improved mechanical properties, pressureless sintered Al_2O_3/SiC nanocomposites also have shown improvements in mechanical properties (Anya and Roberts 1996, 1997, Jeong and Niihara K, 1997, Stearns et al. 1992). Processing of Al_2O_3 with 1, 2.5 and 5 vol% SiC nanocomposites by pressureless sintering has been reported to have significantly improved mechanical and thermal shock resistance properties (Jhao et al. 1993). Researchers have also succeeded to process other oxide and non-oxide based ceramic nanocomposites, such as Al_2O_3/Si_3N_4, Al_2O_3/TiC, mullite/SiC, B_4C/SiC, B_4C/TiB_2, SiC/amorphous SiC, Si_3N_4/SiC and so on by usual powder metallurgical techniques such as pressureless sintering, hot-pressing and hot isostatic pressing. The differences in fundamental mechanisms are very clear between solid-state sintering and liquid-phase sintering. Solid-state sintering processes were specifically applied for the processing of oxide based nanocomposite materials and liquid-phase sintering processes have been applied to the processing of non-oxide based nanocomposite materials (Niihara 1991).

1.3.2 Effect of the use of dopants and other additives

Systematic experimental investigations are still necessary for the use of dopants and other additives for the enhancement of mechanical properties of thermally resistant ceramic nanocomposites. Several investigators have studied the effect of the use of dopants for the enhancement of thermal stability of ceramic materials by controlling the grain boundary mobility during sintering of densified ingredients (Chen and Chen 1994, 1996a, Nowick *et al.* 1979). The effects of dopants Mg^{2+}, Ca^{2+}, Sr^{2+}, Sc^{3+}, Yb^{3+}, Y^{3+}, Gd^{3+}, La^{3+}, Ti^{4+}, Zr^{4+} and Nb^{5+} on the grain boundary mobility of dense CeO_2 have been investigated from $1270^\circ C$ to $1420^\circ C$. Parabolic grain growth was observed in all instances. Together with atmospheric effects, the results support the mechanism of cation interstitial transport being the rate-limiting step. A strong solute drag effect has been demonstrated for diffusion-enhancing dopants such as Mg^{2+} and Ca^{2+}, which, at high concentrations, can nevertheless suppress grain boundary mobility. Severely undersized dopants (Mg, Sc, Ti and Nb) have a tendency to markedly enhance grain boundary mobility, probably due to the large distortion of the surrounding lattice that apparently facilitates defect migration. Overall, the most effective grain growth inhibitor at 1.0% doping is Y^{3+}, while the most potent grain growth promoter is either Mg^{2+} (e.g. 0.1%) or Sc^{3+} at high concentration (greater than 1.0%) (Chen and Chen 1996b).

1.4 Thermal properties of particular ceramic nanocomposites

1.4.1 Al_2O_3, MgO and mullite based nanocomposites

Conventional ball milling techniques have been used for homogeneous mixing of nano-size oxide and non-oxide powder particles with highly pure media for making nanocomposites. The dried mixture powders were hot-pressed under N_2 or Ar atmosphere. Natural kaolin and Al_2O_3 powders were selected for the natural mullite/SiC system. Natural mullite/SiC nanocomposites were prepared by reaction-sintering mixtures of kaolin, Al_2O_3 and SiC powders are processed at $1500^\circ C$ to $1700^\circ C$. The fabrication processes of Al_2O_3/Si_3N_4, Al_2O_3/SiC and MgO/SiC nanocomposites were presented by Niihara (1991).

The variation of Vickers hardness with temperature for the $Al_2O_3/$ 16 vol% Si_3N_4 nanocomposite was noticed in the reported findings. The temperature dependence of the fracture strength for $Al_2O_3/5$ vol% SiC and $MgO/30$ vol% SiC nanocomposites was also reflected in the characterization. Sudden slope changes in the reported figures correspond to the brittle-to-ductile transition temperature (BDTT). The improvement of BDTT by nano-size Si_3N_4 dispersion reaches approximately $450^\circ C$. Tremendous improvements in BDTT were also observed for the Al_2O_3/SiC and $MgO/$ SiC nanocomposites. This improvement in high-temperature hardness and BDTT must be due to the pinning of dislocations by the nano-sized SiC and Si_3N_4 dispersions. The observed improvement in hardness from degradation at high temperatures suggests that nano-size dispersions within matrix grains may improve high-temperature creep behaviour by dislocation pinning with SiC particles.

1.4.2 Si_3N_4/SiC based nanocomposites

Among the available high-temperature material systems currently being developed, fibre-reinforced ceramic matrix nanocomposites (FRCMNCs) have attracted great attention for aerospace applications. Compared with monoilithic ceramics, FRCMNCs exhibit remarkable damage tolerance, with a fracture toughness several times higher than most monolithic ceramics. Defect sensitivity, weak thermal shock resistance and lack of toughness continue to be the two main technical hurdles that exclude monolithic ceramics from aerospace turbine engines. However, high-toughness FRCMNCs require an oxidation-resistant interface between the fibre and the matrix. Due to a weak interface, cracks propagate through the matrix rather than continuing along the fibre during high-temperature applications. In many early ceramic matrix composite systems, the weak interface between the fibre and the matrix contained carbon. However, the carbon layer is oxidized in oxygen-rich environments at temperatures as low as 400°C. After the carbon is gone, the oxygen reacts with the fibre to form a silica (SiO_2) layer on the surface of the fibre (Bonney and Cooper 1990, Brennan 1986, Cooper and Chung 1987). The SiO_2 layer significantly weakens the fibre and also allows strong bonding to the matrix, which results in a significant decrease in fracture toughness of the FRCMNC. It is thus necessary to establish an oxidation-resistant interface coating over the surface of the fibre before fabrication of a FRCMNC.

Advanced high-temperature material systems are considered key techno-logical tools in the evolution of current aerospace propulsion systems. In the past decade, the aerospace design community has greatly improved understanding of the thermal environment within an aerospace turbine engine, thus allowing for new concepts in aerofoil design, cooling and combustion. However, major advances can only be realized with greatly improved materials and a thorough understanding of their high-temperature mechanical behaviour and performance. One such composite was manu-factured by Kaiser Ceramic Composites, which is a joint venture of Dow Corning (Midland, MI) and Kaiser Aerotech (San Leandro, CA), and is sold under the trade name Sylramic™ 202. The silicon–carbon–oxygen (Si–C–O) fibres used in the composite were ceramic-grade Nicalon™ fibres (Nippon Carbon Co., Tokyo, Japan). The fibres were supplied in the form of an eight-harness satin weave (8HSW) cloth. Individual sections of cloth were placed in a CVD furnace and a boron-containing coating was applied to the fibres. Once coated, a total of eight sections of cloth were prepregged with a mixture of polymer, fillers and solvent. The individual sections of prepregged cloth were then placed in a vacuum bag using a hand lay-up procedure and a warp aligned stacking sequence. Once in the bag, the assembly was warm moulded and cured in an autoclave to produce a flat tile. Further curing was followed by pyrolysis under nitrogen at temperatures $> 1000°C$. During pyrolysis, the polymer was pyrolyzed to an amorphous SiNC ceramic matrix. After pyrolysis, the assembly was reinfiltrated with polymer solution and then pyrolyzed again; this procedure was repeated several times to increase the matrix density. Test coupons were then cut from the tiles using a laser, reinfiltrated with polymer solution and pyrolized several times to densify the cured matrix further. Because the porosity in this composite was predominantly open, this may suggest that either additional infiltrations with the polymeric precursor would have been advisable or the viscosity or molecular weight of the final precursor was too high to penetrate and fill the open pores effectively. Typical distributions of fibre, porosity and filler of the transverse cross-section of the nanocomposites were shown in the optical micrographs by the investigators. Large pores in the 8HSW Nicalon/SiNC nanocomposites were also shown by the investigators in optical micrographs (Lee et al. 1998).

Tensile tests have demonstrated that this FRCMNC exhibits excellent strength retention up to 1100°C. The room-temperature fatigue limit was 160 MPa, 80% of the room-temperature tensile strength. In the high-temperature tension tests, each specimen was heated to the test temperature for 15 min and then held there for about 20 min to allow the specimen to equilibrate. When the 20 min soak was complete, the load was applied. A high-temperature extensometer with alumina rods was used to measure strain. All high-temperature tension, creep rupture and fatigue tests followed this heating procedure (Butkus et al. 1992).

For the fabrication of thermal shock resistant Si_3N_4 based nanocomposites, an amorphous Si–C–N precursor powder was prepared by the reaction of $[Si(CH_3)_2]_2NH + NH_3 + N_2$ in a vapour phase gas system at 1000°C and by heat treatment in N_2 atmosphere at 1300°C for 4 h to stabilize the powder. The amorphous precursor powder thus obtained was almost spherical and homogeneous, and the average particle size was 0.2 μm. This Si–C–N amorphous precursor powder actually converts to a mixture of Si_3N_4 and SiC particles when heated at high temperatures. The amorphous Si–C–N powder with various carbon contents was mixed with 8 wt% Y_2O_3 as the sintering aid in a plastic bottle using Si_3N_4 balls and ethanol for over 10 h. The dried mixtures were hot-pressed at 1700–1800°C in N_2 atmosphere. In the Si_3N_4/SiC nanocomposites, the grain morphology of Si_3N_4 was strongly influenced by the SiC dispersions, depending on the volume fraction of SiC. Up to approximately 25 vol% SiC, the growth of elongated Si_3N_4 grains was accelerated by the SiC dispersion. As compared with monolithic Si_3N_4 prepared under the same conditions, the nanocomposites were composed of more uniform and homogeneous elongated grains. On the other hand, above 25 vol% SiC, the growth of elongated grains decreased with the increase of SiC content, and equiaxed and fine Si_3N_4 grains were developed.

Nano–nano composites were also studied for the Si_3N_4/SiC system, including nano-size SiC particles of over approximately 40 vol% SiC (Niihara 1991). It was reported that elongated Si_3N_4 grains were formed by the solution–diffusion–reprecipitation process during the liquid-phase sintering. From this growth mechanism of elongated Si_3N_4 grains and the microstructural change observed in this system, it is reasonable to think that the nano-size SiC particles dispersed within the Si_3N_4 grains act as nuclei for β-Si_3N_4 growth in the solution–diffusion–reprecipitation process. The decrease of elongated grains observed at higher SiC content can thus be explained by the excessive nuclei site of β-Si_3N_4 in the above-mentioned processes. In other words, intragranular SiC particles are thought to be trapped in the Si_3N_4 grains during the solution–diffusion–reprecipitation process in the sintering of Si_3N_4. The mechanical properties of Si_3N_4/SiC nanocomposites will be expected to improve by this morphology change of Si_3N_4 grains (Tani *et al.* 1986). For $Si_3N_4/32$ vol% SiC nanocomposites, fracture toughness and strength were improved with the increase of nano-size SiC content.

It has been reported that grain boundary sliding and/or cavitation are responsible for the high-temperature strength degradation of the monolithic ceramics. The fracture strength of monolithic Si_3N_4 ceramics degrades suddenly above 1200°C because of grain boundary sliding and/or cavitation caused by softening of the grain boundary impurity phase. However, Si_3N_4/SiC nanocomposites did not exhibit strong degradation of strength up to 1400°C. The fracture strength at 1400°C was over 1000 MPa and about 900 MPa even at 1500°C for the 32 vol% SiC nanocomposite in which the nano-size SiC particles were dispersed not only within the Si_3N_4 matrix but also at the grain boundaries (Evans and Blumenthal 1983).

Nano-size SiC particles located at the grain boundaries of the Si_3N_4 matrix make a direct bonding with the matrix in spite of grain boundary impurity phases, and then improve the high-temperature mechanical properties and also probably the oxidation resistance. Some Si_3N_4/SiC nano/nano composites have exhibited superplastic deformation at approximately 1600°C. Superplasticity was provided by controlling the sintering conditions. The important points are to retain the α-Si_3N_4 phase and to incorporate the SiC particles at approximately 40 vol% (Niihara 1991).

1.4.3 Al$_2$O$_3$/SiC nanocomposites

Hard nanoparticles usually have a higher sintering temperature than that of the matrix, so sintering temperature increases with increasing hard-particulate content. In Al$_2$O$_3$/SiC systems, just 5 vol% SiC incorporation can evidently hinder the densification process. Nearly full densities attained by hot-pressing were achieved at 1600°C for 5 vol% SiC, at 1700°C for 11 vol% SiC and at 1800°C for up to 33 vol% SiC, while 1400–1500°C was needed for Al$_2$O$_3$ monolithic ceramics (Niihara 1991). At the same time, matrix grain growth was dramatically inhibited owing to the pinning action of dispersed particles. The sintering temperature for Al$_2$O$_3$ was 1500°C, in which the grain size grew up to 2.6 µm with uneven distribution. Al$_2$O$_3$ grain sizes in 5 vol% and 10 vol% SiC dispersed composites were 1.6 µm and 1.4 µm, respectively, with homogeneous distribution, even though sintering temperatures reached 1700°C or higher (Gao *et al.* 2004).

In general, particles disperse according to their grain size and the type of matrix. For Al$_2$O$_3$/SiC composites, finer particles disperse within the matrix grains and larger particles disperse at the grain boundaries, with the critical grain size typically 200 nm. Niihara (1991) considered that the improved toughness was mainly attributed to the residual stress that results from different thermal expansion coefficients of two phases. In Al$_2$O$_3$/SiC systems, the tensile hoop stress, thought to be over 1000 MPa around nanoparticles within the matrix grains, was generated from the large thermal expansion mismatch. Thus, the material may be toughened primarily by crack deflection and bridging due to nano-sized SiC particles within the matrix grains. However, it was argued by other researchers that the incorporation of SiC into Al$_2$O$_3$ affects toughness to a very limited degree. Hoffman *et al.* (1996) and Ferroni and Pezzotti (2002) found that in Al$_2$O$_3$/5 vol% SiC systems, cracks were not distinctly deflected or bridged by nano-sized SiC particles, and no *R*-curve existed in Al$_2$O$_3$/5 vol% SiC nanocomposites. Zhao *et al.* (1993) considered that the dramatically improved toughening mentioned by Niihara was actually attributed to surface compressive stresses.

The remarkable refinement of matrix grains by the dispersions was associated with sub-grain boundary formation in the matrix grains. Sub-grain boundaries were found to be formed due to the pinning and pile-up of dislocations by intragranular hard particles, which were generated in the matrix during cooling from the sintering temperature by the highly localized thermal stress within and/or around the hard particles caused from the thermal expansion mismatch between the matrix and the dispersions. This thermal expansion mismatch, on the other hand, causes residual compressive stresses at the matrix grain boundaries. Both effects strengthen the composites. The sub-grain boundaries were more extensive for Al$_2$O$_3$/

5 vol% SiC nanocomposites after annealing at 1300°C, and then the fracture strength was further improved from 1050 MPa to 1550 MPa (Yongli 2006).

1.4.4 Al_2O_3/SWCNT nanocomposites

Multifunctional composites of polycrystalline alumina ceramics and single-wall carbon nanotubes (SWCNTs) have been shown to possess unique grain boundary structures (Grossman 1972, Holand et al. 1983, Kawamura 1979, Niihara 1986). These hierarchical structures comprise a three-dimensional network of two-dimensional nets made up of random one-dimensional SWCNTs bundles, embedded within the alumina grain boundaries (Uno et al. 1991). Composites with such unprecedented structures exhibited unusual mechanical properties, including enhanced high-temperature creep resistance (Uno et al. 1992). SWCNTs having an elongation to failure of 20–30% with a combination of stiffness (Young's modulus of 1.5 TPa) show a tensile strength well above 100 GPa. The flexibility of SWCNTs is remarkable and the bending may be fully reversible up to a critical angle as large as 110°. It has been observed that Al_2O_3/10 vol% SWCNTs nanocomposites with uniformly distributed SWCNTs at the grain boundaries show creep resistance over two orders of magnitude larger relative to pure Al_2O_3 of the same grain size (Katagiri et al. 1987). This dramatic increase of creep resistance with partial obstruction of grain boundary sliding by the presence of SWCNTs at the grain boundaries was understood by the investigators to be a new class of high-temperature resistant nanocomposites. In this study, the effect of uniformity of distribution of the SWCNTs at alumina grain boundaries on the creep behaviour of Al_2O_3/SWCNTs nanocomposites was investigated. Several uniaxial compression creep tests were performed to study the high-temperature mechanical properties and thermal stability during deformation of the composites, using a method and equipment described elsewhere (Nose and Fujii 1987) at temperatures of 1300–1350°C and in the stress (σ) range 26–256 MPa. The temperatures used were high enough for diffusion processes to occur, but sufficiently low to avoid grain growth. An argon (Ar) gas atmosphere was used to prevent oxidation of the SWCNTs.

Zhan et al. (2003a, 2003b) fabricated fully dense nanocomposites of SWCNTs with a nanocrystalline alumina matrix at sintering temperatures as low as 1150°C by spark plasma sintering (SPS). The introduction of SWCNTs led to a refinement of grain size. Most of the α-Al_2O_3 grains were in the nanocrystalline range, around 200 nm. The fracture toughness of the Al_2O_3/5.7 vol% SWCNTs nanocomposites was more than twice that of pure alumina and there was almost no decrease in hardness. A toughness of 9.7 MPa.m$^{1/2}$, nearly three times that of pure alumina, was achieved in

$Al_2O_3/10\,vol\%$ SWCNT nanocomposites when sintered under identical conditions.

1.4.5 Al_2O_3/ZrO_2 nanocomposites

The phase diagram shows that the miscibility of Al_2O_3 and ZrO_2 phases is very similar. There is also evidence that the grain boundary structure is stable due to the presence of two phases. It was felt necessary to have these two phases in one microstructure of nano-size. Cotton and Mayo (1996) claimed that no increase in toughness occurs in ZrO_2 nanocomposites with density values close to theoretical unless the materials are heat-treated such that grains become susceptible to a phase transformation. Recent studies have indicated that a different mechanism (rather than phase transformation) of toughening must be operating. Ferroelectric domain switching is responsible for the great increase of toughness of Al_2O_3/ZrO_2 nanocomposites, which is now well established as a mechanism for enhanced toughness without undergoing transformation in ZrO_2. Physically speaking, ferroelasticity is similar to ferromagnetism or ferroelectricity. Drawing the analogy, ferroelasticity can be characterized by the existence of permanent strain and an energy dissipating hysteresis loop between the stress and strain axes. In such materials, new domains or twins can be nucleated depending on the state of stress.

From the literature on superplastic deformation of metals, it can be concluded that nano–nano composites in which the constituent phases have similar grain sizes are preferred. Indeed, fine-grained (submicron but larger than nanocrystalline) Al_2O_3/ZrO_2 nanocomposites have been superplastically deformed. The average value of toughness in HIP sintered Al_2O_3/ZrO_2 nanocomposites was calculated to be $8.38\,MPa.m^{1/2}$. A conventional ceramic material cracks substantially under such a load. This means that the nano–nano composites were actually deforming plastically under load (Bhaduri and Bhaduri 1997). A combination of very rapid sintering at a heating rate of $500°C/min$ and a sintering temperature as low as $1100°C$ for 3 min by the SPS technique, and mechanical milling of the starting γ-Al_2O_3 nanopowder via a high-energy ball milling (HEBM) process, can also result in a fully dense nanocrystalline alumina matrix ceramic nanocomposite. The grain sizes for the matrix and the toughening phase were 96 and 265 nm, respectively. With regard to toughening, a great improvement in fracture toughness ($8.9\,MPa.m^{1/2}$) was observed in the fully dense nanocomposites, nearly three times as tough as pure nanocrystalline alumina (152 nm, $3.03\,MPa.\,m^{1/2}$) (Zhan et al. 2003c).

1.4.6 Metal nanoparticle dispersed nanocomposites

To improve mechanical properties, particularly fracture strength and toughness for high-temperature applications, many attempts have been made to provide ceramic matrix composites incorporating second-phase dispersions such as particulates, platelets, whiskers or fibres (Becher and Wei 1984, Claussen et al. 1977, Prewo et al. 1986, Uchiyama et al. 1986). In this case, ceramic/ceramic systems are the most active field of ceramic matrix composite research. On the other hand, there are also some investigations regarding ceramic/metal composites which have incorporated secondary metal phase dispersions such as tungsten, molybdenum, titanium, chromium, nickel, etc. (Breval et al. 1985, Cho et al. 1980, Hing 1980, McHugh et al. 1966, Naerheim 1986, Rankin et al. 1971). However, for both types of composites, fracture strength and toughness have not been improved simultaneously. This is mainly due to the fact that the addition of second-phase dispersions generally causes an enlargement of the flaw size in the composites. Therefore, these composites have advanced from a micro-order dispersion to a nano-order dispersion, especially in the ceramic/ceramic systems in an attempt to alleviate this problem.

Ceramic/metal nanocomposites consisted of an oxide ceramic and either refractory metal such as in the Al_2O_3/W, Al_2O_3/Mo and ZrO_2/Mo systems or a metal with a low melting point such as Al_2O_3/Ni, Al_2O_3/Cr, Al_2O_3/Co, Al_2O_3/Fe and Al_2O_3/Cu systems. These composites were fabricated by hot-pressing the matrix and oxide powders (Ji and Yeomans 2002, Nawa et al. 1994a, 1994b, Oh et al. 2001, Sekino and Niihara 1995, 1997, Sekino et al. 1997). In recent years, nanocomposites, in which nanometre-sized metal particles are dispersed within the ceramic matrix grains and at the grain boundaries, have shown significant improvements in mechanical properties, such as fracture strength, hardness and creep resistance, even at high temperatures (Niihara 1991, Niihara et al. 1986, 1988b). In the first approach, Al_2O_3 has been well examined as a structural ceramic and has been studied in ceramic/ceramic nanocomposite systems such as Al_2O_3/SiC (Niihara et al. 1986, 1988b).

1.4.7 Al_2O_3/Mo nanocomposites

Molybdenum particles are one of the refractory metals and have a lower thermal expansion coefficient than the Al_2O_3 matrix. A conventional powder metallurgical technique was used to prepare Al_2O_3/Mo nanocomposites. The nanocomposites were made by using α-Al_2O_3 powder as the matrix and different vol% of molybdenum powder as particulate reinforcement (5, 7.5, 10, 15 and 20 vol%). The powder mixtures were ball milled using zirconia milling media in acetone for 24 h. These slurries were dried and passed through a 250 μm screen. The mixtures were then hot-pressed in carbon dies of 40 nm diameter. The hot-pressing conditions were 1400–1700°C with an applied pressure of 30 MPa for 1 h under a vacuum of less than 10^{-4} Torr (1 Torr = 133.322 Pa). The fracture strength of $Al_2O_3/$7.5 vol% Mo nanocomposites hot-pressed at 1400°C exhibited a maximum value of 884 MPa, which was 1.5 times larger than that of the monolithic Al_2O_3 prepared under the same conditions.

When the hot-pressing temperature was increased to 1700°C, strength decreased to around 450 MPa, which corresponded to grain growth of the Al_2O_3 matrix. Thus, in Al_2O_3/Mo nanocomposites, strengthening was assumed to be mainly attributed to the inhibition of grain growth of the Al_2O_3 matrix in the presence of nano-sized molybdenum particles. Many voids were observed in the $Al_2O_3/7.5$ vol% Mo nanocomposites hot-pressed at any temperature. Because of such voids, large molybdenum particles seemed to be pulling out from the grain boundaries or triple junctions of the Al_2O_3, which was observed in the microstructure. These results suggest that the bonding strength of the interfaces between the Al_2O_3 grains and molybdenum particles is not strong. On the other hand, the fracture mode of monolithic Al_2O_3 was mainly intergranular, but in the case of a large grain size, a transgranular fracture mode was observed.

When hot-pressed at a lower sintering temperature of 1400°C, the fracture toughness for the Al_2O_3/Mo nanocomposites was around 4.4 MPa.m$^{1/2}$; nevertheless, a significant improvement in fracture toughness was observed with increasing molybdenum content and hot-pressing temperatutre. Fracture toughness of the $Al_2O_3/20$ vol% Mo nanocomposites hot-pressed at 1700°C exhibited a maximum value of 7.6 MPa.m$^{1/2}$, which was 1.8 times larger than that of the monolithic Al_2O_3 hot-pressed at 1600°C (Nawa et al. 1994b). The $Al_2O_3/5$ vol% Mo nanocomposites exhibited maximum value of 19.2 GPa when hot-pressed at 1400°C. This hardening is supposed to be accounted for by the contribution of the fine-grained microstructure of the Al_2O_3 matrix (Sargent and Page 1978) and/or the intergranular nano-structure of Al_2O_3/Mo nanocomposites, in which nanometre-sized molyb-denum particles are dispersed within the Al_2O_3 grains and at the grain boundaries. In fact, the monolithic Al_2O_3 used in this nanocomposite showed a negative dependence of grain size on hardness and the nanometre-sized metal particle will itself cause hardening (Wierenga et al. 1984). The investigators characterised samples of Al_2O_3/Mo nanocomposites to show the variation of Vickers hardness with molybdenum content; it was found that the hardness values of monolithic and molybdenum polycrystals were 18.5 and 2.1 GPa, respectively. For nanocomposites containing 5–15 vol% Mo hot-pressed at lower sintering temperatures, higher values of hardness were observed than those predicted by the rule of mixtures.

1.4.8   Tetragonal zirconia polycrystal (TZP)/molybdenum (Mo) nanocomposites

To achieve either high strength or high toughness in composites, tetragonal zirconia polycrystals (TZPs), stabilized with Y_2O_3 (Y-TZP) or CeO (Ce-TZP), have been studied (Gupta *et al.* 1978, Masaki 1986). The Y-TZP ceramic containing 2–3 N mol% Y_2O_3 exhibited enhanced strength above 1200 MPa; however, toughness was approximately in the range 5–6 MPa m$^{1/2}$. On the other hand, Ce-TZP ceramic containing 7–12 mol% CeO_2 exhibited high toughness in excess of 10 MPa.m$^{1/2}$, but the strength of 600 to 800 MPa was still modest in comparison with that of Y-TZP. To improve modest strength or toughness, TZP-based composites incorporating Al_2O_3 have been investigated. In Y-TZP containing 30 vol% Al_2O_3, the strength increased in excess of 2400–3000 MPa for isostatically hot-pressed specimens, while the toughness decreased appreciably with increasing Al_2O_3 content (Shikata *et al.* 1990, Tsukuma *et al.* 1985). In the same way, Ce-TZP containing 50 vol% Al_2O_3 showed an improvement in strength of 900 MPa, but the toughness decreased from 20 to 5.5 MPa.m$^{1/2}$ with increasing Al_2O_3 content (Tsukuma *et al.* 1988). A tradeoff between high strength and high toughness is still unsolved for both types of monolithic TZP and TZP/Al_2O_3 composite systems.

With regard to the strength–toughness relationship, Swain and Rose (1986) proposed a mechanism for the limitation of strength in transformation-toughened zirconia, pointing out that maximum strength is limited by the critical stress that induces the tetragonal to monolithic transformation. To overcome these problems, several investigators tried another approach for preserving high strength in the region of high toughness – using ductile metal particles instead of brittle ceramic particles as a secondary phase. It was expected that the metal phase would result in the inherent improvement in toughness, but most of the ceramic matrix composites incorporating metal dispersions such as W, Mo, Ti, Cr, Ni, etc. failed to produce the successful results (Breval *et al.* 1985, Davidge 1979, Jenkins *et al.* 1989, McMurtry *et al.* 1987, Ohji *et al.* 1994, Tuan and Brook 1992). This is mainly due to the fact that the addition of second-phase dispersions generally causes an enlargement of the flaw size in the composites and so this system failed to improve both strength and toughness simultaneously.

Several investigators have studied nanocomposites in which nanometre-sized particles were dispersed within ceramic matrix grains and/or at the grain boundaries to eliminate strength-degrading flaws (Mizutani *et al.* 1997, Nawa *et al.* 1992, Siegel *et al.* 2001). They tried to develop ceramic/metal nanocomposites by selecting Y_2O_3-stabilized TZP (3 mol% Y_2O_3) as the ceramic matrix material and refractory molybdenum (Mo) metal powder as the dispersing particles. The powder mixtures, containing 10, 20, 30, 40 and 50 vol% Mo, were ball milled in acetone for 24 h. The slurries were dried and passed through a 250 μm screen. The mixtures were then hot-pressed in a carbon die at 1400, 1500 and 1600°C with an applied pressure of 30 MPa for 1 h in vacuum of less than 10^{-4} Torr. These nanocomposites possess a novel microstructural feature composed of a mutual intragranular nanostructure in which either nanometre-sized Mo particles or equivalent sized zirconia particles were located within the zirconia grains or Mo grains, respectively. A simultaneous improvement in strength and toughness was achieved, overcoming the strength–toughness tradeoff relation (Nawa et al. 1994b). Scanning electron micrograph (SEM) and transmission electron micrograph (TEM) images clearly indicated improved features of the strength properties of the nanocomposites. Submicron-sized Mo particles were dispersed at the grain boundaries of the fine-grained/submicron-sized zirconia matrix. In addition, a number of extremely fine Mo particles were observed mainly within the zirconia grains. The variations of fracture strength and toughness with Mo content for the 3Y-TZP/Mo composites containing up to 50 vol% Mo were presented by the investigators. The strength increased with increasing Mo content and obtained a maximum value of 1795 MPa at 40 vol% Mo. Moreover, the toughness increased dramatically with increasing Mo content above 30 vol% and reached a maximum value of 18.0 MPa.m$^{1/2}$ at 50 vol% Mo. A summary of the fracture strength–toughness relation for the 3Y-TZP/Mo nanocomposites containing up to 50 vol% Mo was also presented and previous results on various types of zirconia-toughened ceramics demonstrated by Swain and Rose (1986) were also considered.

1.4.9 WC–ZrO$_2$–Co nanocomposites

Tungsten carbide (WC)-based ceramic–metal systems (cermets) have been used for decades in various engineering applications (e.g. cutting tools, rock drill tips, tools and dies as well as general wear parts). In fact, cemented carbides, which are usually aggregates of tungsten carbide particles bonded with cobalt metal via liquid-phase sintering, are regarded commercially as one of the oldest and most successful powder metallurgy products. Such conventional two-phase (WC–Co) composite materials derive their exceptional combination of mechanical properties, such as elastic modulus of 550 GPa, hardness of 16 GPa and fracture toughness of 12 MPa.m$^{1/2}$, from those of their components, i.e. the hard refractory WC and soft ductile metallic Co (Berger et al. 1997, Bock et al. 1992, Cha et al. 2003a, 2003b, Jia et al. 1998, Kim et al. 1997, 2007a, 2007b, Kolaska 1992, Lenel 1980, Masumoto et al. 1986, Sarin 1981, Shi et al. 2005, Sivaprahasam et al. 2007, Sutthiruangwonga and Mori 2003, Suzuki 1986). For further improvement in the performance of such cermets in high-temperature resistant structural and tribological applications, the last few decades have witnessed an increasing surge toward the development of WC–Co cermets with nanoscale microstructure. Furthermore, with the advent of advanced electric field assisted sintering techniques such as spark plasma sintering (SPS), the development of dense WC-based cermets possessing submicrometre nano-sized WC grains has been pursued in various research laboratories (Cha *et al.* 2003a, 2003b, Kim *et al.* 2007a, 2007b, Shi *et al.* 2005, Sivaprahasam *et al.* 2007). In order to overcome problems such as corrosion/oxidation and softness of the metallic phase at high temperature and low wear properties associated with WC-based cermets, researchers incorporated 6 wt% nano-sized ZrO_2 in submicrometre WC and sintered at 1300°C for 5 min via SPS (Biswas *et al.* 2007, Cha and Hong 2003, Kim *et al.* 2004, 2006, Imasato *et al.* 1995). The role of ZrO_2 in enhancing the densification kinetics was also critically analyzed. In another investigation, it was observed that such WC-based nanocomposites possess excellent wear resistance (Venkateswaran *et al.* 2005). However, a serious drawback remained the poor indentation toughness (6 MPa.m$^{1/2}$) of WC-6 wt% ZrO_2 spark plasma sintered at 1300°C for 5 min. To obtain a good combination of fracture toughness, strength and hardness, slight modifications in the compositional window were made: 3 mol% yttria-stabilized ZrO_2 (3Y-TZP) powders were used as the sintering aid and the possibility of exploiting the transformation toughening effect of ZrO_2 was explored (Basu 2005, Garvie *et al.* 1975, Hannink *et al.* 2000, Mukhopadhyay *et al.* 2007). The presence of ZrO_2 nanoparticles changes the fracture mode from intergranular fracture (for WC–Co cermet) to nearly 100% transgranular fracture for the ZrO_2-containing WC-based nanocomposites. The reasons behind such a change in fracture mode for ceramic nanocomposites in the presence of intragranular nano-sized second-phase particles have been discussed elsewhere (Hansson *et al.* 1993, Limpichaipanit and Todd 2009).

As the fracture energy for cleavage (transgranular) fracture is higher than grain boundary (intergranular) fracture in ceramics, such a change in fracture mode has been reported to result in an improvement of fracture toughness, especially for ceramic nanocomposites in which the matrix grains are equiaxed in nature (Chen and Chen 1994, Limpichaipanit and Todd 2009, Mukhopadhyay and Basu 2007). In addition to transformation toughening and change in fracture mode in the presence of ZrO_2, deflection and bridging of cracks by the second-phase particles (ZrO_2) also contribute to the high fracture toughness of WC-6 wt% ZrO_2 (3 N mol% Y_2O_3) and WC-4 wt% ZrO_2 (3 mol% Y_2O_3)-2 wt% Co nanocomposites.

The mechanical properties of some ceramic nanocomposites are different because of their microstructural characteristics. Such differences in mechanical properties of various nanocomposite systems not only depend on the microstructural scale of the nano-sized reinforcement and mechanical response of individual phases, but they also depend on the method of measurement.

1.5   Interface characteristics of ceramic nanocomposites

The phenomenal progress in materials science and technology has led to very high and critical performance demands on ceramic materials having high-mechanical strength and toughness in high-temperature applications. It is becoming increasingly necessary and sometimes absolutely essential to control parameters such as thermal expansion mismatch between reinforcing agents (fibres, whiskers or nano-phase particles) and matrix materials, creep deformation, grain boundary sliding, crack growth, high-temperature oxidation and densification of materials during the processing of ceramic nanocomposites. To achieve high performance, an excellent balance of properties of each constituent of the nanocomposite and the interaction between component items at the interface are of utmost and equal importance. For the development of high-strength ceramic nanocomposites, it is necessary to build strong interfaces between the reinforcements (either micro- or nano-size) and the nano-size matrix materials. However, for the development of high-toughness ceramic nanocomposites, the interfaces between the reinforcements (either micro- or nano-size) and the nano-size matrix should have a low fracture toughness to propagate a crack through the matrix and deflect along the fibre reinforcement/matrix interface rather than continue through the fibre reinforcements.

In many early ceramic matrix composite (CMC) systems, the weak interface was observed in between the Nicalon-grade woven silicon fibre reinforcement (Si–C–O) and the matrix (Si–N–C) contained carbon. However, such a carbon layer will be oxidized in oxygen-rich environments at temperatures as low as 400°C. After the carbon is gone, the oxygen reacts with the fibre to form a silica (SiO_2) layer on the surface of the fibre (Bonney and Cooper 1990, Brennan 1986, Cooper and Chung 1987). As an alternative to carbon, boron nitride (BN) (Brennan *et al.* 1992, Hanigofsky *et al.* 1991, Kmetz *et al.* 1991, Naslain *et al.*, 1991, Shen *et al.* 1992, 1993, Sigh and Burn 1987, Sun *et al.* 1994, Veltri and Galasso 1990) has been considered as a new fibre coating, and several CMC manufacturers have switched to a BN-containing interface between the fibre and the matrix. Microanalysis of fibre/matrix interfaces has been conducted on as-received material using a fibre push-in technique (Hartman and Ashbaugh 1990, Jero *et al.* 1992, Kerans and Parthasarathy 1991). All the interface data were collected at room temperature using a Micro Measure Machine (Process Equipment Co., Tipp City, OH), which is designed specifically for fibre interface testing. The boron-containing fibre/matrix interphase does not rapidly oxidize (compared with the carbon fibre/matrix interphase) due to the formation of a glass seal in the crack mouth for fast oxygen diffusion paths during oxidation at high temperature. Such a phenomena minimizes oxidation at fibre/matrix interface (Folsom *et al.*, unpublished). It has been observed that higher applied stress levels lead to greater matrix-crack densities during the initial loading period. The increased crack density and higher opening displacements act to accelerate the oxidation and rupture processes. The fibre reinforcement (Nicalon fibre) exhibits several creep characteristics, which are important to the understanding of the creep response observed in investigations of a woven Nicalon/Si–N–C ceramic matrix composite (Lee *et al.* 1988). It has been reported that the microstrucure of Nicalon fibre remains unchanged at temperatures <1200°C (Weber *et al.* 1994). Increasing creep resistance over time has also been observed in composites that contain off-axis plies, such as [0/90] cross-ply laminates. Fibres in the 90° plies increase the axial creep resistance of a composite by decreasing creep flow in the matrix (Xu and Holmes 1993). At temperatures in the range of 900–1000°C, it was reported (Kervadec and Chermant 1992, 1993) that strain accumulation in the CMC reinforced by Nicalon fibres is dominated by damage-induced stress redistribution within the composites, and, because of the relatively low test temperatures, the fibre may still behave elastically in this temperature range. Holmes and Cho (1992) reported that for fatigue tests conducted on a unidirectional Nicalon/CAS CMC at stress levels below the proportional limit (225 MPa), matrix cracks along with interface debonding and sliding along the fibre/matrix interface were observed.

The effect of interfacial characteristics for the improvement of tensile creep properties in ceramic nanocomposites was studied by Nakahira and Niihara (1993). High strength at high temperatures suggests that ceramic nanocomposites would possess good creep resistance. They conducted flexure creep tests for an alumina /17 vol% silicon carbide nanocomposite and monolithic alumina at 1100–1400°C at a stress level of 100 MPa and revealed that the creep rate of the former was four orders of magnitude lower than that of the latter. For both scientific and engineering aspects, it is necessary to identify (1) why (or by which mechanism) the creep resistance is so remarkably improved in the ceramic nanocomposite and (2) how (or in which way) the creep and creep rupture behaviour of the ceramic nanocomposite is different from that of the monolithic material. To deal with these problems, uniaxial tensile testing was considered the most suitable method.

It has been speculated that grain boundary diffusion ability in a silicon carbide–alumina interface of a nanocomposite is significantly lower than that in the alumina–alumina interface (Raj and Ashby 1971). This speculation is supported by direct observation of the interfaces: the alumina–alumina interface contained a markedly wider non-crystalline structure than the silicon carbide–alumina interface. A strong particle–matrix interface in a ceramic nanocomposite is supposedly attributable to a great internal compression stress acting on the interface during the cooling process due to thermal expansion mismatch (Ohji *et al.*, unpublished). Transgranular fracture of nanoparticles in the grain boundary crack propagation in the nanocomposite also suggests rigid bonding of the particle–matrix interface. In tensile creep and creep rupture tests, the creep life of the ceramic nanocomposite was 10 times longer and the creep strain at fracture was 8 times smaller than those of the monolith at 1200°C and 50 MPa. The nanocomposite demonstrated transient creep until failure, while accelerated creep was observed in the monolith. Microscopic characterization suggested the following microstructural evolution during creep: as grain boundary sliding proceeded, intergranular silicon carbide nanoparticle rotated and plunged into the alumina matrix, significantly increasing creep resistance.

Fibre-reinforced ceramic matrix composites are being developed for potential use in gas turbine engines and other industrial applications involving high service temperatures and both static and dynamic loads. Ceramic composites have shown low density, enhanced fracture toughness, higher damage tolerance and excellent thermal stability compared to monolithic ceramics. Critical to the successful implementation of fibre-reinforced ceramic matrix composites in such applications is the presence of a suitably weak fibre/matrix interface that is also resistant to oxidation at high temperatures. High strength and toughness at room temperature have been demonstrated in graphite fibre reinforced composites and SiC fibre reinforced composites with a carbon interfacial layer (Levitt 1973, Philips *et al*, 1972, Prewo and Brennan 1980, 1982). However, both of these materials are susceptible to oxidation during long-term high-temperature exposure. Research has been devoted to the design of suitable interfaces to improve the oxidation resistance of composites, and several approaches have been investigated, including the use of fibre coatings and matrix doping (Brennan 1987, Naslain *et al*. 1991, Rice 1987). Sun *et al*. (1996) selectively designed a composite consisting of a barium magnesium aluminosilicate (BMAS) glass-ceramic matrix reinforced with SiC fibres with a SiC/BN coating. The material exhibited retention of most tensile properties up to 1200°C. Monotonic tensile fracture tests produced ultimate strengths of 230–300 MPa with failure strains of 1%, and no degradation in ultimate strength was observed at 1100 and 1200°C. Tensile fatigue experiments were conducted in which the composite survived 10^5 cycles without fracture at temperatures up to 1200°C (Sun *et al*. 1996). The short- and long-term properties observed in this study were derived largely from the degree to which interface microstructure was controlled by the SiC/BN dual coating. In particular, the BN coating provided the desired weak interface necessary for high loading rate toughness, whereas the SiC overlayer provided a barrier to diffusion and reaction at high temperatures. Both layers were more stable in oxidizing atmospheres than either BN coatings alone or previously studied carbon-rich interfaces (Brennan 1986, Cooper and Chung 1987).

1.6   Superplasticity characteristics of thermal shock resistant ceramic nanocomposites

Superplasticity is defined as the ability of a polycrystalline material to exhibit large elongations at high temperatures and relatively low stresses. Today, from an engineering point of view, the term superplasticity is ascribed to a polycrystalline material pulled out to very high tensile elongations prior to failure with necking-free strain. This phenomenon is usually found in many metals, alloys, intermetallics, composites and ceramics (recently in high-temperature superconductor ceramics) when the grain size is small enough: less than $10\,\mu m$ for metals and less than $1\,\mu m$ for ceramics. When a polycrystalline material is deformed at high temperatures, grain boundary sliding (GBS) takes place in two different ways.

- Deformation takes place due to the flow of point defects in the occurrence of GBS. This is called diffusional creep. In these cases, each individual grain suffers almost the same deformation as that imposed on the specimen and the grains that are nearest neighbours remain nearest neighbours. This is termed Lifshitz GBS.
- During GBS, deformation may be accompanied by intergranular slip throughout adjacent grains, by localized slip adjacent to the boundaries or by diffusional process of point defects. This is termed Rachinger GBS.

The grains retain almost their original size and shape even after large deformations when GBS is accommodated by some of the mechanisms involving dislocation movement or diffusion of point defects. This GBS, as the primary mechanism for deformation, is the basis for the high ductility exhibited by some materials at high temperatures and therefore for their structural superplastic behaviour. In many ceramic-related materials and ceramic composites superplasticity is also said to occur even though the polycrystal is deformed in compression, or in three-or four-point bending conditions, as long as GBS is the primary deformation process (Chokshi 1993, Jimenez-Melendo et al. 1998, Nieh et al. 1991, 1997). The first observation of superplasticity in a $3\,mol\%$ yttria-stabilized tetragonal zirconia polycrystal ceramic (Y-TZP) with a grain size of 0.4 μm was reported by Wakai *et al.* (1986). Since then, a large number of fine-grained polycrystalline ceramics and ceramic composites with superplastic behaviour have been developed (Melendez-Martinez and Dominguez-Rodriguez 2004, Wei Chen and Xue 1990).

Several techniques have been developed to achieve ceramics and composites with a fine microstructure, as follows.

- One technique to suppress grain growth is based upon the inclusion of dispersed phases into the ceramics. In this regard, a multiphase ceramic composite containing 40 vol% ZrO_2, 30 vol% spinel and 30 vol% Al_2O_3 has been sintered and superplastically deformed to 1050% at 1650°C at a strain rate of $0.4 s^{-1}$ (Kim *et al.* 2001). Other zirconia-based composites have also been fabricated with inhibition of grain growth (Wang *et al.* 2003, Yonn and Chen 1990). With this technique, it is possible to deform the ceramics, with their microstructure unchanged, at a temperature at which grain growth would be important in ceramics without dispersed phases.

- Another technique makes use of ultra-fine powders sintered under stress-assisted conditions so that the sintering temperature is reduced (Mayo 1996), for instance hot isostatic pressing, hot-pressing, sinter forging, or techniques with a fast heating and cooling ramp, such as spark plasma sintering (SPS) or microwave sintering, which avoid grain growth by reducing the time or temperature of sintering. Very fast densification has been reported in oxides and Si_3N_4-based ceramics by SPS (Shen *et al.* 2003). Microwaves have also been used to sinter PSZ (Wilson and Kunz 1988) and alumina (Mizuno *et al.* 2004).

- The sintering of Y_2O_3 nanocrystalline ceramics ($d = 60$ μm) has been achieved through a two-step sintering method. The first step is pre-sintering at high temperatures to obtain ceramics with intermediate density values of 70–80%. Secondly, suppression of grain growth is achieved by sintering at lower temperatures than those used in the first step, exploiting the difference in kinetics between grain boundary diffusion, which ultimately controls sintering, grain boundary migration, and grain growth (Chen and Wang 2000).

There are a great number of papers dealing with the influence of grain boundary segregation on superplasticity in Y-TZP. It has been shown that the superplastic flow stress at 1400°C of 3Y-TZP doped with different cations is correlated with the ionic radius of the dopant (Mimurada *et al.* 2001, Nakatani *et al.* 2003). Cations with a smaller ionic size decrease the flow stress, whereas those with large ionic sizes increase the flow stress. The authors suggested that the flow stress is determined by grain boundary diffusivity, which is affected by segregation of the dopant. The same improvement in superplasticity has been reported in a $0.3 \, mol\%$ SiO_2 doped 3Y-TZP with $d = 0.35 \, \mu m$ (Morita *et al.* 2004) and a $0.18 \, mol\%$ Al_2O_3 doped 3Y-TZP with $d = 0.4 \, \mu m$ (Sato *et al.* 1999). This behaviour also seems to account for fine-grained Al_2O_3 with ZrO_2 dopant (Wakai *et al.* 1997, Yoshida *et al.* 1947). In the case of Si_3N_4, glass pockets and thin glass film of thickness about $1 \, \mu m$ often remain at grain boundaries (Clarke 1987). Its plasticity is controlled by the viscosity of the intergranular glassy phase and the solubility of the crystalline phase in the liquid, as reviewed in several contributions (Melendez-Martinez and Dominguez-Rodriguez 2004, Wakai *et al.* 1999, Wilkinson 1988). A solid solution of silicon nitride with some aluminium-based compounds or mixtures forms the so-called α or β-SiAlON compounds, which are superplastic by the addition of secondary glassy phase; for example, Rosenflanz and Chen (1998) reported that Li-doped SiAlON deforms 10 times faster than Si_3N_4. A revision of the mechanical properties of these compounds can be found in the literature (Melendez-Martinez and Dominguez-Rodriguez 2004, Wilkinson 1998). The enhancement of superplastic deformation by an intergranular glass phase was also found to be applicable to liquid-phase sintered SiC (Nagano *et al.* 1999, Wang *et al.* 1997). Wang and Raj (1984) pointed out that superplasticity might occur in liquid-phase sintered Si_3N_4 by diffusional creep enhanced by solution and precipitation of crystals in a Si–O–N liquid phase at the grain boundaries. High ductility in compression has been observed, because cavitation at grain boundaries under tension and subsequent fracture occur readily in the presence of an intergranular liquid phase. Wakai *et al.* (1990) reported superplastic deformation of a covalent crystal composite based on Si_3N_4/SiC nano–nano ceramics, which could be elongated by more than 150% at 1600°C. Superplasticized nanoceramic materials can be applied in many technical areas including aerospace, energy, electronics, biology, etc., which often require complex shapes to be manufactured at low prices. The extensive potential applications, together with the possibility of processing dense ceramics and forming complex structures by superplasticity, have been the driving force for the appearance and fast development of a large number of nano–nano ceramic systems with superplastic capabilities. Several industrial processes in the metal and polymer industries have already made use of these high ductility forming techniques. The processing of dense ceramics includes sheet forming, blowing, stamping, forging and joining. A good example of superplastic forming of different ceramics can be found in Figure 1 of the paper by Wei Chen and Xue (1990). In this figure, flat 1 mm thick discs of Si_3N_4, 2Y-TZP–Al_2O_3, 2Y-TZP–Mullite and 2Y-TZP + 0.3% doped with Mn, Fe, Co, Cu and Zn were stretched with a 6.5 mm radius punch at temperatures and forming times depending on the ceramic used. Sinter forging is a promising technique because densification and net shaping are achieved simultaneously by Venkatachari and Raj (1987). This technique avoids cavities and voids because sinter forging is produced by compression. High strength and high fracture toughness of Si_3N_4 have been achieved by superplastic sinter forging due to the reduction of flaw size and grain alignment (Kondo *et al.* 1999).

1.7 Densification for the fabrication of thermal shock resistant ceramic nanocomposites

Densification of graded powder compacts is a major challenge for achieving ceramic nanocomposites with high mechanical strength. Properly densified ceramic-based nanocomposites have received much attention in various high-temperature structural applications. Some ceramics/ceramic nanocomposites in which the nanometre-size second phase was dispersed within the matrix grains were successfully fabricated using a 1989b conventional powder metallurgical technique by Niihara and colleagues (Niihara *et al.* 1988a, 1989b, 1990a, Yanai *et al.* 1990). They showed that Al_2O_3-, Si_3N_4-, MgO- and B_4C-based nanocomposites exhibited high fracture toughness and strength at room temperature, good creep resistance and no degradation of high-temperature strength. Among these, Al_2O_3/SiC nanocomposites were reported to possess high fracture toughness and strength over 1000 MPa, high reliability and good high-temperature mechanical properties (Niihara 1991, Niihara and Nakahira 1989, 1990b). The densification and grain growth behaviours of Al_2O_3 in a temperature range of 1000–1800°C in N_2 atmosphere during hot-pressing $Al_2O_3/5\,vol\%$ SiC nanocomposites were investigated by Nakahira and Niihara (1992). Fully dense (> 98%) monolithic Al_2O_3 was prepared by them at 1500°C. On the other hand, for $Al_2O_3/5\,vol\%$ SiC nanocomposites, the relative densities exceeded the value of approximately 90% at 1300°C and 95% at 1400°C. $Al_2O_3/5\,vol\%$ SiC nanocomposites hot-pressed at and above 1400°C contained no open pores. The fully dense (> 99%) nanocomposites were prepared at 1600°C. The main challenge associated with powder processing is often densification of the graded powder compact. Sintering rates differ with position and uneven shrinkage may lead to warping and cracking, unless sophisticated sintering techniques are employed. Almost all ceramic/ceramic bulk functionally graded materials (FGMs) are sintered by conventional pressureless sintering (Cichocki and Trumble 1998, Marple and Boulanger 1994, Wu *et al.* 1996) or hot-pressing (Kawai and Wakamatsu 1995, Vanmeensel *et al.* 2004), depending on the sintering properties of the two components. In metal/ceramic FGMs with a continuous metal phase and a discontinuous ceramic phase, the sintering rates are controlled by densification of the metal phase and such FGMs can be densified by conventional sintering methods (Neubrand and Rodel 1997).

In addition to conventional sintering, reactive powder processing, also called combustion synthesis or self-propagating high-temperature synthesis (SHS), can be used if the target compounds can be synthesized from the starting powder mixture (Stangle and Miyamoto 1995). This process comprises a rapid and exothermic chemical reaction to simultaneously synthesize some or all of the constituent phases in the FGM and densify the component.

More advanced techniques, such as spark plasma sintering (SPS) or pulsed electric current sintering (PECS) can also be used for FGM fabrication (Tokita 1999). These are pressure-assisted sintering methods in which a high current is pulsed through a die/punch/sample set-up, which can be compared with that of conventional hot-pressing. The large current pulses generate spark plasmas, a spark impact pressure and Joule heating. The sintering mechanism and mechanical properties of the sintered compacts show characteristics different from conventional pressure-assisted sintering processes and parts. This technique offers significant advantages for various kinds of new materials and consistently produces a dense compact in a shorter sintering time and with a finer grain size than conventional methods. With a spark plasma system, large ceramic/metal bulk FGMs (~100 nm across) can be homogeneously densified in a short time with heating and holding times totalling less than one hour. Among the SPS systems are WC-based materials (WC/Co, WC/Co/steel, WC/Mo), ZrO_2-based composites (ZrO_2/steel, ZrO_2/TiAl, ZrO_2/Ni), Al_2O_3/TiAl, etc. (Tokita 1999). To obtain full densification and to overcome the problem of metal softening in WC-based nanocomposites using the SPS technique, 6 wt% nano-sized ZrO_2 was incorporated in submicrometre WC. This led to near theoretical densification (> 99%) at a considerably lower sintering temperature of 1300°C and holding time of 5 min P. The role of ZrO_2 in enhancing the densification kinetics was also critically analyzed by Tokita. In other investigations, it was observed that WC-based nanocomposites possess excellent wear resistance (Biswas *et al.* 2007, Venkateswaran *et al.* 2005).

Microwave sintering is another promising technique for ceramic/metal FGMs to avoid the difficulty of inequality of the shrinkage rate. As a newly developed sintering technique, microwave sintering used microwave irradiation to heat ceramics or ceramic-based composite compacts (Gerdes and Willert-Porada 1994, Willert-Porada *et al.* 1995, Zhao *et al.* 2000). The mechanism of microwave heating is based on the dielectric loss of the ceramic phases involved, resulting in a volumetric heating technique in which heat is generated by the compact itself.

1.8   Test methods for the characterization and evaluation of thermal shock resistant ceramic nanocomposites

The performance characteristics and nature of ceramic nanocomposites depend on well-defined processing routes to achieve a specific microstructure and detailed characterization of the microstructural features is thus extremely important. It is particularly critical in identifying the role of the microstructure in defining the final bulk properties. The microstructural features in nanocomposites that have been linked to bulk properties include the matrix grain size, the reinforcing particle size, its distribution and location (grain boundaries or occluded within the matrix grains), segregation at the various interfaces and residual stress fields. Various test methods have been used for the evaluation and characterization of ceramic nanocomposites, and are discussed in the following subsections.

1.8.1   Scanning electron microscopy (SEM) analysis

The fracture surfaces of composite samples can be analyzed by observing crack propagation in the matrix phase, interface zones, residual porosity, grain sizes, etc. from scanning electron micrographs.

1.8.2   X-ray diffraction (XRD) analysis

XRD spectra of ceramic nanocomposites sintered at different temperatures are extremely useful. Samples sintered at a particular temperature indicate structural behaviour as either amorphous or crystalline in nature. Such an indication is of immense help to researchers aiming to improve the properties of the resultant materials by optimizing the rate of sintering temperature. XRD spectra can also be used to verify the toughness characteristics of ceramic nanocomposites. XRD study does not indicate any phase transformation occurring during a 24 h high-energy ball milling (HEBM) period even though it is longer than the reported minimal time for complete transformation (10 h). It is very interesting to note that the width of XRD for HEBM γ-Al_2O_3 nanopowder is much greater than that for the starting nanopowder without HEBM. The residual stress induced by HEBM is likely to be responsible for the wider XRD peak. Moreover, HEBM can lead to high green density due to pore collapse from high compressive and shear stresses during milling.

1.8.3   Optical microscopy analysis

An extensive network of the porosity of sintered samples can be obtained using an optical microscope. Dark-filled areas reveal the pores to have a crystalline surface texture and the denser (darker) and more porous (lighter) regions of specimen can be readily distinguished. Also, a degree of subsurface detail can be revealed by optical microscopy analysis. Optical microscopy is carried out in reflected light with Nomarski differential interference contrast (DIC) and dark-field modes on a suitable microscope. The fibre volume fraction can be estimated for all samples. Matrix cracks around fibres arising from residual stresses can also be observed and tend to reach the specimen surface via dense matrix regions. Circumferential crack patterns may indicate residual stress arising from fibre/matrix thermal expansion mismatches.

1.8.4   Indentation techniques

The hardness and fracture toughness of materials are determined using indentation techniques. For example, quadrant-disc specimens of Al_2O_3/SiC nanocomposites and Al_2O_3 were cut from a sintered disc (30 mm in diameter and 3.2 mm thick) with a 1 μm polished finish on both sides. Indentation was performed using a Vickers hardness testing machine (Model A.V.K.-C2 Mitutoyo Corp., Kawasaki, Japan). Indentation loads of 20, 50, 100 and 200 N with a holding time of 15 s for each indentation were used. The diagonal, d, and radial crack length, c, with at least five indentations at each load were measured using optical microscopy (Nikon, Inc., Tokyo, Japan). The hardness, H was related to the diagonal, d, of the indentation and the contact load, P, by:

$$H = 1854.4\left(P/d^2\right) \qquad [1.11]$$

The fracture toughness, K_{ic}, was determined using (Anstis *et al.* 1981):

$$K_{ic} = 0.016(E/H)^{1/2}\left(P/c^{3/2}\right) \qquad [1.12]$$

where E is the elastic modulus and c is the radial crack length measured from the indentation centre.

1.8.5   Indentation thermal shock tests

The indentation thermal shock technique developed by Anderson and Rowcliffe (1996, 1998) is used to study the thermal shock and thermal fatigue behaviour of ceramic nanocomposites. In this technique, the thermal shock resistance is measured by studying the propagation of median/radial cracks around a Vickers indentation after single and repeated quenching. The critical thermal shock temperature difference, of the material can be defined with reference to the number of propagating cracks and the amount of crack extension (Maensiri and Roberts 2002).

1.8.6 Indentation fatigue after thermal shock tests

Indentation fatigue tests (Li and Reece 2000, Reece and Guiu 1991a, 1991b, Takakura and Horibe 1992a, 1992b, Vaughan *et al.* 1987) (repeated indentations on the same site) were conducted on thermally shocked specimens of pure Al_2O_3 and $Al_2O_3/5$ vol% SiC nanocomposite sintered at $1700°C$ using a Vickers profile indenter with applied loads of 5, 100, 200, 300 and 400 N. The holding time of each indentation was 10 s, and the interval between successive indentations was 30 s. The number of indentation cycles needed to produce lateral chipping was noted in each case and the fracture surfaces within the chipping area were characterized using SEM. The observed superior thermal shock resistance of the nanocomposites could be due to an increase in thermal conductivity and/or a simultaneous decrease in the thermal expansion coefficient in the nanocomposite, caused by the addition of SiC.

1.8.7 Non-destructive test (NDT) methods

Two acoustic NDT methods may be applicable to ceramic nanocomposites. The first of these, acoustic emission (AE), is simply the monitoring of stress waves generated by a dynamic process occurring within or on the surface of a material by means of a sensitive transducer, usually of the piezoelectric type. The dynamic process of most interest as far as ceramic nanocomposites are concerned is the growth of microcracks and, indeed, AE has been successfully used to monitor microcracking during mechanical testing (Aeberli and Rawlings 1983, Dalgleish *et al.* 1980) and thermal shock treatments (Thompson and Rawlings 1988). It should be noted that AE can only detect active defects (i.e. growing microcracks) and is therefore suitable for the continuous monitoring of damage occurring during production or service but will not detect flaws in any post-production or post-service assessment of a component that is not stressed.

   On the other hand, the acousto-ultrasonic (AU) technique is capable of detecting flaws in post-production and post-service ceramic nanocomposites. In the AU method, stress waves are introduced into the component under examination by means of a pulser and, after travelling through the component, the waves are detected by another transducer. The signals from the transducer are analyzed in a similar manner to that employed in AE. Thus AU monitoring assesses the general condition of the volume of material between the pulser and the receiving transducer and does not attempt to locate individual flaws. It follows that AU may be particularly appropriate for non-destructive testing of ceramic nanocomposites as the defects are usually small, numerous and dispersed: it can be used for detecting flaws in ceramic nanocomposites, as has been shown for the sensitivity of degree of crystallization of a glass-ceramic (De *et al.* 1987) and for the evaluation of percentage porosity in sintered glass and carbon-bonded carbon fibre based composites (Aduda and Rawlings 1990).

1.9   Conclusions

The design principles for the fabrication of different types of high-performance thermal shock resistant ceramic nanocomposites with improved mechanical properties have been highlighted in this chapter. Pertinent factors such as interface characteristics, densification methods and the role of nano-size particulate dispersion in the development of thermal shock resistant and flame retardant nanoceramic materials have also been reviewed. Various test methods for the characterization and evaluation of ceramic nanocomposites have been briefly covered.

To obtain dense bulk ceramic nanocomposites, it is essential to decrease either the sintering temperature or retention time at the highest point, or to employ hot-pressing, hot isostatic pressing, gas pressure sintering or a fast consolidation technique such as microwave sintering or spark plasma sintering. The overall conclusion from the studies on ceramic nanocomposites reported in this chapter can be summarized as follows.

1. A composite formation with nanostructured coatings based on ceramic powders is very attractive for cutting tools and wear-resistant applications. Pulsed electric current sintering can be used for the fabrication of such nanocomposites.

2. Solid-state sintering processes are specifically applied for the processing of oxide based nanocomposite materials and liquid-phase sintering processes are applied for the processing of non-oxide based nanocomposite materials (Niihara 1991).

3. Monolithic Al_2O_3 and MgO exhibit significant degradation of strength at high temperatures, whereas Al_2O_3/Si_3N_4, Al_2O_3/SiC and MgO/SiC nanocomposites show notable improvementes in high-temperature strength up to 1000°C.

4. Si_3N_4/SiC nanocomposites do not exhibit significant degradation of strength up to 1400°C. The fracture strength at 1400°C was over 1000 MPa and about 900 MPa even at 1500°C for a 32 vol% SiC nanocomposite in which nano-sized SiC particles were dispersed not only within the Si_3N_4 matrix but also at the grain boundaries.

5. Researchers have observed that $Al_2O_3/10\,vol\%$ SWCNT nanocomposites with uniformly distributed SWCNTs at the grain boundaries show creep resistance over two orders of magnitude larger than pure Al_2O_3 of the same grain size (Katagiri *et al.* 1987). This dramatic increase in creep resistance, due to partial obstruction of grain boundary sliding by the presence of SWCNTs at the grain boundaries, was considered a new class of high-temperature resistant nanocomposite.

6. To obtain full densification and to overcome the problem of metal softening in WC-based nanocomposites using the SPS technique, $6\,wt\%$ nano-sized ZrO_2 was incorporated into submicrometre WC. This led to near theoretical densification ($> 99\%$) at a considerably lower sintering temperature of $1300°C$ and holding time of 5 min. The role of ZrO_2 in enhancing densification kinetics was also critically analyzed in this investigation (Tokita 1999). WC-based nanocomposites have also been shown to demonstrate excellent wear resistance (Biswas *et al.* 2007, Venkateswaran *et al.* 2005).

7. The toughness of 3Y-TZP/Mo nanocomposites containing up to $50\,vol\%$ Mo increased dramatically to a maximum value of $18.0\,MPa.m^{1/2}$.

8. Superplasticized nanoceramic materials can be applied in many technical areas including aerospace, energy, electronics, biology, etc., which often require complex shapes to be manufactured at low cost. Extensive potential applications and the possibility of processing dense ceramics and forming complex structures by superplasticity have been the driving force for the appearance and fast development of a large number of nano–nano ceramic systems with superplastic capabilities. The superplastic deformation of a covalent crystal composite based on Si_3N_4/SiC nano–nano ceramics, which could be elongated by more than 150% at $1600°C$, has been reported (Wakai *et al.* 1990).

1.10   Future trends

Further research into enhancing the supertoughness, serviceability, thermoforming and machinability of ceramic nano–nano composites is still required. Additionally, multifunctional ceramic nano–nano composites need to be developed by incorporating nano-size particles or fibres as reinforcement in the nano-matrix; this would integrate both thermal shock resistant properties and some electrical, magnetic, optical and oxidation resistant functions. The viability of some structural ceramic nanocomposites with strong potential for future applications is now summarized.

1.10.1 Polymer-derived ceramic nanocomposites

Intense efforts have been made to develop and use new materials for high-temperature applications due to environmental and economical reasons (Belmonte 2006). Higher efficiencies in energy production and decreasing harmful emissions (e.g. CO_2 or NO_x) can, for instance, be achieved by increasing the operating temperature of engines or turbines as well as by reducing the weight of their components. Within this context, ceramic nanocomposites have been shown to be promising candidates for use under extreme conditions as they exhibit good high-temperature properties such as high strength, creep resistance, thermal shock resistance and stability in oxidative and corrosive environments. Future ceramic nanocomposites will need to have some crucial properties such as oxidation resistance, no phase transformation from ambient to operating temperature, chemical stability, low volatility, high resistance to creep deformation, sufficient toughness at ambient temperature and thermal shock resistance. Polymer-derived SiOC/HfO_2 ceramic nanocomposites have been prepared by Ionescu et al. (2010) via chemical modification of a commercially available polysilsesquioxane by hafnium tetra (n-butoxide). The modified polysilsesquioxane-based materials were cross-linked and subsequently pyrolyzed at $1100°C$ in an argon atmosphere to obtain SiOC/HfO_2 ceramic nanocomposites. Annealing experiments at temperatures between 1300 and $1600°C$ were performed and the annealed materials were investigated with respect to chemical composition and microstructure. This ceramic nanocomposite will be able to exhibit a remarkably improved thermal stability up to $1600°C$ in comparison with hafnia-free silicon oxycarbide.

1.10.2 Sol–gel synthesized HfO_2–TiO_2 nanostructured films

Hafnium titanate ceramics are low thermal expansion materials with a high refractoryness (Carlson et al. 1977, Lynch and Morosin 1972, Ruh et al. 1976). Hafnium titanate films are generating increasing interest because of their potential application as high-k dielectric materials for the semiconductor industry. Kidchob et al. (2008) investigated sol–gel processing as an alternative route to obtain hafnium titanate thin films. Hafnia-titania films of different compositions were synthesized using $HfCl_4$ and $TiCl_4$ as precursors. The HfO_2–TiO_2 system with $50\,mol\%$ HfO_2 allowed the formation of a hafnium titanate film after annealing at $1000°C$. The film exhibited a homogeneous nanocrystalline structure and a monoclinic hafnium titanate phase that has never been obtained before in a thin film. The homogeneously distributed nanocrystals had an average size of $50\,nm$. Different compositions, with higher or lower hafnia contents, have been used for the production of anatase crystalline films after annealing at 1000°C.

1.10.3 Superstrong and tough ceramic nanocomposites

This chapter has highlighted the concept of developing superstrong and tough fire retardant ceramic nanocomposites. Much previous research has aimed at improving the fracture toughness of microcomposites. With ceramic nanocomposites, it is possible to manufacture superstrong, super-plastic, supertough, super chemical resistant and thermal shock resistant materials by hybridization of microcomposites and nanocomposites and using nano–nano processing. Further research on platelet-reinforced nanocomposites, whisker-reinforced nanocomposites and nano–nano composites with high aspect ratio ceramic nanofibres in a nano matrix is needed for future severe applications.

1.11   Sources of further information and advice

This chapter has described various aspects of ceramic nanotechnology with special emphasis on thermal shock resistance and flame retardancy properties. The references, review articles, web sites and books listed below give more information on the fabrication, properties and possible applications of ceramic nanocomposites.

- Cherradi N *et al.* (1994) Worldwide trends in functional gradient materials research and development, *Compos. Eng.*, 4(8), 883–894.
- Dresselhaus MS, Dresselhaus G and Eklund PC, editors (1996) *Science of fullerenes and carbon nanotubes.* San Diego, CA: Academic Press.
- Erdogan F (1995) Fracture mechanics of functionally graded materials, *Mater. Res. Soc. Bull.*, 20(1), 43–44.
- Harris PJF (1999) *Carbon nanotubes and related structures – new materials for the twenty-first century.* Cambridge: Cambridge University Press.
- http://fgmdb.nal.go.jp
- http://fgmdb.nal.go.jp/e_whatsfgm.html
- http://ncn.f2g.net/
- http://ninas.mit.edu/lexcom/www/multitherm.html
- http://www.nanotube.msu.edu/
- Kashiwagi T (2007) *Flame retardant mechanism of the nanotubes-based nanocomposites.* Gaithersburg, MD: National Institute of Standards and Technology. Final report. NIST GCR 07-912.
- Li H *et al.* (2000) Experimental investigation of the quasi-static fracture of functionally graded materials, *Int. J. Solid. Struct.*, 37, 3715–3732.

- Markworth AJ *et al.* (1995) Modelling studies applied to functionally graded materials, *J. Mater. Sci.*, 30, 2183–2193.
- Miyamoto Y, Kaysser WA, Rabin BH, Kawasaki A and Ford RG (1999) *Functionally graded materials: design, processing and applications.* Boston/Dordrecht/London: Kluwer.
- Morgan AB and Wilkie CA, editors (2007) *Flame retardant polymer nanocomposites.* Hoboken, NJ: John Wiley & Sons.
- Mortensen A and Suresh S (1995) Functionally graded metals and metal-ceramic composites: part I processing, *Int. Mater. Rev.*, 40(6), 239–265.
- *MRS Bulletin* (2003) Volume 28, Issue 3.
- Nelson GL and Wilkie CA, editors (2001) *ACS Symposium Series #797: Fire and polymers: materials solutions for hazard prevention.* Washington, DC: American Chemical Society.
- Neubrand A and Rodel J (1997) Gradient materials: an overview of a novel concept, *Z. Metallk,* 88(5), 358–371.
- Ohring M (1992) *The materials science of thin films.* Boston, MA: Harcourt Brace Jovanovich.
- Suresh S and Mortensen A (1998) *Fundamentals of functionally graded materials.* Cambridge: Woodhead.
- Tilbrook M *et al.* (2005) Crack propagation in graded composites, *Comp. Sci. Tech.*, 65, 201–220.
- U.S. Environmental Protection Agency (2005). *Environmental profiles of chemical flame-retardant alternatives for low-density polyurethane foam, volume 2: chemical hazard reviews.* Washington, DC:US EPA. EPA 742-R-05-002B.

1.12 References

Aduda, A. O. B. and Rawlings, R. D. (1990), Presentation at *CIMTEC World Ceramic Congress*, Montecatini Terme, Italy. Elsevier, New York.

Aeberli, K. E. and Rawlings, R. D. (1983), 'Effect of simulated body environments on crack propagation in alumina', *J. Mater. Sci. Lett.*, 2, 215–220.

Anderson, A. and Rowcliffe, D. J. (1996), 'Indentation thermal shock test for ceramics', *J. Am. Ceram. Soc.*, 79(6), 1509–1514.

Anderson, A. and Rowcliffe, D. J. (1998), 'Thermal cycling of indented ceramic material', *J. Eur. Ceram. Soc.*, 18, 2065–2071.

Anstis, G. R., Chantikul, P., Lawn, B. R. and Marshall, D. B. (1981), 'A critical evaluation of indentation techniques for measuring fracture toughness: 1, Direct crack measurements', *J. Am. Ceram. Soc.*, 64(9), 533–538.

Anya, C. C. and Roberts, S. G. (1996), 'Indentation fracture toughness and surface flaw analysis of sintered alumina/SiC Nanocomposites', *J. Eur. Ceram. Soc.*, 16, 1107–1114.

Anya, C. C. and Roberts, S. G. (1997), 'Pressureless sintering and elastic constants of alumina/SiC nanocomposites', *J. Eur. Ceram. Soc.*, 17, 565–573.

Banerjee, R. and Bose, N. R. (2006), 'Thermal shock of ceramic matrix composites', in Low, I. M. (ed.) *Ceramic Matrix Composites, Microstructure, Properties and Applications*, Woodhead Publishing, Cambridge, 58–98.

Baste, U. (1993), 'Thermal shock and cyclic loading of ceramic parts in stationary gas turbines', in Schneider G. A. and Petzow G. (eds), *Thermal Shock and Thermal Fatigue Behavior of Advanced Ceramics*, Kluwer Academic, Dordrecht, 87–97.

Basu, B. (2005), 'Toughening of yttria-stabilized tetragonal zirconia ceramics', *Int. Mater. Rev.*, 50(4), 239–255.

Becher, P. F. (1981), 'Transient thermal stress behaviour in ZrO_2-toughened Al_2O_3', *J. Am. Ceram. Soc.*, 64(1), 37–39.

Becher, P. F. and Warwick, W. H. (1993), 'Factors influencing the thaermal shock behaviour of ceramics', in Schneider, G. A. and Petzow, G. (eds), *Thermal Shock and Thermal Fatigue Behavior of Advanced Ceramics*, Kluwer Academic, Dordrecht, 37–48.

Becher, P. F. and Wei, G. C. (1984), 'Toughening behaviour in SiC-whisker-reinforced alumina', *J. Am. Ceram. Soc.*, 67, C267–C269.

Becher, P. F., Lewis, D. III, Carman, K. R. and Gonzalez, A. C. (1980), 'Thermal shock resistance of ceramics: size and geometry effects in quench tests', *J. Am. Ceram. Soc. Bull.*, 59(5), 542–545.

Belmonte, M. (2006), 'Advanced ceramic materials for high temperature applications', *Adv. Eng. Mater.*, 8, 693–703.

Berger, S., Porat, R. and Rosen, R. (1997), 'Nanocrystalline materials: A study of WC-based hard metals', *Prog. Mater. Sci.*, 42, 311–320.

Bhaduri, S. and Bhaduri, S. B. (1997), 'Enhanced low temperature toughness of alumina-zirconia nano/nano composites', *Nanostruc. Mater.*, 8(6), 755–763.

Biswas, K., Mukhopadhyay, A., Basu, B. and Chattopadhyay, K. (2007), 'Densification and microstructure development in spark plasma sintered WC-6 wt% ZrO_2 nanocomposites', *J. Mater. Res.*, 22(6), 1491–1550.

Bock, A., Schubert, D. and Lux, B. (1992), 'Inhibition of grain growth on submicron cemented carbides', *J. Powder Metall. Ind.*, 24(1), 20–26.

Bonney, L. A. and Cooper, R. R. (1990), 'Reaction-layer interfaces in SiC fibre reinforced glass-ceramics: a high-resolution scanning transmission electron microscopy analysis', *J. Am. Ceram. Soc.*, 73(10), 2916–2921.

Brennan, J. J. (1986), 'Tailoring multiphase and composite ceramics', *Mater. Sci. Res.*, 20, 549–560.

Brennan, J. J. (1987), 'Interfacial studies of SiC-fibre-reinforced glass-ceramic-matrix composites', Final Report R87-917546-4, ONR Contract N00014-82-C-0096, Oct. 15, 1987.

Brennan, J. J., Nutt, S. R. and Sun, E. Y. (1992), 'Interfacial microstructure and stability of BN coated Nicalon fibre/glass-ceramic matrix composites', in Evans A. G. and Naslain, R. (eds) *Ceramic Transactions, Vol. 58, High Temperature Ceramic Matrix Composites II: Manufacturing and Materials Development*, American Ceramic Society, Westerville, OH, 53–64.

Breval, E., Dodds, G. and Pantano, C. G. (1985), 'Properties and microstructure of

Ni-alumina composite materials prepared by the sol-gel method', *Mater. Res. Bull.*, 20, 1191–1205.

Buljun, S. T., Baldoni, J. G. and Huckabee, M. L. (1987), 'Si_3N_4–SiC composites,' *J. Am. Ceram. Soc. Bull.*, 66, 347–352.

Butkus, L. M., Zawada, P. L. and Hartman, G. A. (1992), 'Fatigue test methodology and results for ceramic matrix at room and elevated temperature', in *ASTM Special Technical Publication 1157*, American Society for Testing and Materials, Philadelphia, PA, 52–68.

Carlson, G. A., Anderson, J. L., Briesmeister, R. A., Skaggs, S. R. and Ruh, R. (1977), 'Coefficient of thermal expansion and dynamic response to pulsed energy deposition in HfO_2-TiO_2 compositions', *J. Am. Ceram. Soc.*, 60(11–12), 508–510.

Cha, S. I. and Hong, S. H. (2003), 'Microstructures of binderless tungsten carbides sintered by spark plasma sintering process', *Mater. Sci. Eng. A*, 356, 381–389.

Cha, S. I., Hong, S. H. and Kim, B. K. (2003a), 'Spark plasma sintering behavior of nanocrystalline WC-10Co cemented carbide powders', *J. Mater. Sci. Eng. A*, 351, 31–38.

Cha, S. I., Hong, S. H., Ha, G. K. and Kim, B. K. (2003b), 'Microstructure and mechanical properties of nanocrystalline WC-10Co cemented carbides', *Scr. Mater.*, 44, 1535–1539.

Chen, I. W. and Wang, X. H. (2000), 'Sintering dense nanocrystalline ceramics without final-stage grain-growth', *Nature*, 404, 168–171.

Chen, P. L. and Chen, I. W. (1994), 'The role of defect interaction in boundary mobility and cation diffusivity of CeO_2', *J. Am. Ceram. Soc.*, 77(9), 2289–2297.

Chen, P. L. and Chen, I. W. (1996a), 'Grain boundary mobility in Y_2O_3: Defect mechanism and dopant effects', *J. Am. Ceram. Soc.*, 79(7), 1801–1809.

Chen, P. L. and Chen, I. W. (1996b), 'Grain growth in CeO_2: dopant effects, defect mechanism, and solute drag', *J. Am. Ceram. Soc.*, 79(7), 1793–1800.

Cheng, C. M. (1951), 'Resistance to thermal shock', *J. Am. Rocket. Soc.*, 21, 147–153.

Cho, S. A., Puerta, M., Cols, B. *et al.* (1980), *Powder Metall. Int.*, 12, 192.

Chokshi, A. H. (1993), 'Superplasticity in fine-grained ceramic and ceramic composites: current understanding and future prospects', *Mater. Sci. Eng. A*, 166, 119–133.

Cichocki, F. R. and Trumble, Jr. K. P. (1998), 'Tailored porosity gradients via colloidal infiltration of compression-moulded sponges', *J. Am. Ceram. Soc.*, 81(6), 1661–1664.

Clarke, D. R. (1987), 'On the equilibrium thickness of intergranular glass phases in ceramic materials', *J. Am. Ceram. Soc.*, 70, 15–22.

Claussen, N. (1985), *Mater. Sci. Eng.*, 71, 23–28.

Claussen, N., Steeb, J. and Pabst, R. F. (1977), 'Effect of induced microcracking on the fracture toughness of ceramics,' *J. Am. Ceram. Soc. Bull.*, 56, 559–562.

Cooper, R. F. and Chung, K. (1987), 'Structure and chemistry of fibre-matrix interfaces in silicon carbide fibre-reinforced glass-ceramic composites: an electron microscopy analysis', *J. Mater. Sci.*, 22, 3148–3160.

Cotton, B. A. and Mayo, M. J. (1996), 'Fracture toughness of nanocrystalline ZrO_2-3 N mol% Y_2O_3 determined by Vickers indentation', *Scr. Mater.*, 34(5), 809–814.

Dalgleish, B. J., Pratt, P. L., Rawlings, R. D. and Fakhr, A. (1980), *Mater. Sci. Eng.*, 45, 9.

Davidge, R. W. (1979), *Mechanical Behavior of Ceramics*, Cambridge University Press, Cambridge.

Davidge, R. W., Brook, R. J., Cambier, F., Poorteman, M., Leriche, A., Sullivan, D. O., Hampshire, S. and Kennedy, T. (1997), 'Fabrication, properties, and modelling of engineering ceramics reinforced with nanoparticles of silicon carbide', *Br. Ceram. Trans.*, 96(3), 121–127.

De, A., Phani, K. K. and Kumar, S. (1987), 'Acousto–ultrasonic study on glass-ceramics in the system MgO–Al$_2$O$_3$–SiO$_2$', *J. Mater. Sci. Lett.*, 6, 17–19.

Evans, A. G. and Blumenthal, W. (1983), in Bradt, R. C., Evans, A. G., Hasselman, D. P. H. and Lange, F. F. (eds) *Fracture Mechanics of Ceramics*, Vol. 5, Plenum, New York, 423–448.

Ferroni, L. P. and Pezzotti, G. (2002), 'Evidence for bulk residual stress strengthening in alumina/SiC nanocomposites', *J. Am. Ceram. Soc.*, 85(8), 2033–2038.

Folsom, C. A., Lee, S. S. and Zawada, L. P. (unpublished) The intermediate temperature embrittlement phenomenon in a Nicalon-fibre-reinforced ceramic matrix composite with boron containing compounds.

Gao, L., Jin, X. H. and Zheng, S. (2004), *Ceramic Nanocomposites*, Chemical Engineering Publishers, Beijing.

Garvie, R., Hannink, H. and Pascoe, R. T. (1975), 'Ceramic steel', *Nature*, 258, 703–708.

Gerdes, T. and Willert-Porada, M. (1994), 'Microwave sintering of metal-ceramic and ceramic-ceramic composites', *Mater. Res. Soc. Symp.*, 347, 531.

Greskovich, C. and Palm, J. A. (1980), Observations on the fracture toughness of β-Si$_3$N$_4$–β-SiC composites', *J. Am. Ceram. Soc.*, 63, 597–599.

Grossman, D. G. (1972), 'machinable glass-ceramics based on tetrasilicic mica', *J. Am. Ceram. Soc.*, 55, 446–449.

Gupta, T. K., Lange, F. F. and Bechtold, J. H. (1978), 'Effect of stress-induced phase transformation on the properties of polycrystalline zirconia containing metastable tetragonal phase', *J. Mater. Sci.*, 13, 1464–1470.

Hanigofsky, J. A., More, K. L., Lackey, W. J., Lee, W. Y. and Freeman, G. B. (1991), 'Composition and microstructure of chemically vapor-deposited-boron nitride, aluminum nitride and boron nitride + aluminum nitride composites', *J. Am. Ceram. Soc.*, 74(2), 301–305.

Hannink, R. H. J., Kelly, P. M. and Muddle, B. C. (2000), 'Transformation toughening in zirconia-containing ceramics', *J. Am. Ceram. Soc.*, 83, 461–487.

Hansson, T., Warren, R. and Wasen, J. (1993), 'Fracture toughness anisotropy and toughening mechanisms of a hot-pressed alumina reinforced with silicon carbide whiskers', *J. Am. Ceram. Soc.*, 76(4), 841–848.

Hartman, G. A. and Ashbaugh, N. E. (1990), 'A fracture mechanics test automation system for a basic research laboratory', in *ASTM Special Technical Publication 1092*, American Society for Testing and Materials, Philadelphia, PA, 95–110

Hing, P. (1980), *J. Sci. Ceram.*, 10, 521.

Hoffman, M., Rodel, J., Sternitzke, M. *et al.* (1996), 'Fracture toughness and subcritical crack growth in alumina/silicon carbide 'nanocomposites', *Frac. Mech. Ceram.*, 12, 179.

Holand, W., Vogel, W., Mortiere, W. J., Duvigneaud, P. H., Naessens, G. and Plumat, E. (1983), 'A new type of phlogopite crystal in machinable glass-ceramics', *Glass Technol.*, 24, 318–322.

Holmes, J. W. and Cho, C. (1992), 'Experimental observations of frictional heating in fibre-reinforced ceramics', *J. Am. Ceram. Soc.*, 75(4), 929–938.

Homeney, J., Vaughn, W. L. and Ferber, M. K. (1987), 'Processing and mechanical properties of SiC–whisker–Al$_2$O$_3$ matrix composites', *J. Am. Ceram. Soc. Bull.*, 66, 333–338.

Hwang, S. L. and Chen, I. W. (1994), 'Reaction hot pressing of α and β SiAlON ceramics', *J. Am. Ceram. Soc.*, 77, 165–171.

Imasato, S., Tokumoto, K., Kitada, T. and Sakaguchi, S. (1995), 'Properties of ultra fine grain binderless cemented carbide RCCFN', *Int. J. Ref. Met. Hard Mater.*, 13, 305–312.

Ionescu, E., Papendorf, B., Kleebe, H. J. and Riedel, R. (2010), 'Polymer-derived silicon oxycarbide/hafnia ceramic nanocomposites, part ii: stability toward decomposition and microstructure evolution at T $>>$ 1000°C', *J. Am. Ceram. Soc.*, 93, 1774–1782.

Izaki, K., Hakkei, K., Ando, K. and Niihara, K. (1980), in Mackenzie, J. M. and D. R. Ulrich (eds) *Ultrastructure Processing of Advanced Ceramics*, John Wiley & Sons, New York, 891–900.

Izaki, K., Nakahira, A. and Niihara, K. (1991), *J. Japan Soc. Powder Powder Metall.*, 38, 357–361.

Jenkins, M. G., Salem, J. A. and Seshadri, S. G. (1989), 'Fracture of a TiB$_2$ particle/SiC matrix composite at elevated temperature', *J. Comp. Mater.*, 23, 77–90.

Jeong, Y. K. and Niihara, K. (1997), 'Microstructure and mechanical properties of pressureless-sintered of Al$_2$O$_3$-SiC nanocomposites', *J. Nanostruct. Mater.*, 9, 193–196.

Jero, P. D., Parthasarathy, T. A. and Kerans, R. J. (1992), 'Measurement of interface properties from push-out tests', *Ceram. Eng. Sci. Proc.*, 13(7–8), 54–63.

Jhao, J., Stearns, L. C., Harmer, M. P., Chan, H. M., Miller, G. A. and Cook, R. F. (1993), 'Mechanical behavior of alumina-silicon carbide nanocomposites', *J. Am. Ceram. Soc.*, 76(2), 503–510.

Ji, Y. and Yeomans, J. A. (2002), 'Processing and mechanical properties of Al$_2$O$_3$-5vol% Cr nanocomposites', *J. Eur. Ceram. Soc.*, 22(12), 1905–2080.

Jia, K., Fischer, B. and Gallios, T. (1998), 'Microstructure, hardness and toughness of nanostructured and conventional Wc-Co composites', *J. Nanostruct. Mater.*, 10(5), 875–891.

Jimenez-Melendo, M., Dominguez-Rodriguez, A. and Bravo-Leon, A. (1998), 'Superplastic flow in fine-grained yttria-stabilized zirconium polycrystals: constitutive equation and deformation mechanisms', *J. Am. Ceram. Soc.*, 81, 2761–2776.

Jones, R. H., Giancarlie, L., Hasegawa, A., Katoh, Y., Kohyama, A., Riccardi, B., Snead, L. L. and Weber, W. J. (2002), 'Promise and challenges of SiCf/SiC composites for fusion energy applications', *J. Nucl. Mater.*, 307–311, 1057–1072.

Kastritseas, C., Smith, P. and Yeomans, J. (2006), 'Thermal shock of ceramic matrix composites', in Low, I. M. (ed.) *Ceramic Matrix Composites, Microstructure, Properties and Applications*, Woodhead Publishing, Cambridge, 400–433.

Katagiri, G., Ishida, H., Ishitani, A. and Masaki, T. (1987), 'The stress-induced transformation by fracture in Y_2O_3-containing tetragonal zirconia polycrystals', *Mater. Res. Soc. Symp. Proc.*, 78, 43–49.

Kawai, C. and Wakamatsu, S. (1995), 'Synthesis of a functionally gradient material based on C/C composites using an electro-deposition method', *J. Mater. Sci. Lett.*, 14, 467–469.

Kawamura, S. (1979), 'Glass-ceramics' (in Japanese), in Sakka, S., Sakaino, T. and Takahashi, K. (eds) *Glass Handbook*, Asakuma-shoten, Tokyo, 197–200.

Kerans, R. J. and Parthasarathy, T. A. (1991), 'Theoretical analysis of the fibre pullout and pushout tests', *J. Am. Ceram. Soc.*, 74(7), 1585–1596.

Kervadec, D. and Chermant, J. L. (1992), 'Some aspects of the morphology and creep behavior of a unidirectional SiC-MLAS material', *Frac. Mech. Ceram.*, 10, 459–471.

Kervadec, D. and Chermant, J. L. (1993), 'Viscoelastic deformation during creep of ID SiC/MLAS composite', in *High Temperature Ceramic Matrix Composites, Proceeding of the 6th European Conference on Composite Materials* (Sept. 20–24, 1993), Cambridge University Press, Cambridge, 649–657.

Kidchob, T., Falcaro, P., Schiavuta, P., Enzo, S. and Innocenzi, P. (2008), 'Formation of monoclinic hafnium titanate thin films via the sol-gel method', *J. Am. Ceram. Soc.*, 91(7), 2112–2116.

Kim, B. K., Ha, G. H. and Lee, D. W. (1997), 'Sintering and microstructure of nanophase WC/Co hard metals', *J. Mater. Process. Technol.*, 63, 317–321.

Kim, B. N., Hiraga, K., Morita, K. and Sakka, Y. (2001), 'A high-strain-rate superplastic ceramics', *Nature*, 413, 288–291.

Kim, H. C., Shon, I. J., Yoon, J. K., Lee, S. K. and Munir, Z. A. (2006), 'One step synthesis and densification of ultra fine WC by high-frequency induction combustion', *Int. J. Ref. Met. Hard Mater.*, 24, 202–209.

Kim, H. C., Shon, I. J., Garay, J. E. and Munir, Z. A. (2004), 'Consolidation and properties of binderless sub-micron WC by field activated sintering', *Int. J. Ref. Met. Hard Mater.*, 22, 257–264.

Kim, H. C., Shon, J. J., Yoon, J. K. and Doh, J. M. (2007a), 'Consolidation of ultra fine WC and WC-Co hard materials by pulsed current activated sintering and its mechanical properties', *Int. J. Ref. Met. Hard Mater.*, 25(1), 46–52.

Kim, H. C., Jeong, I. K., Shon, J. J., Ko, I. Y. and Doh, J. M. (2007b), 'Fabrication of WC-8 wt% Co hard materials by two-rapid sintering processes', *Int. J. Ref. Met. Hard Mater.*, 25(4), 336–340.

Kmetz, M. A., Laliberte, J. M., Willis, W. S., Suib, S. L. and Galasso, F. S. (1991), 'Synthesis, characterization, and testing strength of CVI SiC/BN/SiC composites', *Ceram. Eng. Sci. Proc.*, 12(9–10), 2161–2174.

Kolaska, H. (1992), 'The dawn of the hardmetal age', *J. Powder Metall. Int.*, 24(5), 311–314.

Kondo, N., Suzuki, Y. and Ohji, T. (1999), 'Superplastic sinter-forging of silicon nitride with anisotropic microstructure formation', *J. Am. Ceram. Soc.*, 82, 1067–1069.

Lange, F. F. (1973), 'Effect of microstructure on strength of Si_3N_4–SiC composite system', *J. Am. Ceram. Soc.*, 56, 445–450.

Lange, F. F. (1974), *J. Ann. Rev. Mater. Sci.*, 4, 365–390

Lange, F. F. (1982), 'Transformation toughening. Part 1: Size effects associated with the thermodynamics of constrained transformations; Part 2: Contribution to fracture toughness; Part 3: Experimental observations in the ZrO_2–Y_2O_3 system; Part 4: Fabrication , fracture toughness and strength of Al_2O_3–ZrO_2 composites; Part 5: Effect of temperature and alloy on fracture toughness', *J. Mater. Sci.*, 17, 225–263.

Lee, S. S., Zawada, P. L. Staehler, M. J. and Folsom, A. C. (1998), 'Mechanical behavior and high-temperature performance of a woven Nicalon^{TM}/Si-N-C ceramic-matrix composite', *J. Am. Ceram. Soc.*, 81(7), 1797–1811.

Lenel, F. B. (1980), *Powder Metallurgy Principles and Applications*, Metal Powder Industries Federation, Princeton, NJ, 383.

Levitt, S. R. (1973), 'High-strength graphite fibre/lithium aluminosilicate composites', *J. Mater. Sci.*, 8, 793–806.

Li, M. and Reece, M. J. (2000), 'Influence of grain size on the indentation-fatigue behaviour of alumina', *J. Am. Ceram. Soc.*, 83(4), 967–970.

Limpichaipanit, A. and Todd, R. I. (2009), 'The relationship between microstructure, fracture, and abrasive wear in Al_2O_3/SiC nanocomposites and microcomposites containing 5 and 10 % SiC', *J. Eur. Ceram. Soc.*, 29, 2841–2848.

Lu, T. J. and Fleck, N. A. (1998), 'The thermal shock resistance of solids', *Acta. Mater.*, 46, 4755–4768.

Lynch, R. W. and Morosin, B. (1972), 'Thermal expansion, compressibility and polymorphism in hafnium and zirconium titanates', *J. Am. Ceram. Soc.*, 55(8), 409–413.

Maensiri, S. and Roberts, S. G. (2002), 'Thermal shock resistance of sintered alumina/silicon carbide nanocomposites evaluated by indentation techniques', *J. Am. Ceram. Soc.*, 85(8), 1971–1978.

Manson, S. S. (1966), *Thermal Stress and Low-cycle Fatigue*, McGraw-Hill, New York.

Marple, B. R. and Boulanger, J. (1994), 'Graded casting and materials with continuous gradients', *J. Am. Ceram. Soc.*, 77(10), 2747–2750.

Masaki, K. (1986), 'Mechanical properties of toughened ZrO_2–Y_2O_3 ceramics', *J. Am. Ceram. Soc.*, 69, 638–640.

Masumoto, Y., Takechi, K. and Imasato, S. (1986), 'Corrosion resistance of cemented carbide', *J. Nippon Tungsten Rev.*, 19, 26–30.

Mayo, M. J. (1996), 'Processing of nanocrystalline ceramics from ultrafine particles', *Int. Mater. Rev.*, 41, 85–115.

McHugh, C. O., Whalen, T. J. and Humenik, M. (1966), 'Dispersion – strengthened aluminium oxide', *J. Am. Ceram. Soc.*, 49, 486–491.

McMurtry, C. H., Boecker, W. D. G., Seshadri, S.G., Zanghi, J. S. and Garnier, J. E. (1987), 'Microstructure and material properties of SiC-TiB$_2$ particulate composites', *J. Am. Ceram. Soc. Bull.*, 66, 325–329.

Melendez-Martinez, J. J. and Dominguez-Rodriguez, A. (2004), 'Creep of silicon nitride', *Prog. Mater. Sci.*, 49, 19–107.

Mimurada, J., Nakano, M., Sasaki, K., Ykuhara, Y. and Sakuma, T. (2001), 'Effect of cation doping on the superplastic flow in yttria-stabilized tetragonal zirconia polycrystals', *J. Am. Ceram. Soc.*, 84, 1817–1821.

Mizuno, M., Obata, S., Takayama, S., Ito, S., Kato, N., Hirai, T. and Sato, M.

(2004), 'Sintering of alumina by 2.45 GHZ microwave heating', *J. Eur. Ceram. Soc.*, 24, 387–391.

Mizutani, T., Kusunose, T., Sando, M. *et al.* (1997), 'Fabrication and properties of nano-sized BN-particulate dispersed Sialon ceramics', *Ceram. Eng. Sci. Proc.*, 18(4B), 669–677.

Morita, K., Hiraga, K. and Kim, B. N. (2004), 'Effect of minor SiO_2 addition on the creep behaviour of superplastic tetragonal ZrO_2' *Acta Mater.*, 52, 3355–3364.

Mukhopadhyay, A. and Basu, B. (2007), 'Consolidation-microstructure-property relationships in bulk nanoceramics and ceramic nanocomposites: a review', *Int. Mater. Rev.*, 52(5), 257–288.

Mukhopadhyay, A, Basu, B., Bakshi, S. D. and Mishra, S. K. (2007), 'Pressureless sintering of ZrO_2-ZrB_2 composite: microstructure and properties', *Int. J. Ref. Metal. Hard Mater.*, 25, 179–188.

Munakata, H., Hayashi, T., Suzuki, H., Saito, H. (1986), 'Pressureless sintering of with Y_2O_3 and Al_2O_3', *J. Mater. Sci.*, 21, 3501–3508.

Munz, D. and Fett, T. (1999), *Ceramics: Mechanical Properties, Failure Behavior, Matrials Selection*, Springer, New York.

Naerheim, Y. (1986), *Powder Metall. Int.*, 18, 158.

Nagano, T., Gu, H., Shinoda, Y., Zhan, D., Mitomo, M. and Wakai, F. (1999), 'Tensile ductility of liquid-phase sintered β-silicon carbide at elevated temperatures', *Mater. Sci. Forum*, 304–306, 507–512.

Nakahira, A. and Niihara, K. (1992), 'Sintering behaviors and consolidation process for Al_2O_3/SiC nanocomposites', *J. Ceram. Soc. Japan*, 100(4), 448–453.

Nakahira, A. and Niihara, K. (1993), 'Microstructure and high temperature mechanical properties for Al_2O_3/SiC nanocomposites', *95th Annual Meeting of the American Ceramic Society, Cincinnati, Ohio* (April 18–22, 1993), Paper SXIV-26-93.

Nakahira, A., Niihara, K. and Hirai, T. (1987), *J. Yogvo-Kyokai-Shi*, 94, 767–772.

Nakatani, K., Nagayama, H., Yoshida, H., Yamamoto, T. and Sakuma, T. (2003), 'The effect of grain-boundary segregation on superplastic behaviour in cation-doped 3Y-TZP', *Scr. Mater.*, 49, 791–795.

Naslain, R., Dugne, O., Guette, A., Sevely, J., Brosse, C. R., Rocher, J. P. and Cotteret, J. (1991), 'Boron nitride interphase in ceramic matrix composites', *J. Am. Ceram. Soc.*, 74(10), 2482–2488.

Nawa, M., Sekino, T. and Niihara, K. (1992), *J. Japan. Soc. Powder Powder Metall.*, 39, 1104.

Nawa, M., Sekino, T., Niihara, K. (1994a), 'Fabrication and mechanical behavior of Al_2O_3/Mo nanocomposites', *J. Mater. Sci.*, 29(12), 3185.

Nawa, M., Yamazaki, K., Sekino, T. *et al.* (1994b), 'A new type of nanocomposite in tetragonal zirconia polycrystal-molybdenum system', *J. Mater. Lett.*, 20(5–6), 299–304.

Neubrand, A. and Rodel, J. (1997), 'Gradient materials: an overview of a novel concept', *Z. Metallk.*, 88(5), 358–371.

Nieh, T. G., Wadsworth, J. and Wakai, F. (1991), 'Recent advances in superplastic ceramics and ceramic composites', *Int. Mater. Rev.*, 36, 146–161.

Nieh, T. G., Wadsworth, J. and Sherby, O. D. (1997), *Superplasticity in Metals and Ceramics*, Cambridge University Press, Cambridge.

Niihara, K. (1986), 'Brittleness improvement of ceramic materials', *Bull. Ceram. Soc. Japan*, 21, 581–589.

Niihara, K. (1989), *Electronic Ceramics*, 9, 44–48.

Niihara, K. (1990), *J. Japan Soc. Powder Powder Metall.*, 37, 348–351.

Niihara, K. (1991), 'New design concept of structural ceramics-ceramic nanocomposites', *J. Ceram. Soc. of Japan*, 99(10), 974–982.

Niihara, K. and Hirai, T. (1986), *J. Scramikkusu*, 26, 598–604.

Niihara, K. and Nakahira, A. (1988), *Proceedings of MRS Meeting of Advanced Composites*, Plenum, New York, 129–134.

Niihara, K. and Nakahira, A. (1989), *Proceedings of MRS Meeting of Advanced Composites*, Plenum, New York, 193–200.

Niihara, K. and Nakahira, A. (1990a), *Proceedings of 7th Korea–Japan Seminar on New Ceramics*, 128–133.

Niihara, K. and Nakahira, A. (1990b), in Vincenzini, P. (ed.) *Advanced Structural Inorganic Composites*, Elsevier, New York, 637–664.

Niihara, K., Nakahira, A., Uchiyama, T. and Hirai, T. (1986), in Bradt, R. C., Evans, A. G., Hasselman, D. P. H. and Lange, F. F. (eds) *Fracture Mechanics of Ceramics*, Vol. 7, Plenum, New York, 103–116.

Niihara, K., Hirano, T., Nakahira, A. and Izaki, K. (1988a), *Proceedings of MRS Meeting on Advanced Materials*, Tokyo, 107–112.

Niihara, K., Nakahira, A., Sasaki, G. and Hirabayashi, M. (1988b), *Proceedings of MRS Meeting on Advanced Materials*, Tokyo, 129.

Niihara, K., Hirano, T., Nakahira, A., Suganuma, K. and Izaki, K. (1989a), *J. Japan Soc. Powder Powder Metall.*, 36, 243–247.

Niihara, K., Nakahira, A., Ueda, H. and Sasaki, H. (1989b), *Proceedings of 1st Japan International SAMPE Symposium*, 1120–1125.

Niihara, K., Izaki, K. and Kawakami, T. (1990a), 'Interfaces in Si_3N_4–SiC nanocomposite', *J. Mater. Sci. Lett.*, 9, 598–599.

Niihara, K., Izaki, K. and Nakihara, A. (1990b), *J. Japan Soc. Powder Powder Metall.*, 37, 352–356.

Niihara, K., Suganuma, K. and Izaki, K. (1999), *J. Mater. Sci.*, 9, 112–116.

Nishida, T., Shino, H., Yamauchi, H. and Nishikawa, T. (1987), *J. Zairyo*, 36, 17–21.

Nose, T. and Fujii, T. (1988), 'Evaluation of fracture toughness for ceramic materials by a single-edge precracked-beam method', *J. Am. Ceram. Soc.*, 71, 328–333.

Nowick, A. S., Wang, D. Y., Park, D. S. and Griffith, J. (1979), 'Oxygen-ion conductivity and defect structure of CeO_2', in Vashishta, P. Mundy, J. N. and Shenoy, G. K. (eds) *Fast Ion Transport in Solids*, Elsevier, Amsterdam, 673–681.

Oh, S. T., Lee, J. S., Sekino, T. *et al.* (2001), 'Fabrication of Cu dispersed alumina nanocomposites using Al_2O_3/CuO and Al_2O_3/Cu-nitrate mixtures', *Scr. Mater.*, 44, (8–9), 2117–2120.

Ohji, T., Nakahira, A., Hirano, T. and Niihara, K. (unpublished) Inhibition mechanism of creep in alumina silicon carbide nanocomposites.

Ohji, T., Nakahira, A., Hirano, T. and Niihara, K. (1994), 'Tensile creep behavior of alumina/SiC nanocomposite', *J. Am. Ceram. Soc.*, 77, 3259–3262.

Ohnabe, H., Masaki, S., Onozuka, M., Miyahara, K. and Sasa, T. (1991), 'Potential application of ceramic matrix composites to aero-engine components', *Composites*, 30A, 489–496.

Philips, D. C., Sambell, R. A. and Bowen, D. H. (1972), 'The mechanical properties of carbon-fibre-reinforced Pyrex glass', *J. Mater. Sci.*, 7, 1454–1464.

Prewo, K. M. and Brennan, J. J. (1980), 'High-strength silicon carbide fibre-reinforced glass-matrix composites', *J. Mater. Sci.*, 15, 463–468.

Prewo, K. M. and Brennan, J. J. (1982), 'Silicon carbide fibre-reinforced glass-ceramic-matrix composites exhibiting high strength and toughness', *J. Mater. Sci.*, 17, 2371–2383.

Prewo, K. M., Brennan, J. J. and Layden, G. K. (1986), 'Fibre reinforced glasses and glass–ceramics for high performance applications', *Am. Ceram. Soc. Bull.*, 65, 305–338.

Raj, R. and Ashby, M. F. (1971), 'Grain boundary sliding and diffusional creep', *Metall. Trans.*, 2(4), 1113–1127.

Rankin, D. T., Stiglich, J. J., Petrak, D. R. *et al.* (1971), 'Hot pressing and mechanical properties of Al_2O_3 with an Mo-dispersed phase', *J. Am. Ceram. Soc.*, 54, 277–281.

Reece, M. and Guiu, F. (1991a), 'Repeated indentation method for studying cyclic fatigue in ceramics', *J. Am. Ceram. Soc.*, 73(4), 1004–1013.

Reece, M. and Guiu, F. (1991b), 'Indentation fatigue of high-purity alumina in fluid environments', *J. Am. Ceram. Soc.*, 74(1), 148–154.

Rice, R. W. (1987), *BN Coating of Ceramic Fibres for Ceramic Fibre Composites*, U. S. Patent No. 4642271, Feb. 10, 1987.

Rosenflanz, A. and Chen, I. W. (1998), 'Classical superplasticity of SiAlON ceramics', *J. Am. Ceram. Soc.*, 81, 713–716.

Ruh, R., Hollenberg, G. W., Charles, E. G. and Patel, V. A. (1976), 'Phase relations and thermal expansion in the system HfO_2-TiO_2', *J. Am. Ceram. Soc.*, 59(11–12), 495–499.

Sargent, P. M. and Page, T. F. (1978), *Proc. Br. Ceram. Soc.*, 26, 209.

Sarin, V. K. (1981), 'Cemented carbide cutting tools', in Chin, D. Y. (ed.) *Advances in Powder Technology*, ASM Material Science, Louisville, KY, 253–287.

Sato, E., Morioka, H., Kuribayashi, K. and Sudararaman, D. (1999), 'Effect of small amount of alumina doping on superplastic behavior of tetragonal zirconia'. *J. Mater. Sci.*, 34, 4511–4518.

Sekino, T. and Niihara, K. (1995), 'Microstructural characteristics and mechanical properties for Al_2O_3/metal nanocomposites', *J. Nanostruc. Mater.*, 6(5–8), 663–666.

Sekino, T. and Niihara, K. (1997), 'Fabrication and mechanical properties of fine-tungsten-dispersed alumina-based composites', *J. Mater. Sci.*, 32(15), 3943–3949.

Sekino, T., Nakajima, T., Satoru, U. *et al.* (1997), 'Reduction and sintering of a nickel-dispersed-alumina composite and its properties', *J. Am. Ceram. Soc.*, 80 (5), 1139–1148.

Shalek, P. D., Petrovic, J. J., Hurley, G. F. and Gac, F. G. (1986), 'Hot pressed SiC whisker/Si_3N_4 matrix composites', *J. Am. Ceram. Soc. Bull.*, 65, 351–356.

Shen, L., Tan, B. J., Suib, S. L. and Galasso, F. S. (1992), 'Dip-coating of BN on Nicalon and C-Nicalon fibres', *Mater. Res. Soc. Symp. Proc.*, 249, 227.

Shen, L., Willis, W. S., Galasso, F. S. and Suib, S. L. (1993), 'Coating of BN interfaces on ceramic yarns from boric acid and ammonia', *Ceram. Eng. Sci. Proc.*, 14(7–8), 556–562.

Shen, Z., Peng, H. and Nygren, M. (2003), 'Formidable increase in the superplasticity of ceramics in the presence of an electric field', *Adv. Mater.*, 15, 1006–1009.

Shi, L. X., Shao, G. Q., Duan, X. L., Zh. Yuan, R. and Lin, H. H. (2005), 'Mechanical properties, phases and microstructure of ultrafine hardmetals prepared by WC-6.29Co nanocrystalline composite powder', *J. Mater. Sci. Eng. A.*, 392, 335–339.

Shikata, R., Urata, Y., Shiono, T. and Nishikawa, T. (1990), *J. Japan. Soc. Powder Powder Metall.*, 37, 357.

Siegel, R. W., Chang, S. K., Ash, B. J. *et al.* (2001), 'Mechanical behaviour of polymer and ceramic matrix nanocomposites', *Scr. Mater.*, 44, 2061–2064.

Sigh, R. N. and Brun, M. K. (1987), 'Effect of boron nitride coating on fibre-matrix interactions', *Ceram. Eng. Sci. Proc.*, 8(7–8), 634–643.

Sivaprahasam, D., Chandrasekar, S. B. and Sundaresan, R. (2007), 'Microstructure and mechanical properties of nanocrystalline WC-12 Co consolidated by spark plasama sintering', *Int. J. Ref. Met. Hard Mater.*, 25(2), 144–152.

Stangle, G. C. and Miyamoto, Y. (1995), 'FGM fabrication by combustion systhesis', *MRS Bull.*, XX(1), 52–53.

Stearns, L. C., Jhao, J. and Harmer, M. P. (1992), 'Processing and microstructural development in Al_2O_3-SiC nanocomposites', *J. Eur. Ceram. Soc.*, 10, 473–477.

Suganuma, K., Sasaki, G., Fujita, T. and Niihara, K. (1991), *J. Japan Soc. Powder Powder Metall.*, 38, 374–377.

Sun, E. Y., Nutt, S. R. and Brennan, J. J. (1994), 'Interfacial microstructure and chemistry of SiC/BN dual-coated Nicalon-fibre-reinforced glass-ceramic matrix composites', *J. Am. Ceram. Soc.*, 77(5), 1329–1339.

Sun, E. Y., Nutt, S. R. and Brennan, J. J. (1996), 'High-temperature tensile behavior of a boron nitride-coated silicon carbide-fibre glass-ceramic composite', *J. Am. Ceram. Soc.*, 79(6), 1521–1529.

Sutthiruangwonga, S. and Mori, G. (2003), 'Corrosion properties of Co-based cemented carbides in acidic solutions', *Int. J. Ref. Met. Hard Mater.*, 21, 135–145.

Suzuki, H. (1986), 'Cemented carbide and sintered hard materials', *J. Maruzen*, 262.

Swain, M. V. and Rose, L. R. F. (1986), 'Strength limitations of transformation-toughened zirconia alloys', *J. Am. Ceram. Soc.*, 69, 511–518.

Takakura, E. and Horibe, S. (1992a), 'Indentation fatigue of silicon carbide and silicon nitride', *Mater. Trans. JIM*, 32(5), 495–500.

Takakura, E. and Horibe, S. (1992b), 'Fatigue damage in ceramic materials caused by repeated indentation', *J. Mater. Sci.*, 27, 6151–6158.

Tani, E., Umebayashi, K., Kishi, K., Kobayasugi, E. and Nishijima, K. (1986), 'Gas pressure sintering of Si_3N_4 with concurrent addition of Al_2O_3 and 5 wt % rare earth oxide: high fracture toughness Si_3N_4 with fibre-like structure', *Am. Ceram. Soc. Bull.*, 65, 1311–1315.

Thompson, I. and Rawlings, R. D. (1988), *Institute of Ceramics*, 42, 15.

Tiegs, T. N. and Bechere, P. F. (1987), 'Sintered Al_2O_3–SiC–whisker composites', *J. Am. Ceram. Soc. Bull.*, 66, 339–342.

Tokita, M. (1999), 'Development of large-size ceramic/metal bulk FGM fabricated by spark plasma sintering', *Mater. Sci. Forum*, 308–311, 83–88.

Tsukuma, K., Ueda, K. and Shimada, M. (1985), 'Strength and fracture toughness of isostatically hot pressed composites of Al_2O_3 and Y_2O_3–partially stabilized ZrO_2', *J. Am. Ceram. Soc.*, 68, C4–C5.

Tsukuma, K., Takahata, T. and Shiomi, M. (1988), in *Advances in Ceramics, Vol. 24, Science and Technology of Zirconia III*, American Ceramic Society, Westerville, OH, 721–728.

Tuan, W. H. and Brook, R. J. (1992), 'Processing of alumina/nickel composites', *J. Eur. Ceram. Soc.*, 10, 95–100.

Uchiyama, T., Niihara, K. and Hirai, T. (1986), *J. Yogvo Kyokaishi*, 94, 756.

Uchiyama, T., Inoue, S. and Niihara, K. (1989), in Somiya, S. and Inomata, K. (eds) *Silicon Carbide Ceramics*, Rokakuho, Uchida, Tokyo, 193–200.

Uno, T., Kasuga, T., Nakayama, S. and Tsutsumi, S. (1991), 'High-strength calcium-mica containing glass-ceramics', in Sakka, S. and Soga, N. (eds) *Proceedings of International Conference on Science & Technology of New Glasses*, Ceramic Society of Japan, Tokyo, 335–340.

Uno, T., Kasuga, T., Nakayama, S. and Ikushima, A. J. (1992), 'Nanocomposite machinable glass-ceramics', in *Proceedings of 16th International Glass Congress, Vol. 4*, Madrid, 73–78.

Vanmeensel, K., Anne, G., Jiang, D., Vleugels, J. and Van Der Biest, O. (2004), 'Homogeneous and functionally graded Si_3N_4-TiCN composites shaped by electrophoretic deposition', *Silic. Indust.*, Special Issue, 69(7–8), 233–239.

Vaughan, D. A. J., Guiu, F. and Dalmau, M. R. (1987), 'Indentation fatigue of alumina', *J. Mater. Sci. Lett.*, 6, 689–691.

Veltri, R. D. and Galasso, F. S. (1990), 'Chemical-vapor-infiltrated silicon nitride, boron nitride and silicon carbide matrix composites', *J. Am. Ceram. Soc.*, 73(7), 2137–2140.

Venkatachari, K. R. and Raj, R. (1987), 'Enhancement of strength through sinter forging', *J. Am. Ceram. Soc.*, 70, 514–520.

Venkateswaran, T., Sarkar, D. and Basu, B. (2005), 'Tribological properties of WC-ZrO_2 nanocomposites', *J. Am. Ceram. Soc.*, 88(3), 691–707.

Wahi, R. P. and Ilschner, B. (1980), 'Fracture behaviour of composites based on Al_2O_3-TiC', *J. Mater. Sci.*, 15, 875–885.

Wakai, F., Sakaguchi, S. and Matsuno, Y. (1986), 'Superplasticity of yttria-stabilized tetragonal ZrO_2 polycrystals', *Adv. Ceram. Mater.*, 1, 259–263.

Wakai, F., Kodama, Y., Sakaguti, S., Murayama, N., Izaki, K. and Niihara, K. (1990), 'A superplastic covalent crystal composite', *Nature*, 344, 421–423.

Wakai, F., Nagano, T. and Iga, T. (1997), 'Hardening in creep of alumina by zirconium segregation at the grain boundary', *J. Am. Ceram. Soc.*, 80, 2361–2366.

Wakai, F., Kondo, N and Shinoda, Y. (1999), 'Ceramics superplasticity', *Curr. Opin. Sol. Sta. Mater. Sci.*, 4, 461–465.

Wang, C. M., Mitomo, M. and Emoto, H. (1997), 'Microstructure of liquid phase sintered superplastic silicon carbide ceramics', *J. Mater. Res.*, 12, 3266–3270.

Wang, H. and Singh, R. N. (1994), 'Thermal shock behaviour of ceramics and ceramic composites', *Int. Mater. Rev.*, 39(6), 228–244.

Wang, J., Taleff, E. M. and Kovar, D. (2003), 'High-temperature deformation of Al_2O_3/Y-TZP particulate composites', *Acta Mater.*, 51, 3571–3583.

Wang, J. G. and Raj, R. (1984), 'Mechanism of superplastic flow in a fine-grained ceramic containing some liquid-phase', *J. Am. Ceram. Soc.*, 67, 399–409.

Weber, C. H., Lofvander, J. P. A. and Evans, A. G. (1994), 'Creep anisotrophy of a continuous-fibre-reinforced SiC/CAS composite', *J. Am. Ceram. Soc.*, 77(7), 1745–1752.

Wei Chen, I. and Xue, L. A. (1990), 'Development of superplastic structural ceramics', *J. Am. Ceram. Soc.*, 73, 2585–2609.

Wierenga, P. E., Dirks, A. G. and Broek, J. J. (1984), 'Ultramicrohardness experiments on vapour deposited films of pure metals & alloys', *J. Thin Solid Films*, 119, 375–382.

Wilkinson, D. S. (1998), 'Creep mechanisms in multiphase ceramic materials', *J. Am. Ceram. Soc.*, 81, 275–299.

Willert-Porada, M., Gerdes, T. and Borchert, R. (1995), 'Application of microwave processing to preparation of ceramic and metal-ceramic FGM', in Ilschner, B. and Cherradi, N. (eds) *Proceedings of 3rd International Symposium on Structural and Functional Gradient Materials*, Presses Polytechniques et Universitaires Romandes, Lausanne, 15–21.

Wilson, J. and Kunz, S. M. (1988), 'Microwave sintering of partially stabilized zirconia', *J. Am. Ceram. Soc.*, 71, C40–C41.

Wu, C. C. M., Kahn, M. and Moy, W. (1996), 'Piezoelectric ceramics with functional gradients: a new application in material design', *J. Am. Ceram. Soc.*, 79(3), 809–813.

Wu, X. and Holmes, J. W. (1993), 'Tensile creep and creep-strain recovery behavior of silicon carbide fibre/calcium aluminosilicate matrix ceramic composites', *J. Am. Ceram. Soc.*, 76(10), 2695–2700.

Yanai, T., Nakahira, A., Suganuma, K. and Niihara, K. (1990), *J. Japan Soc. Powder Powder Metall.*, 37, 571–574.

Yongli, L. (2006), 'Nanophase ceramic composites', in Low, I. M. (ed.) *Ceramic Matrix Composites, Microstructure, Properties and Applications*, Woodhead Publishing, Cambridge, 243–259.

Yonn, C. K. and Chen, I. W. (1990), 'Superplastic flow of two-phase ceramics containing rigid inclusions-zirconia/mullite composites', *J. Am. Ceram. Soc.*, 73, 1555–1565.

Yoshida, H., Okada, K., Ikuhara, Y. and Sakuma, T. (1997), 'Improvement of high-temperature creep resistance in fine-grained Al_2O_3 by Zr^{4+} segregation in grain-boundaries'. *Phil. Mag. Lett.*, 76, 9–14.

Zhan, G. D., Kuntz, J. D., Wan, J. *et al.* (2003a), 'Single-wall carbon nanotubes as attractive toughening agents in alumina-based nanocomposites', *Nature Mater.*, 2, 38–42.

Zhan, G. D., Kuntz, J. D., Wan, J. *et al.* (2003b), 'Plasticity in nanomaterials', *Mater. Res. Soc. Symp. Proc.*, 740–741.

Zhan, G. D., Kuntz, J., Wan, J., Garay, J. and Mukherjee, A. K. (2003c), 'A novel processing route to develop a dense nanocrystalline alumina matrix (<100 nm) nanocomposite material', *J. Am. Ceram. Soc.*, 86(1), 200–202.

Zhao, C., Vleugels, J., Vandeperre, L. and Van Der Biest, O. (2000), 'Cylindrical Al_2O_3/TZP functionally graded materials by EPD', *Br. Ceram. Trans.*, 99(6), 284–287.

Zhao, J., Stearns, L. C., Harmer, M. P., Chan, H. M. Miller, G. A. and Cook, R. F. (1993), 'Mechanical behavior of alumina-silicon carbide nanocomposites', *J. Am. Ceram. Soc.*, 76, 503–510.

2

Magnetic properties of ceramic nanocomposites

D. D. MAJUMDER, D. D. MAJUMDER and
S. KARAN, Institute of Cybernetic Systems and Information
Technology, Kolkata, India

DOI: 10.1533/9780857093493.1.51

Abstract: Nanocomposites are often found in nature, as a multiphase solid material where one of the phases has one, two or three dimensions of less than 100 nm, or structures having nano-scale repeat distances between the different phases that make up the material. In this chapter the magnetic properties of ceramic nanocomposites are presented, along with structures which differ markedly from that of the component materials. It is emphasized that, in the case of nanocomposites, where the main part of the volume is occupied by ceramics (i.e. a chemical compound from the group of oxides, nitrides, borides or silicates, among others), further systematic study will be necessary to improve their optical, electrical and magnetic properties, as well as their tri-biological, corrosion resistance and other protective properties.

Key words: nanocomposites, nano fabricated films, electron tunnelling, magnetic data storage, colossal magnetoresistance, double quantum dots (DQD), soft magnetic materials.

2.1 Introduction

Molecular nanotechnology [1–3] involves developing structures at the molecular scale, designed and built atom by atom. The purpose of this chapter is to provide a technical overview of magnetic nanocomposites [4–26] and to detail current trends in their development and application. The science and technology relating to nanostructures is an interdisciplinary area of research and development that has been growing in past decades. Among these nanoscale materials are nanocomposites. These often have properties

51

that are different from conventional microscale composites and can be synthesized using simple and inexpensive techniques.

Over the last 15 years or so, the development of magnetic nanocomposites [27] has been critical in the field of information technology, bio-medical informatics and telecommunication. The seminal book by Majumdar and Das [27] suggested different potential approaches to providing large memory capacity materials for computer technology, including magnetic drum, magnetic core, magnetic cell, magnetic bubble, semiconductor and optical memory technologies. The authors predicted that any technology has an ultimate limit, where either further improvement is not possible or the cost of such improvement exceeds the return.

The areal data density of magnetic media, especially hard disk memory, has increased at an astonishing rate over the last three decades (30% per year for 1970–1990, 60% per year since 1990 and up to 100% per year in recent years). Extrapolating the 60–100% growth rate leads to bit sizes that clearly cannot be achieved using current technology. In recent years, increasing data storage density has meant decreasing the number of magnetic grains needed to store a bit of data from a thousand down to a few hundred, as well as decreasing the grain size. Today, the highest density hard disk drives store up to 120 billion bits (120 Gb) per square inch of disk surface. As the bit density increases, all components of the system – the media, read head, write head, tracking system, detection and error correction electronics, etc. – become more difficult to design and construct.

All magnetic data recording systems are currently based on thin sheets containing small, single-domain, magnetically decoupled particles, which switch independently. The granularity of the medium produces noise, which becomes unacceptable at small bit sizes. The media noise is now the primary factor that limits ultrahigh-density magnetic recording systems. Reducing the grain size, which reduces noise, is a challenge, because there is a critical size of the thermal instability of magnetization. To solve this problem, the particles should have a strong magnetic anisotropy. The single-domain status of the particles is important, because it prevents undesirable magnetization reversal through domain wall motion. All current commercial disk drive systems (excluding optical memories) operate using continuous granular media magnetized in-plane, called longitudinal recording. The magnetic recording industry has an enormous investment in longitudinal recording technology and is understandably very reluctant to move to an alternative technology.

A number of disk drive companies have commissioned a study, managed by the National Storage Industry Consortium (NSIC), to determine the limit on conventional longitudinal magnetic recording. The study group concluded that the extension of continuous longitudinal recording to 100 Gb/inch was not impossible. The study concluded that a grain diameter of 8 nm, with an appropriately adjusted anisotropy field (6000 Oe), would meet both thermal stability and signal/noise requirements. Media with an average grain size of 8 nm should not be extraordinarily hard to generate, but the model imposes another, very demanding, requirement on the grain size; namely that the grains should be monodisperse. As current technologies reach their limit, nanoscale materials seem to have the greatest potential for further developments in high-density magnetic data storage.

2.2   Magnetic nanocomposites

Imagine being able to record 100 movies on a disk the size of a CD. This is the potential of holographic data storage based on developments in nanotechnology. Nanoscale structures are an intermediate form of matter, which fills the gap between atoms/molecules and bulk materials. These types of structures frequently exhibit unusual physical and chemical properties which differ from those observed in bulk three-dimensional materials. One way forward in the field of high-density data storage systems is to utilize magnetic nanoparticles embedded in various matrices.

Magnetic nanocomposites are generally composed of ferromagnetic particles [28, 29] of nanometer grained size distributed either in a non-magnetic or magnetic matrix. The size and distribution of the particles play an important role in determining the properties of these materials. Generally, the electrical and magnetic behavior of nanostructured systems is governed by both the intrinsic properties of the nanostructures and their interactions with the matrix. Thus, the magnetic behavior of the system can be controlled by the size, shape, chemical composition and structure of the nanoparticles, and/or by the nature of the matrix in which they are embedded. However, the preparation of stable magnetic nanoscale materials that contain uniformly distributed nanoparticles (in a specified range of sizes of nanoparticles with a narrow size distribution) using conventional techniques is rather difficult. The fabrication and future utilization of such types of nanosized systems requires the use of novel technological processes.

The critical temperature of magnetic particles is an important property, and it is difficult to calculate the magnetization of nanomaterials without knowledge of the critical temperature. A variational cumulant expansion (VCE) [100–102] method has been developed for the calculation of the critical point of magnetic films. The methodology is, in principle, capable of dealing with a crystal of any lattice structure and of any geometric shape. The critical point T_c for thin films of cubic lattices has been investigated with simple cubic (sc), body-centered-cubic (bcc) and face-centered-cubic (fcc) structures as a function of the number of monolayers L grown along various directions such as $<110>$, $<111>$ and $<100>$. It has been found that T_c is

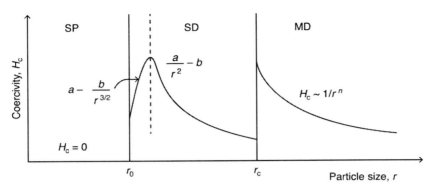

2.1 Overview of the size dependence of coercivity exhibited by magnetic particles.

higher for planes containing the highest population of nanoparticles in the film, provided that the crystal film contains the same number of monolayers but is grown in different directions. Thus, the highest T_c is found along the <100>, <110> and <111> directions in the sc, bcc and fcc structures, respectively. The common feature is that the most populated plane corresponds to the most favorable growth direction and has the highest T_c value. Needless to say, the film tends to a bulk in any case as L approaches infinity.

This chapter now moves on to review the properties of magnetic nanoparticles such as coercivity, superparamagnetism and electrical transport.

2.3   Size-dependent magnetic properties

Magnetic nanoparticles exhibit a number of magnetic properties that can be attributed to the reduced dimensions of the particles. These include coercivity and superparamagnetism.

2.3.1 Coercivity

The coercivity [41–56] of magnetic nanoparticles has a striking dependence on their size, as shown in Fig. 2.1. The coercivity H_c is zero below the superparamagnetic (SP) particle size limit r_0, single-domain (SD) behavior (SD) is shown between r_0 and the single-domain limit r_c, and multi-domain (MD) behavior for $r > r_c$.

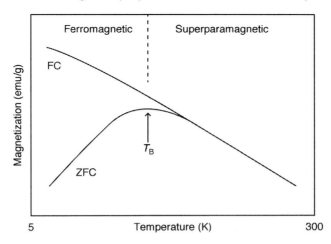

2.2 Schematic diagram of zero-field-coded (ZFC) and field-coded (FC) magnetization curves as a function of temperature taken in an applied field *H*. Arrow indicates blocking temperature, T_B.

2.3.2 Superparamagnetism

Very fine ferromagnetic particles have very short relaxation times, even at room temperature, and behave superparamagnetically [57–69]. Their behavior is paramagnetic but their magnetization values are typical of ferromagnetic substances. The individual particles have normal ferromagnetic movements but very short relaxation times, enabling them to rapidly follow directional changes in an applied field. Superparamagnetism is characterized by two significant features. Firstly, there is no hysteresis, which means both retentivity and coercivity are zero. Secondly, magnetization curves measured at different temperatures superimpose when magnetization (M) is plotted as a function of field (H)/temperature (T), as is shown in Fig. 2.2. This also demonstrates that superparamagnetism can be destroyed by cooling. The temperature at which this occurs is called the blocking temperature (T_B) and is dependent linearly on volume and the magnitude of the crystal field anisotropy.

For a particle of constant size below the blocking temperature T_B, the magnetization will be stable and shows hysteresis for those particles which have a relaxation time for demagnetization of longer than 100 s. For uniaxial particles (using the same criterion for stability):

$$T_B = \frac{KV}{25k_B} \qquad [2.1]$$

where K = anisotropy constant, V = volume of the particle and k_B = Boltzmann's constant (1.38×10.23 J K).

If one considers Ni as an example with an anisotropy constant $K = 4.5 \times 103$ J m^{-3}, then for a particle size of 20 mm, the particle will show a blocking temperature (T_B) at -55 K using equation 2.1. Below T_B, the magnetization will be relatively stable and will show ferromagnetic behavior. Above T_B, the thermal energy will be sufficient to suppress ferromagnetic behavior and the particles thus become superparamagnetic.

2.4   Colossal magnetoresistance (CMR)

Colossal magnetoresistance [103–109] is a property of some materials, mostly manganese-based perovskite oxides, that enables them to dramatically change their electrical resistance in the presence of a magnetic field. The magnetoresistance of conventional materials enables changes in resistance of up to 5%, but materials featuring CMR may demonstrate resistance changes by very high orders of magnitude.

CMR was initially discovered in mixed-valence perovskite manganites in the 1950s by G. H. Jonker and J. H. van Santen [133]. In the initial double-exchange model to explain the phenomenon, the spin orientation of adjacent Mn-moments is associated with kinetic exchange of e_g-electrons. Consequently, alignment of the Mn-spins by an external magnetic field causes higher conductivity. However, the double-exchange model did not adequately explain the high insulating resistivity above the transition temperature. One later model is the so-called half-metallic ferromagnetic model, which is based on spin-polarized band structure calculations using the local spin-density approximation (LSDA) of the density functional theory (DFT) where separate calculations are carried out for spin-up and spin-down electrons. The half-metallic state is concurrent with the existence of a metallic majority spin band and a non-metallic minority spin band in the ferromagnetic phase.

This model can be contrasted with the Stoner model of itinerant ferromagnetism [134–138]. In the Stoner model, a high density of states at the Fermi level makes the non-magnetic state unstable. With spin-polarized calculations on covalent ferromagnets, the exchange-correlation integral in the LSDA-DFT takes the place of the Stoner parameter. The density of states at the Fermi level does not play a special role. A significant advantage of the half-metallic model is that it does not rely on the presence of mixed valency as does the double-exchange mechanism and it can therefore explain the observation of CMR in stoichiometric phases like pyrochlore $Tl_2Mn_2O_7$. Microstructural effects have also been investigated for polycrystalline samples and it has been found that the magnetoresistance is often dominated by the tunneling of spin-polarized electrons between grains, giving rise to an intrinsic grain-size dependence to magnetoresistance.

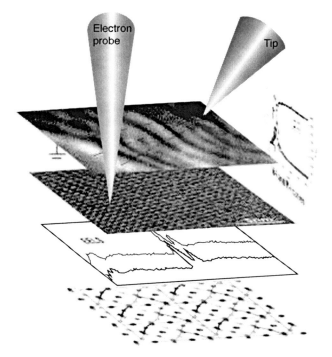

2.3 Electron probe in imaging technology.

However, a full quantitative understanding of the CMR effect remains elusive and it is still the subject of significant research.

In trying to understand the mechanism behind CMR, scientists have used an electron probe to make images and collect other data while using a scanning tunneling microscope tip to apply current or an electric field to the sample. In Fig. 2.3, the first layered image of black lines shows polaron waves, which propagate during the application of the current. Fine dots in the second layer are the individual atoms, while the periodic dot-clusters show the electron ordered state. The graph of electron energy loss spectroscopy (EELS) reveals bonding-electron excitation. The bottom layer is a structural model of the crystal lattice and the vertical graph shows the electric resistance (I–V curve) of the crystal when current is applied.

Experiments at the U.S. Department of Energy's Brookhaven National Laboratory have shed new light on some materials' ability to dramatically change their electrical resistance in the presence of an external magnetic or electric field [139–141]. The Brookhaven scientists studied crystalline perovskite manganites that had been doped with extra charge carriers – electrons or 'holes' (the absence of electrons) – using various state-of-the-art electron microscopy techniques. In an unprecedented experiment, the scientists used a scanning tunneling microscope that was built inside an electron microscope to apply an electric stimulus to the sample while observing its response at the atomic scale.

Using this technique, they obtained, for the first time, direct evidence that a small electric stimulus can distort the shape of the crystal lattice and also cause changes in the way charges travel through the lattice. The lattice distortions accompanied the charge carrier as it moved through the lattice, producing a particle-like excitation called a polaron. Polarons can be pictured as a charge carrier surrounded by a 'cloth' of the accompanying lattice vibrations. This research showed polarons melting and reordering – that is, undergoing a transition from solid to liquid to solid again – in response to the applied current. This has been identified as the key mechanism for CMR. The technique has also allowed scientists to study polaron behavior. Static long-range ordering of polarons forms a polaron solid, which represents a new type of charge and orbital ordered state. The related lattice distortions connect this phenomenon to colossal resistance effects, and suggest ways of modifying charge density and electronic interactions at the vicinity of electric interfaces and electrodes.

Colossal resistance effects could result in miniaturization of electric circuits that operate at lower power. Research on CMR has had a direct impact on the development of new electronic and spintronic devices (devices that use a combination of electron spin and charge). Such devices include new forms of 'non-volatile' computer memory, i.e. memory that can retain stored information even when not powered, such as resistive random access memory (RRAM).

As we have noted, magnetoresistance [70–74] in materials is of enormous technological importance, particularly in the design of magnetic memory systems, since these materials can be used as read heads for hard disks, magnetic storage and sensing devices. The effectiveness of these materials is directly related to the percentage change of resistance in an external magnetic field. Magnetoresistance is defined as:

$$\mathrm{MR}(T) = \frac{\rho(H, T) - \rho(0, T)}{\rho(0, T)} \qquad [2.2]$$

where $\rho(H, T)$ and $\rho(0, T)$ are the resistivity at a given temperature T in the presence and absence of magnetic field H, respectively.

Perovskite manganites $RE_{1-x}A_xMnO_3$ (RE = trivalent rare earth element such as La, Pr, Sm, Gd, etc. and A = divalent alkaline earth ion such as Ca, Sr, Ba, etc.) have attracted considerable interest in recent years because of their CMR behavior. In the perovskite structure, (RE, A) elements occupy the A-site position (corner of a cube) and manganese

2.4 Schematic illustration of the double-exchange mechanism in manganites.

occupies the B-site position (body center of a cube). All the face-centered positions are occupied by oxygen. At low temperatures, the resistivity is metallic both in magnitude and its dependence on temperature. With an increase in temperature, it increases up to a temperature T_{MI} (called the metal–insulator transition temperature) beyond which it decreases, having a negative temperature coefficient of resistance while maintaining a large magnitude. The metal–insulator transition at this temperature T_{MI} is usually accompanied by a ferromagnetic to paramagnetic transition.

The coexistence of metallic conductivity and ferromagnetic coupling in these materials at low temperatures has been explained in terms of a double-exchange mechanism. This mechanism involves the excitation of d electrons from the Mn cation with the highest number of such electrons (lower valency, Mn^{3+} in the present case) into an overlapping anion orbital (O^{2-} ion in this case), with the transfer of one anion p electron to the other cation (Mn^{4+} in the present case). This type of electron hopping is schematically shown in Fig 2.4. Simultaneous transfer of an electron from Mn^{3+} to O^{2-} and from O^{2} to Mn^{4+} is called double exchange.

Figure 2.5 shows the electronic phase diagram of $La_{1-x}Ca_xMnO_3$ (a CMR oxide) showing the compositional stability of different phases. The Curie temperature is maximized at $x = 3/8$ according to Hwang and co-workers [142–144], contrary to the $x = 0.30$ believed by many to be the most optimal compositions for ferromagnetism. Electrical transport in these materials exhibits both insulating and metallic behavior depending on the transition temperature T_{MI}. These transport mechanisms are discussed in brief in the next section.

2.5   Electrical transport/resistivity

The resistivity as a function of temperature and/or material composition is shown in Fig. 2.6. The curve can be described as consisting of two different regions. In the low-temperature regime, the resistivity increases with temperature. In the high-temperature regime, the resistivity decreases with increasing temperature. The resistivity has a metallic behavior ($dp/dT > 0$) below the peak and a semiconducting (insulator) behavior ($dp/dT < 0$) above. The conduction mechanism in the manganites exhibits both metallic and insulating behavior depending on composition and temperature. In a

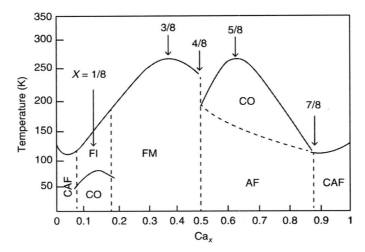

2.5 Electronic phase diagram of La$_{1-x}$Ca$_x$MnO$_3$ showing the compositional stability of different phases: FM, ferromagnetic metal; FI, ferromagnetic insulator; AF, antiferromagnetism; CAF, canted AF; and CO, charge/orbital ordering.

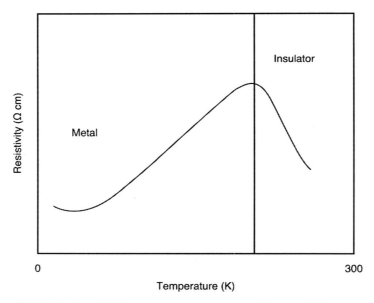

2.6 Typical resistivity versus temperature and composition.

particular composition range, manganites undergo insulator to metal transition on cooling [75–91].

2.5.1 Insulating region

Temperature causes electrons to be excited into the conduction band and hence resistivity is considered as a thermally activated process. Jonker and Van Santen [133] measured the resistivity of $LaMnO_3-AMnO_3$ (where A = Ca, Sr and Ba) and found that the resistivity plotted as $\log \rho$ versus $1/T$ was linear, showing thermally activated behavior given by the relation:

$$\rho(T) = A \exp\left(\frac{E_o}{k_B T}\right) \qquad [2.3]$$

where E_o is the activation energy, k_B is Boltzmann's constant and A is a constant.

An alternative electrical transport mechanism in the insulating region is due to the formation of polarons (a strong coupling between an electron and phonons). Polaron-mediated hopping behavior can be described as:

$$\rho(T) = BT \exp\left(\frac{E_o}{k_B T}\right) \qquad [2.4]$$

where B is a measure of ideal conductivity at elevated temperatures and depends on polaron concentration. The activation energy for polaron-mediated conduction can be described using the following equation proposed by Yeh *et al.* [145]:

$$\rho(T) = \rho_0 T\alpha \exp\left(\frac{E_o}{k_B T}\right) \qquad [2.5]$$

where ρ_0 is the residual resistivity at $T = 0$, α equal to 1.6 identifies the behavior as non-adiabatic small polaron hopping in most cases and α equals 1 in some cases. Other research suggests that variable range hopping (VRH) best describes electronic transport. VRH has been suggested by Mott to describe transport at low temperatures when the electronic states are localized near the Fermi energy [146–151]:

$$\rho(T) = \rho_0 \exp\left(\frac{E_o}{k_B T}\right)^{1/4} \qquad [2.6]$$

Coey *et al.* [152–155] found that this kind of expression best fitted their data on a variety of films. To summarize, there is clear evidence in the paramagnetic insulating region of activated behavior, but there is no agreement on the exact form of temperature dependence.

2.5.2 Metallic region

In the metallic region the resistivity has been found to be quite well described by

$$\rho(T) = \rho_0 + CT^2 + DT^n \qquad [2.7]$$

where C is the electron–electron scattering coefficient and D is the electron–phonon or electron–magnon scattering coefficient. The value of n has been predicted to be 5 for electron–phonon scattering while it has a value of 4.5 for electron–magnon scattering.

2.6   Spin-dependent single-electron tunneling phenomena

Scientific researchers are continuously searching for new technological tools to comprehend and manipulate the basic elements available in nature. They develop experiments and theories that can seem at first sight highly abstract, virtually useless and far from meaning anything to our society. Strikingly, it is precisely this attitude of exploring the 'extremes of what is possible' that has induced the most significant technological breakthroughs to impact our society. One important example is the discovery of a means to record the magnetic properties of atomic nuclei, which is now widely applied as an imaging tool in hospitals. Rapid developments in the computer industry are based on the invention of the transistor, which was a result of fundamental interest in the nature of electrons at the interface between a metal and a semiconductor. Finally, the boost in data storage of the last decade, induced by the discovery of a phenomenon called giant magnetoresistance, was based on research on magnetic layers only one nanometer thick.

These developments have relied heavily on knowledge about the most fundamental laws of physics, called quantum mechanics. Quantum mechanics predicts how electrons move in materials, what processes lead to light emission and the sources of magnetism. At this small scale, the laws of nature exhibit some peculiar properties that contradict usual physical laws. Examples of this are particles that start behaving like waves and the striking manifestation of superposition states, where a particle can exist at multiple positions at the same time. Most surprisingly of all, it has been found that two particles can share a connection, called entanglement, even if they are separated by a very long distance.

It is now possible to control these phenomena in very small systems like a single atom or electron. This level of control makes it possible to address important questions about the stability of quantum superpositions and entangled states and how these are affected by measurements or interactions with the environment. It is still an open question to what extent macroscopic systems can be in superposition states and, if not, where and why the transition to the macroscopic world occurs. Currently, experiments are being developed to study these phenomena. Examples of such experiments are the observation of interference of large molecules with themselves, coupling a small well-defined quantum system with a macroscopic system like a tiny mirror, and the preparation of large numbers of photons in a superposition state. In parallel to addressing these fundamental issues, the question is also raised as to whether this high level of control over small quantum objects can form a useful basis for technological applications. It is too early to say whether applications will fully exploit the fundamental resources available in nature, but major conceptual breakthroughs have already been achieved. For example, entangled states have proven a valuable resource for novel quantum communication protocols, such as quantum cryptography, which is fundamentally unbreakable.

2.6.1 Molecular magnets as qubits: spin tunneling

A molecular magnet is a molecule that is typically ferromagnetic or antiferromagnetic with an isolated spin center. Physical candidates for an ideal molecular magnet are evaluated based on three criteria:

- the total spin of the isolated system is high
- they are typically large molecules with little intermolecular interaction, utilizing the $1/r^3$ dipole–dipole interaction
- they exhibit high magnetic anisotropy.

By far the most studied molecules have been Mn_{12} and Fe_8, each being ferromagnetic with high magnetic anisotropy and total spin $S = 10$. The resulting spin states of both Mn_{12} and Fe_8 can be modeled as a double-well potential, which becomes the basis for further analysis of molecular magnets.

The most promising use of molecular magnets for both data storage and computation is through the exploitation of the quantum phenomenon of spin tunneling. The concept of the intrinsic angular momentum of an electron, later referred to as its spin, was introduced by the Dutch physicists George Uhlenbeck and Samuel Goudsmit in 1925. In retrospect, the electron's spin was first observed in 1922 by the Stern–Gerlach experiment. Stern and Gerlach found that a bundle of neutral silver particles is split into two beams due to the interaction with an inhomogeneous magnetic field, and explained this by two quantized states of the angular momentum.

A better theoretical explanation of the Stern–Gerlach experiment was given in 1927 by Wolfgang Pauli, by taking into account the intrinsic angular momentum of electrons, their spin. The spin of an electron is a pure quantum mechanical property, which cannot be described by classical mechanics. It is represented by the spin quantum number, which can take the values $1/2$ or $-1/2$ in the case of an electron. These two eigenstates are usually referred to as spin-down and spin-up.

The foundation of spin tunneling lies in the high magnetic anisotropy, S_z, of the system and the resulting double-well potential, which can be perturbed with both longitudinal, H_z, and transverse, H_x, magnetic fields to induce degenerate states that can be manipulated as qubits.

The application of an external longitudinal magnetic component to the system results in a perturbation of the double potential, while the presence of an external transverse field is responsible for the spin tunneling effect. By assuming low temperatures (~1 K) to minimize spin–phonon interactions, the single-spin Hamiltonian for molecules like Mn_{12} and Fe_8 becomes:

$$\tilde{H} = \tilde{H}_a + \tilde{H}_Z + \tilde{H}' \qquad [2.8]$$

$$\tilde{H}_a = -AS_z^2 - BS_z^4 \qquad [2.9]$$

$$\tilde{H}_Z = g\mu_B A_z S_z \qquad [2.10]$$

$$\tilde{H}' = g\mu_B A_x S_x \qquad [2.11]$$

where equations 2.9, 2.10 and 2.11 are the anisotropic, Zeeman (longitudinal external field) and transverse field components of the Hamiltonian, respectively. It is sufficient to attribute the presence of spin tunneling to the non-commutability of the H' term with the remaining S_z elements of the Hamiltonian. Hence, each eigenstate $|M\rangle$ is not stationary and will tunnel through the anisotropic barrier when the state is degenerate with $|-M\rangle$. It is important to note that even in a zero transverse field environment, Mn_{12} has been shown to still exhibit spin tunneling due to violations of transverse-anisotropic selection rules, indicating the inherent presence of a transverse component. One fascinating aspect of this phenomenon is its appearance in magnetic hysteresis (see Fig. 2.7).

2.6.2   The Kondo effect and controlled spin entanglement in coupled double-quantum-dots (DQDs)

Manipulation of the electron spin degree of freedom for future device applications is of increasing importance in the field of quantum mechanics. In particular, spintronics is an area of research that seeks to develop future-generation spin-based devices. On a more fundamental level, the spin degree of freedom plays a major role in diverse, strongly correlated systems, such as

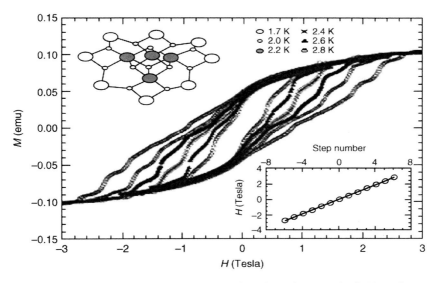

2.7 Magnetization of Mn$_{12}$ as a function of magnetic field at six different temperatures. The stepped magnetic hysteresis shows relaxation of the field due to magnetization (spin) tunneling.

high-T_c superconductors, 2-d metal–insulator transition at large r_s (ratio of Coulomb to kinetic energy) and Kondo systems.

The Kondo effect in the fully controllable semiconducting double-quantum-dot (DQD) system is described below. In particular, this section addresses the following two areas of interest.

• The DQDs presented represent the first experimental realization of the two-impurity Kondo system, which has been studied theoretically for nearly two decades.

• Direct evidence for the controlled entanglement of quantum dot spins is presented, one on each dot, into a spin-singlet state.

These findings have important implication for quantum computation applications. Similar studies and results have been reported recently in the work of Craig *et al.* [156], though with a different coupled quantum dot scheme.

In the past decade, the semiconducting quantum dot system has proven to be a versatile and ideal system for studying correlated effects involving the spin degree of freedom. Under appropriate conditions, an isolated spin on a quantum dot closely models an isolated spin impurity and, through its coupling to conduction electrons in the leads connected to the dot, can give rise to a rich variety of correlated behaviors. As a result of the tunability of the semiconducting quantum dot system, it is possible to continuously vary a number of physical parameters to produce qualitatively distinct behaviors, or to systematically map out the behavior of a particular phenomenon of interest. Many interesting results have been obtained in individual dots, including the even–odd effect in the filling of a dot with electrons, Kondo physics under various conditions and mixed valence behavior. In DQDs, many authors have investigated the splitting between symmetric and antisymmetric combinations of the dot wave functions and molecular bonding–antibonding behavior.

2.6.3 The DQD as a model two-impurity spin system

In an idealized scenario neglecting intra-dot exchange and correlation, the filling of the quantum dot levels proceeds in an even–odd fashion, as successively higher energy orbital levels are filled first with one excess electron, that is paired when the next electron is added. Pauli exclusion ensures that the total spin of the quantum dot thus alternates between $S = 1/2$ and $S = 0$, depending on whether the total number of electrons is odd or even. Because the filling of the odd electron involves the next higher orbital level, the energy required has an added contribution of the level spacing, ΔE, in addition to the Coulomb charging energy, U. This leads to an even–odd (or odd–even) alternation of the spacing in the addition spectrum. Within this scenario, when the number of electrons is odd and the dot spin is $1/2$, the quantum dot behaves much like a spin-1/2 magnetic impurity imbedded in a metal. In this semiconductor context, the advantage is that the coupling to the Fermi sea of electrons, which reside in the connecting electrical leads, can be tuned in a controlled fashion.

A successful model, which captures much of the correlated spin physics under the situation described above, is given by the Anderson Hamiltonian equation:

$$H = H_{\mathrm{L}} + H_{\mathrm{R}} + H_{\mathrm{QD}} + H_{\mathrm{T}} \qquad [2.12]$$

$$H_{\mathrm{L(R)}} = \sum_{k,\sigma \in \mathrm{L(R)}} \varepsilon_k c_{k\sigma}^* c_{k\sigma} \qquad [2.13]$$

$$H_{\mathrm{QD}} = \sum_{m,\sigma} \varepsilon_{m\sigma} d_{m\sigma}^* d_{m\sigma} + U \sum_{m>m'} n_{m\uparrow} n_{m'\downarrow} \qquad [2.14]$$

$$H_{\mathrm{T}} = \sum_{k,m,\sigma \in \mathrm{L(R)}} (V_{km} c_{k\sigma}^* d_{m\sigma} + \mathrm{h.c.}) \qquad [2.15]$$

$$\Gamma = \frac{\Gamma_{\mathrm{L}} \Gamma_{\mathrm{R}}}{\Gamma_{\mathrm{L}} + \Gamma_{\mathrm{R}}} \qquad [2.16]$$

$$\Gamma_{L(R)}^{m}(\omega) = 2\pi \sum_{k \in L(R)} |V_{km}|^2 \delta(\omega - \varepsilon_k) \qquad [2.17]$$

where H_L, H_R are the Hamiltonians for the left (top) and right (bottom) leads, H_{QD} is the Hamiltonian for the quantum dot and H_T the dot–lead interaction. Within the quantum dot, $\varepsilon_{m\sigma}$ represents the orbital energy levels and Γ is the level broadening due to coupling to the leads.

This model exhibits the required even–odd filling. In the case of an odd total number of electrons, the last excess electron occupies the next higher orbital level, which is the lowest unoccupied level, yielding a net spin of 1/2 for the quantum dot as desired. In this situation of an excess odd electron with spin 1/2, the Anderson Hamiltonian gives rise to different scenarios depending on the ratio of the energy of last occupied orbital measured, from the chemical potential of the leads, $|-\varepsilon_0|$, to the level broadening due to coupling to the leads, Γ, i.e. $|-\varepsilon_0|/\Gamma$. When this ratio lies between 0 and 0.5, charge fluctuates readily on and off of the quantum dot and the system is in the mixed valence regime. For a ratio in excess of 0.5, charge is essentially quantized on the dot and Kondo physics becomes relevant. Under this condition, the Anderson Hamiltonian can be mapped into the Kondo Hamiltonian, familiar in the context of a magnetic impurity imbedded in a host metal:

$$H = \sum_{i,j,s} t_{ij} c_{is}^* c_{js} + J \sum_i \mathbf{s}_{ci} \cdot \mathbf{s}_{fi} \quad T_K \propto \exp(-1/\rho J) \qquad [2.18]$$

The corresponding Kondo energy scale in the quantum dot case is given by $E_K = k_B T_K$:

$$T_K = (U\Gamma)^{1/2} \exp\left(\frac{-\pi[|-\varepsilon_0|(U + \varepsilon_0)]}{\Gamma U}\right) \qquad [2.19]$$

Going from the individual quantum dot to the coupled DQD case involves an additional dot–dot tunnel-coupling term:

$$H_{\text{dot-dot}} = t \sum_\sigma (d_{L\sigma}^* d_{R\sigma} + \text{h.c.}) \qquad [2.20]$$

Here we are taking the simplest case of full symmetry between the two (left/right or top/bottom) dots, in their electronic energy level structure and level splitting ΔE, on site Coulomb repulsion U and level broadening Γ, for respective couplings to the left (top) and right (bottom) leads. Only the coupling term between the excess, last odd electron on each dot needs be included as the relevant coupling term. The inter-dot tunnel coupling, t, gives rise to an effective antiferromagnetic coupling, $J = 4t^2/U$, between the two excess spins. According to theoretical analysis, the following scenarios arise as t/Γ and J/T_K are varied.

Assuming U to be the largest energy scale as is usually the case in

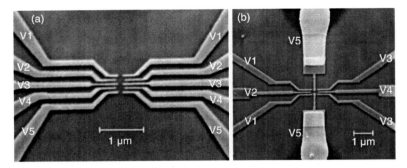

2.8 SEM micrographs of (a) series-coupled and (b) parallel-coupled DQDs.

experiment, when the ratio $t/\Gamma < 1$, this problem maps onto the two-impurity Kondo problem discussed by Jones and co-workers [157, 158], characterized by the antiferromagnetic coupling J and Kondo scale T_K, but with an additional term due to the tunnel-coupling, which breaks the symmetry in the even and odd channels. The basic behavior is similar to the scenario in the two-impurity Kondo problem where competition between Kondo and antiferromagnetic correlations leads to a continuous phase transition (or crossover) at a critical value of the coupling ratio, $J/T_K \approx 2.5$. However, the non-Fermi liquid quantum critical point at $J/T_K \sim 2.5$ is not accessible due to breakage of the symmetry between the even and odd channels, and a crossover behavior is expected instead.

If t is tuned to $t > \Gamma$ before the antiferromagnetic transition point can be reached, the system undergoes a continuous transition from a separate Kondo state of individual spins on each dot (atomic-like) to a coherent bonding–antibonding superposition of the many-body Kondo states of the dots (molecular-like). Both the antiferromagnetic state and coherent bonding state exhibit a double-peaked Kondo resonance in the differential conductance versus source–drain bias and involve entanglement of the dot spins into a spin-singlet. Therefore they are likely to be closely related.

The parallel-coupled case has only recently been analyzed in a model without inter-dot tunnel coupling where the antiferromagnetic coupling occurs via electrostatic coupling, yielding a discontinuous transition. Two DQD geometries were studied: series-coupled and parallel-coupled. Whereas the series-coupled geometry is more likely to be relevant for quantum computation applications, the parallel-coupled geometry is well suited for studying the quantum phase transition (crossover) in the two-impurity Kondo problem.

Figure 2.8 shows the device patterns for the series- and parallel-coupled DQDs. The devices are basically complex, multi-gate field effect transistors, which are operated in the quantum regime. The bright lines in the electron micrograph ending in finger-like features represent metallic gates, which are utilized to drive away electrons residing 90 nm below the top surface, at the $GaAs/Al_xGa_{1-x}As$ heterojunction, with the application of negative gate voltages. The 'pincher' gates, V1 and V5 in Fig. 2.8(a), and V1 and V3 in Fig. 2.8(b), control the coupling to the leads, setting the scale for the level broadening, Γ, due to coupling to the leads, while the pincher gates V3 (V5 for the parallel case) control the inter-dot tunneling, t. Plunger gates V2 and V4 control separately the number of electrons on each dot.

The real difficulty in the operation of such complex devices arises from the close proximity of all the gates. Each quantum dot has a lithographic size of ~180 nm. The close proximity gives rise to sizable capacitive coupling between all gates, as well as to the electron gas puddles, which make up the quantum dots. Changing one gate voltage will inevitably affect the charge on all other gates and in the puddles at the same time. To be able to tune the device parameters over a significant range (e.g. the inter-dot tunneling matrix t) without affecting other characteristics (e.g. electron number), it is necessary to experimentally 'diagonalize' the capacitance matrix. Practically speaking, this means that in order to tune t, in addition to varying the main gate voltage at V5, it is necessary to map out other gates, some of which need to be adjusted in an amount proportional to the change in V3 (V5), which in effect means that it is necessary to tune a linear combination of different gates, with V3 (V5) being the dominant one.

2.6.4 Spin-dependent single-electron tunneling effects in epitaxial Fe nanoparticles

Fe/MgO/Fe nanoparticle /MgO/Co double-tunnel junctions have been prepared by molecular beam epitaxy for current perpendicular-to-plane transport measurements on submicrometer-sized pillars. Microstructural observations indicate that the samples exhibit a fully epitaxial layered structure with sharp and flat interfaces including well-defined separated Fe nanoparticles between the barriers. The introduction of asymmetric MgO tunnel barriers (i.e. with different thicknesses), in the double junction leads to a clear observation of a Coulomb staircase and associated tunnel magnetoresistance oscillations. Estimation of the capacitance of the system indicates that these transport phenomena are due to charging effects of the magnetic particles.

2.6.5 Coulomb promotion of spin-dependent tunneling

An important phenomenon is the transport of spin-polarized electrons through a magnetic single-electron transistor (SET) in the presence of an external magnetic field. Assuming the SET to have a nanometer size central island with a single-electron level we find that the interplay on the island between coherent spin-flip dynamics and Coulomb interactions can make the Coulomb correlations promote rather than suppress current through the device.

2.7   Applications: cobalt-doped nickel nanofibers as magnetic materials

By developing a fuller understanding of their properties, materials science has produced magnetic materials far more powerful than those available only a few decades ago. For instance, ferromagnetic metal nanostructures reveal physical and chemical properties that are characteristic of neither the atom nor the bulk counterparts. Quantum size effects and the large surface area of magnetic nanoparticles dramatically change some of the magnetic properties and exhibit superparamagnetic phenomena and quantum tunneling of magnetization because each particle might be considered as a single magnetic domain. Consequently, some metal nanoparticles such as Fe, Co, and Ni have been given much attention for use in various applications such as electronic, optical and mechanical devices, magnetic recoding media, catalysis, superconductors, ferrofluids, magnetic refrigeration systems and contrast enhancement in magnetic resonance imaging carriers for drug delivery.

Research suggests that the magnetic properties of those materials are highly dependent on particle shape. For instance, the coercivity of nickel nanofibers at room temperature was about 100 times the magnitude of the bulk material. One-dimensional (1D) magnetic nanomaterials are expected to have interesting properties, as the geometrical dimensions of the material become comparable to key magnetic length scales, such as the exchange length or the domain wall width. This has resulted in the fabrication of nanoscale magnetic logic junctions with ferromagnetic nanowires as building blocks. Magneto-optical switches have been prepared using suspensions of ferromagnetic nanowires. Among the 1D nanoshapes, nanofibers have considerable importance because of their long axial ratio characteristic, making them good candidates for nanodevices and nano-membranes.

Electrospinning is the most popular technique utilized in the production of functional nanofibers because of its simplicity, low cost and high yield. Metal base nanofibers are produced by the electrospinning of a sol–gel composed of a metal precursor and an accordant polymer. In the field of pure metal nanofibers, the electrospinning process has been exploited to synthesize Co, Cu, Fe and Ni in a nanofibrous shape by calcination of electrospun metal precursor/polymer nanofiber mats in a hydrogen atmosphere. However, calcination in an argon atmosphere has recently been introduced as a safe, economically preferable and more effective alternative strategy to produce silver, nickel and cobalt nanofibers.

Among the common ferromagnetic metals, cobalt has distinct magnetic properties. The incorporation of cobalt nanoparticles in nickel nanofibers results in better magnetic properties compared with the bulk. Electrospinning of a colloidal solution rather than a sol–gel (which is widely utilized in the conventional electrospinning technique) has been suggested as a novel strategy to prepare cobalt nanoparticles/ nickel acetate/ poly(vinyl alcohol) and nanofiber mats. Calcination of the dried electrospun mats in an argon atmosphere at 700°C for 5 h leads to the production of cobalt-doped nickel nanofibers. Overall, the magnetic properties studied point to an improvement in the magnetic parameters of synthesized cobalt-doped nickel nanofibers compared with pristine ones.

Generally, the electrospun solution is either a polymer(s) dissolved in a proper solvent or metallic precursor/polymer solution. The distinct feature of these solutions is that they have to be completely miscible. In other words, in the case of a metallic precursor, it should be soluble in a suitable solvent because it has to hydrolyze and polycondensate in the final precursor/polymer mixture to form the gel network.

2.7.1 Assessing the magnetic properties of cobalt-doped nickel nanofibers

Synthesized cobalt-doped nickel nanofibers revealed better magnetic properties compared with pristine ones, using a gravimetric analyzer (TGA, Pyris1, PerkinElmer Inc., USA). Information about the phase and crystallinity was obtained by using a Rigaku x-ray diffractometer (XRD, Rigaku, Japan) with Cu KR (λ) (1.540 Å) radiation over the Bragg angle ranging from 30 to 100°. High-resolution images and selected area electron diffraction patterns were obtained with a transmission electron microscope (TEM, JEOL JEM-2010, Japan) operated at 200 kV.

The magnetic properties of the nanofibers were evaluated using commercial superconducting quantum interface device (SQUID) magnetometery. The nanofibers were weighed and then placed into capsules in an inert gas environment. After this, the capsules were sealed with paraffin wax to prevent air oxidation of the nanofibers. The masses of the pristine and cobalt-doped nickel nanofibers were 11.48 and 5.26 mg respectively.

The mechanism of the spin-reorientation transition (SRT) in the Ni/Fe/Ni/W(110) system has been investigated using in situ low-energy electron microscopy, x-ray magnetic circular dichroism measurements and first-principles electronic structure calculations [110–125]. It was found that the growth of Fe on a flat Ni film on a W(110) crystal resulted in the formation of nanosized particles, instead of a uniform monolayer of Fe as commonly assumed. This interfacial nanostructure leads to a change of the system's dimensionality from 2D to 3D, which simultaneously weakens the dipolar interaction and enhances the spin–orbit coupling in the system and drives the observed SRT.

2.8   Applications: amorphous soft magnetic materials

The newest addition to the class of soft magnetic materials is not crystalline, but amorphous and nanocrystalline. Amorphous magnets were first fabricated in 1967 from their liquid states by means of a rapid-quenching technique. Amorphous magnets exhibit a wide range of new phenomena, in contrast to those of their crystalline counterparts where the periodicity of constituent atoms by and large determines the overall magnetic properties. Atoms in an amorphous magnet are distributed randomly, resulting in a disordered state and physicists are struggling to explain the magnetic order resulting from the complete absence of a specific crystal structure.

Nanocrystalline magnetic materials have been known for over 20 years since the discovery of Finemet by Yoshisawa and co-workers in 1988. However, it is only the recent developments in complex alloys and their metastable amorphous precursors that have revolutionized the field of soft magnetism. Nanocrystalline magnetic materials are commonly produced by the partial crystallization of its amorphous precursors. The microstructure of these materials consists of nanosized ferromagnetic materials embedded in an amorphous matrix. The matrix phase must also be ferromagnetic to facilitate the exchange coupling between the nanoparticles. As a result of this coupling, they often exhibit vanishing magnetocrystalline anisotropy.

Since soft magnetic properties are strongly related to the crystalline anisotropy, the exchange interaction in nanocrystalline magnetic materials often results in an improvement of their soft magnetic properties. Interesting properties of nanocrystalline magnetic materials are a consequence of the effects induced by the nanocrystalline structure. They include interface physics, the influence of grain boundaries, the averaging of magnetic anisotropy by exchange interactions, the decrease in exchange length and the existence of a minimum two-phase structure in such materials.

Amorphous and nanocrystalline materials have been investigated for soft magnetic applications such as in transformers and inductive devices. The crucial parameters that decide their applicability in such devices are high induction, high permeability and high Curie temperature. As well as these requirements, these materials must also possess good mechanical properties and corrosion resistance. In achieving such goals the key issues include alloy chemistry, structure and, most importantly, the ability to tailor the microstructure. Therefore, materials selected for soft magnetic applications must be optimized in terms of their intrinsic and extrinsic magnetic properties, as well as their morphology. Intrinsic magnetic properties such as saturation magnetic induction B_s and Curie temperature T_C are determined by alloy composition and crystal structure. Permeability μ, which is an extrinsic property, is usually determined by chemistry, crystal structure, microstructure and morphology (shape).

In particular, alloys with small magnetocrystalline anisotropies and magnetostrictive coefficients give rise to excellent soft magnetic properties. Alloys for soft magnetic applications may be single phase (Type I, amorphous) or bi-phasic materials (Type II, nanocrystalline) with a nanocrystalline ferromagnetic phase and a residual amorphous phase at the grain boundaries. The Type II nanocrystalline alloys possess:

- high resistivity (50–80 $\mu\Omega$cm)
- low magnetocrystalline anisotropy
- increased mechanical strength.

With properties such as these, nanocrystalline alloys have great potential as soft magnetic materials. The most common compositions for soft magnetic applications either in the amorphous or in the nanocrystalline state are metal–metalloid based (Fe, Co, Ni)–(Si, B) alloys with small additions of Mn, Nb, C and, for the nanocrystalline case, Cu. This alloy system has a good glass-forming ability and is easily accessible by rapid solidification as a thin ribbon in large-scale production.

There has been extensive research in amorphous and nanocrystalline materials in melt-spun ribbon form, which exhibit excellent magnetic properties: large saturation magnetostriction, high saturation magnetization, low anisotropy energies and low coercivity. These factors have made soft magnetic ribbon materials excellent candidates for sensors and actuator devices. Despite their excellent magnetic properties, the as-cast melt-spun ribbons suffer from high randomly orientated stresses, which give rise to a complicated domain structure. However, it is well established that in the stress relieved or magnetically annealed state, they exhibit excellent soft magnetic properties. The disadvantages of these ribbon materials are the difficulty of incorporation into submillimeter dimensional devices and, most importantly, there is no suitable means of bonding such materials onto microfabricated structures.

Ribbon materials are currently bonded to larger devices using epoxy resin. It has been found that the optimized domain structure obtained by magnetic annealing is disturbed on curing the epoxy resin, which induces stress in the ribbon. Since metallic glasses are widely used for sensor applications, a thin-film form of this material would be of great interest for integrating thin-film sensors with today's microelectronics. This can be realized by depositing thin films of this material onto suitable substrates. This not only allows miniaturization of the sensor elements, as dictated by technological demands for smaller and smaller electrical components, but also enables the same microfabrication technologies to be used in production of both electronic and magnetic devices especially for applications like MEMS, NEMS, etc. This makes it commercially more attractive because of the reduced costs and the applicability to a wider range of systems.

2.9   Applications: assembly of magnetic nanostructures

Arrays of magnetic nanostructures can find applications in high-density recording media and magnetic random access memory (MRAM). An important challenge is to assemble these magnetic nanostructures in an effective and controllable way. Several strategies have been developed for the growth of nanostructured magnetic materials, including nanolithography-based methods, solution-based approaches and template-based methods. Some of these methods, however, require high temperatures and special conditions, while in other cases they demand complex and time-consuming procedures. For instance, in template-assisted growth of nanostructures, the selection of suitable catalysts and templates is not straightforward, and the removal of templates and the stabilization of unsupported nanostructures are crucial issues that may compromise the structural and physical properties. The ability to create ordered arrays of well-defined and periodic nanostructures in an accurate, fast and inexpensive fashion would be of great interest for future applications.

The hierarchical self-assembly of nanoscale building blocks (nanoclusters, nanowires, nanobelts and nanotubes) is a technique for building functional electronic and photonic nanodevices. Fractal structures are common in nature across all length scales – from self-assembled molecules, to the shapes of coastlines, to the distribution of galaxies and even to the 3D shapes of clouds. On the nanoscale, dendritic fractals are one type of hyperbranched structure that are generally formed by hierarchical self-assembly under non-equilibrium conditions. Investigation of hierarchically self-assembled fractal patterns in chemical systems has shown that the distinct size, shape and chemical functionality of such structures make them promising candidates for the design and fabrication of new functional nanomaterials, but it is challenging to develop simple and novel synthetic approaches for building hierarchically self-assembled fractal architectures of various systems (see Fig. 2.9 and Fig. 2.10).

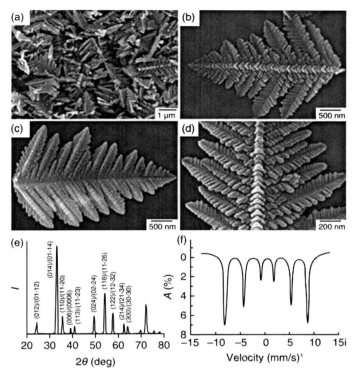

2.9 Electron microscopy images and chemical characterization of a-Fe$_2$O$_3$ fractals synthesized with a K3[Fe(CN)6] concentration of 0.015 m at 1408 C. (a) Low-magnification SEM image of fractals showing the high yield and good uniformity. (b) SEM image of a single a-Fe$_2$O$_3$ fractal taken from one side. (c) SEM image of a single a-Fe$_2$O$_3$ fractal taken from the other side. (d) Higher magnification image of a single a-Fe$_2$O$_3$ fractal showing striking periodic corrugated structures on the main trunk. (e) XRD pattern of the sample confirming formation of a pure a-Fe$_2$O$_3$ phase; I = intensity. (f) Mössbauer spectrum of the sample recorded at room temperature; A = absorption.

Oblique-angle vapor deposition offers advantages in the fabrication of nanostructures over large areas. Vapor atoms arrive on the substrate at oblique angles relative to the surface normal of the substrate. The evaporant nucleates on the substrate and the region behind the nucleus does not receive any further vapor because of the shadowing by the nucleus. Therefore, vapor will be deposited only onto the nucleus. This gives rise to the formation of isolated columnar structures. This technique requires no templates, relatively low temperatures and less harmful chemicals in nanostructure fabrication.

The preparation of thin films of Fe–Ni amorphous alloys by a simple thermal evaporation technique has also been investigated. The focus is on

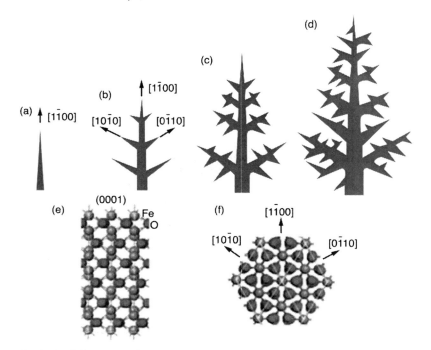

2.10 (a–f) Proposed formation process of the micro-pine dendrite structure by fast growth along the six crystallographically equivalent directions.

Fe–Ni due to their excellent soft magnetic properties. Their saturation magnetization and magnetostriction is high compared with Co-based amorphous alloys, making them interesting candidates for magnetic shielding devices due to their higher permeability. The microstructure and surface morphology of Fe–Ni films are modified using ion beam irradiation and thermal annealing. The evolution of the surface with parameters such as ion fluence, vapor deposition time, annealing temperature and substrate surface roughness has been studied using atomic force microscopy (AFM).

It has been found that the film's surface roughness and microstructure have a significant influence on magnetic properties such as coercivity, remanence, saturating field, demagnetizing field and spin reorientation transition. Films have been structurally characterized using glancing angle x-ray diffraction (GAXRD), transmission electron microscopy (TEM) and scanning electron microscopy (SEM). The surface morphology of the films was probed using atomic force microscopy (AFM) and scanning tunneling microscopy (STM). Imaging of the magnetic domains has been carried out using magnetic force microscopy (MFM). Compositions of the films were determined using energy dispersive x-ray spectroscopy (EDS) and x-ray photoelectron spectroscopy (XPS). The magnetic properties of samples have been characterized employing vibrating sample magnetometer (VSM) and ferromagnetic resonance (FMR) techniques.

Nanocrystalline Fe–Ni thin films have been prepared by the partial crystallization of vapor-deposited amorphous precursors. The microstructure was controlled by annealing the films at different temperatures. X-ray diffraction, transmission electron microscopy and energy dispersive x-ray spectroscopy investigations showed that the nanocrystalline phase was that of Fe–Ni. Grain growth was observed with an increase in annealing temperature. X-ray photoelectron spectroscopy observations showed the presence of native oxide layer on the surface of the films. Magnetic studies using a vibrating sample magnetometer showed that coercivity has a strong dependence on grain size. This is attributed to the random magnetic anisotropy characteristics of the system. The observed coercivity dependence on grain size is explained using modified random anisotropy model.

Glancing angle x-ray diffraction studies showed that the irradiated films retain their amorphous nature. The topographical evolution of the films under swift heavy ion bombardment was probed using atomic force microscopy and it was noticed that surface roughening was taking place with ion beam irradiation. Magnetic measurements using a vibrating sample magnetometer showed that the coercivity of the films increased with an increase in ion fluence. The observed coercivity changes are correlated with topographical evolution of the films under swift heavy ion irradiation.

Investigations on the effect of thermal annealing on the surface roughness of the Fe–Ni thin found that surface smoothing of the film occurs at higher annealing temperature. Ferromagnetic resonance measurements revealed that the demagnetizing field along the in-plane direction decreased with annealing temperature while that along the out-of plane direction increased. This resulted in a transition of surface magnetization direction from out-of-plane to in-plane. The observed modifications of the magnetic properties are correlated with annealing induced surface modification in Fe–Ni thin films.

Atomic force microscopy was used to study the evolution of the surface of these columnar thin films with deposition time. It has been found that the root mean square roughness increased with deposition time but showed a less significant change at longer deposition times. The separation between the nanostructures increased sharply during the initial stages of growth and the change was less significant at higher deposition time. These results suggest that, during initial stages of growth, self-shadowing dominates and as deposition time increases surface diffusion also plays an active role in the growth process.

It is important to note the influence of substrate surface roughness on the structural and magnetic properties of obliquely deposited amorphous nanocolumns of Fe–Ni. Experiments showed that the surface roughness of the substrate determines the morphology of the columnar structures and this in turn has a profound influence on the magnetic properties. Nucleation of Fe–Ni nanocolumns on a smooth silicon substrate was at random, while that on a rough glass substrate was defined by irregularities on the substrate surface. It has been found that magnetic interaction between nanocolumns prepared on a silicon substrate was due to their small inter-column separation. Well-separated nanocolumns on a glass substrate resulted in exchange isolated magnetic domains.

The size, shape and distribution of nanocolumns can be tailored by appropriately choosing the surface roughness of the substrate, which will find potential applications in thin-film magnetism. In a technique for measuring the magnetostriction in amorphous Fe–Ni thin film an optical fiber long-period grating (LPG) was used. An LPG consists of a periodic modification of the refractive index of the core of an optical fiber. For these gratings, the energy typically couples from the fundamental guided mode to the discrete forward-propagating cladding mode. When tensile stress is applied to the optical fiber LPG the periodic spacing changes and reversibly causes the coupling wavelength to shift. This provides a sensitive mechanism to measure the stress/strain and also the magnetostriction of a material attached to the fiber grating. Fe–Ni thin films were coated on a LPG. The magnetostriction of the films was determined from the change in the peak position of the attenuation band when a magnetic field was applied along the axis of the fiber. Results indicate that LPGs in combination with Fe–Ni thin films can act as a potential candidate in the field of magnetostrictive sensors.

Magnetic fine particles of Fe–Co and Fe–Ni alloys encapsulated by graphitic carbon (C) have been synthesized by annealing Fe_2O_3, Co_3O_4 and NiO with carbon powders at 1273 K in a nitrogen atmosphere. Saturation magnetizations of the Fe–Co and Fe–Ni particles were 122–150 and 7.8–105 A m^2/kg, respectively. X-ray diffraction measurement showed that the Fe–Co was composed of two phases with face-centered-cubic (fcc) and body-centered-cubic (bcc) phases, while the Fe–Ni was only a fcc phase. Lattice constants of both particles depended on their composition, which suggested that they were alloys. This was confirmed by Mössbauer spectroscopy. Electron microscope images and electron energy loss spectroscope spectra revealed that these particles, with diameters of 100–200 nm, were encapsulated by C layers with a thickness of several tens of nanometers.

Magnetic nanomaterials, fabricated by various methods, have attracted a great deal of attention for their technological applications, such as magnetic storage media, sensors and magnetic random access memory (MRAM). In these nanomaterials, each ferromagnetic region is a single domain in which the magnetic spins are aligned in one direction. Thus every ferromagnetic region represents a single-domain magnet, which exhibits unique magnetic properties for further applications, such as high squareness and high coercivity. Among these nanomaterials, nanowires are of intense interest.

Many methods for the fabrication of nanowires have been developed. Among these methods, template synthesis was regarded as quite useful because it is extremely general with regard to the types of materials that can be prepared. Many ferromagnetic nanowires of Fe, Co, Ni and their alloys are fabricated by this method. However, these nanowires often lose their high coercivity due to demagnetization by neighboring magnetic domains. In order to overcome this drawback, the non-magnetic layer (NM) is inserted between each ferromagnetic layer (FM), to form an FM/NM/ FM ... periodic structure. The magnetic properties of these multi-layer nanowires can be varied according to the thickness of FM and NM layers.

Highly ordered composite nanowires with multi-layer Ni/Cu and NiFe/ Cu have been fabricated by pulsed electrodeposition onto nanoporous alumina membrane. The diameter of wires can be easily varied by the pore size of alumina, ranging from 30 to 100 nm. The applied potential and the duration of each potential square pulse determines the thickness of the metal layers. Nanowires have been characterized using transmission electron microscopy (TEM), magnetic force microscopy (MFM), and vibrating sample magnetometer (VSM) measurements (see Fig. 2.11 to Fig. 2.15). The MFM images indicate that every ferromagnetic layer separated by Cu layer was present as single isolated domain-like magnet.

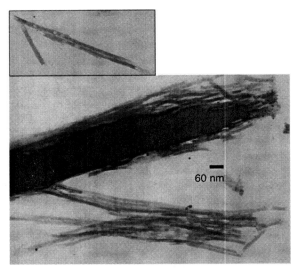

2.11 TEM micrograph of a bundle of Ni/Cu nanowires with a diameter of 40 nm. The inset shows the bamboo-like structure.

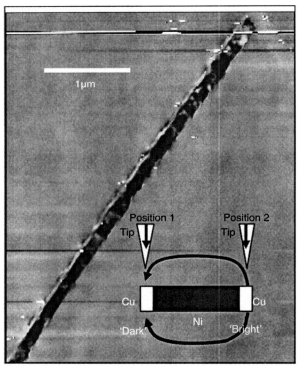

2.12 Lift-mode MFM graphs of Ni/Cu nanowires. The deposition times of Ni and Cu are 20 s (»200 nm) and 30 s (»50 nm) respectively.

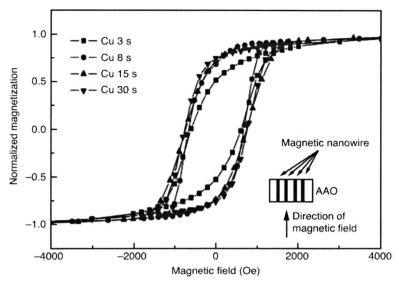

2.13 Hysteresis loops of 35 nm Ni/Cu wires with different deposition time of Cu and constant deposition time (20 s) of Ni (»200 nm). The applied field is perpendicular to the membrane.

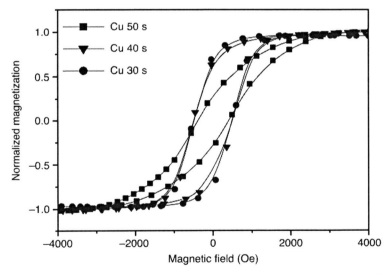

2.14 Hysteresis loops of 45 nm NiFe/Cu wires with different deposition times of Cu and constant deposition time (10 s) of NiFe alloy. The applied field is perpendicular to the membrane (parallel to the wires).

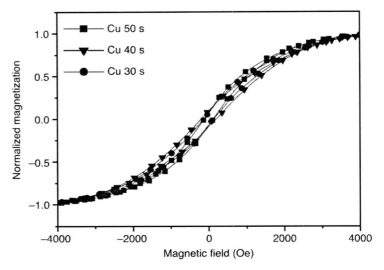

2.15 Hysteresis loops of 45 nm NiFe/Cu wires with different deposition times of Cu and constant deposition time (10 s) of NiFe alloy. The applied field is parallel to the membrane (perpendicular to the wires).

2.10   References and further reading

1. D. Dutta Majumder, *et al.*, 'Nano-materials: science of bottom-up and top-down nanotechnology education – a paradigm shift,' *IETE Tech Rev*, vol. 24, p. 16, 2007.
2. D. D. Majumder, *et al.*, 'Characterization of gold and silver nanoparticles using it's color image segmentation and feature extraction using fuzzy C-means clustering and generalized shape theory,' in *Communications and Signal Processing (ICCSP), 2011 International Conference*, pp. 70–4, 2011.
3. D. Dutta Majumder, *et al.*, 'Synthesis and characterization of gold nanoparticles – a fuzzy mathematical approach,' *Pattern Recognition and Machine Intelligence*, vol. 6744, S. Kuznetsov, *et al.*, Eds., Springer, Berlin, 2011, pp. 324–32.
4. H. Zeng, *et al.*, 'Exchange-coupled nanocomposite magnets by nanoparticle self-assembly,' *Nature*, vol. 420, pp. 395–8, Nov 28 2002.
5. P. Zhang, *et al.*, 'Synthesis of mesoporous magnetic Co-NPs/carbon nanocomposites and their adsorption property for methyl orange from aqueous solution,' *J Colloid Interface Sci*, vol. 389, pp. 10–15, Jan 1 2013.
6. M. Y. Razzaq, *et al.*, 'Memory-effects of magnetic nanocomposites,' *Nanoscale*, vol. 4, pp. 6181–95, Sep 28 2012.
7. J. C. Serna, *et al.*, 'Magnetic properties of nanocomposites formed by magnetic nanoparticles embedded in a non-magnetic matrix: a simulation approach,' *J Nanosci Nanotechnol*, vol. 12, pp. 4979–83, Jun 2012.
8. Q. Wang, *et al.*, 'Superparamagnetic magnetite nanocrystals-graphene oxide nanocomposites: facile synthesis and their enhanced electric double-layer capacitor performance,' *J Nanosci Nanotechnol*, vol. 12, pp. 4583–90, Jun 2012.

9.   I. V. Kityk, *et al.*, 'Photoinduced spectra for magneto electric (1-x)BiFeO(3)(x) CuFe(2)O(4) nanocomposites,' *Spectrochim Acta A Mol Biomol Spectrosc*, vol. 97, pp. 695–8, Nov 2012.

10. M. R. Nabid, *et al.*, 'Preparation of new magnetic nanocatalysts based on TiO (2) and ZnO and their application in improved photocatalytic degradation of dye pollutant under visible light,' *Photochem Photobiol*, vol. 89, pp. 24–32, Jul 21 2012.

11. K. Ni, *et al.*, 'Magnetic catechol-chitosan with bioinspired adhesive surface: preparation and immobilization of omega-transaminase,' *PLoS One*, vol. 7, p. e41101, 2012.

12. D. Huang, *et al.*, 'Development of magnetic multiwalled carbon nanotubes as solid-phase extraction technique for the determination of p-hydroxybenzoates in beverage,' *J Sep Sci*, vol. 35, pp. 1667–74, Jul 2012.

13. S. Li, *et al.*, 'Fabrication of magnetic Ni nanoparticles functionalized water-soluble graphene sheets nanocomposites as sorbent for aromatic compounds removal,' *J Hazard Mater*, vol. 229–230, pp. 42–7, Aug 30 2012.

14. W. Wei and X. Qu, 'Extraordinary physical properties of functionalized graphene,' *Small*, vol. 8, pp. 2138–51, Jul 23 2012.

15. Y. L. Dong, *et al.*, 'Graphene oxide-Fe3O4 N magnetic nanocomposites with peroxidase-like activity for colorimetric detection of glucose,' *Nanoscale*, vol. 4, pp. 3969–76, Jul 7 2012.

16. C. Ding, *et al.*, 'Electrochemiluminescent determination of cancer cells based on aptamers, nanoparticles, and magnetic beads,' *Chemistry*, vol. 18, pp. 7263–8, Jun 4 2012.

17. O. Cintra e Silva Dde, *et al.*, 'Successful strategy for targeting the central nervous system using magnetic albumin nanospheres,' *J Biomed Nanotechnol*, vol. 8, pp. 182–9, Feb 2012.

18. P. Papaphilippou, *et al.*, 'Multiresponsive polymer conetworks capable of responding to changes in pH, temperature, and magnetic field: synthesis, characterization, and evaluation of their ability for controlled uptake and release of solutes,' *ACS Appl Mater Interfaces*, vol. 4, pp. 2139–47, Apr 2012.

19. P. Thanikaivelan, *et al.*, 'Collagen based magnetic nanocomposites for oil removal applications,' *Sci Rep*, vol. 2, p. 230, 2012.

20. S. Srivastava, *et al.*, 'Magnetic-nanoparticle-doped carbogenic nanocomposite: an effective magnetic resonance/fluorescence multimodal imaging probe,' *Small*, vol. 8, p. 1099, Apr 10 2012.

21. H. Zarei, *et al.*, 'Magnetic nanocomposite of anti-human IgG/COOH-multiwalled carbon nanotubes/Fe(3)O(4) as a platform for electrochemical immunoassay,' *Anal Biochem*, vol. 421, pp. 446–53, Feb 15 2012.

22. D. Gopi, *et al.*, 'Synthesis and spectroscopic characterization of magnetic hydroxyapatite nanocomposite using ultrasonic irradiation,' *Spectrochim Acta A Mol Biomol Spectrosc*, vol. 87, pp. 245–50, Feb 15 2012.

23. G. Jie, *et al.*, 'Quantum dots-based multifunctional dendritic superstructure for amplified electrochemiluminescence detection of ATP,' *Biosens Bioelectron*, vol. 31, pp. 69–76, Jan 15 2012.

24. M. Yin, *et al.*, 'Magnetic self-assembled zeolite clusters for sensitive detection and rapid removal of mercury(II),' *ACS Appl Mater Interfaces*, vol. 4, pp. 431–7, Jan 2012.

25. J. Yue, *et al.*, 'Deposition of gold nanoparticles on beta-FeOOH nanorods for detecting melamine in aqueous solution,' *J Colloid Interface Sci*, vol. 367, pp. 204–12, Feb 1 2012.
26. M. Darbandi, *et al.*, 'Bright luminescent, colloidal stable silica coated CdSe/ZnS nanocomposite by an in situ, one-pot surface functionalization,' *J Colloid Interface Sci*, vol. 365, pp. 41–5, Jan 1 2012.
27. J. D. D. Dutta Majumdar and J. Das, *Digital Computer Memory Technology*, John Wiley & Sons (Asia) Pte Ltd, 1980.
28. A. Nacev, *et al.*, 'The behaviors of ferro-magnetic nano-particles in and around blood vessels under applied magnetic fields,' *J Magn Magn Mater*, vol. 323, pp. 651–68, Mar 1 2011.
29. S. K. Misra, *et al.*, 'A 236-GHz Fe EPR study of nano-particles of the ferromagnetic room-temperature semiconductor Sn(1-x)Fe(x)O(2)(x = 0.005),' *Appl Magn Reson*, vol. 36, pp. 291–5, Dec 1 2009.
30. F. A. Gallagher, *et al.*, 'Magnetic resonance imaging of pH in vivo using hyperpolarized 13C-labelled bicarbonate,' *Nature*, vol. 453, pp. 940–3, Jun 12 2008.
31. S. Engel, *et al.*, 'Colour tuning in human visual cortex measured with functional magnetic resonance imaging,' *Nature*, vol. 388, pp. 68–71, Jul 3 1997.
32. R. B. Tootell, *et al.*, 'Visual motion aftereffect in human cortical area MT revealed by functional magnetic resonance imaging,' *Nature*, vol. 375, pp. 139–41, May 11 1995.
33. M. S. Albert, *et al.*, 'Biological magnetic resonance imaging using laser-polarized 129Xe,' *Nature*, vol. 370, pp. 199–201, Jul 21 1994.
34. G. A. Press, *et al.*, 'Hippocampal abnormalities in amnesic patients revealed by high-resolution magnetic resonance imaging,' *Nature*, vol. 341, pp. 54–7, Sep 7 1989.
35. J. B. Aguayo, *et al.*, 'Nuclear magnetic resonance imaging of a single cell,' *Nature*, vol. 322, pp. 190–1, Jul 10 1986.
36. H. A. Neumann, *et al.*, 'Treatment of human clonogenic tumor cells and bone marrow progenitor cells with bleomycin and peplomycin under 40.5 degrees C hyperthermia in vitro,' *Eur J Cancer Clin Oncol*, vol. 25, pp. 99–104, Jan 1989.
37. N. L. Trevaskis, *et al.*, 'Correction to Targeted drug delivery to lymphocytes: a route to site-specific immunomodulation?,' *Mol Pharm*, vol. 8, p. 2484, Dec 5 2011.
38. A. Jain, *et al.*, 'Mannosylated solid lipid nanoparticles as vectors for site-specific delivery of an anti-cancer drug,' *J Control Release*, vol. 148, pp. 359–67, Dec 20 2010.
39. B. Polyak and G. Friedman, 'Magnetic targeting for site-specific drug delivery: applications and clinical potential,' *Expert Opin Drug Deliv*, vol. 6, pp. 53–70, Jan 2009.
40. Y. Morimoto, *et al.*, 'Biomedical applications of magnetic fluids. I. Magnetic guidance of ferro-colloid-entrapped albumin microsphere for site specific drug delivery in vivo,' *J Pharmacobiodyn*, vol. 3, pp. 264–7, May 1980.
41. A. Namai, *et al.*, 'Hard magnetic ferrite with a gigantic coercivity and high frequency millimetre wave rotation,' *Nat Commun*, vol. 3, p. 1035, 2012.
42. L. Zhu, *et al.*, 'Multifunctional L1(0)-Mn(1.5)Ga films with ultrahigh coercivity,

giant perpendicular magnetocrystalline anisotropy and large magnetic energy product,' *Adv Mater*, vol. 24, pp. 4547–51, Aug 28 2012.

43. P. Granitzer, *et al.*, 'Porous silicon/Ni composites of high coercivity due to magnetic field-assisted etching,' *Nanoscale Res Lett*, vol. 7, p. 384, 2012.

44. R. Fernandez, *et al.*, 'Microstructural enhancement of high coercivity L1(0)–FePt films for next-generation magnetic recording media,' *J Nanosci Nanotechnol*, vol. 11, pp. 3889–93, May 2011.

45. Z. X. Wang, *et al.*, 'Synchronized time- and high-field-resolved all-optical pump-probe magneto-optical setup based on a strong alternating magnetic field and its application in magnetization dynamics of high coercivity magnetic medium,' *Rev Sci Instrum*, vol. 82, p. 034703, Mar 2011.

46. M. Urbaniak, *et al.*, 'Domain-wall movement control in Co/Au multilayers by He(+)-ion-bombardment-induced lateral coercivity gradients,' *Phys Rev Lett*, vol. 105, p. 067202, Aug 6 2010.

47. T. N. Narayanan, *et al.*, 'The synthesis of high coercivity cobalt-in-carbon nanotube hybrid structures and their optical limiting properties,' *Nanotechnology*, vol. 20, p. 285702, Jul 15 2009.

48. J. Escrig, *et al.*, 'Geometry dependence of coercivity in Ni nanowire arrays,' *Nanotechnology*, vol. 19, p. 075713, Feb 20 2008.

49. J. N. Wang, *et al.*, 'Synthesis of carbon encapsulated magnetic nanoparticles with giant coercivity by a spray pyrolysis approach,' *J Phys Chem B*, vol. 111, pp. 2119–24, Mar 1 2007.

50. F. Iskandar, *et al.*, 'High coercivity of ordered macroporous fept films synthesized via colloidal templates,' *Nano Lett*, vol. 5, pp. 1525–8, Jul 2005.

51. S. Maheswaran, *et al.*, 'Phosphonate ligands stabilize mixed-valent {Mn(III) (20-x)Mn(II)x} clusters with large spin and coercivity,' *Angew Chem Int Ed Engl*, vol. 44, pp. 5044–8, Aug 12 2005.

52. H. P. Liang, *et al.*, 'Ni-Pt multilayered nanowire arrays with enhanced coercivity and high remanence ratio,' *Inorg Chem*, vol. 44, pp. 3013–5, May 2 2005.

53. L. Pust, *et al.*, 'Domain-wall coercivity in ferromagnetic systems with nonuniform local magnetic field,' *Phys Rev B Condens Matter*, vol. 54, pp. 12262–71, Nov 1 1996.

54. F. Cebollada, *et al.*, 'Angular dependence of coercivity in Nd–Fe–B sintered magnets: Proof that coherent rotation is not involved,' *Phys Rev B Condens Matter*, vol. 52, pp. 13511–8, Nov 1 1995.

55. X. C. Kou, *et al.*, 'Coercivity mechanism of sintered Pr17Fe75B8 and Pr17Fe53B30 permanent magnets,' *Phys Rev B Condens Matter*, vol. 50, pp. 3849–60, Aug 1 1994.

56. D. I. Paul and A. Cresswell, 'Single-domain crystal-anisotropy-dominated coercivity,' *Phys Rev B Condens Matter*, vol. 48, pp. 3803–9, Aug 1 1993.

57. N. Paunovic, *et al.*, 'Superparamagnetism in iron-doped CeO(2-y) nanocrystals,' *J Phys Condens Matter*, vol. 24, p. 456001, Nov 14 2012.

58. S. W. Kim, *et al.*, 'Superparamagnetism of Cu2Se nanoparticles,' *J Nanosci Nanotechnol*, vol. 12, pp. 5880–3, Jul 2012.

59. B. Ghosh, *et al.*, 'Observation of superparamagnetism to flux closure behaviour in ZnO nanoparticle agglomerates,' *J Phys Condens Matter*, vol. 24, p. 366002, Sep 12 2012.

60. R. J. Tackett, *et al.*, 'Evidence of low-temperature superparamagnetism in Mn3O4 nanoparticle ensembles,' *Nanotechnology*, vol. 21, p. 365703, Sep 10 2010.
61. M. Knobel, *et al.*, 'Superparamagnetism and other magnetic features in granular materials: a review on ideal and real systems,' *J Nanosci Nanotechnol*, vol. 8, pp. 2836–57, Jun 2008.
62. D. Brinzei, *et al.*, 'Photoinduced superparamagnetism in trimetallic coordination nanoparticles,' *J Am Chem Soc*, vol. 129, pp. 3778–9, Apr 4 2007.
63. D. Magana, *et al.*, 'Switching-on superparamagnetism in Mn/CdSe quantum dots,' *J Am Chem Soc*, vol. 128, pp. 2931–9, Mar 8 2006.
64. M. Mikhaylova, *et al.*, 'Superparamagnetism of magnetite nanoparticles: dependence on surface modification,' *Langmuir*, vol. 20, pp. 2472–7, Mar 16 2004.
65. S. R. Shinde, *et al.*, 'Co-occurrence of superparamagnetism and anomalous hall effect in highly reduced cobalt-doped rutile TiO2-delta films,' *Phys Rev Lett*, vol. 92, p. 166601, Apr 23 2004.
66. S. I. Woods, *et al.*, 'Direct investigation of superparamagnetism in Co nanoparticle films,' *Phys Rev Lett*, vol. 87, p. 137205, Sep 24 2001.
67. M. E. McHenry, *et al.*, 'Superparamagnetism in carbon-coated Co particles produced by the Kratschmer carbon arc process,' *Phys Rev B Condens Matter*, vol. 49, pp. 11358–63, Apr 15 1994.
68. A. Slawska-Waniewska, *et al.*, 'Superparamagnetism in a nanocrystalline Fe-based metallic glass,' *Phys Rev B Condens Matter*, vol. 46, pp. 14594–7, Dec 1 1992.
69. R. S. de Biasi and A. A. Fernandes, 'Ferromagnetic resonance evidence for superparamagnetism in a partially crystallized metallic glass,' *Phys Rev B Condens Matter*, vol. 42, pp. 527–9, Jul 1 1990.
70. C. Wan, *et al.*, 'Geometrical enhancement of low-field magnetoresistance in silicon,' *Nature*, vol. 477, pp. 304–7, Sep 15 2011.
71. N. H. Pham, *et al.*, 'Electromotive force and huge magnetoresistance in magnetic tunnel junctions,' *Nature*, vol. 458, pp. 489–92, Mar 26 2009.
72. Z. H. Xiong, *et al.*, 'Giant magnetoresistance in organic spin-valves,' *Nature*, vol. 427, pp. 821–4, Feb 26 2004.
73. M. M. Parish and P. B. Littlewood, 'Non-saturating magnetoresistance in heavily disordered semiconductors,' *Nature*, vol. 426, pp. 162–5, Nov 13 2003.
74. N. Manyala, *et al.*, 'Magnetoresistance from quantum interference effects in ferromagnets,' *Nature*, vol. 404, pp. 581–4, Apr 6 2000.
75. H. Pfau, *et al.*, 'Thermal and electrical transport across a magnetic quantum critical point,' *Nature*, vol. 484, pp. 493–7, Apr 26 2012.
76. D. Porath, *et al.*, 'Direct measurement of electrical transport through DNA molecules,' *Nature*, vol. 403, pp. 635–8, Feb 10 2000.
77. K. A. Grossklaus, *et al.*, 'Electrical transport in ion beam created InAs nanospikes,' *Nanotechnology*, vol. 23, p. 315301, Aug 10 2012.
78. S. Kim, *et al.*, 'Role of self-assembled monolayer passivation in electrical transport properties and flicker noise of nanowire transistors,' *ACS Nano*, vol. 6, pp. 7352–61, Aug 28 2012.
79. S. N. Mohammad, 'Quantum-confined nanowires as vehicles for enhanced electrical transport,' *Nanotechnology*, vol. 23, p. 285707, Jul 20 2012.

80. A. W. Tsen, *et al.*, 'Tailoring electrical transport across grain boundaries in polycrystalline graphene,' *Science*, vol. 336, pp. 1143–6, Jun 1 2012.

81. S. W. Yoon, *et al.*, 'Effects of Pt junction on electrical transport of individual ZnO nanorod device fabricated by focused ion beam,' *J Nanosci Nanotechnol*, vol. 12, pp. 1466–70, Feb 2012.

82. A. B. Mostert, *et al.*, 'Role of semiconductivity and ion transport in the electrical conduction of melanin,' *Proc Natl Acad Sci U S A*, vol. 109, pp. 8943–7, Jun 5 2012.

83. J. Dolinsek, 'Electrical and thermal transport properties of icosahedral and decagonal quasicrystals,' *Chem Soc Rev*, vol. 41, pp. 6730–44, Oct 21 2012.

84. E. K. Lee, *et al.*, 'Large thermoelectric figure-of-merits from SiGe nanowires by simultaneously measuring electrical and thermal transport properties,' *Nano Lett*, vol. 12, pp. 2918–23, Jun 13 2012.

85. S. A. Thomas, *et al.*, 'Neutron diffraction and electrical transport studies on the incommensurate magnetic phase transition in holmium at high pressures,' *J Phys Condens Matter*, vol. 24, p. 216003, May 30 2012.

86. B. A. Assaf, *et al.*, 'Modified electrical transport probe design for standard magnetometer,' *Rev Sci Instrum*, vol. 83, p. 033904, Mar 2012.

87. A. Hamam and R. R. Lew, 'Electrical phenotypes of calcium transport mutant strains of a filamentous fungus, Neurospora crassa,' *Eukaryot Cell*, vol. 11, pp. 694–702, May 2012.

88. B. Aissa, *et al.*, 'Electrical transport properties of single wall carbon nanotube/polyurethane composite based field effect transistors fabricated by UV-assisted direct-writing technology,' *Nanotechnology*, vol. 23, p. 115705, Mar 23 2012.

89. S. Takagi, *et al.*, 'Gating electrical transport through DNA molecules that bridge between silicon nanogaps,' *Nanoscale*, vol. 4, pp. 1975–7, Mar 21 2012.

90. K. Bapna, *et al.*, 'Implication of local moment at Ti and Fe sites for the electrical and magneto-transport properties of degenerate semiconducting Ti(1)-xFexO(2)-d epitaxial films,' *J Phys Condens Matter*, vol. 24, p. 056004, Feb 8 2012.

91. A. Casaca, *et al.*, 'Electrical transport properties of CuS single crystals,' *J Phys Condens Matter*, vol. 24, p. 015701, Jan 11 2012.

92. H. Fischer, *et al.*, 'Hemocompatibility of high strength oxide ceramic materials: an in vitro study,' *J Biomed Mater Res A*, vol. 81, pp. 982–6, Jun 15 2007.

93. M. Hamid, *et al.*, 'Synthesis of isostructural cage complexes of copper with cobalt and nickel for deposition of mixed ceramic oxide materials,' *Inorg Chem*, vol. 45, pp. 10457–66, Dec 25 2006.

94. N. Trabandt, *et al.*, 'Limitations of titanium dioxide and aluminum oxide as ossicular replacement materials: an evaluation of the effects of porosity on ceramic prostheses,' *Otol Neurotol*, vol. 25, pp. 682–93, Sep 2004.

95. L. W. Brinkmann, 'Endosseous implantation procedures for non-metallic aluminum oxide ceramic implant materials,' *Phillip J Restaur Zahnmed*, vol. 3, pp. 281–97, Nov 1986.

96. T. Gabor, 'Chemico-physical study of oxide layers on the margins of metal-ceramic materials,' *Cesk Stomatol*, vol. 77, pp. 423–7, Dec 1977.

97. A. Breustedt, *et al.*, 'Comparative polarisation optical studies of ceramic materials reinforced with aluminum oxide' *Stomatol DDR*, vol. 24, pp. 199–201, Mar 1974.

98. R. S. Saksena, *et al.*, 'Petascale lattice-Boltzmann studies of amphiphilic cubic liquid crystalline materials in a globally distributed high-performance computing and visualization environment,' *Philos Transact A Math Phys Eng Sci*, vol. 368, pp. 3983–99, Aug 28 2010.

99. K. Nakamura, *et al.*, 'Parameter estimation of in silico biological pathways with particle filtering towards a petascale computing,' *Pac Symp Biocomput*, vol. 14, pp. 227–38, 2009.

100. R. J. Liu and T. L. Chen, 'Variational cumulant expansion for the Heisenberg model with the critical temperature determined to third order,' *Phys Rev B Condens Matter*, vol. 50, pp. 9169–73, Oct 1 1994.

101. Y. Xia, *et al.*, 'Equivalence of the linked-cluster expansion and the variational cumulant expansion in the high-temperature limit,' *Phys Rev B Condens Matter*, vol. 46, pp. 3125–7, Aug 1 1992.

102. C. I. Tan and X. T. Zheng, 'Variational-cumulant expansion in lattice gauge theory at finite temperature,' *Phys Rev D Part Fields*, vol. 39, pp. 623–635, Jan 15 1989.

103. E. J. Wildman, *et al.*, 'Colossal magnetoresistance in Mn2+ oxypnictides NdMnAsO(1–x)F(x),' *J Am Chem Soc*, vol. 134, pp. 8766–9, May 30 2012.

104. C. Sen, *et al.*, 'First order colossal magnetoresistance transitions in the two-orbital model for manganites,' *Phys Rev Lett*, vol. 105, p. 097203, Aug 27 2010.

105. X. Zhou, *et al.*, 'Colossal magnetoresistance in an ultraclean weakly interacting 2D Fermi liquid,' *Phys Rev Lett*, vol. 104, p. 216801, May 28 2010.

106. A. M. Goforth, *et al.*, 'Magnetic properties and negative colossal magnetoresistance of the rare earth Zintl phase EuIn2As2,' *Inorg Chem*, vol. 47, pp. 11048–56, Dec 1 2008.

107. S. Weber, *et al.*, 'Colossal magnetocapacitance and colossal magnetoresistance in HgCr2S4,' *Phys Rev Lett*, vol. 96, p. 157202, Apr 21 2006.

108. K. Kuepper, *et al.*, 'The x-ray magnetic circular dichroism spin sum rule for 3d (4) systems: Mn(3+) ions in colossal magnetoresistance manganites,' *J Phys Condens Matter*, vol. 24, p. 435602, Oct 31 2012.

109. H. Sakurai, *et al.*, 'Unconventional colossal magnetoresistance in sodium chromium oxide with a mixed-valence state,' *Angew Chem Int Ed Engl*, vol. 51, pp. 6653–6, Jul 2 2012.

110. M. Rybicki and I. Zasada, 'Noncollinear spin reorientation transition in S = 1 ferromagnetic thin films,' *J Phys Condens Matter*, vol. 24, p. 386005, Sep 26 2012.

111. C. Lee, *et al.*, 'Spin reorientation in the square-lattice antiferromagnets RMnAsO (R = Ce, Nd): density functional analysis of the spin-exchange interactions between the rare-earth and transition-metal ions,' *Inorg Chem*, vol. 51, pp. 6890–7, Jun 18 2012.

112. A. Marcinkova, *et al.*, 'Iron spin-reorientation transition in NdFeAsO,' *J Phys Condens Matter*, vol. 24, p. 256007, Jun 27 2012.

113. S. H. Park, *et al.*, 'Canted antiferromagnetism and spin reorientation transition in layered inorganic-organic perovskite (C6H5CH2CH2NH3)2MnCl4,' *Dalton Trans*, vol. 41, pp. 1237–42, Jan 28 2012.

114. S. A. Nikitin, *et al.*, 'Giant rotating magnetocaloric effect in the region of spin-reorientation transition in the NdCo(5) single crystal,' *Phys Rev Lett*, vol. 105, p. 137205, Sep 24 2010.

115. T. Slezak, *et al.*, 'Noncollinear magnetization structure at the thickness-driven spin-reorientation transition in epitaxial Fe films on W(110),' *Phys Rev Lett*, vol. 105, p. 027206, Jul 9 2010.

116. R. Fromter, *et al.*, 'Imaging the cone state of the spin reorientation transition,' *Phys Rev Lett*, vol. 100, p. 207202, May 23 2008.

117. J. Choi, *et al.*, 'Magnetic bubble domain phase at the spin reorientation transition of ultrathin Fe/Ni/Cu(001) film,' *Phys Rev Lett*, vol. 98, p. 207205, May 18 2007.

118. K. Y. Wang, *et al.*, 'Spin reorientation transition in single-domain,' *Phys Rev Lett*, vol. 95, p. 217204, Nov 18 2005.

119. J. Hong, *et al.*, 'Manipulation of spin reorientation transition by oxygen surfactant growth: a combined theoretical and experimental approach,' *Phys Rev Lett*, vol. 92, p. 147202, Apr 9 2004.

120. L. M. Garcia, *et al.*, 'Orbital magnetic moment instability at the spin reorientation transition of Nd2Fe14B,' *Phys Rev Lett*, vol. 85, pp. 429–32, Jul 10 2000.

121. E. Y. Vedmedenko, *et al.*, 'Magnetic microstructure of the spin reorientation transition: a computer experiment,' *Phys Rev Lett*, vol. 84, pp. 5884–7, Jun 19 2000.

122. G. Garreau, *et al.*, 'Spin-reorientation transition in ultrathin Tb/Co films,' *Phys Rev B Condens Matter*, vol. 53, pp. 1083–6, Jan 15 1996.

123. H. Fritzsche, *et al.*, 'Angular dependence of perpendicular magnetic surface anisotropy and the spin-reorientation transition,' *Phys Rev B Condens Matter*, vol. 49, pp. 15665–8, Jun 1 1994.

124. V. L. Sobolev, *et al.*, 'Spin-reorientation phase transition in Nd2CuO4 in an external magnetic field: Unusual manifestations of magnetoelastic coupling,' *Phys Rev B Condens Matter*, vol. 48, pp. 3417–22, Aug 1 1993.

125. Z. Q. Qiu, *et al.*, 'Asymmetry of the spin reorientation transition in ultrathin Fe films and wedges grown on Ag(100),' *Phys Rev Lett*, vol. 70, pp. 1006–9, Feb 15 1993.

126. M. R. Safari, *et al.*, 'Accuracy verification of magnetic resonance imaging (MRI) technology for lower-limb prosthetic research: utilising animal soft tissue specimen and common socket casting materials,' *Sci World J*, vol. 2012, p. 156186, 2012.

127. Y. Yang, *et al.*, 'Portable magnetic tweezers device enables visualization of the three-dimensional microscale deformation of soft biological materials,' *Biotechniques*, vol. 51, pp. 29–34, Jul 2011.

128. R. Shakya, *et al.*, 'Amphiphilic and magnetic properties of a new class of cluster-bearing [L2Cu4(mu4-O)(mu2-carboxylato)4] soft materials,' *Chemistry*, vol. 13, pp. 9948–56, 2007.

129. J. P. Butler, *et al.*, 'Measuring surface-area-to-volume ratios in soft porous materials using laser-polarized xenon interphase exchange nuclear magnetic resonance,' *J Phys Condens Matter*, vol. 14, pp. L297–304, Apr 8 2002.

130. S. Yin, *et al.*, 'Local structure evolution of FexNi77-xCu(1-)Nb2P14B6 soft magnetic materials by mechanical alloying,' *J Synchrotron Radiat*, vol. 8, pp. 889–91, Mar 1 2001.

131. A. Vazquez and O. Sotolongo-Costa, 'Dynamics of a domain wall in soft-

magnetic materials: barkhausen effect and relation with sandpile models,' *Phys Rev Lett*, vol. 84, pp. 1316–9, Feb 7 2000.

132. R. Kucharkowski, *et al.*, 'Determination of boron in amorphous and nanostructured soft magnetic alloys, ribbons and layered materials by ICP atomic emission spectrometry,' *Anal Bioanal Chem*, vol. 355, pp. 256–60, Jun 1996.

133. G. H. Jonker and J. H. Van Santen, 'Ferromagnetic compounds of manganese with perovskite structure,' *Physica*, vol. 16, pp. 337–349, 1950.

134. T. A. Kaplan, *et al.*, 'Non-Stoner continuum in the double exchange model,' *Phys Rev Lett*, vol. 86, pp. 3634–7, Apr 16 2001.

135. M. P. Gokhale and D. L. Mills, 'Spin excitations of a model itinerant ferromagnetic film: Spin waves, Stoner excitations, and spin-polarized electron-energy-loss spectroscopy,' *Phys Rev B Condens Matter*, vol. 49, pp. 3880–93, Feb 1 1994.

136. R. Saniz and S. P. Apell, 'Model calculation of Stoner-excitation cross sections in spin-polarized crystals,' *Phys Rev B Condens Matter*, vol. 48, pp. 3206–11, Aug 1 1993.

137. I. I. Lo, *et al.*, 'Pressure dependence of the de Haas-van Alphen effect in $ZrZn2$: Deviations from the Stoner-Wohlfarth model,' *Phys Rev Lett*, vol. 62, pp. 2555–58, May 22 1989.

138. P. M. Marcus and V. L. Moruzzi, 'Stoner model of ferromagnetism and total-energy band theory,' *Phys Rev B Condens Matter*, vol. 38, pp. 6949–53, Oct 1 1988.

139. http://dx.doi.org/10.1016/S0022-3093(02)01640-X

140. R. Pool, 'Brookhaven chemists find new fusion method: Firing clusters of heavy water molecules into a deuterium target produces fusion at a surprisingly high rate, but the practical applications are unclear,' *Science*, vol. 245, pp. 1448–9, Sep 29 1989.

141. J. Xiao, *et al.*, 'A supramolecular solution to a long-standing problem: The 1,6-polymerization of a triacetylene,' *Angew Chem Int Ed Engl*, vol. 39, pp. 2132–35, Jun 16 2000.

142. H. Y. Hwang, *et al.*, 'Spin-polarized intergrain tunneling in $La2/3Sr1/3MnO_3$,' *Phys Rev Lett*, vol. 77, pp. 2041–44, Sep 2 1996.

143. H. Y. Hwang, *et al.*, 'Lattice effects on the magnetoresistance in doped $LaMnO_3$,' *Phys Rev Lett*, vol. 75, pp. 914–17, Jul 31 1995.

144. H. Y. Hwang, *et al.*, 'Pressure effects on the magnetoresistance in doped manganese perovskites,' *Phys Rev B Condens Matter*, vol. 52, pp. 15046–49, Dec 1 1995.

145. N. C. Yeh and C. C. Tsuei, 'Quasi-two-dimensional phase fluctuations in bulk superconducting $YBa_{2}Cu_{3}O_{7}$ single crystals,' *Phys Rev B*, vol. 39, pp. 9708–11, 1989.

146. H. Liu, *et al.*, 'Mott and Efros-Shklovskii variable range hopping in CdSe quantum dots films,' *ACS Nano*, vol. 4, pp. 5211–6, Sep 28 2010.

147. E. Medina, *et al.*, 'Magnetoconductance anisotropy and interference effects in variable-range hopping,' *Phys Rev B Condens Matter*, vol. 53, pp. 7663–72, Mar 15 1996.

148. R. Rosenbaum, 'Crossover from Mott to Efros-Shklovskii variable-range-

hopping conductivity in InxOy films,' *Phys Rev B Condens Matter*, vol. 44, pp. 3599–603, Aug 15 1991.

149. S. S. Yan, *et al.*, 'Spin-dependent variable range hopping and magnetoresistance in Ti(1-x)Co(x)O(2) and Zn(1-x)Co(x)O magnetic semiconductor films,' *J Phys Condens Matter*, vol. 18, pp. 10469–80, Nov 22 2006.

150. M. Caban-Acevedo, *et al.*, 'Synthesis, characterization, and variable range hopping transport of pyrite (FeS(2)) nanorods, nanobelts, and nanoplates,' *ACS Nano*, vol. 7, pp. 1731–9, Feb 26 2013.

151. V. P. Arya, *et al.*, 'Effect of magnetic field on Mott's variable-range hopping parameters in multiwall carbon nanotube mat,' *J Phys Condens Matter*, vol. 24, p. 245602, Jun 20 2012.

152. J. M. Coey, *et al.*, 'Magnetic stabilization and vorticity in submillimeter paramagnetic liquid tubes,' *Proc Natl Acad Sci USA*, vol. 106, pp. 8811–7, Jun 2 2009.

153. J. M. Coey, *et al.*, 'Ferromagnetism of a graphite nodule from the Canyon Diablo meteorite,' *Nature*, vol. 420, pp. 156–9, Nov 14 2002.

154. J. M. Coey, *et al.*, 'Giant transverse hysteresis in an asperomagnet,' *Phys Rev B Condens Matter*, vol. 41, pp. 9585–87, May 1 1990.

155. J. M. Coey, *et al.*, 'Donor impurity band exchange in dilute ferromagnetic oxides,' *Nat Mater*, vol. 4, pp. 173–9, Feb 2005.

156. N. J. Craig, *et al.*, 'Tunable nonlocal spin control in a coupled-quantum dot system,' *Science*, vol. 304, pp. 565–7, Apr 23 2004.

157. B. A. Jones, *et al.*, 'Mean-field analysis of two antiferromagnetically coupled Anderson impurities,' *Physical Review B*, vol. 39, pp. 3415–18, 1989.

158. I. I. Affleck, *et al.*, 'Conformal-field-theory approach to the two-impurity Kondo problem: Comparison with numerical renormalization-group results,' *Phys Rev B Condens Matter*, vol. 52, pp. 9528–46, Oct 1 1995.

3

Optical properties of ceramic nanocomposites

R. BANERJEE and J. MUKHERJEE,
Central Glass & Ceramic Research Institute, India

DOI: 10.1533/9780857093493.1.92

Abstract: Nanoscale constituents in nanocomposites possess excellent optical properties that differ from the macroscale properties. This chapter focuses on the optical properties of nanoscale materials incorporated in glass and ceramics, especially transmittance, absorption, non-linearity and luminescence. The fluorescence property of carbon nanotube–glass composite is a new observation which has hitherto not been studied. All these properties make them a potential material for optical sensors, ultraviolet–infrared shielding windows and other biological applications.

Key words: nanocomposites, glass, ceramics, optical properties.

3.1   Introduction

The term 'composite' originally arose in engineering when two or more materials were combined in order to rectify some shortcoming of a particularly useful component. In other words, a composite material can be defined as a heterogeneous mixture of two or more homogeneous phases that have been bonded together. The phases can be of essentially any material class, including metal, polymer, ceramic, carbon. In many cases, the dimensions of one of the phases of a composite material are small, say between 10 nm and a few micrometers, and under these conditions that particular phase has properties rather different from those of the same material in the bulk form.

When the microstructural scale falls in the range of a few nanometers, such a material is referred to as a 'nanocomposite'. In mechanical terms, nanocomposites differ from conventional composite materials due to the exceptionally high surface to volume ratio of the reinforcing phase and/or its

92

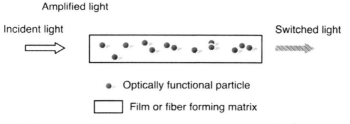

3.1 Schematic illustration of optical nanocomposite.[1]

exceptionally high aspect ratio. Nanocomposite materials show great promise as they can provide the necessary stability and processability for many important applications. Due to their nanoscale constituents they also exhibit optical and electronic properties that differ from the corresponding macroscopic properties. There is a wide range of potential applications of nanocomposite materials. Among them are electrical and optical sensors, dispersions, coatings and novel optical glasses.

The applications of optical materials and the need for novel optically functional and transparent materials are expanding. In addition to optical needs such as switching and amplification, the materials must be integrated into existing structures such as waveguides and optical fibers.[1] In general, the principle in the construction of an optical composite involves the intimate mixing of optically functional materials within a processable matrix. This type of composite is schematically shown in Fig. 3.1, where the enclosing matrix imparts processability in film or fiber forms and the small particles possess the desirable optical properties.[1]

Examples of incorporated phases include quantum-confined semiconductors, solid-state lasers, small molecules and polymers. Matrix materials can be polymers, copolymers, polymer blends, glasses or ceramics. Using such a composite structure, nanocomposites have been formed with non-linear optical and laser amplification properties, among others.[2,3] In these types of composites, optical scattering must be avoided; this results from a mismatch between the refractive index of the matrix and that of the particles. Refractive index mismatch is not so important in the case of small particles (typically <25 nm), but for larger particles the refractive index of the matrix and the particles must be carefully matched to avoid scattering.[1]

Since the 1970s there has been considerable progress in the sol–gel technique for the manufacture of glasses, glass–ceramics and ceramics. This method has been successfully used for a variety of products ranging from bulk glasses[4, 15] and optical fibers, to special coatings, ultra-pure powders and multifunctional materials. The sol–gel processed transparent porous matrix also offers an exciting potential as a host matrix for doping optically active organic molecules.[6] This matrix exhibits many properties of inorganic glasses as well as some unique characteristics that can lead to novel developments in photonics. Avnir *et al.*[7,8] demonstrated for the first time that organic molecules can be added to a sol–gel matrix and thin films. Since then, many groups around the world have introduced organic materials into a variety of inorganic matrices via the sol–gel process. Sol–gel processed materials have been used in the areas of solid-state lasers and in platforms for chemical and biosensors.[9,10] The sol–gel technique has also been used to fabricate non-linear optically active composites for applications in optical telecommunications.[11] Organically modified and sol–gel processed materials were shown to exhibit excellent non-linear optical $\chi^{(2)}$ and $\chi^{(3)}$ properties. The sol–gel method also provides a more convenient route to prepare luminescent glass optical fibers doped with rare earth (RE) ions,[12,13] waveguides[14,15] and lasers.[16]

To date, there are only a few sol–gel derived materials that meet the materials quality required for fabricating devices. One of the major problems has been to produce useful bulk materials with controlled doping (organic and inorganic) at the nanoscale. Another problem has been the fabrication of a low-loss optical fiber with active species incorporated within it. Sol–gel approaches show promising results in producing useful materials for photonics. They include: (i) RE-doped glasses; (ii) multiphasic nanostructured composites, which combine the merits of inorganic glass and an organic polymer; and (iii) optical fibers with active molecules incorporated within the fiber matrix and dispersed homogeneously for sensing or lasing applications.

3.2   Optical properties of ceramic nanocomposites

Many of the interesting optical properties of ceramic nanocomposites, including absorption, fluorescence, luminescence and non-linearity, may be studied by incorporating semiconductor nanoparticles into polymer, glass or ceramic matrix materials. Nanocomposite structures have been used to create optically functional materials, and in all these systems, very small particle sizes (see Fig. 3.2) enhance the optical properties while the matrix materials act to stabilize the particle size and growth. Laser-active composites can be made by incorporating ceramic nanoparticles of solid-state laser materials into polymer matrix materials. This structure allows the formation of solid-state laser amplifying films, which would traditionally be very difficult to make. Nanocomposite structures provide a new method to improve the processability and stability of materials with interesting optical properties. The applications of such composites are extremely broad, ranging from solid-state amplifier films to transparent magnets.[1]

In nanocomposites, light scattering is remarkably reduced compared to composites with larger particles. This renders nanocomposite materials

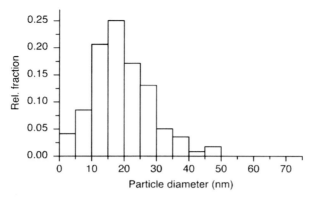

3.2   Particle size distribution of a typical poly(ethyleneoxide)-PbS nanocomposite.[17]

attractive for optical studies and applications. The refractive index of some nanocomposite has been found to depend linearly on the volume fraction of the particles. When the particles become very small, the refractive index changes.[17] In optical applications, the refractive index is a key feature. Typical refractive indices for organic polymers are between 1.3 and 1.7.[8,15] Isotropic refractive indices around 2.1 have also been found (e.g. for poly (thiophene) or aromatic poly(imide)s[19–21]) and these values are extraordinarily high for polymers. The theoretical lower limit of the refractive index of organic polymers has been estimated to be close to 1.29.[22] So far, an isotropic refractive index above 2.5 ('ultrahigh' refractive index[23]) or below 1.25 ('ultralow' refractive index[24]) has never been reported for a pure polymer. Inorganic materials cover a more extended range of refractive indices (Table 3.1). In technical applications, the refractive index of inorganic fillers is commonly between 1.4 and 1.7,[25,26] and is occasionally up to 2.7 (TiO_2)[25,27]). By incorporating of inorganic colloids with extreme refractive indices in organic polymers, an attempt has been made to obtain composite materials with refractive indices outside the typical range of polymers and hitherto unknown for polymer composites. High refractive index polymers have many applications, ranging from anti-reflection coatings for solar cells to high refractive index lenses.

After transmission of visible light through composites comprising a transparent polymer matrix with embedded particles, the intensity loss by scattering is substantially reduced for small particles, i.e. particles with diameters below 50–100 nm (nanoparticles) (Fig. 3.2).[17] Similarly, infrared (IR) windows are typically made from single-crystal ceramics as single-crystal windows have excellent optical properties, but they cannot withstand the mechanical demands of some applications, leading to catastrophic failures. Ceramic nanocomposites can be used in making IR transparent

Table 3.1   Refractive index of selected inorganic materials with extreme refractive indices at different wavelengths

	Refractive index		
	413.3 nm	619.9 nm	826.6 nm
Os	4.05	3.98	2.84
W	3.35	3.60	3.48
Si crystalline	5.22	3.91	3.67
Si amorphous	4.38	4.23	3.86
Ge	4.08	5.59–5.64	4.65
GaP	4.08	3.33	3.18
GAs	4.51	3.88	3.67
InP	4.40	3.55	3.46
InAs	3.20	4.00	3.71
InSb	3.37	4.19	4.42
PbS	3.88	4.29	4.50[a]
PbSe	1.25–3.00	3.65–3.90	4.64
PbTe	1.0–1.8	6.4	3.8
Ag	0.173	0.131	0.145
Au	1.64	0.194	0.188
Cu	1.18	0.272	0.260

[a]At 885.6 nm

3.3   Infrared transparent ceramic dome.

windows (see Fig. 3.3) that are mechanically robust, which can be used as protection for optical detectors and sensors in military applications. As a consequence, related materials (nanocomposites) have found particular interest in optical studies. Polymer inorganic nanocomposites with extreme refractive indices and anisotropic optical behavior have also been prepared. The optical properties of nanocomposites have attracted attention in many fields, for example in the areas of photoconductivity, non-linear optics and transparent magnetic materials. Further, nanocomposites can act as transparent ultraviolet (UV) absorbing layers or they may exhibit extreme refractive indices or dichroic effects in polarized light.

3.3   Transmittance and absorption

In nanocomposites, particular attention should be paid to the intensity loss of transmitting light by scattering, which varies with particle size. Transparency is determined by reflection and absorption, as described by[28]:

$$\frac{I}{I_0} = (1 - R)^2 e^{-\gamma h} \tag{3.1}$$

where I is the intensity of the transmitted light and I_0 that of the incident light, R is the reflection, γ is the extinction coefficient (including scattering and absorption components) and h is the optical path length. For transparency, the conductive filler particles themselves must be transparent and the scattering at the interface between the filler particles and the matrix must be as small as possible. Scattering can be minimized either by matching the refractive indices of the transparent conductive filler particles and the polymer matrix or by reducing one dimension of the conductive filler particles which thereby becomes much smaller than the wavelength of visible light. Since matching of the refractive indices of a polymer and transparent ceramic conductors may not be possible, nanosized conductive particles can be used, where the concentration of the conductive particles is kept as low as possible. Transparent conductive coatings have been fabricated by combining nanosized antimony-doped tin oxide particles and a polymer (photographic-grade gelatin or poly(n-butyl methacrylate) and poly(n-butyl acrylate) lattices).

Sun et al.[28] fabricated transparent, conductive composite coatings from suspensions of poly(vinyl acetate-acrylic) (PVAc-co-acrylic) copolymer lattices (50–600 nm) and nanosized antimony-doped tin oxide particles (~15 nm). The suspensions were deposited as coatings onto poly(ethylene terephthalate) substrates and dried at 50°C. They investigated the optical transmittance and scattering behavior of the coatings and compared them with PVAc-co-acrylic coating. They found that composite coatings had lower transparency because of the Rayleigh scattering. The transparency of the composite coatings was improved by a reduction in the coating thickness. The best transparency for the coatings with a direct-current conductivity of approximately 10^{-2} S/cm was around 85% at a wavelength of 600 nm.

Luo et al.[29] prepared and studied the optical properties of novel transparent Al-doped zinc oxide (AZO)/epoxy nanocomposites. AZO nanoparticles were prepared by calcinations of the precursor (precursor synthesized via the homogeneous precipitation method) at different temperatures. They studied the optical properties, namely visible-light transmittance, UV and IR opaqueness, of the AZO/epoxy nanocomposites as a function of the AZO content by using a UV–VIS spectrophotometer and a UV–VIS–near-infrared (NIR) spectrophotometer (Figs 3.4 and 3.5).

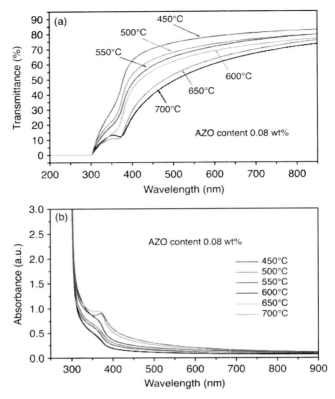

3.4 UV–VIS spectra of AZO/epoxy nanocomposites: (a) transmittance; (b) absorbance.[29]

Their experimental results showed that a nanocomposite with a thickness of about 3.5 mm containing an extremely low content (0.08 wt%) of AZO nanoparticles after calcination at 600°C would have excellent overall optical properties, having UV-shielding efficiency greater than 85% at ~360 nm and good visible transmittance greater than 70% at ~650 nm. Moreover, AZO addition in the composite was found to change the IR light absorptive capability. Therefore, a nanocomposite containing AZO shows enhanced IR shielding efficiency compared to that of the epoxy matrix and the nanocomposite containing pure ZnO nanoparticles. The as-prepared transparent AZO/epoxy nanocomposites have potential for use in a range of engineering applications such as UV- and IR-shielding windows, contact lenses and heat mirrors.

Rare earth (RE) ions are generally incorporated into a glass in their trivalent state. However, some of the RE ions can exist in a divalent state as well. These divalent RE ions show very strong emission because of the dipole-allowed electronic transition. Glasses doped with Sm^{2+} or Eu^{2+}

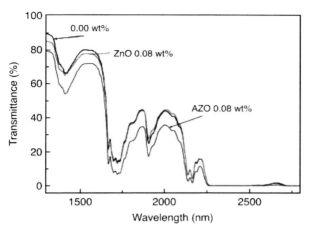

3.5 Transmittance of AZO/epoxy nanocomposites as a function of AZO content.[29]

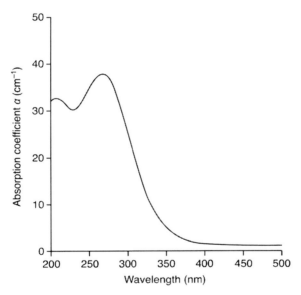

3.6 Absorption spectrum of Eu–Al co-doped silica glass.[30]

have attracted a great deal of attention for applications in memory devices and as phosphor materials. The broad emission of Eu^{2+} ions is host sensitive. Depending on the host matrix, Eu^{2+} ions could emit from the UV to the red region of the electromagnetic spectrum.

It has been shown that Eu^{3+} ions were spontaneously reduced in a sol–gel matrix during densification.[30] The optical absorption spectrum in the 200–

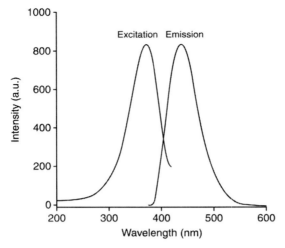

3.7 Excitation and emission spectra of Eu–Al co-doped silica glass.[30]

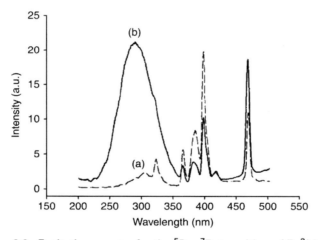

3.8 Excitation spectra for the $^5D_0-^7F_2$ transition of Eu^{3+} ions in (a) SiO_2 and (b) $Al_2O_3–SiO_2$ glasses.[30]

500 nm range of the Eu–Al co-doped glasses densified at 1125°C in air is shown in Fig. 3.6. The spectrum shows a broad peak around 266 nm that is attributed to a transition of the Eu^{2+} ions from 4f ^{7} to 4f ^{6}5d (t_g). The emission and excitation spectra of the Eu–Al co-doped sample at 1125°C are shown in Fig. 3.7. The emission shows a broad intense peak at 435 nm. A qualitative measurement based on the emission intensity and the absorption coefficient of these glasses shows that a maximum of 85% of the Eu^{3+} ions were converted to Eu^{2+} ions. Figure 3.8 shows the excitation spectra for the

Table 3.2 Third-order non-linear optical (NLO) properties measured in nanocomposite materials[1]

	Measured NLO strength	Unit	Part of χ^3 measured[a]	λ (nm)
Composite				
CdS in Nafion	-6.1×10^{-7}	cm^2/W	Im $(\chi^3)/\alpha$	480
CdS in Nafion/NH$_3$	-8.3×10^{-7}	cm^2/W	Im $(\chi^3)/\alpha$	450
Capped CdSe in PMMA	1.2×10^{-5}	cm/W	Im (χ^3)	544–560
CdS$_x$Se$_{1-x}$ glass	1.3×10^{-8}	esu	$[\text{Re}(\chi^3)^2 + \text{Im}(\chi^3)^2]^{0.5}$	532
CdS in sol–gel glass	5×10^{-12}	esu cm	$[\text{Re}(\chi^3)^2 + \text{Im}(\chi^3)^2]^{0.5}/\alpha$	380
PPV in SiO$_2$	3×10^{-10}	esu	$[\text{Re}(\chi^3)^2 + \text{Im}(\chi^3)^2]^{0.5}$	602
PPV in V$_2$O$_5$	6×10^{-10}	esu	$[\text{Re}(\chi^3)^2 + \text{Im}(\chi^3)^2]^{0.5}$	602
GaAS in Vycor glass	-5.6×10^{-12}	cm^2/W	Re (χ^3)	1064
Standard NLO materials				
Fused quartz	8.5×10^{-14}	esu	Re (χ^3)	1064
SF$_6$	8×10^{-13}	esu	Re (χ^3)	1064
CdS	-5×10^{-11}	esu	Re (χ^3)	610

[a]α is the absorption coefficient

$^5D_0$–$^7F_2$ transition of Eu^{3+} ions (614 nm) in densified silica and Al^{3+} co-doped silica matrices. The sharp lines observed between 200 and 500 nm are assigned to the f–f transitions of the Eu^{3+} ions in the glasses. However, a broad band with a peak at 300 nm is observed only in the aluminum co-doped sample that contains both Eu^{2+} and Eu^{3+} ions. The excited Eu^{2+} ions in the 4f ^{6}5d (e$_g$) level can relax to the ground state by transferring energy to the Eu^{3+} ions which, in turn, are excited to the $^5D_2$ level.

3.4   Non-linearity

Non-linear optics is the branch of optics that describes the behavior of light in non-linear media, that is, media in which the dielectric polarization P responds non-linearly to the electric field E of the light (i.e electric polarization is the cause of optical non-linearity), which can be expressed as[31]

$$P_{NL} = \chi^{(2)}EE + \chi^{(3)}EEE + \cdots \qquad [3.2]$$

where P_{NL} is non-linear electric polarization, E is the incident electric field and $\chi^{(2)}$ and $\chi^{(3)}$ are the second- and third-order non-linear optical susceptibility terms, respectively. Theoretically, $\chi^{(2)}$ is zero for materials having inversion symmetry in a region whose size is of the order of the wavelength of an incident beam but $\chi^{(3)}$ has a finite value in all substances (Table 3.2).

Glasses are optically isotropic so they should not have second-order

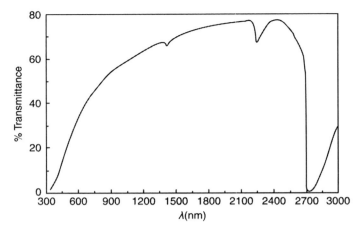

3.9 Optical transmittance spectra of 30KTP–70SiO$_2$ heat treated at 850°C for 2 h.[31]

optical non-linearity. However, photo-induced second harmonic generation (SHG) was observed from GeO$_2$-doped SiO$_2$ glass fibers. Li *et al.*[31] successfully fabricated transparent bulk xKTP-$(1-x)$SiO$_2$ nanocomposites containing nanocrystalline KTiOPO$_4$ (KTP) dispersed in a silica glass matrix by the sol–gel method. They tried to identify the phase structure of KTP/SiO$_4$ nanocomposites and to observe optical SHG in the transparent nanocomposites or translucent glass–ceramics. They studied the optical properties of the nanocomposites and found that these are transparent in the visible range of the electromagnetic spectra. Figure 3.9 shows the optical transmission spectrum of a 30KTP–70SiO$_2$ nanocomposite heat treated at 850°C. For all the samples, the UV band edge was observed near 350 nm, which is the same as KTP crystal. This common value indicates that the UV absorption band of glasses and nanocomposites originates from the Ti–O band. The strong absorption near 2800 nm indicates that the materials contain a large amount of OH$^-$.

The SHG spectra of the glass and nanocomposite heat treated at 800°C for 2 h are shown in Fig. 3.10. As multicomponent glasses do not contain crystallites they do not exhibit SHG but the KTP/SiO$_2$ nanocomposite containing KTP nanocrystals does exhibit SHG. In non-linear optical materials, microscopic SHG structural units must be ordered spatially and in a non-centrosymmetric manner to generate a bulk second-order effect from the non-linear optical materials. This study clearly demonstrates that KTP nanocrystals were formed in the KTP/SiO$_2$ nanocomposite and further indicates that the SHG in the nanocomposites comes from the KTP nanocrystals embedded in the SiO$_2$ glass matrix. Since the KTP nanocrystals are randomly oriented in the SiO$_2$ glass matrix, it should be noted that

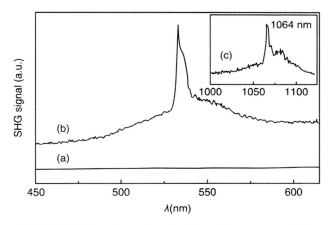

3.10 SHG (532 nm) spectra for (a) multicomponent glass and (b) 30KTP- 70SiO₂ nanocomposite prepared by heating at 750°C; (c) fundamental beam (1064 nm).[31]

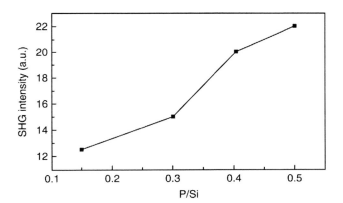

3.11 Dependence of SHG signal intensity from xKTP-$(1-x)$SiO₂ heat treated at 750°C on P/Si ratio.[31]

optical SHG in the KTP/SiO₂ nanocomposites is incoherent. Therefore, oriented nanocomposites may exhibit a strong phase-matched SHG.

Li *et al.*[31] also plotted the dependence of SHG intensity on P/Si ratio in xKTP-$(1-x)$SiO₂ nanocomposites, as shown in Fig. 3.11. The SHG signal intensity increases as a function of P/Si ratio, which suggests that the SHG signal increases as a function of KTP content. Li *et al.* also stated that the base glass is the amorphous portion of the transparent nanocomposite and so it should exhibit negligible SHG. This implies that the nanocomposites are expected to exhibit SHG signal intensity lower than pure KTP powders, depending on the content of ferroelectric crystallites in the glasses. Optical SHG has been reported for transparent silicate and borate glass–ceramics

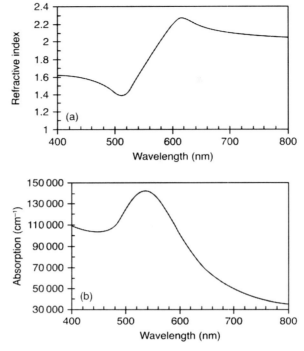

3.12 Spectral variation of the optical linear coefficients for Au:SiO$_2$ thin film of thickness 100 nm and gold volume fraction of 0.2. (a) Refractive index *n*. (b) Absorption coefficient β.[3]

containing crystalline phases that exhibit second-order non-linear optical properties, such as BaB$_2$O$_4$ and LiNbO$_3$.[31]

Third-order non-linearity is also seen in ceramic nanocomposites. Debrus *et al.*[3] studied the third-order non-linear optical properties of Au:SiO$_2$ thin films at the surface plasmon resonance wavelength by the *z*-scan technique using a nanosecond laser (Fig. 3.12). They prepared the films by multilayer deposition using a sputtering technique. For a metal volume fraction equal to 0.2, they obtained a very large negative imaginary part, $Im\chi_{eff}^{(3)} = -3.0 \times 10^{-6}$ esu, which is characteristic of an absorption saturation. The non-linear absorption coefficient β was found to be negative and equal to 1.1×10^{-6} cm/W.

3.5   Luminescence

Luminescence is another important property of nanoparticles and its primary application is in medical, biological or pharmaceutical areas. Semiconducting nanoparticles are currently the most important luminescent

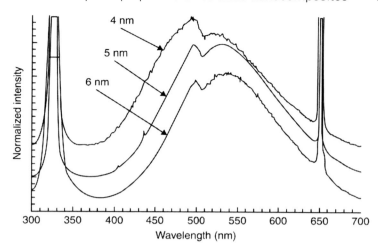

3.13 Emission spectra of ZnO/m-PMMA nanocomposites, influenced by particle size. The spectra show excitation in first order at 325 nm and at 650 nm in second order. The spectra are stacked and normalized.[32]

nanoparticles. With respect to nanoparticles, quantum dots based on sulphides, selenides or tellurides of zinc and cadmium show the best luminescence efficiency but these materials are toxic and carcinogenic and show limited thermodynamic stability against oxidation.

Luminescence of oxide nanoparticles is also subjected to rapid aging caused by the formation of hydroxides at the surface of oxide nanoparticles, which leads to quenching luminescence. This problem can be reduced by coating the surface of the oxide nanoparticles with a polymer to protect it against ambient air or even water. Vollath and Szabó demonstrated that oxide nanoparticles show luminescence provided they are coated with a polymer or ceramic.[32] They synthesized such nanocomposites using the Karlsruhe microwave plasma process and showed that interaction between the ceramic core and the organic coating also leads to luminescence. They found that the carbonyl group attached to the oxide core exhibits strong luminescence and this depends strongly on particle size: with a decrease in particle size the luminescence showed a blue shift and narrowing of emission lines, which predicts a direct correlation between the molecular orbital and energy bands of the ceramic particle. Figure 3.13 shows the luminescence spectra of ZnO/m-PMMA nanocomposites with varying size of zinc oxide nanoparticles. The figure shows a broad emission spectra consisting of two broad overlapping lines. The splitting is a consequence of the interaction between ZnO nanoparticles and their surroundings.

A blue shift was observed for both the overlapping luminescence peaks of the ZnO/m-PMMA nanocomposites with a decrease in nanoparticle size.

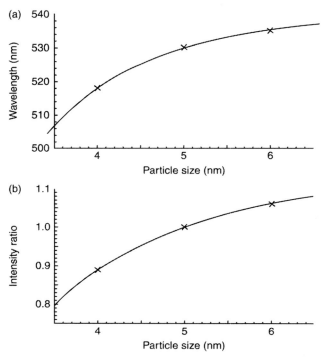

3.14 Influence of particle size on the emission spectra of ZnO. (a) Blue shift of the ZnO emission line with decreasing particle size. The experimental points were fitted with equation 3.3. (b) Intensity ratio of the quantum confinement peak over the oxide–polymer interaction peak. The experimental points follow equation 3.5.[32]

The wavelength of the emission maximum as a function of particle size follows the relationship (Fig. 3.14(a)):[32]

$$\frac{1}{\lambda} = \frac{1}{\lambda_0} + bd^{-3} \qquad [3.3]$$

$$\Delta\frac{1}{\lambda} = bd^{-3} \qquad [3.4]$$

From an experimental fit the values used are $\lambda_0 = 542.7 \pm 0.4\,\text{nm}$ and $b = 5.61 \times 10^{-3} \pm 1 \times 10^{-4}\,\text{nm}^{-2}$; d is the particle diameter in nanometers. The exponent -3 is the same as that found for the blue shift in ZnO related to quantum confinement.

Table 3.3 Influence of particle size on the wavelength and intensity of the emission spectrum of ZnO/m-PMMA nanocomposites

Particle size (nm)	Position of first maximum (nm)	Position of second maximum (nm)	Intensity ratio of second to first maximum
4	495	519	0.89
5	497	530	1.00
6	498	535	1.06

The intensity ratio of peak 2 over peak 1 (I_2/I_1) increases with increasing particle size (Table 3.3):[32]

$$\frac{I_2}{I_1} = A + kd^{-2} \qquad [3.5]$$

where I_1 and I_2 are the intensities of peak 1 and peak 2 respectively and $A = 1.120 \pm 2 \times 10^{-4}$ and $k = -4.9 \pm 5 \times 10^{-3}$ nm^{-2} are the values taken from the experimental fit (Fig. 3.14(b)). The position of the first and the second intensity maximum as a function of the particle size together with the intensity ratio of these two peaks is given in Table 3.3. A strong blue shift with decreasing particle size of the second and a less pronounced one of the first maximum was observed. Additionally, the intensity ratio of the second over the first peak increases with increasing particle size.

The observation discussed above can be described by the relation of the luminescence to the ceramic core and coating. The ceramic core absorbs the existing UV-quanta and the excitation is transferred from the ceramic core to the coating, where the emission occurs. The wavelength and intensity of luminescence of oxide nanocomposites can be tailored by changing the nanoparticle size as the luminescence of these oxide nanocomposites depends on the surface. In the case of semiconducting oxides such as ZnO, a mixture of the quantum confinement mechanism with the surface interaction mechanism is observed. These composites can be used in nanophotonic waveguides for the transport of electromagnetic energy using the Förster mechanism.

The low-temperature luminescence of Cr^{3+} ions in a silicate glass system and in nanocrystalline glass–ceramic composites with embedded β-Ga_2O_3 nanocrystals has been studied.[33] The precursor glass showed a wide range of crystal fields. Due to octahedrally coordinated Cr^{3+} ions located in a weak crystal field, the spectrum of the precursor glass was dominated by a broadband $^4T_2$–$^4A_2$ transition. The spectra of the glass–ceramic nanocomposites showed a crystal-like ^2E–$^4A_2$ emission and demonstrated that Cr^{3+} ions are located within the crystalline environment. It was observed[33] that the host glass matrix nucleated with gallium oxide nanocrystals (i.e. glass–

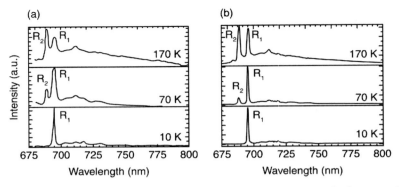

3.15 Luminescence spectra of composite 1 (a) and of composite 2 (b) measured between 10 and 170 K. With temperature rise, the structured R-lines' emissions overlap with the associated Stokes phonon sidebands, which become increasingly noticeable.[33]

ceramic nanocomposites) indicated the prevailing contribution of the crystal-like $^2E-^4A_2$ emissions (R-lines) of Cr^{3+} ions in a strong crystal field. The low-temperature studies demonstrated that below 70 K, the luminescence of Cr^{3+} ions could be demonstrated by sharp R-lines of the $^2E-^4A_2$ transition and above 70 K the Cr^{3+} luminescence can be transformed to a combination of R-lines and their sidebands as well as a small contribution of the $^4T_2-^2A$ transition (Fig. 3.15). The conclusion for the formation of $Ga_2O_3:Cr^{3+}$ nanocrystallites and the relocation of Cr^{3+} ions from the host glass to the nanocrystalline phase comes from the overall modification of the luminescence profiles from the broad $^4T_2-^4A_2$ emission for the as-quenched glass to the narrow R-lines for the glass–ceramic composites. The presented results help in the formulation of a simple distribution model in which most of the Cr^{3+} ions are located within the nanocrystalline phase. The developed nanocrystalline glass–ceramics are a promising new class of Cr^{3+} doped oxide glass-based optically active composite materials.

Sorarù *et al.*[34] prepared erbium-activated SiC/SiO_2 nanocomposites doped with Er^{3+}. The concentrations of Er^{3+} were varied from 1 to 4 mol% and samples were prepared by pyrolysis of sol–gel derived precursors. The gels were obtained from modified silicon alkoxides containing Si–CH$_3$ and Si–H groups. Crack-free, monolithic and transparent SiCO glass discs were obtained after pyrolysis at temperatures in the range 800–1300°C. Emission in the C-telecommunication band was observed at room temperature for all the samples upon continuous-wave excitation at 980 nm. A spectral bandwidth of 48 nm, large enough for wavelength division multiplexing (WDM) application, was observed for all the samples. The shape of the $^4I_{13/2}-^4I_{15/2}$ luminescence band appears to be

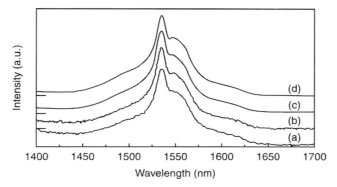

3.16 Photoluminescence spectra relative to the $^4I_{13/2}$–$^4I_{15/2}$ transition of Er^{3+} ions for Er^{3+} activated SiC/SiO_2 nanocomposite pyrolyzed at 1200°C for 1 h and doped with (a) 1 N mol%, (b) 2 mol%, (c) 3 mol% and (d) 4 mol% Er^{3+}. The emission spectra were obtained using a diode laser operating at 980 nm as a pump beam.[14]

independent of the erbium content, at least for Er^{3+} concentration up to 4 mol% (Fig. 3.16). This effect suggests that the SiCO network is flexible enough to accommodate Er^{3+} without appreciable matrix strains. The luminescence spectra do not show evidence of crystalline environment for the Er^{3+} ion, suggesting that at least the majority of erbium ions are accommodated in the glass. The arrow in Fig. 3.17(a) shows the excitation wavelength (980 nm) used to obtain the luminescence spectra shown in Fig. 3.16. The UV edge shifts to longer wavelengths with respect to pure silica glasses with an increase in annealing temperature, resulting in a corresponding decrease in the electronic bandgap energy.

Since the 1990s, the use of fluorescent nanoparticles as indicators in biological applications such as imaging and sensing has dramatically increased. These applications require that the fluorescent nanoparticles are monodisperse, bright, photostable and amenable to further surface modification for the conjugation of biomolecules and/or fluorophores. Aslan *et al.*[35] developed core–shell (silver core–silica shell) (Fig. 3.18) nanoparticles with various shell thicknesses featuring a variety of fluorophores to show the versatility of the core–shell architecture. They demonstrated their applicability for two platform technologies: metal-enhanced fluorescence and single nanoparticle sensing. They also developed three different fluorescent probes: an organic fluorophore (Rh800), a doped lanthanide probe (non-covalently linked) and another organic fluorophore (Alexa 647) covalently linked to the silica shell. When compared with the control sample fluorescent nanoparticles (nanobubbles), fluorescent nano-particles with core–shell architecture yielded up to 20-fold (with Rh800)

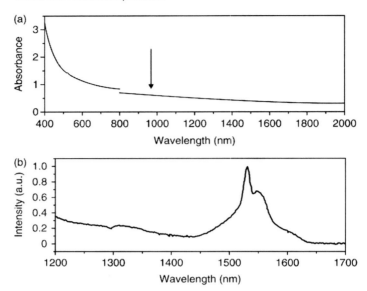

3.17 Optical absorption (a) and photoluminescence (b) spectra for the Er^{3+} activated SiC/SiO$_2$ nanocomposite doped with 1 mol% and submitted to thermal treatment at 1200°C. The emission spectra were obtained using a diode laser operating at 980 nm as a pump beam.[14]

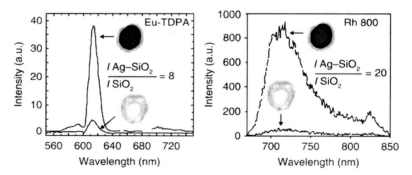

3.18 Fluorescence emission intensity of Eu-TDPA-doped Ag@SiO$_2$ and Rh800-doped Ag@SiO$_2$ and the corresponding fluorescent nanobubbles (control samples), Eu-TDPA-doped SiO$_2$ and Rh800-doped SiO$_2$. The diameter of the Ag is 130 ± 10 nm and the thickness of the shell is 11 ± 1 nm (optimized) for all the samples.[35]

enhancement of the fluorescence signal and a potentially 200-fold increase in particle detectability.

Conducting polypyrrole wide-bandgap semiconducting silica (SiO$_2$) nanocomposites were prepared by polymerizing pyrrole in the presence of colloidal silicon oxide sol.[36] It was found that with an increase in silica

Table 3.4 Pyrrole compositions in samples[36]

Sample	Amount of pyrrole, x (ml)
S1	0.50
S2	0.65
S3	0.80

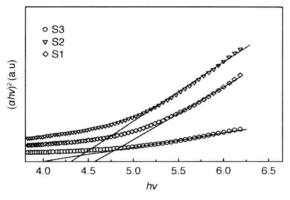

3.19 Plot of $(\alpha h v)^2$ versus $h v$ for different nanocomposite samples (Table 3.4) to determine optical bandgap.[36]

concentration the optical absorption spectra showed that the π–π* transition of polypyrrole shifts from 3.9 eV to 4.58 eV. Samples with three different concentrations of pyrrole were prepared, as shown in Table 3.4.

Generally, the optical bandgap in a semiconductor is determined by plotting absorption coefficients (α) as $(\alpha h v)^{1/m}$ against $h v$ where m represents the nature of the transition and $h v$ is the photon energy; m may have different values, such as 1/2, 2, 3/2 or 3 for allowed direct, allowed indirect, forbidden direct and forbidden indirect transitions, respectively. The optical absorption coefficient α near the absorption edge for direct interband transitions is given by:[37]

$$\alpha = \frac{B(h v - E_g)^{1/2}}{h v} \qquad [3.6]$$

where B is the absorption constant for a direct transition. For an allowed indirect transition, one can plot $(\alpha h v)^2$ against $h v$, as shown in Fig. 3.19, and extrapolate the linear portion of it to $\alpha = 0$ to obtain the corresponding bandgap. The estimated bandgaps are 4.58, 4.30 and 3.90 eV for S1, S2 and S3 respectively. The increase of bandgap with silica concentration implies that the electronic structure of polypyrrole is affected by silica.[36]

3.6   Optical properties of glass–carbon nanotube (CNT) composites

Fluorescence of single-walled carbon nanotube (SWCNT) bundles has become an emerging field due to its wide range of application, especially in fiber optics. We report for the first time[38] SWCNT borosilicate glass composite having strong NIR fluorescence near 0.84–2.03 μm with 325 nm laser excitations. The emission spectrum was 1185 nm wide and contained three main transmission windows for fiber optic telecommunications, which strongly indicates the strong potential of this material in broadband fiber optic telecommunications and fabrication of NIR tunable lasers.

As noted above, the total spectral width of the emission of the composite was found to be 1185 nm. While analyzing the emission spectra for both glass and the composite, it was observed that the fluorescence peaks at 844, 1193, 1562 and 2029 nm were generated due to structural defects of SiO_2.[39] Other emission peaks can be considered to be the bandgap fluorescence of CNT bundles. Fluorescence bands were seen at 1420, 1502, 1635, 1694, 1768, and 1865 nm, together with three shoulders at 954, 1267 and 1318 nm, although such emissions remain elusive in the base glass spectra (Fig. 3.20). The high intensities of the bands show that the SWCNT bundles are made from semiconducting tubes. However, the presence of metallic tubes in a bundle will rapidly quench the emission. Two extremely strong wide emission bands were observed around 1121–1350 nm and 1500–1600 nm, centered at 1267, 1318 and 1562 nm. This is of considerable significance in fiber optic telecommunications in terms of lowest dispersion and best transmission. Broadened spectra having a high intensity at 1318 nm and 1562 nm are very important for the ability to carry large amounts of information at a slightly different wavelength in wavelength division multiplexing systems. Strong emission from the composite at 954 nm in the range of 800–1030 nm is also useful for short-range communication devices like remote controls for audio and video systems.

Researchers have shown that SWCNT fluorescence can be used as a laser source for optical amplification.[40] The total emission of 1185 nm width is well within the tunable range of conventional NIR lasers like Ti:Saphire (460 nm) and InGaAs (161 nm). Full width at half maximum data of the composite, together with the area under the emission curve (σ) of the emission bands are shown in Table 3.5. The data show that the composite is a material with great potential for the fabrication of tunable NIR lasers, which will be useful as an NIR emitter for fiber optic telecommunications (800–1600 nm), in light detection and ranging (Lidar) and infrared countermeasure (IRCM) applications (1800–2100 nm). The fluorescence from semiconducting SWCNT bundles can be explained by two important photophysical phenomena – intertube carrier migration[41] and the excitonic

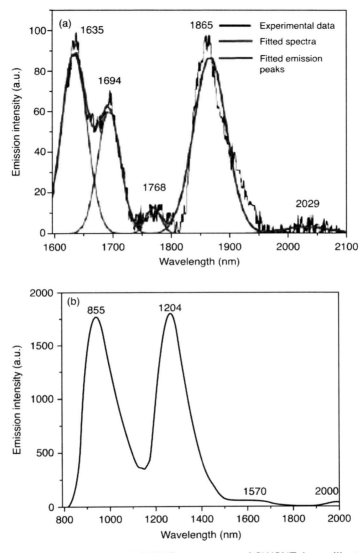

3.20   (a) Near-infrared (NIR) fluorescence of SWCNT–borosilicate glass composite. (b) NIR fluorescence spectrum of the base glass.[38]

effect[42–44] – depending on the intertube separation. In the composite described here, each nanotube within the bundles is separated by an insulating glassy layer. This separation is 3.34–5.01 nm, which is quite a large distance for charge carriers to migrate between the nanotubes. Thus, the fluorescence of the SWCNT bundles in the composite is a result of the

Table 3.5 Proposed radiative emission features of SWCNT and base glass with 325 nm excitation[38]

Single-walled carbon nanotube bundles

Emission wavelength (nm)	Excited to	Emission from	Emission feature	FWHM (nm)	Area under emission curve
954	eh_{33} (8.3)	eh_{11} (8.3)	E-E resonance	164	72276
1267	eh_{44} (8.7) $^{donor}eh_{44}$ (10.5)	eh_{11} (8.7) $^{acceptor}eh_{11}$ (8.7)	E-E resonance EET	53	9282
1318	eh_{44} (13.2) $^{donor}eh_{44}$ (14.0)	eh_{11} (13.2) $^{acceptor}eh_{11}$ (13.2)	E-E resonance EET	84	18700
1420	eh_{44} (9.8) $^{donor}eh_{44}$ (11.6) $^{donor}eh_{44}$ (13.0)	eh_{11} (9.8) $^{acceptor}eh_{11}$ (9.8) $^{acceptor}eh_{11}$ (9.8)	E-E resonance EET EET	66	13752
1502	eh_{44} (13.3) $^{donor}eh_{44}$ (13.5) $^{donor}eh_{44}$ (10.8)	eh_{11} (13.3) $^{acceptor}eh_{11}$ (13.3) $^{acceptor}eh_{11}$ (13.3)	E-E resonance EET EET	37	2509
1635	eh_{55} (13.3) $^{donor}eh_{55}$ (13.5) $^{donor}eh_{55}$ (13.5) $^{donor}eh_{55}$ (13.5)	eh_{11} (14.3) $^{acceptor}eh_{11}$ (14.6) $^{acceptor}eh_{11}$ (14.6) $^{acceptor}eh_{11}$ (14.6)	E-E resonance EET EET EET	43	4785
1694	eh_{55} (13.8) $^{donor}eh_{55}$ (18.1)	eh_{11} (13.8) $^{acceptor}eh_{11}$ (13.8)	E-E resonance EET	38	2921
1768	eh_{55} (12.10) $^{donor}eh_{55}$ (15.5) $^{donor}eh_{55}$ (17.1) $^{donor}eh_{55}$ (16.3)	eh_{11} (12.10) $^{acceptor}eh_{11}$ (12.10) $^{acceptor}eh_{11}$ (12.10) $^{acceptor}eh_{11}$ (12.10)	E-E resonance EET EET EET	25	404
1865	eh_{55} (12.11) $^{donor}eh_{55}$ (14.9) $^{donor}eh_{55}$ (18.4)	eh_{11} (12.11) $^{acceptor}eh_{11}$ (12.11) $^{acceptor}eh_{11}$ (12.11	E-E resonance EET EET	56	6096

Borosilicate base glass

Emission wavelength (nm)	Emission feature	FWHM	Area under emission curve
844	Defects related fluorescence	97	51902
1193	of SiO_2	103	80572
1563		58	2491
2029		82	331

combined effect of exciton–exciton resonance and exciton energy transfer between the adjacent nanotubes. Further detailed discussion is beyond the scope of this book, but will no doubt be addressed in future work.

3.7   References

1.   Beecroft L L and Ober C K (1997) *Chem. Mater.* **9**, 1302–1317.
2.   Li D, Lin Y, Zhang L and Yao X (2000) *J. Non-Cryst. Solids* **261**, 273–276.
3.   Debrus S, Lafait J, May M, Pinçon N, Prot D, Sella C and Venturini J (2000) *J. Appl. Phys.* **88**, 4469–4475.
4.   Kirkbir F, Murata H, Mayers D, Chaudhari S R and Sarkar A (1996) Drying and sintering of sol–gel derived large SiO2 N monoliths. *J. Sol–Gel Sci. Technol.* **6**, 203–217.
5.   Susa K, Matsuyama S I, Satoh S and Suganuma T (1982) New optical fiber fabrication method. *Electron. Lett.* **18**, 499–500.
6.   Nogues J-L R and Moreshead W V (1990) Porous gel–silica, a matrix for optically active components. *J. Non-Cryst. Solids* **121**, 136–142.
7.   Avnir D, Kaufman V R and Reisfeld R (1985) Organic fluorescent dyes trapped in silica and silica–titania thin films by the sol–gel method. Photophysical, film and cage properties. *J. Non- Cryst. Solids* **74**, 395–406.
8.   Avnir D, Levy D and Reisfeld R (1984) The nature of the silica cage as reflected by spectral changes and enhanced photo stability of trapped rodamine 6G. *J. Phys. Chem.* **88**, 5956–5959.
9.   Braun S, Rappoport S, Zusman R, Avnir D and Ottolenghi M (1990) Biochemically active sol–gel glasses: the trapping of enzymes. *Mater. Lett.* **10**, 1–5.
10. Narang U, Bright F V and Prasad P N (1993) Characterization of rhodamine 6G-doped thin sol–gel films. *Appl. Spectrosc.* **47**, 229–234.
11. Burzynski R and Prasad P N (1994) New photonic media prepared by sol–gel process. In: Klein L C (ed.) *Photonics and Nonlinear Optics with Sol–Gel Processed Inorganic Glass: Organic Polymer Composite*, Kluwer, Boston, MA, Ch. 19.
12. Wu F Q, Machewirth D, Snitzer E and Sigel G H Jr (1994) An efficient single mode Nd3 + fiber laser prepared by the sol–gel method. *J. Mater. Res.* **9**, 2703–2705.
13. Wu F Q, Puc G, Foy P, Snitzer E and Sigel G H Jr (1993) Low-loss rare earth doped single-mode fiber by sol–gel method. *Mater. Res. Bull.* **28**, 637–644.
14. Bahtat A, Bouazaoui M, Bahtat M and Mugnier J (1994) Fluorescence of Er^{3+} ions in TiO2 planar waveguides prepared by a sol–gel process. *Opt. Commun.* **111**, 55–60.
15. Benatsou M, Capoen B, Bouazaoui M, Tchana W and Vilcot J P (1997) Preparation and characterization of sol gel derived Er^{3+}: Al_2O_3-SiO_2 planar waveguides. *Appl. Phys. Lett.* **71**, 428–430.
16. Thomas I M, Payne S A and Wilke G D (1992) Optical properties and laser demonstration of Nd-doped sol–gel silica glasses. *J. Non-Cryst. Solids* **151**, 183–194.
17. Caseri W (2000) *Macromol. Rapid Commun.* **21**, 705–722.

18. Brandrup J and Immergut E H (Eds) (1989) *Polymer Handbook*, Wiley, New York, NY.
19. Sugiyama T, Wada T and Sasabe H (1989) *Synth. Met.* **28**, C323–328.
20. Gaudiana R A and Minns R A (1991) *J. Macromol. Sci. A* **28**, 831–842.
21. Rogers H G, Gaudiana R A, Hollinsed W C, Kalyanaraman P S, Manello J S, McGowan C, Minns R A and Sahatjian R (1985) *Macromolecules* **18**, 1058–1068.
22. Groh W and Zimmermann A (1991) *Macromolecules* **24**, 6660–6663.
23. Weibel M, Caseri W, Suter U W, Kiess H and Wehrli E (1991) *Polym. Adv. Technol.* **2**, 75–80.
24. Zimmermann L, Weibel M, Caseri W, Suter U W and Walther P (1993) *Polym. Adv. Technol.* **4**, 1–7.
25. Lee S M (Ed.) (1990) *International Encyclopedia of Composites*, Vol. **2**, VCH, New York, NY.
26. MarK H F, Bikales N M, Overberger C G, Menges G and Kroschwitz J I (Eds) (1987) *Encyclopedia of Polymer Science and Engineering*, Vol. **7**, Wiley, New York, NY.
27. Pfaff G and Reynders P (1999) *Chem. Rev.* **99**, 1963–1981.
28. Sun J, Gerberich W W and Francis L F (2003) J. Polym. Sci.: B Polym. Phys. **41**, 1744–1761.
29. Luo Y-S, Yang J-P, Dai X-J, Yang Y and Fu S-Y (2009) *J. Phys. Chem. C* **113**, 9406–9411.
30. Biswas A, Friend C S and Prasad P N (1999) Spontaneous reduction of Eu^{3+} ion in Al Co-doped sol–gel silica matrix during densification. *Mater. Lett.* **39**, 227–231.
31. Li D, Lin Y, Zhang L and Yao X (2000) *J. Non-Cryst. Solids* **261**, 273–276.
32. Vollath D and Szabó D S (2004) *J. Nanopart. Res.* **6**, 181–191.
33. Lipinska-Kalita K E, Kalita P E, Krol D M, Hemley R J, Gobin C L and Ohki Y (2006) *J. Non-Cryst. Solids* **352**, 524–527.
34. Soraru G D, Zhang Y, Ferrari M, Zampedri L and Rocha Gonçalves R (2004) *J. Euro. Ceram. Soc.* **25**, 277–281.
35. Aslan K, Wu M, Lakowicz J R and Geddes C D (2007) *JACS Commun*, online 01/19/2007.
36. Dutta K and De S K (2006) *Solid State Commun.* **140**, 167–171.
37. Pankove J I (1971) *Optical Processes in Semiconductors*, Prentice Hall, Englewood Cliffs, NJ.
38. Ghosh A, Ghosh S, Das S, Das P K and Banerjee R (2013) *Chem. Phys. Lett.* **570**, 113–117.
39. Xiang W, Liang X and Yang Y (2010) *Glass Phys. Chem.* **36**, 36.
40. Gaufres E, Izard N, Le Roux X, Morini D M, Kazaoui S, Cassan E E and Vivien L (2010) *Appl. Phys. Lett.* **96**, 231105.
41. Torrens O N, Milkie D E, Zheng M and Kikkawa J M (2006) *Nano Lett.* **6**, 231105-1–231105-3.
42. Tan P H, Rozhin A G, Hasan T, Hu P, Scardaci V, Milne W I and Ferrari A C (2007) *Phys. Rev. Lett.* **99**, 137402-1–137402-4.
43. Wang F, Dukovic G, Brus L E and Heinz T F (2005) *Science* **308**, 838.
44. Jiang J, Saito R, Samsonidze G G, Jorio A, Chou S G, Dresselhaus G and Dresselhaus M S (2007) *Phys. Rev. B* **75**, 035407-1–035407-13.

4

Failure mechanisms of ceramic nanocomposites

P. HVIZDOŠ, P. TATARKO, A. DUSZOVA
and J. DUSZA, Slovak Academy of Sciences, Slovakia

DOI: 10.1533/9780857093493.1.117

Abstract: This chapter first builds a basic understanding of structural failure and its determining critical factors. It describes typical fracture origins and modes of crack propagation. It then deals with the concept of reinforcing ceramic nanocomposites. Different strategies for preventing failures are discussed, and the influence of microstructure and secondary nanometric phases on friction and wear properties of some ceramic nanocomposites is described.

Key words: mechanical properties, strength, fracture toughness, damage mechanism, wear.

4.1 Introduction

Due to their crystallographic structure and strong atomic bonds, modern ceramic materials have many excellent properties. These include extremely high hardness and strength, high thermal and chemical stability, high corrosion and wear resistance. However, these strong atomic bonds (usually covalent or ionic) limit the possibility of dislocation movement and plastic deformation. This leads to their well-known weakness – low fracture toughness and poor resistance to crack growth and the resultant high brittleness and lower reliability.

Reliability is closely related to the presence of technological defects within the microstructure (i.e. flaws that arise during processing steps such as mixing and sintering). These flaws include pores and their clusters, impurities, clusters of secondary phases and large grains. Flaws in manufactured parts are difficult to identify and cannot be eliminated easily, thus creating stress concentrators during machining and/or shaping.

117

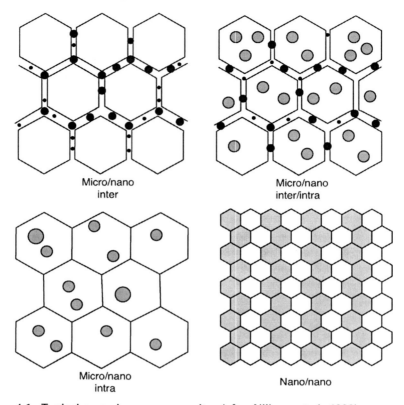

Micro/nano
inter

Micro/nano
inter/intra

Micro/nano
intra

Nano/nano

4.1 Typical ceramic nanocomposites (after Niihara *et al.*, 1993).

One means of overcoming these drawbacks is the preparation of composite materials, where the base ceramic matrix is reinforced by secondary phases in the form of particles, platelets, whiskers or fibres. These additives act in a variety of ways. In general, their role is to dissipate the energy of crack propagation by deflecting and arresting cracks caused by the presence of residual stress fields, crack bridging, crack branching, micro-cracking and other micro-mechanisms.

In recent years, there has been great interest in 'nanoceramics'. The concept of structural ceramic nanocomposites was proposed by Niihara *et al.* (1993) and was based on results achieved during the study of $Si_3N_4 + SiC$ and $Al_2O_3 + SiC$ systems. Nanocomposite ceramics may be defined as composites consisting of more than one solid phase, where at least one of the phases has a nanometric dimension. These nanophases are generally nanoparticles located inside matrix grains and/or at the grain boundaries. Depending on the location of the nanophases, nanocomposites may be classified as inter-type, intra-type, inter/intra-type and nano/nanotype (Fig. 4.1).

In addition to these nanoparticle-reinforced materials, great effort has been made in recent years to develop material systems with nanostructures in the form of fibrous and sheet-like structures (nanofibres, nanotubes and graphene sheets). Among the most promising candidates for new structural and functional nanocomposites are those with carbon-based filamentous nanomaterials (Iijima, 1991) such as carbon nanotubes (CNTs) and carbon nanofibres (CNFs). These have attracted attention due to their outstanding mechanical properties, excellent thermal performance and useful electrical characteristics (high electrical conductivity).

There are two principal aims in the current development of new ceramic/ CNT composites – to improve the mechanical properties of ceramic materials by reinforcement with carbon nanofibres and the development of functionalised ceramics with improved magnetic and electric properties. Studies show that CNTs (both single-wall and multi-wall) should be ideal reinforcing/functionalising elements for composites because of their small size, low density and good electrical and thermal conductivity.

This chapter first builds a basic understanding of structural failure and its determining critical factors. It describes the typical fracture origins found in various types of ceramic nanocomposites and the modes of crack propagation. It then deals with the concept of residual stresses and their use in achieving reinforcing effects due to their ability to influence crack propagation. Additional strategies for preventing failures are discussed and, finally, the influence of microstructure and secondary nanometric phases on friction and wear properties of some ceramic nanocomposites is described.

4.2   Rupture strength

4.2.1 Theoretical tensile strength

The concept of the theoretical tensile strength of a crystal was developed by Orowan (1949). His model assumes that the solid separates into two pieces, with instant separation of the atoms along the direction perpendicular to the applied normal stress. Here, no stress concentrators are considered and the separation takes place simultaneously across the body when a critical stress (i.e. rupture strength) is reached. The energy involved is equal to the energy of the two new surfaces created by the cleavage. So if the surface energy per unit area is γ and the cross-section of the specimen is A, the total energy of the formed surfaces is $2\gamma A$.

According to these assumptions, Orowan's model leads to the following

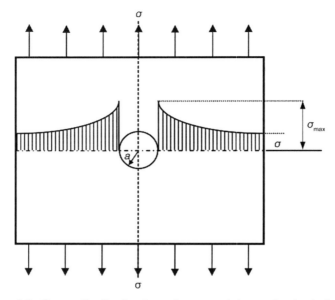

4.2 Stress distribution in a plate containing a circular hole.

relations for surface energy γ and rupture strength σ_f:

$$\gamma = \frac{Ea_0}{\pi^2} \qquad\qquad [4.1]$$

$$\sigma_f = \left(\frac{E\gamma}{a_0}\right)^{1/2} \cong \frac{E}{\pi} \qquad\qquad [4.2]$$

where E is the modulus of elasticity and a_0 is the interatomic distance. The value of γ is difficult to determine, and this model gives most materials unrealistically high values of rupture strength. The reason for this is the presence of structural defects, which act as stress concentrators (Fig. 4.2). The stresses at their tips can easily exceed the theoretical cohesive strength of the material, so leading to crack propagation and failure.

4.2.2 Fracture toughness

Linear elastic fracture mechanics is used for the basic description of crack propagation through a solid brittle material. Its basic assumptions may be summarised as follows.

1. The material behaves like a linear elastic continuum: no non-elastic deformations are considered. This is usually a good approximation of a ceramic material.

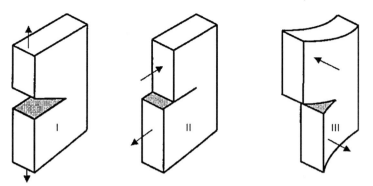

4.3 Crack tip separation modes: I, opening mode; II, sliding mode; III, tearing mode.

2. Cracks have infinitely sharp tips: cracks are considered to be flat (2D) structures with ideally sharp tips, i.e. tip radius of curvature $r \to 0$.
3. Crack borders are free of traction forces: however, in reality traction forces exist in many cases.

According to linear elastic fracture mechanics, the stress near the crack tip is:

$$\sigma_{r\theta} = \frac{K}{(2\pi r)^{1/2}} f(\theta) \qquad [4.3]$$

where r and θ are polar coordinates and K is a constant called the stress intensity factor. By using elasticity theory it can be shown that:

$$K = Y\sigma(\pi a)^{1/2} \qquad [4.4]$$

where σ is the applied stress, a is half the crack length and Y is a constant that depends on the crack opening mode and the geometry of the specimen (the so-called geometric factor). The geometric factor Y is for small notch-like or semicircular cracks, approximately 1.12. For a circular crack embedded in a homogeneous uniaxial tensile stress field, it is $2/\pi$.

The three modes of crack tip separation are shown in Fig. 4.3. Generally, the most critical and most important is mode I, known as the opening mode. The stress intensity factor for this situation is denoted K_{I} (notations of K_{II} and K_{III} are used for other modes). The critical value of K_{I}, at which the crack begins to propagate, is denoted as K_{IC}. This value is an important material parameter and is known as the fracture toughness.

When the value of the fracture toughness K_{IC} (using equation 4.4) is known, the critical strength σ_{f} of a material with certain types of defects (with size of a_{c}) or critical crack length (critical size of defect) can be calculated:

$$\sigma_f = \frac{K_{IC}}{Y(\pi a_c)^{1/2}} \qquad\qquad [4.5]$$

$$a_c = \frac{1}{\pi Y^2}\left(\frac{K_{IC}}{\sigma_f}\right) \qquad\qquad [4.6]$$

Equations 4.5 and 4.6 describe the close relation between fracture strength and size of the critical defects. It may clearly be seen that given a particular value of K_{IC}, the size of microstructural features (whether macro-, micro- or nano-cracks) determines the failure or survival of the stressed material.

4.2.3 Measurement of fracture toughness

The measurement of fracture toughness in ceramics is based on the methods developed for metals. However, in newly developed brittle materials, the choice is restricted as only a limited number of small specimens are available and of practical use. The most common methods are as follows.

Single edge notched beam (SENB) and single edge pre-cracked beam (SEPB)

Both are simple to use. In a rectangular beam, the sharpest possible notch is made on one side (Fig. 4.4(a)). In a SEPB, a crack is created at the tip of the notch, i.e. a situation close to an ideally sharp controlled flaw is created. The beam is loaded in four-point bending mode and the fracture toughness is calculated from the fracture load:

$$K_{IC} = M\frac{F_{max}}{B(W)^{1/2}}\left(\frac{S_1 - S_2}{W}\right)\left(\frac{3(\alpha)^{1/2}}{2(1-\alpha)^{3/2}}\right), \text{where}\quad M = f(\alpha = a/W) \quad [4.7]$$

Double torsion (DT)

The double torsion test (Fig. 4.4(b)) is convenient because the stress intensity K_1 does not depend on the length of the crack for $0.25\ L < a < 0.75\ L$. The crack can therefore be propagated in a stable and controllable manner. The fracture toughness is given by:

$$K_1 = FW_m\left(\frac{3(1+v)}{Wt^3 t_1}\right)^{1/2} \qquad\qquad [4.8]$$

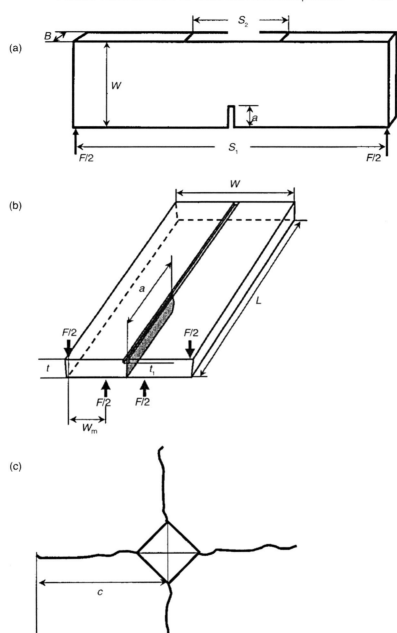

4.4 Fracture toughness measured by (a) SENB, (b) DT and (c) indentation method.

A notch is cut in the brittle material. A crack occurs during loading and grows with F. This method requires only small specimens.

Indentation methods

Indentations can generate cracks in brittle materials and the equilibrium length, c, of these cracks is related to the indentation load and fracture toughness. However, the relationship is not simple and many authors have proposed different formulas. One of the most widely used, which is suitable for very brittle materials, was developed by Anstis *et al.* (1981):

$$K_{IC} = \eta \left(\frac{E}{H}\right)^{1/2} \frac{F}{c^{3/2}}$$

[4.9]

where η is a shape factor (0.016 for a Vickers indentation), E and H are Young's modulus and hardness respectively, F is the indentation load and c is the crack length (Fig. 4.4(c)). These tests are easy to carry out and may be done with very small specimens. Their disadvantages are high scatter and subjective errors. This technique is therefore used mostly for comparative purposes and to obtain preliminary results for new materials.

4.2.4   Measurement of strength

It is difficult to prepare and perform a standard tensile strength test for new brittle materials. The most widely used methods for measuring strength are three- and four-point bending of rectangular bars (Fig. 4.5). The stress across the specimen changes in a linear manner from compression to tension. For this reason, the fracture usually originates either close to, or directly at, the tensile surface. In an elastic beam bent at four points, the bending moment is constant in the inner span (Fig. 4.5(b)). After the fracture load F is reached, the strength can be calculated as:

$$\sigma_c = \frac{3}{2} F \frac{(S_1 - S_2)}{W^2 B}$$

[4.10]

For three-point bending, $S_2 = 0$.

4.2.5   Critical flaw, statistical character of strength in ceramics, Weibull distribution

As shown by equation 4.5, there is a direct relationship between the size of the critical flaw and the resultant measured fracture strength. The measured strength is therefore not a fixed value but depends on the size (and shape) of defects and flaws that result from the technological process of material

(a)

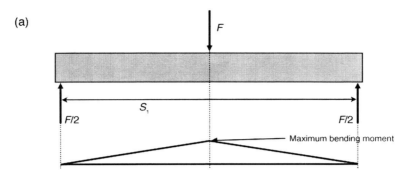

(b)

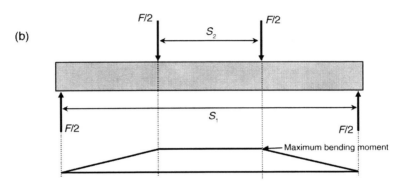

4.5 Flexure test of strength: (a) three-point bending; (b) four-point bending.

preparation and which are statistically distributed throughout the micro-structure. The strength cannot therefore be characterised by one value but must be statistically treated and described by the appropriate distribution.

Weibull (1951) proposed a distribution that expresses the cumulative probability of failure P_f as a function of maximum applied tensile stress, σ, acting in an effective volume of sample V, using three independent variables:

$$P_f = 1 - \exp - \left[\int_V \left(\frac{\sigma - \sigma_u}{\sigma_0} \right) \right]^m dV \qquad [4.11]$$

where m is the Weibull modulus, which characterises the scatter of strength values, σ_0 is a normalising parameter called characteristic strength and σ_u is the threshold stress (i.e. stress under which the failure probability is zero). Weibull's analysis assumes a normal distribution of defect size, that only one population of critical defects is present and that existing defects do not interact among themselves.

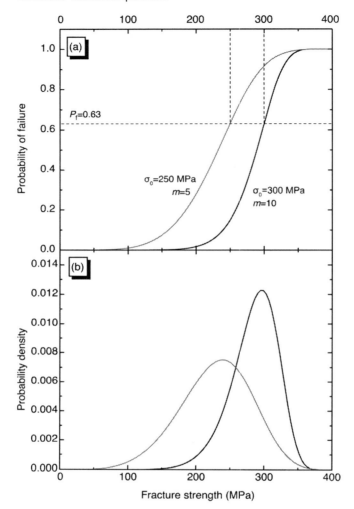

4.6 Examples of two-parameter Weibull distribution. (a) Two
distributions of cumulative probability of failure, P_f, with different
parameters σ_0 and m. (b) Comparison of the corresponding first
derivatives (probability densities) of these distributions.

As may be seen, the probability of failure increases with increasing
volume, because in larger volumes, larger defects are more likely to be
present. Considering two different volumes, V_1 and V_2, and using equation
4.11 the relationship between their respective measured strengths is:

$$\sigma_1/\sigma_2 = (V_2/V_1)^{1/m} \qquad [4.12]$$

This fact must be kept in mind when comparing results obtained by different methods, for instance in the case of results from three-point and four-point bending tests.

Good structural ceramic materials should have high parameters of the Weibull distribution: m is of particular interest as a high value of m means high reproducibility of the strength values and high reliability of the material (the strength distribution if narrow). Traditional ceramics have m values of around 5. In high-quality advanced structural ceramics, m is typically between 10 and 15. Making the defect size distribution narrow and/or ensuring some flaw tolerance in the material are means of improving the Weibull modulus m.

Without additional information, the Weibull distribution (equation 4.11) is difficult to use because it contains many unknown variables. In practice, when testing normalised samples with constant volume, and assuming that σ_u is negligible with respect to characteristic strength (which for most ceramics is possible), the so-called two-parameter Weibull distribution may be used:

$$P_f = 1 - \exp\left[-\left(\frac{\sigma}{\sigma_0}\right)^m\right] \tag{4.13}$$

This is possible to evaluate and therefore is frequently used. Figure 4.6 shows examples of two different two-parameter Weibull distributions for ceramic materials, which illustrate the effect of σ_0 and m on the shape of probability function. The first derivative of P_f with respect to strength, σ, gives the failure probability density (Fig. 4.6(b)) and describes the frequency of the strength values.

The two-parameter Weibull distribution (equation 4.13) can be linearised as:

$$\ln\left[\ln\left(\frac{1}{1 - P_f}\right)\right] = m \ln \sigma - m \ln \sigma_0 \tag{4.14}$$

and in this form its results are typically plotted as shown in Fig. 4.7.

Correct Weibull analysis is important in characterising the strength of ceramics. International standards (e.g. EN 843-5: 2006) require the testing of sets of at least 30 specimens in order to calculate the Weibull parameters with sufficient confidence.

4.3   Fracture origins

As already mentioned, fractures occur when the local stress at particular defects reaches a critical value. The defect becomes unstable and a crack begins to grow. In the absence of plastic deformation, this process does not stop and leads to catastrophic failure. The most important feature in this respect is therefore the critical flaw at the place where the fracture begins,

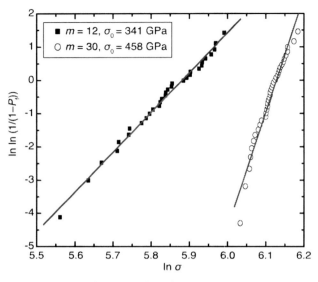

4.7 Linearised Weibull plots for two materials. The parameters of the two two-parameter distributions are shown in the figure.

which is known as the fracture origin. In ceramics, the fracture origin usually has two sources: (i) microstructure imperfections that arise during processing, such as pores, inner residual cracks, non-sintered clusters of secondary phases, impurities and unwanted inclusions, large grains, etc. and (ii) defects from machining and handling (usually surface cracks). While the latter may be quite easily seen and identified, it is necessary to know and minimise the former type of defect to ensure the reliability of the component. These defects are therefore the subject of extensive studies. This is even more important in composite materials, as the processes of milling and mixing two or more starting powder phases leads to various technological defects (e.g. Dusza and Kromp, 2003; Balázsi *et al.*, 2006; Dusza *et al.*, 2009; Csehová *et al.*, 2011).

Pores are one of the most frequently occurring technological flaws, especially macropores and clusters of pores that remain after non-optimal sintering. Typical examples are shown in Fig. 4.8, where macropores in various ceramic and/or non-metallic materials are illustrated and identified as the most common causes of failure in their respective materials. These macropores can be simply void areas where the matrix grains have not bonded adequately (Fig. 4.8(a)), impurities remaining from processing (Fig. 4.8(b)), empty spaces connected to clusters of very large grains in bimodal structures (Fig. 4.8(c)) or other types of macropores (Fig. 4.8(d)). However, residual microporosity may also have a significant influence on structural integrity. An example is illustrated in Fig. 4.9 in which fast spark plasma sintering of monolithic alumina resulted in incomplete bonding of the grains

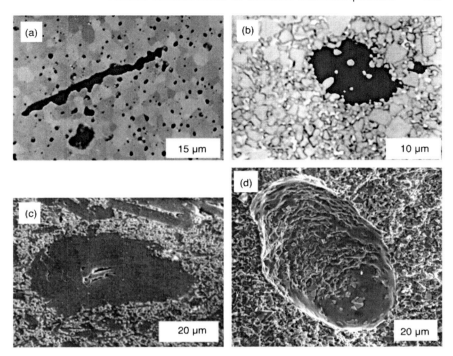

4.8 (a) Macropore in $MoSi_2$. SiO_2 glassy phase in the form of globular particles is also visible. (b) Impurity as a microstructural defect in tungsten carbide–cobalt system. (c) Porosity as a consequence of large grain clustering in bimodal microstructure of Si_3N_4. (d) Macropore as a fracture origin in Si_3N_4 ceramics.

4.9 Micropores at the grain boundaries in alumina prepared by spark plasma sintering (Puchý *et al.*, 2011).

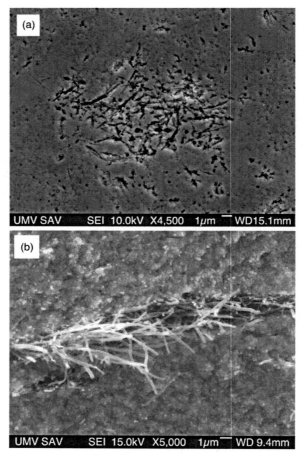

4.10 Clusters of carbon nanofibers (CNFs) in tetragonal zirconia matrix (ZrO_2–2vol% CNF): (a) visible on polished surface; (b) as a fracture origin observed on fracture surface.

where many submicron pores at the grain boundaries could still be found. In this case, these micropores gave rise to a lower than expected fracture structural integrity (Puchý *et al.*, 2011).

One of the most critical aspects governing the reliability of the microstructure in fibrous nanocomposites is the homogeneous dispersion of admixed nanofibres (Balázsi *et al.*, 2006), which has proved difficult to achieve. Insufficient homogenisation leads to fibre clusters of several micrometres in size. Such clusters are often the largest defects in the microstructure (Fig. 4.10) and thus the cause of failure. In other cases they form small islands, such as CNTs in Si_3N_4 (Fig. 4.11), which are smaller than the typical critical defect. They do not become fracture origins or

UMV SAV   SEI 10.0kV X10,000   1μm WD10.1mm

4.11 Clustering of carbon nanotubes (CNTs) in a silicon nitride matrix (Si_3N_4–3vol% CNT) (Hvizdoš *et al.*, 2011).

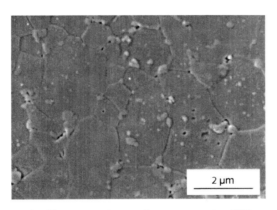

2 μm

4.12 Clusters and porosity connected to SiC nanoparticles in Al_2O_3–SiC nanocomposite.

reduce the strength but their toughening potential is not fully realised, thus the materials containing them do not exhibit a significant improvement in fracture toughness (Balázsi *et al.*, 2006; Hvizdoš *et al.*, 2011).

In nanocomposites with nanoparticles, the microstructures typically consist of fine grains because the secondary-phase particles are introduced in order to hinder the grain growth. Macropores are seldom present in such materials. However, residual pores at grain boundaries and triple-point junctions (Fig. 4.12), and/or secondary particle clusters such as those reported by Csehová *et al.* (2011) are typical defects.

4.4   Crack propagation, toughening mechanisms

In linear elastic fracture mechanics, K_{IC} is a constant (equation 4.4). When a crack starts to propagate, it cannot be stopped and will grow catastrophically across the whole specimen. In order to improve the performance of brittle materials, various strategies have been devised for slowing down and potentially stopping crack growth. In composite systems, the toughening phases act either behind the crack tip, where they slow down further opening of an existing crack by keeping mechanical contact between the crack faces, or in front of the crack tip where, by causing residual stresses, they cause the crack to deviate. By dissipating energy in this way, they serve as pinning sites for arresting cracks. In nanocomposites, the first type is mainly used in systems containing nanofibres and nanotubes. The second type uses hard nanoparticles such as silicon carbide.

4.4.1   Toughening by contact mechanisms

The mechanisms by which elongated structures (whiskers, large grains, nanofibres) can hinder crack propagation are crack deflection, crack bridging, pull-out and the associated friction. These are often observed at fracture lines and fracture surfaces and are schematically illustrated in Fig. 4.13. Duszová *et al.* (2008) observed these mechanisms acting in ZrO_2

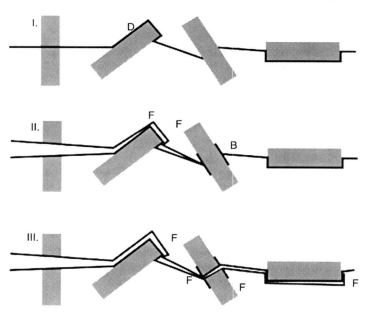

4.13 Toughening mechanisms acting behind the crack tip between the fracture surfaces: D, deflection; F, friction; B, bridging.

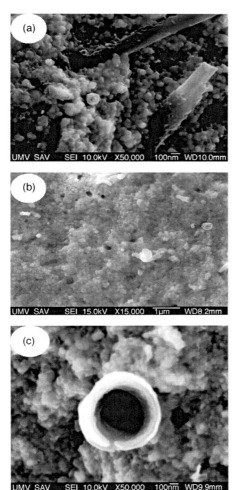

4.14 Characteristic toughening mechanisms in CNF–ZrO$_2$ composites in the form of crack deflection (a) and pull-out (b). In the case of CNFs with a large diameter perpendicular to the fracture surface, telescoping of inner walls from outer walls was observed (c).

ceramics containing carbon nanofibres (Fig. 4.14). Similar behaviour has been documented in a number of works on different types of ceramics reinforced by CNTs (Balázsi *et al.*, 2006; Puchý *et al.*, 2011). Contact mechanisms lead to the development of closing forces which act in the process zone behind the crack tip. This behaviour has been described and modelled in other works (Hsueh and Becher, 1988; Choi *et al.*, 1992; Hvizdoš *et al.*, 1994).

4.4.2   Residual stresses

In their seminal work, Niihara *et al.* (1993) emphasised that of the types
defined in Fig. 4.1, it is the intra-type that is usually expected to improve the
mechanical properties of ceramic nanocomposites, largely due to its ability
to utilise the relatively high residual stresses arising in such structures.

Because of their transgranular fractures, bridging zone mechanisms such
as mechanical or frictional interlocking or pull-out in Al_2O_3–SiC
nanocomposites should be excluded *a priori*. Therefore, only mechanisms
acting ahead of or directly behind the crack tip may be taken as applicable
for these nanocomposites. The spherical shape and small size of the
reinforcing elements means the toughening effect of crack deflection in these
composites is minimal and elastic crack pinning is the only mechanism that
will contribute to toughness.

According to Ohji *et al.* (1998), even when a dispersed particle consists of
a brittle material that fails at a small crack opening distance, the high
strength and the rigid interface with the matrix create an extremely high
shielding stress and a steep increase in fracture resistance. An *R*-curve with a
high tearing modulus may be responsible for the high strength in Al_2O_3–SiC
nanocomposites. The effects of residual stresses are utilised in this composite
to render the pinning mechanism effective (Fig. 4.15).

Internal stresses are generated during the cooling that follows the
fabrication of ceramic nanocomposites. This is due to the difference in the
thermal expansion coefficients of the matrix (α_m) and the nanoparticle (α_p),
which influence the grain and grain boundary strength. The thermal
expansion stress, σ_T, inside a single spherical particle in an infinite matrix is

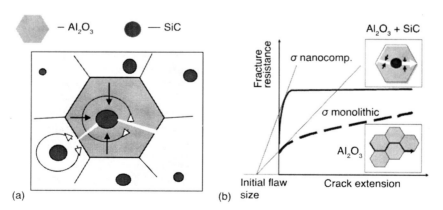

(a)   (b)

4.15 Schematic illustration of (a) crack propagation in Al_2O_3–SiC
composite and (b) *R*-curves of an Al_2O_3–SiC nanocomposite and a
monolithic alumina (Dusza and Šajgalík, 2009).

expressed by the Selsing formula (Selsing, 1961):

$$\sigma_T = \frac{\Delta\alpha\Delta T}{(1 + v_m)/2E_m + (1 - 2v_p)/E_p}$$ [4.15]

where $\Delta\alpha = \alpha_p - \alpha_m$ is the difference in thermal expansion coefficient, $\Delta T = T_{pl.} - T_{room}$, v is Poisson's ratio, E is Young's modulus, and the subscripts m and p represent matrix and particle respectively.

The tangential, σ_{Tt}, and radial, σ_{Tr}, stress distributions in the matrix around the particle are:

$$\sigma_{Tt} = -\frac{\sigma_T}{2}\left(\frac{r}{x}\right)^3$$ [4.16]

and

$$\sigma_{Tr} = \sigma_T\left(\frac{r}{x}\right)^3$$ [4.17]

where r denotes the radius of the particle and x is the radial distance from the particle surface.

In Al_2O_3 with SiC nanoparticles, the compressive stresses inside the SiC nanoparticles may reach several gigapascals, while the tensile stresses around them are of the order of hundreds of megapascals. These calculations have been experimentally confirmed by similar values of average residual stresses in Al_2O_3–SiC nanocomposites. The described residual stresses mean that, during propagation, the crack in Al_2O_3 deflects towards the nearest SiC particle because of the tangential tension around it. With an increase in applied stress, the crack leaves the particle and starts to propagate to the nearest SiC particle.

Fracture morphology

Figures 4.16 and 4.17 compare the crack propagation mechanism in monolithic alumina (Al_2O_3), where it is intergranular, and in nanocomposites where, due to residual stresses, it becomes transgranular. This change in the crack path is a specific feature that differentiates the fracture mechanisms of monolithic alumina and nanocomposites. In alumina, the crack path changes from transgranular to intergranular as the grain size decreases (Vekinis *et al.*, 1990) but remains transgranular in the nanocomposites, even where the grain sizes are very small. Figure 4.18 compares the detailed representative fracture surfaces of nanocomposites containing 5% and 10% SiC. It should be noted, however, that intergranular crack propagation in alumina usually leads to energy dissipation in the wake of the crack by grain bridging/interlocking, and to

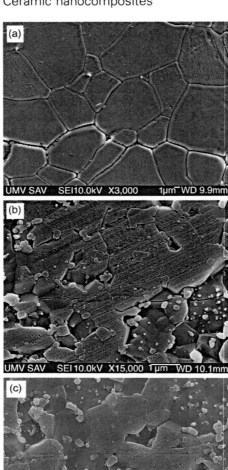

4.16 SEM micrographs of crack propagation: (a) Al_2O_3; (b) Al_2O_3–
5% vol. SiC; (c) Al_2O_3–10% vol. SiC.

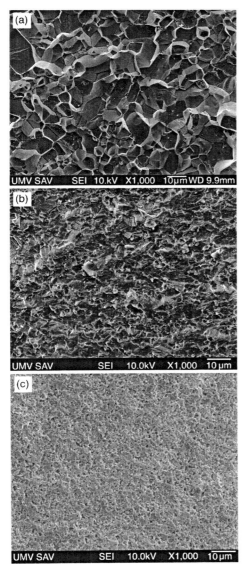

4.17 SEM morphology of fracture surfaces of: (a) Al_2O_3; (b) Al_2O_3–5% vol. SiC; (c) Al_2O_3–10% vol. SiC.

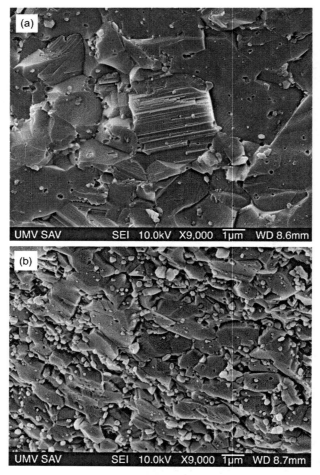

4.18 SEM morphology of (a) Al_2O_3–5% vol. SiC and (b) Al_2O_3–10% vol. SiC.

a strong *R*-curve effect with an overall increase in toughness (Rödel *et al.*, 1990). So the same or higher fracture toughness in the Al_2O_3–SiC nanoceramics must have a different origin in the form of residual stresses around the SiC particles.

4.5   Preventing failures

During the last two decades, various approaches have been used to improve the failure tolerance and reliability of ceramic materials (Fig. 4.19). They may be divided into the following classes.

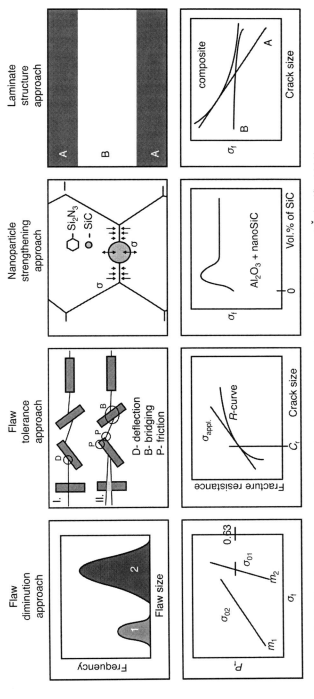

4.19 The four main research areas in advanced ceramic research (Dusza and Šajgalík, 2009).

At room temperature:

- improving the strength level and reducing its scatter, i.e. enhancing reliability by reduction of the critical defect size (improved properties of powders and of powder processing, clean-room manufacturing, etc.) – the flaw diminution approach (Lange, 1989);
- promoting energy-absorbing mechanisms occurring in the frontal process zone ahead of a propagating crack (stress-induced martensitic transformation, microcracking, etc.) – the flaw tolerance approach (Evans and Cannon, 1986);
- promoting localised bridging behind the crack tip (as frictional and mechanical interlocking, or pull-out), by which the flaw tolerance of the material can be improved – the flaw tolerance approach (Evans and Cannon, 1986);
- improving structural reliability by designing novel laminar composites that promote crack deflection at the inter-layer boundaries and/or utilise compressive residual stresses generated during cooling from the sintering temperature due to the differences in thermal expansion coefficients between layers of different composition – the laminar structure approach (Suresh, 1997; Lugovy et al., 1999);
- improving strength by incorporating nano-sized, second-phase particles with different expansion coefficients into the matrix – nanoparticle dispersion strengthening (Sternitzke, 1997).

At high temperature:

- optimising the grain size and the grain boundary phase geometry, composition and crystallinity with the aim of improving high-temperature strength, fatigue and creep resistance (Dusza and Šajgalík, 2005).

Over the last decade, a variety of advanced monolithic and composite ceramics have been fabricated using these approaches and they possess enhanced toughness, flaw tolerance and improved high-temperature properties. In addition to the main/matrix phase and sintering additives, a certain amount of reinforcing phase is present (up to 50 vol. %). These composite ceramics can be divided into the following groups:

- ductile phase reinforced: $Al_2O_3 + Al$, $Al_2O_3 + Nb$, $WC + Co$ etc.
- particle reinforced: $SiC + TiC$, $Al_2O_3 + ZrO_2$, etc.
- whisker/platelet reinforced: $Si_3N_4 + Si_3N_4$-wh., $Si_3N_4 + SiC$-wh., $Al_2O_3 + SiC$-wh., etc.
- *in-situ* reinforced ceramics: Si_3N_4, SiC, etc. (without second reinforcing phase)
- nanocomposites: $Al_2O_3 + SiC$, $Si_3N_4 + SiC$, etc.

- layered/laminated composites: two, three or multiple layers based on oxides, nitrides, etc.

Most of these materials are treated using classical powder processing, sintering, hot-pressing or hot isostatic pressing method, with or without sintering additives in monolithic form or with different volume fractions in the reinforcing phase. During the last decade, these methods have significantly improved control of the microstructure (grain size and shape, defects, flaws and intergranular phases in terms of amount, distribution and microchemistry). This has been achieved through the production of fine and pure powders, the design of the starting composition and the development of controlled powder processing and densification procedures.

4.6   Wear of ceramic nanocomposites

Despite improvements in fracture toughness and strength of modern ceramic materials, they are still predominantly used in applications that make use of their exceptional hardness and chemical and thermal stability. Typical examples are cutting tools, ceramic seals, ball bearings, etc. (i.e. applications without high loads but with intense mechanical contact). Such applications require high wear resistance and good tribological behaviour. There are two advantages in using nanocomposites in this area: firstly, the improvement in fracture toughness and hardness can improve the wear resistance and, secondly, the addition of appropriate secondary phases can lower the friction coefficient due to the development of a self-lubricating effect.

4.6.1   Nanocomposites with carbon nanofilaments

Recently, efforts to develop ceramics (based on alumina, silicon nitride and zirconia) with carbon filamentous phases (carbon nanofibres (CNFs) and nanotubes (CNTs)) have produced interesting results in the area of tribological properties (An *et al.*, 2003; Lim *et al.*, 2005; Balázsi *et al.*, 2006; Yamamoto *et al.*, 2008; Hvizdoš *et al.*, 2010, 2011). It has been shown that in self-mated tribological pairs, the coefficient of friction could be significantly reduced, from values between 0.5 and 0.8 typical for the dry sliding of ceramics, down to 0.2–0.3 (Fig. 4.20) (Hvizdoš *et al.*, 2011). This reduction, however, was generally accompanied by a lowering of the wear resistance (Fig. 4.21) due to a less than optimal microstructure, which results from difficulties with the incorporation of CNTs/CNFs. However, Hvizdoš *et al.* (2011) also found an optimum for fractions of around 5% CNT in Si_3N_4/CNT nanocomposite, where the wear resistance improved, probably

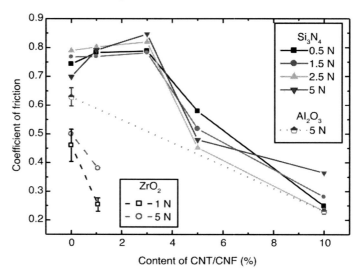

4.20 Coefficient of friction as a function of amount of carbon phases in ceramic matrix composite materials.

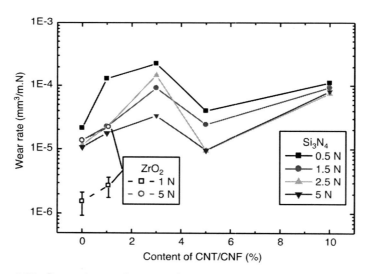

4.21 Dependence of wear resistance on carbon content of experimental materials tested at various load levels.

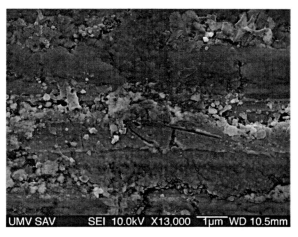

UMV SAV   SEI 10.0kV X13,000   1µm WD 10.5mm

4.22 Detail of the wear track in ZrO₂–CNF composite – pull-out of CNFs and smeared transferred film.

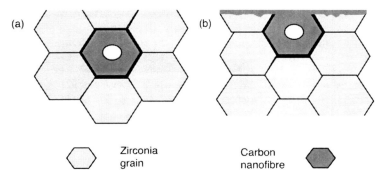

4.23 Schematic representation of wear and creation of transferred carbon film in ZrO₂–CNF nanocomposite: (a) polished surface before the wear test; (b) tribofilm formation during the wear test.

due to significantly reduced friction. In Si_3N_4–10%CNT, it only reached values similar to those for Si_3N_4–1%CNT.

The micro-mechanisms of wear surface damage were identified by observation of the wear tracks of composite materials. Microscopic studies of the worn surface of a ZrO_2–CNF composite showed pulled-out CNFs (Fig. 4.22), suggesting the presence of a smeared layer of graphite and crushed CNFs, the so-called transferred film (Fig. 4.23), which significantly lowers the friction coefficient, even though a relatively low amount of carbon (1.07 wt%) was incorporated into the microstructure. The values of the friction coefficient and wear rates suggest that the CNTs in silicon nitride based composites were much less effective in creating transferred film. This may be due to their greater mechanical stability and better distribution throughout the microstructure. This could make pull-out and destruction in the contact zone more difficult issues.

4.6.2 Nanocomposites with hard nanoparticles

The wear behaviour of composites reinforced with hard nanoparticles differs in character from that of those reinforced with carbon nanofilaments. Tatarko *et al.* (2010) presented a thorough study of silicon nitride based nanocomposites sintered with a wide range of rare earth oxides that were further doped with SiC nanoparticles. In general, the wear behaviour of Si_3N_4-based materials is reported to be controlled by the following mechanisms:

- mechanical- or thermal-induced micro-cracking due to fatigue-assisted stresses resulting from friction at high normal loads;
- tribochemical reaction with water vapour at low temperatures and normal loads ($<400°C$, $<10\,N$);
- tribooxidation at temperatures that depend on the composition and nature of the intergranular phases (Dong and Jahanmir, 1993; Skopp *et al.*, 1995; Wang and Hsu, 1996).

The introduction of SiC nanoparticles into a silicon nitride matrix improves hardness, strength, resistance to creep, oxidation and corrosion in Si_3N_4 ceramics. With regard to tribology, the addition of ceramic particles has two main functions: to increase wear resistance and to improve self-lubricating properties by selective oxidation of these compounds.

In work by Tatarko *et al.* (2010), the wear resistance (measured by the ball-on-disc method of tribology) increased with decreasing size of rare earth cations in Si_3N_4 monoliths as well as Si_3N_4–SiC composite materials sintered with different rare earth oxide additives. Figure 4.24 shows the friction coefficient and specific wear rate of the specimens as a function of the size of rare earth (RE) cations for such materials. The friction coefficient of the materials slightly decreased with a decreasing ionic radius in RE^{3+} from 0.74 to 0.70 in the case of monoliths, and from 0.71 to 0.64 in the case of composites sintered with La_2O_3 and Lu_2O_3, respectively. Similarly, the specific wear rate decreased with a decreasing ionic radius in RE^{3+} from $3.22 \times 10^{-5}\,mm^3/Nm$ to $1.15 \times 10^{-5}\,mm^3/Nm$ for the monolithic Si_3N_4 sintered with La_2O_3 and Lu_2O_3, respectively, and from $1.91 \times 10^{-5}\,mm^3/Nm$ to $0.89 \times 10^{-5}\,mm^3/Nm$ in the case of composites doped with the same additives. As may be seen from Fig. 4.24, the friction coefficient, as well as the specific wear rate of composites, was always lower when compared to the monoliths with the same sintering additives, thus indicating the higher wear resistance of Si_3N_4–SiC composites.

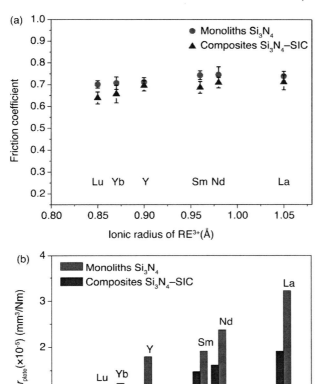

4.24 Influence of different rare earths on (a) the friction coefficient and (b) the specific wear rate of Si_3N_4 monoliths and Si_3N_4–SiC composites.

It is well known that the bonding strength between Si_3N_4 grains and grain boundary phases increases with the decreasing ionic radius of RE^{3+}. Therefore the highest bonding strength in Si_3N_4 containing Lu_2O_3 as a sintering additive restricts dropping of the individual silicon nitride grains during the wear experiment. This fact, together with its high hardness and fracture toughness values, are the reasons why Si_3N_4 doped with Lu exhibited the best wear resistance among all the studied monolithic materials. The addition of SiC nanoparticles in this class of materials leads to lower fracture resistance. However, the composites exhibited higher wear resistance in comparison with that of monolithic Si_3N_4. This is probably due to their higher degree of hardness, the positive effect of which was more significant than the influence of the lower fracture toughness.

The wear mechanisms and tribological characteristics were almost analogous for monolithic Si_3N_4 and Si_3N_4–SiC composite materials. Mechanical wear (microfracture) and tribochemical reaction were found to be the main wear mechanisms for all the studied materials. Examples of the worn surfaces of the tested plate specimens at room temperature are shown in Fig. 4.25. SEM observations of wear tracks revealed that with increasing wear resistance, the worn surface becomes quite smooth, but adherent debris is still observed. Addition of SiC into the Si_3N_4 materials results in a higher incidence of 'islands' of coherent debris. Moreover, the highest amount of coherent debris was observed in material with the highest wear resistance, i.e. for the composite sintered with Lu_2O_3. These 'islands' of coherent debris, which were observed in all studied wear tracks in different volumes, constitute a tribofilm affording some protection to the ceramic surfaces and decreasing the wear coefficient of materials (Fig. 4.26(a) and 4.26(b)). A larger amount of coherent debris was observed in both kinds of material in the case of composites or materials doped with smaller ionic radius RE^{3+}. The tribochemical reactions created SiO_2-based tribofilm in the tested samples, which was partially removed above a critical load, resulting in microfractures in discrete regions due to the propagation of a micro-crack (Fig. 4.26(c) and 4.26(d)). This microfracture, shown in Fig. 4.26(d), has very similar features to the fracture surface of the material after the common bending test.

An increase in fractured areas and in the amount of silicon nitride debris were observed in specimens with lower wear resistance. The fractured areas show a mixture of intergranular and transgranular failure modes in the worn surfaces. According to the mechanisms described for wear in silicon nitride based materials (Dong and Jahanmir, 1993; Skopp et al., 1995; Wang and Hsu, 1996), cracks initiate below the surface and their propagation is assisted by fatigue effects in the mixed intergranular/transgranular mode, leading to microscale delaminations and fragmentation.

Coherent layers formed on the wear surfaces contain large amounts of oxygen (observed by EDX analyses), suggesting that the layers are composed mainly of the oxidation products of silicon nitride. The formation of oxide may be accelerated by the presence of small particles of silicon nitride formed by fracture under the high contact stresses at the wear interface. These particles have a high specific area, which would increase the amount of reaction product formed when compared to a reaction with the ceramic bulk. This would cause some oxidation at room temperature as the silicon nitride surface is likely to be rapidly covered by oxide layers.

Unlike the CNTs/CNFs containing nanocomposites, where the secondary phases reduce the coefficient of friction but also lower the wear resistance,

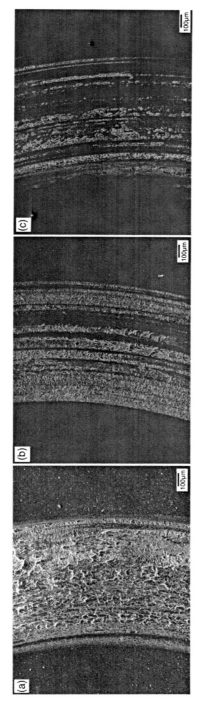

4.25 Worn tracks after test at 22°C of (a) La-doped monolith, (b) La-doped composite and (c) Lu-doped composite (Tatarko et al., 2010).

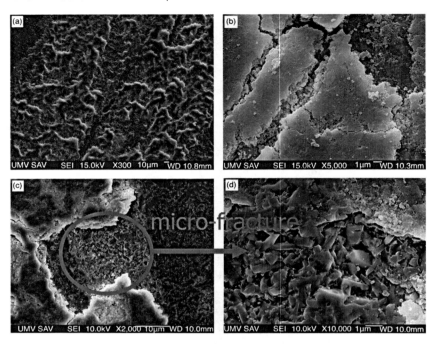

4.26 (a–d) Wear track of Si₃N₄–SiC composite with Yb₂O₃ consisting of tribofilm which is partially removed, resulting in microfracture in discrete regions.

incorporation of SiC nanoparticles leaves the coefficient of friction unchanged while improving the wear resistance. This suggests that further development has the potential to improve the overall tribological performance of ceramic nanocomposites.

4.7   Future trends

Various research areas will be of interest in the future (Basu, 2008):

- multi-function integrated nanocomposite materials, developed in various ceramic systems in an effort to make significant improvements in the mechanical and thermal properties of multi-functional structural ceramic materials for the energy, environment and biomedical sectors
- development of 'intermaterials': a special class of multi-functional material containing inorganic, organic and metallic materials, incorporated at the nano, molecular and atomic levels; new design concepts on molecular/cluster/lattice-level composites, their related results and impacts on future aspects of research
- polymer-derived nanostructured ceramics, including the processing of

nanostructured composite coatings, the use of polymer-derived ceramics to join ceramics and the processing of controlled porosity ceramics (e.g. *in-situ* carbon nanotubes in a ceramic matrix)

- nanostructured ceramics for fuel cell and battery applications, including the development of various state-of-the-art materials for solid oxide fuel cells (SOFC) and nanostructured ceramics that enable the operation of SOFC in a temperature range of 500–650°C. It appears that the doping of nanosized Ag in Li–CoO_2 cathodes increases the electrical conductivity by two to three orders of magnitude with reference to undoped materials. The synthesis of nanocrystalline $Li_4Ti_5O_{12}$ using an aqueous combustion process with alanine fuel will be the subject of future research.

It seems likely that there will be intensive development of ceramic- and carbon-based filler nanocomposites in the near future. In addition to carbon nanotubes and carbon nanofibres, new carbon-based fillers such as graphene and graphene platelets (GPL), also called graphene nanoplatelets (GNP) and/or multi-layer graphene nanosheets (MGN), offer alternatives to the more expensive nanotubes. Currently, there are only a few reports on the use of graphene additives to improve the mechanical and/or functional properties of bulk advanced ceramics.

A graphene nanosheet/alumina composite has been prepared by Wang *et al.* (2011) using spark plasma sintering. The results show that the fracture toughness and conductivity of the graphene nanosheet/alumina composite are about 53% and 13 orders of magnitude higher than those of unreinforced alumina material, respectively. Walker *et al.* (2011) applied aqueous colloidal processing methods to obtain uniform and homogeneous dispersions of GPL and Si_3N_4 ceramic particles which were densified using spark plasma sintering. A fracture toughness of $6.6\,MPam^{0.5}$ was measured for a composite with 1.5 wt% of GPL, which was significantly higher than the value measured for the monolithic silicon nitride with globular grains of α-phase. The observed toughening mechanisms on the fracture surfaces of the nanocomposites are in the form of graphene necking and crack bridging, crack deflection and graphene sheet pull-out. Kun *et al.* (2011) prepared and characterised silicon nitride based nanocomposites with differing amounts of carbon reinforcement in the form of multilayer graphene, graphite nanoplatelets and nano-graphene platelets. They found that both the bending strength and elastic modulus decreased with the addition of carbon-based fillers and that the composite with multi-layered graphene exhibited the highest strength and module of elasticity among the composites.

4.8   Sources for further information

One of the most authoritative sources of information about this topic is *Handbook of Nanoceramics and Their Based Nanodevices*. This is a five-volume set, edited by Tseung-Yuen Tseng (National Chiao Tung University, Taiwan) and Hari Singh Nalwa (Editor-in-Chief, *Journal of Nanoscience and Nanotechnology*, USA), published in January 2009 (3500 pages, ISBN: 1-58883-114-0, US\$ 2999.00).

Another important source is the more general *Handbook of Nanophysics*, a seven-volume work edited by Klaus D. Sattler and published in 2011 by CRC Press, Taylor and Francis Group (Hardback ISBN-978-1-4200-7540-3).

Among the references mentioned in this chapter, the most important are the founding papers by IIjima (1991) and Niihara *et al.* (1993).

Regarding internet interest groups, the very lively and active community site called Nanopaprika is recommended (www.nanopaprika.eu).

4.9   References

An J W, You D H and Lim D S (2003), 'Tribological properties of hot-pressed alumina-CNT composites', *Wear*, 255, 677–681.

Anstis G R, Chantikul P, Lawn B and Marshall D (1981), 'A critical evaluation of indentation techniques for measuring fracture toughness: I. Direct crack measurements', *J Am Ceram Soc*, 64, 533–538.

Balázsi C, Fényi B, Hegman N, Kovér Z, Wéber F, Vértesy Z, Kónya Z, Kiricsi I, Biró LP and Arató P (2006), 'Development of CNT/ Si3N4 composites with improved mechanical and electrical properties', *Composites: Part B*, 37, 418–424.

Basu B (2008), 'Nanoceramics and nanocomposites', *Curr Sci*, 95, 570–571.

Choi S R, Salem J A and Sanders W A (1992), 'Estimation of crack closure stresses for in situ toughened silicon nitride with 8 wt% scandia', *J Am Ceram Soc*, 75, 1508–1511.

Csehová E, Andrejovská J, Limpichaipanit A, Dusza J and Todd R (2011), 'Hardness and indentation load-size effect in Al2O3–SiC nanocomposites', *Kovove Mater*, 49, 119–124.

Dong X and Jahanmir S (1993), 'Wear transition diagram for silicon nitride', *Wear*, 165, 169–180.

Dusza J and Kromp K (2003), 'Fracture and mechanical properties of MoSi2 and MoSi2 + SiC', *Key Eng Mat*, 251–252, 13–18.

Dusza J and Šajgalík P (2005), 'Si3N4 and Al2O3 based ceramic nanocomposites', *Int J Mat and Prod Technol*, 23, 91–120.

Dusza J and Šajgalík P (2009), 'Silicon nitride and alumina-based nanocomposites', *In Handbook of Nanoceramics and their Based Nanodevices*, Vol. 2, American Scientific Publishers, 253–283. ISBN 1-58883-116-7.

Dusza J, Blugan G, Morgiel J, Kuebler J, Inam F, Peijs T, Reece MJ and Puchý V

(2009), 'Hot pressed and spark plasma sintered zirconia/carbon nanofiber composites', *J Eur Ceram Soc*, 29, 3177–3184.

Duszová A, Dusza J, Tomášek K, Blugan G and Kuebler J (2008), 'Microstructure and properties of carbon nanotube/zirconia composite', *J Europ Ceram Soc*, 28, 1023–1027.

EN 843-5: (2006), Advanced technical ceramics – Mechanical properties of monolithic ceramics at room temperature – Part 5: Statistical analysis.

Evans A G and Cannon R M (1986), 'Toughening of brittle solids by martensitic transformations', *Acta Met*, 34, 761–800.

Hsueh C-H and Becher P F (1988), 'Evaluation of bridging stress from R-curve behavior for nontransforming ceramics', *J Am Ceram Soc*, 71, C234–C237.

Hvizdoš P, Kupková M, Rudnayová E and Dusza J (1994), 'Rising fracture resistance of Al_2O_3 and Al_2O_3 + SiC whisker ceramics', *CFI - Berichte der DKG*, 9, 55–64.

Hvizdoš P, Puchý V, Duszová A and Dusza J (2010), 'Tribological behavior of carbon nanofiber–zirconia composite', *Scripta Mat*, 63, 254–257.

Hvizdoš P, Duszová A, Puchý V, Tapasztó O, Kun P, Dusza J and Balázsi C (2011), 'Wear behavior of ZrO_2-CNF and Si_3N_4-CNT nanocomposites', *Key Eng Mat*, 465, 495–498.

Iijima S (1991), 'Helical microtubules of graphitic carbon', *Nature*, 354, 56–58.

Kun P, Tapasztó O, Wéber F and Balázsi C (2011), 'Determination of structural and mechanical properties of multilayer graphene added silicon nitride-based composites', *Ceram Int*, in press; doi:10.1016/j.ceramint.2011.06.051.

Niihara K, Nakahira A and Sekino T (1993), 'New nanocomposite structural ceramics', *Mat Res Soc Symp Proc*, 286, 405–412.

Lange F F (1989), 'Powder processing science and technology of increased reliability', *J Am Ceram Soc*, 72, 3–15.

Lim D S, You D H, Choi H-J, Lim S H and Jang H (2005), 'Effect of CNT distribution on tribological behavior of aluminum-CNT composites', *Wear*, 259, 539–544.

Lugovy M, Orlovskaya N, Berroth K and Kuebler J (1999), 'Macrostructural engineering of ceramic-matrix layered composites', *Compos Sci Technol*, 59, 1429–1437.

Ohji T, Jeong Y K, Choa Y H and Niihara K (1998), 'Strengthening and toughening mechanisms of ceramic nanocomposites', *J Am Ceram Soc*, 81, 1453–1460.

Orowan E (1949), 'Fracture and strength of solids', *Rep Prog Phys*, 12, 185.

Puchý V, Dusza J, Hvizdoš P, Inam F and Reece M J (2011), 'Mechanical and electrical properties of Al_2O_3–CNT nanocomposites', *Chem. Papers*, 105, 635–637.

Rödel J, Kelly J F and Lawn B R (1990), 'In-situ measurements of bridged crack interfaces in the scanning electron microscope', *J Am Ceram Soc*, 73, 3313–3318.

Selsing J (1961), 'Internal stresses in ceramics', *J Am Ceram Soc*, 44, 419.

Skopp A, Woydt M and Habig H (1995), 'Tribological behavior of silicon nitride materials under unlubricated sliding between 22°C and 1000°C', *Wear*, 181–183, 571.

Sternitzke M (1997), 'Structural ceramic nanocomposites', *J Eur Ceram Soc*, 17, 1061–1082.

Suresh S (1997), 'Modeling and design of multi-layered and graded materials', *Prog Mater Sci*, 42, 243–251.

Tatarko P, Kašiarová M, Dusza J, Morgiel J, Šajgalík P and Hvizdoš P (2010), 'Wear resistance of hot-pressed Si3N4/SiC micro/nano composites sintered with rare-earth oxide additives', *Wear*, 269, 867–874.

Vekinis G, Ashby M F and Beaumont P W R (1990), 'R-curve behaviour of Al₂O₃ ceramics', *Acta Met*, 38, 1151–1162.

Walker L S, Marotto V R, Rafiee M A, Koratkar N and Corral E L (2011), 'Toughening in graphene ceramic composites', *ACSNano*, 5, 3182–3190.

Wang K, Wang Y, Fan Z, Yan J and Wei T (2011), 'Preparation of graphene nanosheet/alumina composites by spark plasma sintering', *Mat Res Bull*, 46, 315–318.

Wang Y and Hsu S M (1996), 'Wear and wear transition mechanisms of ceramics', *Wear*, 195, 112–122.

Weibull W (1951), 'A statistical distribution function of wide applicability', *J Appl Mech - Trans ASME* 18, 293–297.

Yamamoto G, Hashida T, Adachi K and Takagi T (2008), 'Tribological properties of single-walled carbon nanotube solids', *J Nanosc Nanotech*, 8, 2665–2670.

5

Multiscale modeling of the structure and properties of ceramic nanocomposites

V . T O M A R , Purdue University, USA

DOI: 10.1533/9780857093493.1.153

Abstract: One of the most recent developments in ceramics has been the distribution of multiple phases in a ceramic composite at the nanoscopic length scale. An advanced nanocomposite microstructure such as that of polycrystalline silicon carbide (SiC)–silicon nitride (Si_3N_4) nanocomposites contains multiple length scales with grain boundary thickness of the order of 50 nm, SiC particle sizes of the order of 200–300 nm and Si_3N_4 grain sizes of the order of 0.8–1.5 μm. Designing the microstructure of such a composite for a targeted set of material properties is, therefore, a daunting task. Since the microstructure involves multiple length scales, multiscale analyses based material design is an appropriate approach for such a task. With this view, this chapter presents an overview of the current state of the art and work performed in this area.

Key words: ceramic nanocomposites, microstructure, multiple length scales, multiscale modeling.

5.1   Introduction

Over the past half century, ceramics have received significant attention as candidate materials for use as structural materials under conditions of high loading rates, high temperature, wear, and chemical attack that are too severe for metals. However, the inherent brittleness of ceramics has prevented their wide use in different applications. Significant scientific effort has been directed towards making ceramics more flaw-tolerant through design of their microstructures by:

- incorporation of fibers or whiskers that bridge the crack faces just behind the crack tip
- designing microstructures with elongated grains that act as bridges between crack faces just behind the crack tip
- incorporating second phase particles, which deflect the crack, making it travel a more tortuous path
- incorporating secondary phases that undergo stress-induced volume expansion that forces the crack faces together.

However, one of the most recent developments has been the distribution of multiple phases in a ceramic composite at the nanoscopic length scale. Owing to the prevalence of nanoscopic features, such composites are referred to as ceramic nanocomposites.

The definition of nanocomposite material has broadened significantly to encompass a large variety of systems such as one-dimensional (1D), two-dimensional (2D), three-dimensional (3D) and amorphous materials, made of distinctly dissimilar components and mixed at the nanometer scale. The general class of nanocomposite organic/inorganic materials is a fast-growing area of research. Reducing the sizes of structural features in materials leads to a significant increase in the portion of surface/interface atoms. The surface/interface energy essentially controls the properties of a solid of such type. Interfaces provide a means to introduce non-homogeneity in the material. This non-homogeneity acts as a significant modification of both thermal and mechanical properties of the composites. Selective mixing of materials in a highly tailored morphology with a high percentage of interface area leads to materials with enhanced properties. The properties of nanocomposite materials depend not only on the properties of their individual parents but also on their morphology and interfacial character-istics.

Nanocomposites find their use in various applications because of the improvements in the properties over the simpler structures. As an example, for components used in a gas turbine engine, a lifetime up to 10 000 h and a retained strength of ~300 MPa at a temperature of 1400°C have been postulated, together with negligible creep rate. Furthermore, at elevated temperatures, the material must exhibit high resistance to thermal shock, oxidation, and subcritical crack growth. Ceramic nanocomposites have been shown to be extremely important for such future applications. Advanced bulk ceramic composite materials that can withstand high temperatures (> 1500°C) without degradation or oxidation can also be used for applications such as structural parts of motor engines, catalytic heat exchangers, nuclear power plants, and combustion systems, besides their use in fossil energy conversion power plants. These hard, high-temperature-stable, oxidation-resistant ceramic composites and coatings are also in demand for aircraft and spacecraft applications.

One such material system in this class of composites, silicon carbide/ silicon nitride (SiC/Si_3N_4) composites, has been shown to perform very well under high-temperature oxidizing conditions. Interest in such nanocomposites started with the experiments of Niihara [1] who reported large improvements in both the fracture toughness and the strength of materials by embedding nanometer range (20–300 nm) particles within a matrix of larger grains and at the grain boundaries (GBs). A 200% improvement in both strength and fracture toughness, better retention of strength at high temperatures, and better creep properties were observed. An advanced nanocomposite microstructure such as that of polycrystalline silicon carbide (SiC)–silicon nitride (Si_3N_4) nanocomposite contains multiple length scales with GB thickness of the order of 50 nm, SiC particle sizes of the order of 200–300 nm and Si_3N_4 grain sizes of the order of 0.8 to 1.5 µm [2]. Designing the microstructure of such a composite (and similar others such as TiN–Si_3N_4, SiC–Al_2O_3, SiC–SiC, graphene/CNT + SiC, and carbon fiber + SiC nanocomposites) for a targeted set of material properties is, therefore, a daunting task. Since the microstructure involves multiple length scales, multiscale analyses based material design is an appropriate approach for such a task.

5.2   Multiscale modeling and material design

A multiscale modeling paradigm is shown in Fig. 5.1. Atomistic analyses at the nanoscale can impart important information about the effect of critical features such as a grain boundary (GB), an interface, or a triple junction, etc. on mechanical deformation behavior of a small nanoscale (~ few nanometers) sample. In multiscale modeling such information is used to formulate macroscale (> a few micrometers) material models for understanding microstructure-dependent deformation behavior of a material sample such as the one shown in Fig. 5.1. Appropriate mathematical models of microstructure property relations allow one to relate performances such as fracture toughness, ultimate strength, fatigue lifetime, etc. to key material microstructure parameters like volume fraction, particle size, and phase composition.

Since a typical nanoscale test sample is much smaller and is subjected to varied surroundings in a typical microstructure (e.g. Fig. 5.1), the incorporation of nanoscale information in macroscale models is subjected to statistical uncertainty. If a complex microstructure is to be designed for a targeted set of properties, it is important that such uncertainties be correctly quantified and incorporated within a robust material design framework. The development of a variable fidelity model management framework that can

Molecular   Continuum

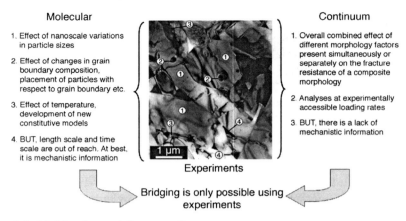

1. Effect of nanoscale variations in particle sizes

2. Effect of changes in grain boundary composition, placement of particles with respect to grain boundary etc.

3. Effect of temperature, development of new constitutive models

4. BUT, length scale and time scale are out of reach. At best, it is mechanistic information

1. Overall combined effect of different morphology factors present simultaneously or separately on the fracture resistance of a composite morphology

2. Analyses at experimentally accessible loading rates

3. BUT, there is a lack of mechanistic information

Experiments

Bridging is only possible using experiments

5.1 Multiscale modeling paradigm.

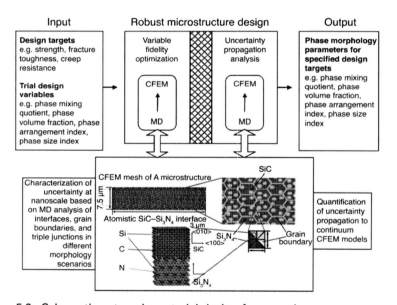

5.2 Schematic petascale material design framework.

incorporate material behavior analyses at multiple length scales in a design optimization framework has been reported [3–6].

Figure 5.2 details the process flow of a petascale multiphysics model management tool for multiscale material design. Deployed on a petascale machine, the design tool developed in this research, that integrates atomistic and mesoscale analyses using a variable fidelity model management framework, will facilitate a significant reduction in nanomaterials' development cost and time with a simultaneous increase in the possible different

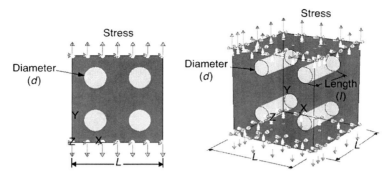

5.3 High- and low-fidelity models for CFCC nanocomposites.

combinations of individual composite material phases to achieve desired material performance. The model management framework [3, 4], besides managing the models and scales, is also well suited to control hierarchical parallelism. The natural hierarchy is molecular dynamics (MD) within the cohesive finite element method (CFEM) within design under uncertainty, using a mixed programming model SHMEM^{TM} by SGI for CFEM and MPI for MD and the uncertainty modeling. Both MD and the uncertainty quantification (via quasi-Monte Carlo integration) can use 1000 processors, and CFEM 10, so 1000 uncertainty quantification groups of 10 CFEM groups of 1000 HMC processors is 10^7 processors, nearing exascale.

Material design analyses of the model system have been performed to understand the morphology-related parameters that must be controlled for optimal targeted set of properties. The application of the design tool is focusing on the continuous fiber ceramic composite (CFCCs) models of SiC–Si$_3$N$_4$ nanocomposites (Fig. 5.3). The second phase (circles and cylinders) are the SiC fibers that have higher elastic modulus and higher creep resistance (E) but lower yield stress and fracture toughness, than that of the primary Si$_3$N$_4$ phase. The problem is to design the most suitable CFCC, with maximum strength and creep resistance for a set of external temperatures T, where the number of design variables will depend on whether the simulation tests are run on the 2D or 3D model. The design variables to be considered in the nanocomposite design optimization problem, for the 2D model, are the fiber diameter (d) and the external temperature (T). For the 3D model, the design variables to be considered are the fiber diameter (d), the length of fibers (l) and the external temperature (T). The problem definition in standard form is:

$$\left. \begin{array}{l} \text{minimize: } f(d,\, l,\, T) = \{-\sigma_U(d,\, l,\, T),\ \varepsilon_c(d,\, l,\, T)\} \\ \text{subject to: } d_{min} \le d \le d_{max} \quad l_{min} \le l \le l_{max} \text{ and } T = 1500°C \end{array} \right\} [5.1]$$

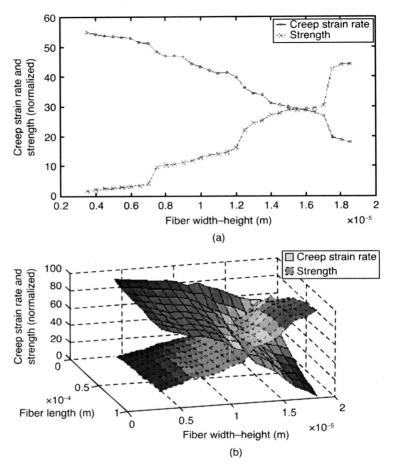

5.4 (a) Strength and creep strain rate at 1500°C as a function of the design variable width–height (*d*) for the 2D low-fidelity model. (b) Strength and creep strain rate at 1500°C as a function of the design variables width–height (*d*) and length of fibres (*l*) for the 3D high-fidelity model.

Figure 5.4 illustrates normalized (0–100) function values for the strength and creep strain rate as a function of design variables for the high-fidelity model (3D) and low-fidelity model (2D). Figure 5.4(a) shows an increase in the CFCC strength and a corresponding decrease in the creep strain rate as the design variable *d* increases. Similarly, for the high-fidelity model, Fig. 5.4(b) shows an increase in the CFCC strength and a corresponding decrease in the creep strain rate as the design variables *d* and *l* increase.

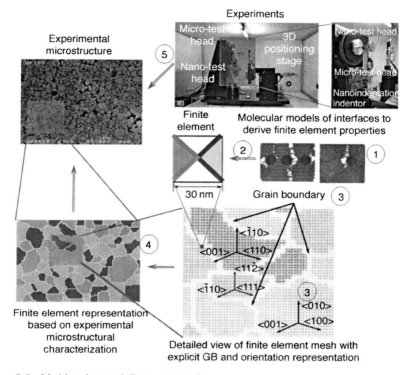

5.5 Multiscale modeling approach.

5.3   Multiscale modeling approach

The multiscale analyses (at nanometer and micrometer length and time scales) are based on a combination of CFEM and MD based techniques as shown in Fig. 5.5. In such analyses we extract information between multiple phases of a material based on molecular simulations. Such information is usually not available in experiments, but is incorporated in continuum models and later compared and validated with experiments.

Several studies have shown that the incorporation of SiC nanoparticles in a Si_3N_4 matrix improves high-temperature creep resistance while maintaining or improving the room-temperature strength [e.g. 1, 2, 7–14]. Following this observation, innovative processing approaches using polymer-derived amorphous Si–C–N powders have been developed to produce *in-situ* SiC nanoparticle reinforced composites [11, 14–19]. While the material system is an excellent candidate for high-temperature applications, a careful analysis of the effect of the underlying morphology on the overall high-temperature behavior of the material system is lacking. Such studies exist for composites in which the reinforcement size is in the several-micron range and, in these studies, both the reinforcement size and volume fraction significantly affect the fracture toughness and strength [20]. A fundamental understanding of the effect of nanosized SiC reinforcements on the mechanical behavior of Si_3N_4 matrix composites is required before further attempting to improve the properties of these composites by varied morphological alterations in experiments.

Tomar [21, 22] reported on the effect of morphological variations in second-phase SiC particle placement and GB strength on the dynamic fracture strength of $SiC–Si_3N_4$ nanocomposites using continuum analyses based on a mesoscale cohesive finite element method (CFEM). It was found that high-strength and relatively small sized SiC particles act as stress concentration sites in the Si_3N_4 matrix, leading to intergranular Si_3N_4 matrix cracking as a dominant failure mode. However, as a result of a significant number of nanosized SiC particles being present in microsized Si_3N_4 matrix, the SiC particles invariantly fall in wake regions of microcracks, leading to significant structural strength. This mechanism was further examined using 3D molecular dynamics (MD) simulations of crack propagation in $SiC–Si_3N_4$ nanocomposites with cylindrical SiC inclusions.

In the case of $SiC–Si_3N_4$ nanocomposites, MD analyses have also revealed that the second-phase particles act as significant stress raisers in the case of single crystalline Si_3N_4 phase matrix, affecting the strength significantly. However, the particles' presence does not have a significant effect on the mechanical strength of bicrystalline or nanocrystalline Si3N4 phase matrices. The strength of the $SiC–Si_3N_4$ nanocomposite structures showed an uncharacteristic correlation between GB thickness and temperature. The strength showed a decrease with increase in temperature for structures having thick GBs having diffusion of C, N, or Si atoms. However, for structures with no appreciable GB thickness (no diffusion of C, N, or Si atoms), due to particle clustering and increase in $SiC–Si_3N_4$ interfacial strength with temperature, the strength improved with an increase in temperature. Figure 5.6 shows snapshots of fracture propagation analyses in such nanocomposites obtained using the CFEM. Current research work focuses on obtaining experimental images of the ceramic nanocomposites developed by collaborators, developing nanoscale CFEM meshes on such images, and performing failure analyses using the combination of MD and CFEM techniques.

As noted earlier, high-strength and relatively small sized SiC particles act as stress concentration sites in the matrix, leading to intergranular Si_3N_4 matrix cracking as a dominant failure mode. CFEM analyses also revealed that the SiC nanosized particles invariantly fall in wake regions of microcracks, leading to significant mechanical strength. This finding was confirmed in the MD analyses that revealed that particle clustering along the

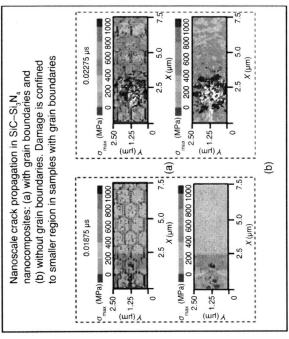

Nanoscale crack propagation in SiC–Si$_3$N$_4$ nanocomposites: (a) with grain boundaries and (b) without grain boundaries. Damage is confined to smaller region in samples with grain boundaries

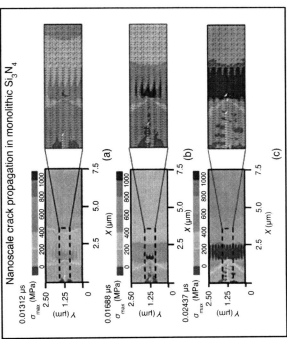

5.6 (a–c) Snapshots of mesoscale crack propagation and damage propagation in SiC–Si$_3$N$_4$ nanocomposites.

GBs significantly increases the strength of these nanocomposites. While some nanocomposite morphologies have sharply defined SiC–Si_3N_4 interfaces [e.g. 23], other nanocomposite morphologies have diffusion of C, N, or Si atoms at the interfaces [e.g. 24].

Classical MD replaces a comprehensive quantum mechanical treatment of interatomic forces with a phenomenological description in the form of an interatomic potential. MD has been used recently to achieve nm/cycle fatigue crack extension rate similar to that observed in experiments [e.g. 25]. The MD simulation results of deformation twinning in two-dimensional nanocrystalline Al with grain sizes from 30 nm to 90 nm by Yamakov and co-workers [26] have been found to be in close agreement with experimental observations reported by Liao et al. [27]. MD simulations have proven to provide phenomenological trends on deformation mechanisms of nanocrystalline materials in agreement with experiments [e.g. 28, 29]. Atomistic analyses of the nanocomposite mechanical strength as a function of phase morphology are relatively new and have focused on a very limited set of issues [e.g. 30–34]. Both SiC and Si_3N_4 have been individually analyzed in atomistic simulations for different mechanical strength related issues [e.g. 35–38]. However, SiC–Si_3N_4 nanocomposite morphologies are analyzed in this work for the first time.

5.4   The cohesive finite element method (CFEM)

Molecular dynamics (MD) simulations are performed using a well-established nanocomposite molecular dynamics simulation framework [32, 39]. The CFEM analyses are performed using the analyses criterion developed and reported in [40]. The CFEM is based on continuum mechanical foundations. In the CFEM meshes, each phase is modeled with hyperelastic constitutive behavior based on available experimental evidence [41–44]. In the absence of experimental information, the distribution of the crystalline orientation of the Si_3N_4 and SiC phases is neglected. In the MD morphologies, crystalline orientations are explicitly considered.

The complete CFEM framework has been described earlier [21, 22], so it is only briefly described here. A Lagrangian finite deformation formulation is used to account for the finite strains and rotations in crack tip regions. The CFEM simulations are carried out under plane strain assumption. Although the discussion in the presented research focuses on tensile loading, compression and contact can also be dealt with within this framework [48]. An irreversible bilinear cohesive law is used [21, 22]. Fracture energy per unit cohesive surface area is the same as the fracture energy of the material, Φ_0 (Table 5.1). The damage in cohesive surfaces is tracked through a parameter Φ_d, which is a function of the extent of the separation of cohesive surfaces. $\Phi_d = \Phi_0$ when surfaces have separated to cause fracture. Φ_d is

Table 5.1 Bilinear cohesive law parameters for different phases in microstructures (phase volume fractions $(V_f)_{SiC} = 0.2$ $(V_f)_{Si3N4} = 0.7$ $(V_f)_{GB} = 0.1$)

Component	Φ_0 (N/m)	T^{max} (GPa)	Δ (Nm)	E (GPa)	N	P (kg/m^3)
SiC (SC)	19.53	1.02	38.3	449	0.16	3215
Si$_3$N$_4$ (SN)	191.5	2.3	166.5	210	0.22	2770
Grain boundary (G)	238.7	2.38	200.6	200	0.16	4000
SC-G	19.53	1.02	38.3	—	—	—
SC-SN	19.53	1.02	38.3	—	—	—
SN-G	191.5	2.3	166.5	—	—	—
Homogenized (H)	127.8	2.03	125.9	256.8	0.202	2982
H-SC	19.53	1.02	38.3	—	—	—
H-SN	127.8	2.03	125.9	—	—	—
H-GB	127.8	2.03	125.9	—	—	—

partly converted into the surface energy and partly spent on causing damage in the material adjacent to crack surfaces through microcrack formation that is not based on a pre-specified criterion. A unique damage parameter can be defined to phenomenologicaly track the progressive softening of cohesive surfaces interspersed throughout the composite microstructure. This parameter D is defined such that:

$$D = \frac{\Phi_d}{\Phi_0} \qquad [5.2]$$

Note that $0 \le D \le 1$, with $D = 0$ indicating fully recoverable interfacial separation and $D = 1$ signifying complete separation or total fracture. In the numerical analysis carried out by Tomar and co-workers [46, 47], and in the presented research, D is used as a state variable quantifying the degree of the damage, providing a phenomenological measure for failure analysis.

5.4.1 Experiment-based calculation of the bilinear cohesive law parameters

Overall, five parameters are needed to specify the cohesive behavior, including the maximum tensile strength T_{max}, the critical separations Δ_{nc} and Δ_{lc}, characteristic separation η_0 and α. Note that only four of these parameters are independent since $\alpha = \Delta_{nc}/\Delta_{lc}$. Calibration of these parameters is an important aspect in the implementation of the CFEM model. T_{max} is commonly assumed to be a fraction of the Young's modulus (1/100 to 1/1000) [48]. The critical separations Δ_{nc} and Δ_{lc} are usually obtained by equating the area under the cohesive relation to the formation energy per unit area of the corresponding fracture surface. In this regard, experimental efforts have been reported [49, 50]. The value of α is typically obtained from the ratio between the tangential and normal energy release rates [54].

In this study, the approach for parameter selection as described by Xu and Needleman [48] is used. The value of characteristic separation η_0 is taken as 0.001 [50]. GBs in the nanocomposites have a glassy structure consisting of densification aids such as Y_2O_3 and other rare earth oxides such as samarium, gadolinium, dysprosium, erbium, and ytterbium (verified using the TEM observations on the Si_3N_4 phase [52]). Experimental data on the fracture properties and strength of GBs are not available. The GBs' chemical composition is a complex and uncharacterized combination of different compounds such as Y_2O_3, MgO, etc. with glassy structure. Accordingly, the fracture properties for GBs cannot be specified based on chemical composition. However, experiments for polycrystalline Si_3N_4 have shown that the presence of GBs results in lowering of mechanical strength owing to GB sliding (attributed to glassy structure) and an increase in fracture strength owing to the crack-deflection effect (roughly of the order of 5%) [53]. Accordingly, the GBs are arbitrarily assigned 5% higher fracture strength than that of the Si_3N_4 matrix. An increase in fracture strength for structural and glassy materials is often accompanied with a reduction in elastic modulus. Accordingly, while arbitrary assigning 5% higher fracture toughness to GBs than Si_3N_4 phase, the elastic modulus is made smaller than that of the Si_3N_4 phase by 5%. The immediate effect is that the simulations are qualitative in nature. With the availability of experimental measurements on GB properties, more realistic properties of GBs can be incorporated to increase the accuracy of the simulation predictions.

Because the GBs have finite widths, there are three phases (GBs, SiC, and Si3N4) in the microstructures analyzed. The cohesive parameters are calculated using experimental information on the elastic moduli and Φ_0 [1, 41–44, 54–57]. Homogenized properties are calculated using volume weighted averaging. The values for Φ_0 are obtained based on surface energy release rate measurements during fracture experiments on bulk SiC and Si_3N_4 reported in the literature. An interface between any two of the three phases is assigned the cohesive properties corresponding to the weaker phase. Table 5.1 shows the material properties for analyzing microstructures shown in Fig. 5.6.

5.4.2 CFEM problem setup

Microstructures analyzed using CFEM are shown in Fig. 5.6. Since a given unique set of phase morphology defining parameters (such as the location of SiC particles in the current research) corresponds to multiple sets of phase morphologies, three different random samples corresponding to each microstructural representation shown in Fig. 5.6 are used to characterize the material behavior. The microstructures have Si_3N_4 grain size approximately equal to 1.2 μm and SiC particle size approximately equal to 200 nm. The volume fraction of the SiC phase is fixed at 20%. The average GB width is approximately 120 nm. Accordingly, the microstructures with GBs have the approximate GB volume fraction of 10%.

To track complex crack/microcrack patterns, arbitrary crack paths and crack branching, cohesive surfaces are specified along all finite element boundaries as an intrinsic part of the finite element model. All cohesive surfaces serve as potential crack paths in the microstructure; therefore, fracture inside each microstructural phase and along interphase boundaries can be explicitly resolved (for more details see [40], [46] and [47]). Accordingly, the analyses are able to take into account the intergranular as well as intragranular fracture.

The finite element meshes used have a uniform structure with 'cross-triangle' elements of equal dimensions arranged in a quadrilateral pattern (Fig. 5.7). This type of triangulation is used since it gives the maximum flexibility for resolving crack extensions and arbitrary fracture patterns [40]. Because of the computational limitations and the requirement that stress wave reflections do not interfere with the analyses results [40], the microstructures are embedded in a uniform finite element mesh (see the mesh surrounding the microstructure in Fig. 5.7). The uniform mesh has elements with higher size increasing the overall size of the sample to delay the stress wave reflection and minimize its effect on dynamic fracture while simultaneously reducing the computational load. Since the crack propagation is limited to the microstructural window whose size had been analyzed in a previous research [40], the results are unaffected by the presence of the uniform mesh.

The dimensions for the microstructural region (7.5 μm × 30 μm) are limited by the memory sizes of the supercomputers used (a 48 processor Opteron Linux cluster in this work). These regions are much larger than the length scales involved in all microstructures. Thus, reasonable representations of the microstructures are achieved. Material outside the micro-structure window is assumed to be homogeneous and assigned effective properties representative of those for the SiC–Si_3N_4 ceramic composite. Computations are carried out for side-cracked samples under tensile loading. The length of the initial crack is $a_i = 9.0$ μm. Tensile loading is applied by imposing velocity boundary conditions along the upper and the lower edges of the specimen in the direction shown in Fig. 5.7. The boundary velocity V_0 (0.5 m/sec and 2 m/sec) is imposed on the bottom and top edges with a linear ramp from zero to V_0 in the initial phase of loading. This represents the loading of the pre-crack by a tensile wave with a stress amplitude of $\sigma = \rho\, C_L\, V_0$ (ρ is homogeneous material density and C_L is the

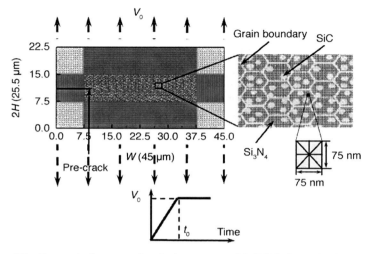

5.7 Dynamic fracture simulation setup with FEM discretization for performing the CFEM simulations.

homogeneous material longitudinal wave velocity) (14.3 MPa and 57.3 MPa at $V_0 = 0.5$ m/sec and 2 m/sec, respectively) and a linear ramp from zero to that value in the initial phase of loading. The specimen is stress free and at rest initially. Conditions of plain strain are assumed to prevail.

For the CFEM mesh and the bilinear cohesive law used, convergence analyses of the dependence of mesh element size on simulation results have been performed [40]. Based on those analyses, we require the characteristic finite element size, h, to satisfy: 750 nm $> h \geq 30$ nm. The upper limit is based on the minimum cohesive zone size estimate based on the properties listed in Table 5.1 [40]. The lower limit is based on the elimination of material softening because of the use of bilinear law with finite initial stiffness [40]. The characteristic element size of 75 nm satisfies the convergence criterion. The characteristic size corresponds to four finite elements (Fig. 5.7). Accordingly, there are 542 000 elements in each analyzed finite element mesh.

5.5   Molecular dynamics (MD) modeling

The MD simulations described here were carried out under 3D conditions (using periodic boundary conditions (PBCs)). The focus of the analyses is on understanding the deformation mechanisms in the composites and on delineating the factors affecting their strength. The microstructures analyzed are shown in Fig. 5.8. Six different atomistic morphologies were analyzed: M1 is a 10 nm sized cubic single crystal Si_3N_4 block; M2 a 10 nm sized cubic bi-crystalline Si_3N_4 block; M3 is a microstructure based on M1 with a 3 nm

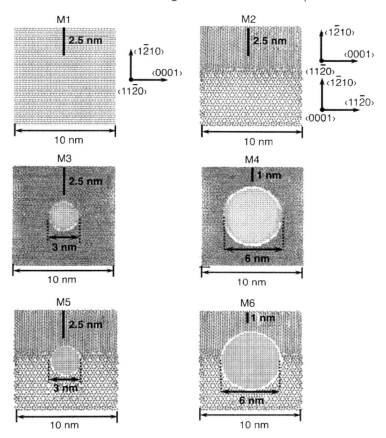

5.8 Atomistic microstructures analyzed for crack propagation.

sized cylindrical inclusion; M4 is based on M1 with 6 nm sized cylindrical inclusion; M5 is based on M2 with 3 nm sized cylindrical inclusion; M6 is based on M2 with 6 nm sized cylindrical inclusion. Because of the imposition of PBCs, microstructures M3 to M6 represent composites with second-phase SiC inclusions arranged periodically. Figure 5.8 also shows crack lengths in the analyzed nanocomposites.

The structures shown in Fig. 5.8 are three-dimensional. The structures were formed by placing SiC cylindrical particles in Si_3N_4 blocks with holes. Setting up of the $SiC-Si_3N_4$ interface in the structures can critically affect the internal stresses. In order to relieve any artificial internal stresses, first Si_3N_4 blocks with holes were prepared and equilibrated without the presence of the corresponding SiC cylindrical particles. SiC particles that are to be inserted in the Si_3N_4 holes were separately equilibrated as clusters. After equilibration, the Si_3N_4 block and the corresponding SiC particle were put together to form a SiC–Si$_3$N$_4$ composite. Then the composite structure was again equilibrated. A gap of 2 Å was ensured between the SiC particles and the corresponding Si$_3$N$_4$ block's hole in order to prevent additional buildup of internal stresses. However, changing the gap did not influence the observed trends and results.

The composites were analyzed for crack propagation as well as for mechanical deformation without an initial crack. The mechanical deformation methodology is based on earlier work by the authors [32, 39]. The initial crack in the nanocomposites was defined by switching of interatomic interactions along the plane shown in Fig. 5.8. After that, mechanical deformation was applied and crack propagation visually followed and analyzed. This methodology for analyzing crack propagation is based on earlier published work [28, 58]. Important elements of the MD framework are now described.

5.5.1 Interatomic potential for SiC–Si$_3$N$_4$ material system

Classical MD simulations of Si$_3$N$_4$ + SiC material systems require an interatomic potential to describe Si–Si, Si–N, Si–C, N–N, C–C, and N–C interactions. The potential should be fitted to the properties of Si$_3$N$_4$, SiC and to an approximation to the interfacial transitions between these components. The silicon nitride family consists of two members, α and β. The higher symmetry β phase has a hexagonal lattice (space groups C_{6h}^2, $N176$) with a primitive cell containing two Si$_3$N$_4$ formula units ($a = 7.606$ Å, $c = 2.909$ Å). The lower symmetry α phase is trigonal (space groups C_{3v}^4, $N159$) and has a primitive cell nearly twice as large ($a = 7.746$ Å, $c = 5.619$ Å) with twice as many atoms. Both structures consist of a SiN$_4$ tetrahedron forming a 3D network with each N corner common to three tetrahedra. Each NSi3 polyhedra has a pyramid-like form in the α phase, while the β phase can be regarded as being built of planar N2Si3 and nearly planar N1Si3 triangles with each Si corner common to four triangles that are oriented perpendicular to the direction (001) or along it. The interatomic potentials for the Si$_3$N$_4$ material system have been developed elsewhere [37, 59–61]. Fang *et al.* [62] carried out investigations of phonon spectrum and thermal properties in cubic-Si$_3$N$_4$.

Of all the above approaches, the potential of Ching *et al.* [61] was chosen to model β-Si$_3$N$_4$ because of its simplicity and ability to enable large-scale MD simulations. β-Si$_3$N$_4$ was chosen because of its relative abundance in comparison to the α form. Different polytypes of SiC exist at ambient pressure, which are differentiated by the stacking sequence of the tetrahedrally bonded Si–C bilayers [63]. Among these polytypes, β-SiC (cubic-SiC) is of much interest for its electronic properties. In particular, in recent years many theoretical and experimental studies have been carried out to investigate the different properties and possible applications of this material.

The majority of interatomic potentials developed for SiC material system focus on describing the material properties of β-SiC [e.g. 64–68]. We choose Tersoff's bond order potential to model interatomic interactions in SiC. At the interface of SiC and Si_3N_4, we need to be able to describe Si–C, Si–Si, C–N, and Si–N interactions. Tersoff's potential is useful only for describing the bulk Si–Si, C–C, and Si–C interactions [69]. Accordingly, a pair potential form similar to that used for Si_3N_4 is sought to be used for interfacial interactions at the SiC–Si_3N_4 interfaces. We use the of potentials Marian et al. [70] to describe Si–N, Si–Si interactions, the potential of Vincent and Merz [71] to describe C–N interactions, and the potential of Jian et al. [72] to describe Si–C interactions at the SiC–Si_3N_4 interfaces.

5.5.2 High-performance computing and mechanical deformation algorithm for MD simulations

MD simulations are performed using a modified version of a scalable parallel code, DL_POLY 2.14 [73, 74]. Electrostatic calculations using the code can be carried out for charged as well as neutral systems using well-established algorithms [e.g. 75–77]. The code has been modified and tested on a system of 1 000 000 atoms for a model ceramic matrix composite (Al + Fe_2O_3) material system and is benchmarked for scalable high-performance classical MD simulations for large atomic ensembles with millions of atoms [73, 78].

The simulations primarily focus on obtaining virial stress versus strain relations and visual atomistic deformation information in order to delineate the deformation mechanisms. In previous uniaxial quasistatic deformation analyses of nanocrystalline Cu by Schiøtz et al. [79], strain was calculated by recording the changes in positions of individual atoms. The average virial stress was calculated at every step in order to obtain the stress–strain relations. A modified version of this approach is used here. An alternative method to obtain uniaxial stress–strain curves is to record strain–time curves at several values of applied stress and then deduce the stress–strain relations [80]. Spearot et al. [81] used both methods and found that the modified method [82] works better because it controls the applied strain and closely emulates controlled displacement experiments. The modifications to the method of Schiøtz et al. [79] include the use of a combination of the algorithms for NPT and NVT ensembles. Alternating steps of stretching and equilibration at constant temperature are carried out to approximate uniaxial quasistatic deformation. Initially, the system is equilibrated at a specified temperature (300 K). During equilibration, NVT equations of motion are used to relax the pressure on structure in all three directions. During stretching, the MD computational cell is stretched in the loading direction using a modified version of the NPT equations of motion [83]. NPT equations ensure that the structure has lateral pressure relaxed to atmospheric values during deformation [32, 39]. In this algorithm, the rate of change of a simulation cell volume, $V(t)$, is specified using a barostat friction coefficient parameter η such that:

$$\frac{d\eta(t)}{dt} = \frac{1}{Nk_B T_{ext} \tau_P^2} V(t)(P - P_{ext}) \tag{5.3}$$

and

$$\frac{dV(t)}{dt} = 3\eta(t)V(t) \tag{5.4}$$

where P is the instantaneous pressure, P_{ext} is the externally applied pressure, N is total number of atoms in the system, k_B is the Boltzmann constant, T_{ext} is the external temperature, and τ_P is a specified time constant for pressure fluctuations. For a given cross-sectional area, the specification of η in equation 5.4 is equivalent to specifying strain rate for the change in simulation cell length. Further, for a given η, equation 5.3 can be modified as:

$$\frac{d\eta(t)}{dt} = \frac{1}{Nk_B T_{ext} \tau_P^2} V(t)(P - P_{ext}) - \gamma\eta \tag{5.5}$$

In equation 5.5, the term $\gamma\eta$ acts as a damping coefficient for reducing fluctuations in pressure during the stretching of the simulation cell. During the simulations, the system is initially equilibrated at $T_{ext} = 300$ K. After equilibration, the computational cell is stretched in the loading direction using $\eta = 0.1$ psec^{-1}. The values of $\gamma = 0.5$ and $P_{ext} = 1$ atmospheric pressure are used.

The values for η and γ are calculated in trial calculations that focused on achieving the best balance between simulation time (low η results in long simulation times and vice versa) and pressure fluctuations (high γ results in excessive pressure damping with increase in residual stresses along periodic boundaries). In the analyses reported in this chapter, the MD equilibration time in between the periods of stretching is chosen as 2.0 psec.

5.6   Dynamic fracture analyses

As pointed out earlier, simulations focus on comparing the CFEM analyses of the effect of the second-phase SiC particles with the MD analyses. MD results are analyzed using visuals as well as virial stress–strain relations.

Virial stress represents internal resistance offered by a material to an externally imposed load [84]. Since virial stress is an internal quantity, it is also an indicator of the increase in the stress concentration caused by the presence of an inhomogeny in an otherwise homogenous body. This characteristic is used to analyze the effect of the cracks in the presented analyses. The focus of the analyses is on developing an understanding of the correlations between particle size, particle placement with respect to a bicrystalline interface and the strength against crack propagation.

Figure 5.9 shows the stress field in a monolithic Si_3N_4 sample of the same size as that of the CFEM meshes shown in Fig. 5.7 as the crack propagates. As shown, the crack tip stress field is resolved using the CFEM mesh used in the research. As shown in the inset, the fine size of the mesh does not allow extreme distortion of the finite elements. In order to understand the stress distribution and damage evolution, the damage parameter (D (equation 5.2)) and the maximum principal stress (σ_{max}) distributions are analyzed in all microstructures. As explained earlier, $0 \leq D \leq 1$, with $D = 0$ indicating fully recoverable interfacial separation and $D = 1$ signifying complete separation or fracture. The results for damage distribution in all three types

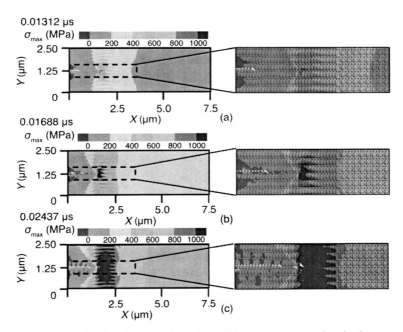

5.9 Stress distribution as a function of damage progression in the case of bulk Si_3N_4 at the loading rate of 2 m/sec at time (a) 0.01312 μsec, (b) 0.01688 μsec, and (c) 0.02437 μsec. The close-up inset in each plot shows how the crack tip stress field is resolved and elements distorted as crack progresses.

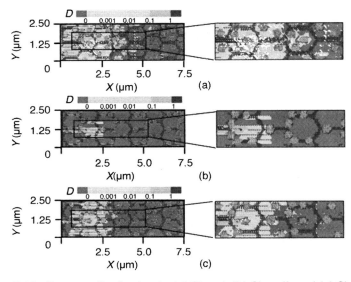

5.10 Damage distribution in (a) Class I, (b) Class II, and (c) Class III microstructures at the loading rate of 2 m/sec and time $t = 0.01875\,\mu$sec. Close-up insets offer higher resolution view of the damage region. Broken particles have distorted 'crumbled' appearance. The intact particles have near circular appearance. Corresponding fragmented GBs show up as zigzag lines in the diffused damage region. These lines are outlines of the fragmented GB elements.

of microstructures at time $t = 0.01875\,\mu$sec (Fig. 5.10) at $V_0 = 2$ m/sec are now discussed.

A concentrated stress field similar to homogeneous materials near the propagating crack tip is seen in all three microstructures. The distribution of stress around the crack tip is strongly affected by the distribution of second-phase SiC particles with respect to Si_3N_4 GBs. The presence of SiC particles inside Si_3N_4 grain interior in the case of Class I and Class III microstructures results in stress concentration over and in excess of the crack tip stresses that are seen in the case of Class II microstructure where the SiC particles are placed solely along GBs. In the case of the Class II microstructure, near crack tip stresses are concentrated in a very small region. This is contrary to the case of the Class I microstructure where the near crack tip stress fields are spread over a larger microstructural region. Owing to the SiC particle stress concentration, intergranular microcracks originating in the microstructure can also be seen. Stress distribution in the Class III microstructure lies in between that of the Class I and Class II microstructures. The effect of the difference in the fracture and bulk properties of GBs and Si_3N_4 matrix on stress distribution is insignificant in comparison to the effect of the second-phase particle placement. GBs are

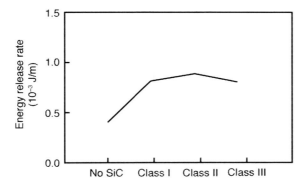

5.11 Mean energy release rate values as a function of microstructure at loading velocity of 2 m/sec. No SiC is a control microstructure with no SiC particles.

shown to restrict the damage to grain interior. However, SiC particles primarily determine how soon the structure develops microcracks.

Self-reinforced Si_3N_4 is an important example of the high-toughness structural ceramics first manufactured during 1970s. Embedding nanosized SiC particles in a Si_3N_4 matrix is one proven approach to improve its mechanical properties. It is also verified here by comparing the average energy release rate of all the microstructures with that of a control microstructure that contains no SiC particles at all (Fig. 5.11). As shown, a significant increase in the fracture resistance is observed. For the first time, Niihara [1] reported almost 200% improvements in both the fracture toughness and the strength of $SiC–Si_3N_4$ nanocomposites by embedding nanometer size (20–300 nm) particles within a matrix of larger grains and at the Si_3N_4 GBs. Since then, a number of experiments focusing on manufacturing $SiC–Si_3N_4$ nanocomposites have been reported [2, 53]. So far it has been very difficult to control the processing routes and conditions to place SiC particles selectively along GBs. However, it is possible to place second-phase particles in a combination that places them along GBs as well as in grain interiors near GBs, which may ultimately result in very high strength $SiC–Si_3N_4$ nanocomposites [14]. Such mechanisms have been observed for other SiC particle reinforced nanocomposites [85, 86].

Figure 5.12 shows stress–displacement plots for atomistic microstructures M1 to M6 in order to compare the internal stresses when an initial crack is present in all microstructures with the internal stresses when the initial crack is absent in all the microstructures. Figures 5.13 and 5.14 show visuals of crack propagation as a function of total displacement for microstructures M1, M2, M4, and M6. The results from Figs 5.12 to 5.14 are now collectively discussed.

As shown in Fig. 5.12(a), microstructures M1 and M2 have the same

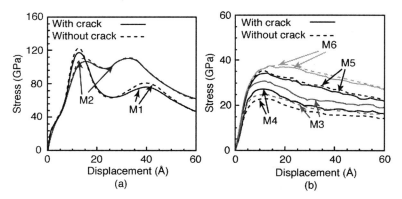

5.12 Virial stress versus displacement plots for (a) microstructures M1 and M2 and (b) microstructures M3, M4, M5, and M6.

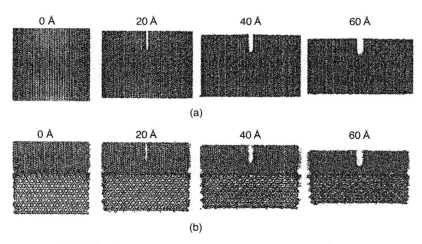

5.13 Crack propagation as a function of displacement for microstructure (a) M1 and (b) M2.

stress–displacement profile with or without initial cracks. Microstructure M2 has an interface, as shown in Fig. 5.8, which leads to lower peak stress values. However, overall, the area under the stress–strain curve in the case of M2 is higher. This indicates that the presence of an interface leads to an increase in toughness for Si_3N_4. The presence of an initial crack does not have any significant effect on the stress–strain profile for both M1 and M2. This may be attributed to the long-range Coulumbic interactions that are not affected by the switching off of interatomic interactions in the crack plane.

As noted earlier, virial stress is an internal quantity signifying the resistance of atomistic system to external load or pressure. In this case,

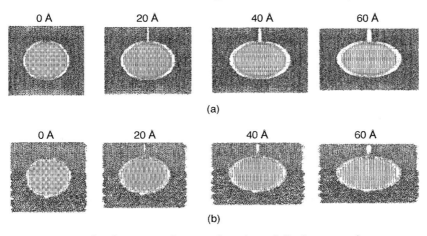

5.14 Crack propagation as a function of displacement for microstructure (a) M4 and (b) M6.

owing to the presence of Coulombic interactions, the presence of nanosized small cracks does not result in appreciable stress concentration. That leads to the stress profiles for a system with and without crack being almost the same. A secondary peak can also be observed in all stress–strain curves in Fig. 5.12(a). This secondary peak may correspond to an internal reorganization in all the microstructures. However, atomistic visualization did not reveal any such changes in the structural features. This is also attributable to the complex hexagonal structure of Si_3N_4, which is extremely difficult to analyze visually. Overall, an important finding from the plots in Fig. 5.12(a) is that the presence of interfaces results in an increase in the overall toughness of the material.

In Fig. 5.12(b), stress–strain plots for M3, M4, M5, and M6 are shown. A direct comparison with the plots in Fig. 5.12(a) reveals that the presence of second-phase SiC particles results in significant weakening of M1 and M2. The effect of the initial crack is also found to be significant. This is apparent by looking at the differences between the curves with initial crack and the ones without initial cracks. It has been pointed out [21] that high strength and relatively small sized SiC particles act as stress concentration sites in Si_3N_4 matrix, leading to intergranular Si_3N_4 matrix cracking as a dominant failure mode. However, as a result of a significant number of nanosized SiC particles being present in the microsized Si_3N_4 matrix, the SiC particles invariantly fall in wake regions of microcracks, which results in significant structural strength. The observations here confirm the first point – that the presence of SiC particles will significantly weaken M1 and M2 against deformation and fracture. A comparison of deformation without initial crack and with initial crack for microstructures M3 and M4 reveals that, in the absence of an interface in Si_3N_4, the presence of an initial crack will lead to higher internal stresses. The curves also show that internal stresses will decrease with an increase in particle size and the corresponding reduction in the interparticle spacing (owing to the imposed PBCs). The difference between the stresses with and without initial cracks can be explained by a significant stress concentration caused due to the presence of second-phase particles and their interactions with the propagating initial crack. This stress concentration reduces with an increase in particle size, leading to lower internal stresses overall.

In the case of microstructures M5 and M6 with SiC inclusion now placed along a Si_3N_4 interface, a reversal in trend regarding the effects of particle size and initial crack is observed. As was observed earlier in the case of M1 and M2, the presence of a crack does not result in an appreciable stress concentration in M5 and M6. A combined effect of the change in particle size and the presence of an interface is now that the internal stresses rise with an increase in particle size and a reduction in particle spacing. This result is in direct contrast to earlier observations of a reduction in internal stresses with increase in particle size for microstructures M3 and M4. The difference is now the presence of the interface. The interface, while imparting higher toughness for a given particle size, is also a site for structural discontinuity. The presence of second-phase particles with a reduction in spacing between the particles will lead to a stronger material when particles are present along the interface. The trend is reversed when particles are present in the grain interior.

Overall, the combination of observations from Figs 5.12, 5.13, and 5.14 reveals that the effect of interfaces is to strengthen the material by raising toughness. The presence of second-phase particles will lead to a reduction in strength. The presence of nanosized cracks in an ionic material does not offer significant stress concentration until aided by the presence of second-phase inclusions. The toughening of the nanocomposite in the presence of a combination of interfaces and second-phase particles is strongly dependent upon whether the second-phase particles are placed along interfaces or in grain interiors. In addition, the strengthening is also dependent upon the size and relative spacing between particles, irrespective of whether the particles are in grain interiors or at the interfaces. The trend, however, is different for interfaces and grain interiors.

5.7   Conclusions

This chapter has CFEM and MD based analyses to understand the effect of second-phase SiC particles on the strength of SiC–Si_3N_4 nanocomposites. CFEM analyses can explicitly account for the effect of nanoscale inclusions on the strength of nanocomposites. However, the CFEM cannot account for the effect of nanoscale size changes and nanoscale position changes on the strength of nanocomposites. MD analyses can be very useful to reveal this information, along with information on the effect of interfaces and the mechanism of deformation for certain specific orientations. Overall, CFEM analyses demonstrated that, irrespective of the location of second-phase particles, the final failure mode in all microstructures is brittle fragmentation with initial microcracks forming and propagating mainly in the Si_3N_4 matrix. Later parts of brittle fragmentation include GB and SiC fragmentation. Second-phase particles have two important effects. If they are present in front of the crack tip they weaken the microstructure because of the stress concentration caused by them. However, if they are present near GBs, they cause a crack bridging effect with an increase in strength of the microstructure.

MD analyses also reveal that the second-phase particles act as significant stress raisers, resulting in strength reduction of single-crystal and bi-crystalline Si_3N_4 blocks by a factor of almost two times. The stress concentration increases almost 1.5 times with doubling the size of SiC particles. With smaller SiC particles, the interfacial boundary in the bi-crystalline Si_3N_4 block acts as a stress reliever. However, with an increase in the size of SiC particles and with a decrease in the spacing between adjacent SiC particles, the presence of an interfacial boundary results in significant internal stress rise. This indicates that the placement of SiC particles along interfacial boundaries will not always lead to strengthening of a SiC–Si_3N_4 nanocomposite. Overall, MD analyses confirm the CFEM findings concerning the effect of second-phase SiC particles on SiC–Si_3N_4 nanocomposite strength. In addition, the analyses also indicate that the strengthening of a nanocomposite by placing second-phase particles along grain boundaries is only possible for a selective few second-phase particle sizes with interparticle spacing being another important factor.

5.8   References

1. Niihara, K., New design concept for structural ceramics – Ceramic nanocomposites. *J. Ceram. Soc. Jpn: Centennial memorial issue*, 1991, 99(10): 974–982.
2. Weimer, A.W. and Bordia, R.K., Processing and properties of nanophase SiC/Si3N4 composites. *Composites Part B*, 1999, 30: 647–655.
3. Gano, S.E., Agarwal, H., Renaud, J.E., and Tovar, A., Reliability based design using variable fidelity optimization. *Struct. Infrastruct. Engg*, 2006, 2(3–4): 247–260.
4. Gano, S.E., Renaud, J.E., and Sanders, B., Variable fidelity optimization using a kriging based scaling function. *10th AIAA/ISSMO Multidisciplinary Analysis and Optimization Conference*, Albany, New York, 2004.
5. Mejia-Rodriguez, G., Renaud, J.E., and Tomar V., A variable fidelity model

management framework for designing multiphase materials. *ASME J. Mech. Des.*, 2007, 130: 091702-1–13.

6. Mejia-Rodriguez, G., Renaud, J.E., and Tomar V., A methodology for multiscale computational design of continuous fiber SiC–Si$_3$N$_4$ ceramic composites based on the variable fidelity model management framework. *3rd AIAA Multidisciplinary Design Optimization Specialist Conference*, Honolulu, Hawaii, 2007.

7. Besson, J.L., Mayne, M., Bahloul-Hourlier, D., and Goursat, P., Nanocomposites Si$_3$N$_4$–SiCN: Influence of SiC nanocomposites on the creep behavior. *Key Engg. Mater.*, 1997, 132–136: 1970–1973.

8. Buljan, S.T., Baldoni, J.C., and Huckabee, M.L., Si$_3$N$_4$–SiC composites. *Am. Ceram. Soc. Bull.*, 1987, 66(2): 347–352.

9. Chedda, M.S., Flinn, B.D., Leckie, R., and Bordia, R.K., Effect of sub-micron sized reinforcements on the high temperature behavior of Si$_3$N$_4$ composites. *Ceram. Trans.*, 2000, 103: 223–236.

10. Lange, F.F., Effect of microstructure on strength of Si$_3$N$_4$–SiC composite system. *J. Am. Ceram. Soc.*, 1973, 56(9): 445–450.

11. Niihara, K., Izaki, K., and Kawakami, T., Hot-Pressed Si$_3$N$_4$–32%SiC nanocomposites from amorphous Si–C–N powder with improved strength above 1200°C. *J. Mater. Sci. Lett.*, 1990, 10: 112–114.

12. Rendtel, A., Hubner, H., Herrman, M., and Schubert, C., Silicon nitride/silicon carbide nanocomposite materials: II, hot strength, creep, and oxidation resistance. *J. Am. Ceram. Soc.*, 1998, 81(5): 1109–1120.

13. Sternitzke, M., Review: Structural ceramic nanocomposites. *J. Eur. Ceram. Soc.*, 1997, 17: 1061–1082.

14. Wan, J., Duan, R.-G., Gasch, M.J., and Mukherjee, A.K., Highly creep-resistant silicon nitride/silicon carbide nano-nano composites. *J. Am. Ceram. Soc.*, 2006, 89(1): 274–280.

15. Bill, J. and Aldinger, F., Precursor-derived covalent ceramics. *Adv. Mater.*, 1995, 7: 775–787.

16. Kroke, E., Li, Y.-L., Konetschny, C., Lecomte, E., Fasel, C., and Riedel, R., Silazane derived ceramics and related materials. *Mat. Sci. Engg.: R: Reports*, 2000, 26(4–6): 197–199.

17. Riedel, R., Seher, M., and Becker, G., Sintering of amorphous polymer-derived Si, N and C containing composite powders. *J. Eur. Ceram. Soc.*, 1989, 5: 113–122.

18. Riedel, R., Streker, K., and Petzow, G., In-situ polysilane-derived silicon carbide particulates dispersed in silicon nitride composite. *J. Am. Ceram. Soc.*, 1989, 72(11): 2071–2077.

19. Wan, J., Gasch, M.J., and Mukherjee, A.K., Silicon carbonitride ceramics produced by pyrolysis of polymer ceramic precursors. *J. Mater. Res.*, 2000, 15: 1657–1660.

20. Choi, H.-J., Cho, K.-S., and Lee, J.-G., R-curve behavior of silicon-nitride-titanium nitride composites. *J. Am. Ceram. Soc.*, 1997, 10(10): 2681–2684.

21. Tomar, V., Analyses of the role of the second phase SiC particles in microstructure dependent fracture resistance variation of SiC–Si$_3$N$_4$ nanocomposites. *Model. Sim. Mater. Sci. Eng.*, 2008, 16: 035001.

22. Tomar, V., Analyses of the role of grain boundaries in mesoscale dynamic

fracture resistance of SiC–Si$_3$N$_4$ intergranular nanocomposites. *Eng. Fract. Mech.*, 2008, 75: 4501–4512.

23. Bill, J., Kamphowe, T.W., Mueller, A., Wichmann, T., Zern, A., Jalowieki, A., Mayer, J., Weinmann, M., Schuhmacher, J., Mueller, K., Peng, J., Seifert, H.J., and Aldinger, F., Precursor-derived Si–(B–)C–N ceramics: thermolysis, amorphus state, and crystallization. *Appl. Organometal. Chem.*, 2001, 2001 (15): 777–793.

24. Jalowiecki, A., Bill, J., Aldinger, F., and Mayer, J., Interface characterization of nanosized B-doped Si$_3$N$_4$/SiC ceramics. *Composites Part A*, 1996, 27A: 721.

25. Farkas, D., Willemann, M., and Hyde, B., Atomistic mechanisms of fatigue in nanocrystalline metals. *Phys. Rev. Lett.*, 2005, 94: 165502.

26. Yamakov, V., Wolf, D., Phillpot, S.R., and Gleiter, H., Deformation twinning in nanocrystalline Al by moleculardynamics simulation. *Acta Mater.*, 2002, 50: 5005–5020.

27. Liao, X.Z., Zhou, F., Lavernia, E.J., He, D.W., and Zhua, Y.T., Deformation twins in nanocrystalline Al. *Appl. Phys. Lett.*, 2003, 83(24): 5062–5064.

28. Abraham, F.F., How fast can cracks move? A research adventure in materials failure using millions of atoms and big computers. *Adv. Phys.*, 2003, 52(8): 727–790.

29. Kadau, K., Germann, T.C., Lomdahl, P.S., and Holian, B.L., Microscopic view of structural phase transitions induced by shock waves. *Science*, 2002, 296: 1681.

30. Dionald, W.A.R.D., Curtin, W.A., and Yue, Q., Mechanical behavior of aluminium–silicon nanocomposites: A molecular dynamics study. *Acta Mater.*, 2006, 54(17): 4441–4451.

31. Song, M. and Chen, L., Molecular dynamics simulation of the fracture in polymer-exfoliated layered silicate nanocomposites. *Macromol. Theory Sim.*, 2006, 15(3): 238–245.

32. Tomar, V. and Zhou, M., Analyses of tensile deformation of nanocrystalline α-Fe2O3 + fcc-Al composites using classical molecular dynamics. *J. Mech. Phys. Solids*, 2007, 55: 1053–1085.

33. Zeng, Q.H., Yu, A.B., and Lu, G.Q., Molecular dynamics simulations of organoclays and polymer nanocomposites. *Int. J. Nanotechnol.*, 2008, 5(2–3): 277–290.

34. Zeng, Q.H., Yu, A.B., Lu, G.Q., and Standish, R.K., Molecular dynamics simulation of organic-inorganic nanocomposites: Layering behavior and interlayer structure of organoclays. *Chem. Mater.*, 2003, 15: 4732–4738.

35. Tsuruta, K., Totsuji, H., and Totsuji, C., Neck formation processes of nanocrystalline silicon carbide: A tight-binding molecular dynamics study. *Phil. Mag. Lett.*, 2001, 81(5): 357.

36. Tsuruta, K., Totsuji, H., and Totsuji, C., Parallel tight-binding molecular dynamics for high-temperature neck formation processes of nanocrystalline silicon carbide. *Mater. Trans.*, 2001, 42(11): 2261.

37. Mirgorodsky, A.P., Baraton, M.I., and Quintard, P., Lattice dynamics and prediction of pressure-induced incommensurate instability of a β-Si$_3$N$_4$ lattice with a simple mechanical model. *Phys. Rev. B*, 1993, 48(18): 13326–13332.

38. Lidorikis, E., Bachlechner, M.E., Kalia, R.K., Nakano, A., Vashishta, P., and Voyiadjis, G.Z., Coupling length scales for multiscale atomistic-continuum

simulations: Atomistically induced stress distributions in Si/Si_3N_4 nanopixels. *Phys. Rev. Lett.*, 2001, 87(8): 086104.

39. Tomar, V. and Zhou, M., Tension-compression strength asymmetry of nanocrystalline α-Fe_2O_3 + fcc-Al ceramic–metal composites. *Appl. Phys. Lett.*, 2006, 88: 233107 (1–3).

40. Tomar, V., Zhai, J., and Zhou, M., Bounds for element size in a variable stiffness cohesive finite element model. *Int. J. Num. Meth. Engg*, 2004, 61: 1894–1920.

41. Klopp, R.W. and Shockey, D.A., The strength behavior of granulated silicon carbide at high strain rates and confining pressure. *J. Appl. Phys.*, 1991, 70(12): 7318–7326.

42. Holmquist, T.J. and Johnson, G.R., Response of silicon carbide to high velocity impact. *J. Appl. Phys.*, 2002, 91(9): 5858–5866.

43. Walker, J., Analytically modeling hypervelocity penetration of thick ceramic targets. *Int. J. Impact Engg.*, 2003, 29(1–10): 747–755.

44. Loubens, A., Rivero, C., Boivin, P., Charlet, B., Fortunier, R., and Thomas, O., Investigation of local stress fields: Finite element modeling and high-resolution X-ray diffraction. *MRS Proc.*, 2005, 875: 0.83, doi: 10, 1577/PROC-875-08.3.

45. Minnaar, K., *Experimental and numerical analysis of damage in laminate composites under low velocity impact loading.* 2002. PhD Thesis, Georgia Institute of Technology, Atlanta, GA.

46. Tomar, V. and Zhou, M., Deterministic and stochastic analyses of dynamic fracture in two-phase ceramic microstructures with random material properties. *Eng. Fract. Mech.*, 2005, 72: 1920–1941.

47. Zhai, J., Tomar, V., and Zhou, M., Micromechanical modeling of dynamic fracture using the cohesive finite element method. *J. Engg Mater. Tech.*, 2004, 126: 179–191.

48. Xu, X.P. and Needleman, A., Numerical simulations of fast crack growth in brittle solids. *J. Mech. Phys. Solids*, 1994, 42: 1397–1434.

49. Sorensen, B.F. and Jacobsen, T.K., Determination of cohesive laws by the J integral approach. *Engg Frac. Mech.*, 2003, 70: 1841–1858.

50. Cornec, A., Scheider, I., and Schwalbe, K.-H., On the practical application of the cohesive zone model. *Engg Frac. Mech.*, 2003, 70: 1963–1987.

51. Espinosa, H.D., Dwivedi, S., and Lu, H.-C., Modeling impact induced delamination of woven fiber reinforced composites with contact/cohesive laws. *Comp. Meth. Appl. Mech. Engg*, 2000, 183: 259–290.

52. Niihara, K., Suganuma, K., Nakahira, A., and Izaki, K., Interfaces in Si_3N_4–SiC nanocomposites. *J. Mater. Sci. Lett.*, 1990, 9: 598–599.

53. Ajayan, P.M., Schadler, L.S., and Braun, P.V., *Nanocomposite Science and Technology*, 2003: Wiley-VCH.

54. Schwetz, K.A., Kempf, T., Saldsleder, D., and Telle, R., Toughness and hardness of LPS-SiC and LPS-SiC based composites. *Ceram. Engg Sci. Proc.*, 2004, 25(3): 579–588.

55. Messier, D.R. and Croft, W.J., *Silicon Nitride*, 1982: Army Research Laboratory, AMMRC TR 82-42.

56. Liu, X.-J., Huang, Z.-Y., Pu, X.-P., Subn, X.-W., and Huang, L.-P., Influence of planetary high-energy ball milling on microstructure and mechanical

properties of silicon nitride ceramics. *J. Am. Ceram. Soc.*, 2005, 88(5): 1323–1326.

57. Blugan, G., Hadad, Y.M., Janczak-Rusch, J., Kuebler, J., and Graulez, T., Fractography, mechanical properties, and microstructure of commercial silicon nitride–titanium nitride composites. *J. Am. Ceram. Soc.*, 2005, 88(4): 926–933.

58. Latapie, A. and Farkas, D., Molecular dynamics investigation of the fracture behavior of nanocrystalline α-Fe. *Phys. Rev. B*, 2004, 69: 134110–8.

59. Wendel, J.A. and Goddard, W.A., The Hessian biased force field for silicon nitride ceramics: Predictions of thermodynamic and mechanical properties for α- and β-Si_3N_4. *J. Chem. Phys.*, 1992, 97(7): 5048–5062.

60. Mota, F.d.B., Justo, J.F., and Fazzio, A., Hydrogen role on the properties of amorphous silicon nitride. *J. Appl. Phys.*, 1999, 86(4): 1843–1847.

61. Ching, W.-Y., Xu, Y.-N., Gale, J.D., and Ruehle, M., Ab-initio total energy calculation of α- and β-silicon nitride and the derivation of effective pair potentials with application to lattice dynamics. *J. Am. Ceram. Soc.*, 1998, 81 (12): 3189–3196.

62. Fang, C.M., Wijs, G.A.d., Hintzen, H.T., and With, G.d., Phonon spectrum and thermal properties of cubic Si_3N_4 from first-principles calculations. *J. Appl. Phys.*, 2003, 93(9): 5175–5180.

63. Morkoc, H., Strite, S., Gao, G.B., Lin, M.E., Sverdlov, B., and Burns, M., Large-band-gap SIC, Ill-V nitride, and II-VI ZnSe-based semiconductor device technologies. *J. Appl. Phys.*, 1994, 76(3): 1363–1398.

64. Tersoff, J., Empirical interatomic potential for silicon with improved elastic properties. *Phys. Rev. B*, 1988, 38: 9902–9905.

65. Tersoff, J., Modeling solid-state chemistry: Interatomic potentials for multicomponent systems. *Phys. Rev. B (Rapid Comm.)*, 1989, 39(8): 5566–5568.

66. Tersoff, J., Carbon defects and defect reactions in silicon. *Phys. Rev. Lett.*, 1990, 64: 1757–1760.

67. Huang, H., Ghoniem, N.M., Wong, J.K., and Baskes, M.I., Molecular dynamics determination of defect energetics in β-SiC using three representative empirical potentials. *Model. Sim. Mater. Sci. Engg*, 1995, 1995 (3): 615–627.

68. Noreyan, A., Amar, J.G., and Marinescu, I., Molecular dynamics simulations of nanoindentation of β-SiC with diamond indentor. *Mater. Sci. Engg B*, 2005, 117: 235–240.

69. Ma, Y. and Garofalini, S.H., Application of the Wolf damped Coulomb method to simulations of SiC. *J. Chem. Phys.*, 2005, 122: 094508(1–5).

70. Marian, C.M., Gastreich, M., and Gale, J.D., Empirical two-body potential for solid silicon nitride, boron nitride, and borosilazane modifications. *Phys. Rev. B*, 2000, 62(5): 3117–3124.

71. Vincent, J. and Merz, K.M., A highly portable parallel implementation of AMBER using the Message Massing Interface standard. *J. Comp. Chem.*, 1995, 11: 1420–1427.

72. Jian, W., Kaiming, Z., and Xide, X., Pair potentials for C–C, Si–Si and Si–C from inversion of the cohesive energy. *J. Phys.: Cond. Matter*, 1994, 6: 989–996.

73. Tomar, V., Atomistic modeling of the Al + Fe_2O_3 material system using classical molecular dynamics. In *Mechanical Engineering*, 2005: Georgia Institute of Technology, Alanta. P. 295.

74. Smith, W., Yong, C.W., and Rodger, P.M., DL_POLY: Application to molecular simulation. *Mol. Sim.*, 2002, 28(5): 385–471.
75. Ding, H.-Q., Karasawa, N., and Goddard III, W.A., Atomic level simulations on a million particles: The cell multipole method for Coulomb and London nonbond interactions. *J. Chem. Phys.*, 1992, 97(6): 4309–4315.
76. Wolf, D., Reconstruction of NaCl surfaces from a dipolar solution to the Madelung problem. *Phys. Rev. Lett.*, 1992, 68(22): 3315–3318.
77. Darden, T.A., York, D.M., and Pedersen, L.G., Particle mesh Ewald. An *N*.log (*N*) method for Ewald sums in large systems. *J. Chem. Phys.*, 1993, 98: 10089–10092.
78. Tomar, V. and Zhou, M., Classical molecular-dynamics potential for the mechanical strength of nanocrystalline composite fcc-Al + α-Fe$_2$O$_3$. *Phys. Rev. B*, 2006, 73: 174116 (1–16).
79. Schiøtz, J., Di Tolla, F.D., and Jacobsen, K.W., Softening of nanocrystalline metals at very small grain sizes. *Nature*, 1998, 391: 561–563.
80. Van Swygenhoven, H. and Caro, A., Plastic behavior of nanophase Ni: A molecular dynamics computer simulation. *Appl. Phys. Lett.*, 1997, 71(12): 1652–1654.
81. Spearot, D.E., Jacob, K.I., and McDowell, D.L., Nucleation of dislocations from [001] bicrystal interfaces in aluminum. *Acta Mater.*, 2005, 53: 3579–3589.
82. Schiøtz, J., Vegge, T., Di Tolla, F.D., and Jacobsen, K.W., Atomic-scale simulations of the mechanical deformation of nanocrystalline metals. *Phys. Rev. B*, 1999, 60: 11971–11983.
83. Melchionna, S., Ciccotti, G., and Holian, B.L., Hoover NPT dynamics for systems varying in shape and size. *Mol. Phys.*, 1993, 78(3): 533–544.
84. Zhou, M., A new look at the atomic level virial stress. On continuum-molecular system equivalence. *Proc. Royal Soc. London A*, 2003, 459: 2347–2392.
85. Ohji, T., Jeong, Y.-K., Choa, Y.-H., and Niihara, K., Strengthening and toughening mechanisms of ceramic nanocomposites. *J. Am. Ceram. Soc.*, 1998, 81(6): 1453–1460.
86. Perez-Regueiro, J., Pastor, J.Y., Llorca, J., Elices, M., Miranzo, P., and Moya, J.S., Revisiting the mechanical behavior of aluminum/silicon carbide nanocomposites. *Acta Mater.*, 1998, 46(15): 5399–5411.

Part II

Types

6

Ceramic nanoparticles in metal matrix composites[*]

F . HE, Purdue University, USA

DOI: 10.1533/9780857093493.2.185

Abstract: This chapter introduces research conducted on metal matrix
nanocomposites (MMNCs). The chapter reviews the material system
used for MMNCs and explains the principles to choose the
reinforcement for metal matrix composites. The mechanical properties of
MMNCs fabricated by different processes are summarized and models
used to predict and describe the effect of different strengthening
mechanisms on MMNCs are introduced. Different approaches used to
produce MMNCs are reviewed and categorized.

Key words: metal matrix nanocomposites (MMNCs), mechanical
properties, mechanical alloying, ultrasonic aided casting.

6.1   Introduction

Metal matrix composites (MMCs) have been studied since the early 1960s
and were developed to improve the mechanical properties of metallic
materials. A typical MMC consists of a ductile metal matrix and rigid
ceramic reinforcement. It retains its metallic properties (high ductility,
machinability, and toughness) while adding ceramic properties (high
strength, thermal stability, and modulus) (Varadan *et al.*, 2010). MMCs
are used as structural material in the aerospace, sport and automotive
industries because of the material's high strength to weight ratio. The

---

[*] Part of Section 6.5 was adapted from He, F., Han, Q., Jackson, M. J., 2008. Nanoparticulate
reinforced metal matrix nanocomposites – a review. *International Journal of Nanoparticles.* 1(4).
pp. 301–309. Used with permission from Inderscience. Inderscience retains copyright of the
original material.

category MMC encompasses a wide range of scales and microstructures (Davis and Ward, 1993).

Currently, MMCs can be classified into three categories:

- particle reinforced MMCs
- short fibre or whisker reinforced MMCs
- continuous fibre or sheet reinforced MMCs.

Particle reinforced MMCs are of the most interest since they are isotropic, easier to manufacture, and often lower cost. The properties of particle reinforced MMCs are greatly affected by particle size and distribution. Research on particle reinforced MMCs has shown that tensile strength and ductility decrease with increasing reinforcement size. Nanosized particles are predicted to enhance the properties of MMCs more than their microsized counterparts, creating metal matrix nanocomposites (MMNCs) (Ajayan *et al.*, 2003). Nanoparticle and carbon nanotube (CNT) reinforced MMNCs are the most commonly researched types.

MMNC research did not attract as much interest as ceramic and polymer matrix nanocomposites until recently, mostly due to the inferior mechanical properties reported so far. Despite some improvements in hardness and wear resistance (Kim *et al.*, 2007; Laha *et al.*, 2009), the overall properties did not achieve predicted values due to the existence of defects, such as porosity and agglomeration of nanoparticles. However, the potential development of a novel nanocomposite with excellent mechanical and unique physical properties has generated an increasing amount of research interest in recent years.

MMNCs have shown numerous different characteristics compared to composites with microsize reinforcement. When the dimension of a material approaches the nanoscale, new properties emerge due to size confinement, quantum phenomena and Coulomb blockage. Varadan *et al.* (2010) summarized the special properties of nanomaterials, including their structural, thermal, chemical, mechanical, magnetic, optical, electronic and biological properties. When it comes to MMNCs, a couple of characteristics should be noted.

- When the reinforcement scales down to nanosize, due to the agglomeration of nanoparticles, the fabrication of MMNCs is much more difficult than the production of MMCs.
- The reinforcement mechanism is different in MMNCs and MMCs. Adding microsize reinforcement has successfully improved some mechanical properties of a metal matrix. The strengthening mechanism is more due to the hard reinforcement to stop the movement of dislocation. Due to the small reinforcement size, MMNCs are reinforced

through different mechanisms, which will be explained later in the chapter.

- When ceramic particles approach nanosize, their thermal properties change due to the increase in surface energy and different interatomic spacing.
- When the reinforcing phase scales down to nanometersize, it leads to difficulties in material testing technology. TEM instead of SEM is commonly used for particles observation. Motta *et al.* (2004) illustrated different methods to observe the distribution of nanoparticles in Cu/Al_2O_3 nanocomposites.

However, even with all these difficulties, MMNCs have shown other promising characteristics that still attract researchers in this field. For example, the ductility of the metal matrix is expected to remain the same while other properties of MMNCs would be enhanced considerably (e.g. improved strength and creep properties, better elevated temperature performance, and better machinability (Cao *et al.*, 2008a)). Also low-volume nanoparticles can help keep the thermal and electrical conductivity of the MMC whilst improving the mechanical properties. This character is essential for functional composites such as Cu/Al_2O_3, used as electronic packaging material. He *et al.* (2008) successfully produced nanodiamond (ND)/copper nanocomposite, which maintained the electrical conductivity of copper whilst increasing the mechanical properties of the metal matrix. Furthermore, nanoparticles in a metal matrix will affect the solidification and heat treatment process, which will have a strong effect on micro-structure and the mechanical properties of the MMNC. A good example is that, with the existence of nano SiC particles, the Mg matrix grain size is refined, which will lead to balanced mechanical properties, as reported by Cao *et al.* (2008a). The assumption is that nanoparticle could act as a nucleate during the solidification process. Also, the presence of nanopar-ticles can suppress the softening processes (i.e. dislocation annihilations) and pinned grain boundaries, resulting in a more equiaxed grain structure in the composite. The dispersion of Al_2O_3 particles into a copper matrix produces Cu/Al_2O_3 nanocomposites with excellent resistance to high-temperature annealing while maintaining high thermal and electrical conductivities (Motta *et al.*, 2004).

The most commonly used metals in MMNC research are aluminum, copper, magnesium, titanium, and iron. The reinforcing constituent is normally ceramic, including borides, carbides, nitrides, oxides, and their mixtures. Mechanical, chemical, and physical properties of both matrix and reinforcement should be considered when selecting the material system (Ibrahim *et al.*, 1991). For *in-situ* production processes, the thermal stability of different ceramic particles is the main factor that determines the reinforcement used in the system. For *ex-situ* manufacturing processes, a successful material system depends on good wettability between the ceramic and metal area and the surface bonding between the two materials. Different metal matrix and reinforcement systems used in research are summarized in this chapter.

Experiments have been conducted to fabricate, observe reinforcement/ matrix interfaces, and improve the mechanical and physical properties of MMNCs. Despite the many publications in this field, limited achievement has been made to date to improve matrix properties without the attenuation of some other properties. This chapter, summarizes these achievements and limitations.

The majority of current research on MMNCs focuses on process development. The MMNC fabrication process can be categorized into *in-situ* and *ex-situ*. In *in-situ* MMCs, the reinforcements are formed by chemical reactions between elements or between elements and compounds during the manufacturing process. The advantages of *in-situ* methods are: the formation of reinforcements that are thermodynamically stable; the reinforcing particles surfaces are clean, resulting in strong interfacial bonding; finer particle size; a more uniform distribution in the matrix (Shehata *et al.*, 2009). For *ex-situ* manufacturing, the advantage is potentially being able to produce bulk material. Due to the poor wettability between a metal and ceramic reinforcement, and also because nanoparticles tend to agglomerate due to their electrostatic and van der Waals forces in the liquid mixing process, a limited amount of MMNCs have reached their predicted mechanical properties.

Based on all the recent research on MMCs, critical issues that need to be solved to successfully fabricate MMNCs are as follows.

- *Dispersion.* The distribution of reinforcing particulates has to be uniform to achieve predicted mechanical properties. However, this is extremely difficult because nanoparticles tend to agglomerate. Particle cluster will be extremely harmful for MMNCs.
- *Reactivity.* The research shows that some nanoparticle reactivity decreases as size increases (Ramakrishna and Ghosh, 2003). The effect of particle size on the reactivity of nanoparticles may lead to interesting chemical reactions between the interface of the metal matrix and ceramic reinforcement for *in-situ* and *ex-situ* MMNC manufacturing processes.
- *Thermal stability.* When particles scale down to nanosize, the thermal stability of ceramic particles can be affected (Park and Lee, 2009).
- *Wettability.* Wettability between ceramic particles and metal is poor. Copper coating and acid treatment have been used to improve the wettability between CNTs and metals (Neubauer *et al.*, 2010). A cost-

effective method is needed to improve the wettability between ceramic reinforcement and metals.

• *Cost.* In order to get a uniform dispersion of nanoparticles, a large amount of energy input is needed. The cost of manufacturing MMNCs needs to be reduced for further application in industry.

• *Bulk material for structure material usage.* Due to the high energy density input needed, manufacturing bulk MMNC for structure material fabrication is limited.

6.2   Material selection

Compatibility between the reinforcement and the matrix is the most important factor to build a successful MMC system. Matrices used are metals including copper, magnesium and aluminum. All of these metals have also been commonly used for microsize particulate reinforced MMC. Most of the reinforcements used are low-dimensional materials, ceramic particles and CNTs.

6.2.1   Reinforcement material

Ibrahim *et al.* (1991) summarized the selection criteria for ceramic reinforcement, which include:

• elastic modulus
• tensile strength
• density
• melting temperature
• thermal stability
• coefficient of thermal expansion
• size and shape
• compatibility with matrix material
• cost
• availability in the market.

With the matrix material determined, the above factors should be evaluated to choose a suitable ceramic reinforcement.

Of all the possible candidates, CNTs are the most popular reinforcement used in MMNCs due to their rapidly developing fabrication technology. CNT/ceramic and CNT/polymer nanocomposites have successfully enhanced the mechanical properties of the matrix. CNT/metal systems have made less improvement due to inferior mechanical properties, mostly caused by the defects formed in bulk metals during the fabrication process. However, the unique mechanical and physical properties of CNTs still make them the most promising reinforcements for creating metal matrix structural

Table 6.1 Properties of reinforcement

Reinforcement	Elastic modulus (GPa)	Knoop hardness	Density (10^3 kg/m^3)	Expansivity (10^{-6}/°C)
Al_2O_3	400–537	2100	3.96	7.92
AIN	310	N/A	3.26	4.84
B_4C	448	N/A	2.52	6.08
CNT	910–1260	N/A	~1.2	1
SiC	410–700	2480	3.2	4.5
Si_3N_4	1900	~2500	3.2	2.9
TiB_2	648	~1800	4.5	8.0

Source: Kelly, 1985; Ibrahim *et al.*, 1991; Salvetat *et al.*, 1999.

composites with unique characters (Laha *et al.*, 2009). To develop a continuous interface and homogenous distribution of CNTs in a metal matrix, pre-treatment of the CNT is needed. Acid treatment, copper coating and other different surface modification processes have been successfully used to pre-treat CNTs (Daoush *et al.*, 2009).

Other reported MMNCs are ceramic particle/metal nanocomposites. The ceramic particles used are carbides (e.g. SiC, B_4C), nitrides (e.g. Si_3N_4, AlN), oxides (e.g. Al_2O_3, SiO_2), and boride (e.g. TiB_2), similar to those used for microsized reinforcement in MMCs (Ibrahim *et al.*, 1991). TiB_2 is well known for its high rigidity and hardness, high mechanical strength and good electrical and thermal conductibility. The intermetallic nanometer-sized phase also acts as reinforcement (e.g. Al_3Ti in an Al matrix; Hsu *et al.*, 2006). Selected properties of commonly used reinforcements are listed in Table 6.1.

6.2.2 Matrix selection

The list of matrix materials is extensive. Most of the metal matrices used for MMNCs are light metals used as structural materials, such as aluminum, titanium and magnesium, but copper and nickel-based superalloys have also been considered for specific applications. With the addition of nanosize particulates, the mechanical properties of metal matrices are expected to improve.

6.2.3 Matrix–ceramic interface

The matrix–ceramic interface determines load transfer and distribution of the MMNCs during deformation and therefore affects elastic modulus and strength. Promoting wettability, controlling chemical interactions, and minimizing oxide formation are necessary to maximize interfacial bond strength for *in-situ* processes, as the reinforcements are produced from chemical reaction. The only issue for the *in-situ* process is to control the chemical reaction by controlling process parameters. For the *ex-situ* process, especially for casting of MMNCs, improving the interface by improving the poor wettability between ceramic particulates and liquid metal is a determinant factor for the success of the process. Ibrahim *et al.* (1991) summarized three factors that can be improved to increase wettability:

- increasing the surface energy of the solid
- decreasing the solid–liquid interfacial energy
- decreasing the surface tension of the liquid metal.

Cao *et al.* (2008a) reported that applying ultrasonic vibration to liquid magnesium can produce a chemically bonded interface between Mg alloy and SiC. Lan *et al.* (2004) theorized that ultrasonic vibration breaks the gas and silica layer and makes a clean SiC surface to accelerate the wetting of SiC and aluminum. Hashim *et al.* (2001a, 2001b) summarized the methods used to improve the wettability between SiC particles and molten aluminum alloy. These methods include:

- the addition of alloying elements to the matrix
- coating the ceramic particles
- pre-treating the ceramic particles.

The wetting behavior of CNTs and liquid metals has been studied in detail. It is reported that surface tension, with a cut-off limit between 100 and 200 mN/m, is the determining factor for wetting. In this case, aluminum (surface tension of 865 mN/m), copper (1270 mN/m), and iron (1700 mN/m) would not be easily wetted on the surface of multi-walled carbon nanotubes (MWNTs) (Daoush *et al.*, 2009).

The metal–ceramic interface is affected by the process used to fabricate MMNCs. Nanocomposite particles can be fabricated with mechanical alloying and sintered later using other methods. This process will be introduced later in the chapter. Other chemical methods are also used to get a better metal–ceramic interface. Shehata *et al.* (2009) used *in-situ* technology to produce Al_2O_3/Cu nanocomposite powders. He *et al.* (2008) discovered a chemical method to co-deposit copper and nanodiamond. Daoush *et al.* (2009) reported electroless deposition to form a copper coating on CNTs and successfully fabricated CNT/Cu nanocomposite powder. The different chemical methods listed in Table 6.2 can produce thermal, stable and clean metal–ceramic interfaces.

Table 6.2 Methods used to make stable ceramic/matrix interfaces

Methods used to prepare particles	Reinforcement/ matrix	Forming method	Reference
Disintegrated melt deposition (DMD)	CNT/ZK60A	DMD + hot extrusion	Paramsothy et al. (2011b)
	Si_3N_4/ZK60A	DMD + hot extrusion	Paramsothy et al. (2011a)
	Al_2O_3/Mg	DMD + hot extrusion	Hassan et al. (2008)
	Y_2O_3/Mg	DMD + hot extrusion	Hassan (2011)
N/A	SiC/Mg	Ultrasonic-aided casting	Cao et al. (2008a)
	SiC/Mg	Ultrasonic-aided casting	Nie et al. (2011)
Electroless copper deposition	CNT/Cu	Electroless copper deposition + spark plasma sintering	Daoush et al. (2009)
Co-deposition	Nanodiamond/Cu	Co-deposition + hot pressing	He et al. (2008)
In-situ	Al_2O_3/Cu	*In-situ* produced particle + cold pressing	Shehata et al. (2009)
	TiB_2/Cu	*In-situ* from Cu–Ti alloy, Ti, B_2O_3 and carbon powder	Tu et al. (2002)
Mechanical alloying	Al_2O_3/Mg	Mechanical alloyed particle + cold compacted + hybrid microwave sintering + hot extrusion	Prasad et al. (2009)
	Al_2O_3/$A_3$56	Mechanical milling + friction stir processing	Mazaheri et al. (2011)
	Carbon black/Al	BP-ECAP (back pressure, equal-channel angular pressing)	Goussous et al. (2009)
	Si_3N_4/Al	Mechanical mixing + hot press + hot extrusion	Ma et al. (1996)
	CNT/Al	Ball mill + plasma spray forming	Laha et al. (2009)

6.3   Physical and mechanical properties of metal matrix nanocomposites (MMNCs)

6.3.1   Properties of MMNCs with different fabrication methods

Adding microsize ceramic particles to a metal matrix has proven to successfully improve strength, stiffness, and creep resistance. However, due to the large size of the particles, the ductility of metal matrix is always reduced. A uniform dispersion of nanoparticles provides a good balance between the strengtheners and interparticle spacing effects and is predicted to maximize the yield strength and still retain ductility. Therefore, MMNCs with nanoparticles acting as strengtheners have been the most promising approach to produce the desired mechanical properties in materials (i.e. enhanced hardness, Young's modulus, 0.2% yield strength and ultimate tensile strength (UTS)) (Tjong, 2007).

Ma *et al.* (1996), Cao *et al.* (2008a), Hassan *et al.* (2008) and Hassan (2011) reported that all the mechanical properties, including ductility, were improved by adding nanoparticles into the metal matrix. All the reported experiments show adding ceramic nanoparticles will improve elastic modulus, hardness, and UTS. However, the majority of the experiments indicated that by adding nanoparticles, ductility is reduced, mostly due to process defects.

Ma *et al.* (1996) compared the effect of nanosized and microsized reinforcement in their research. In the experiment, the 1 vol% Si_3N_4 (15 nm)/Al nanocomposite exhibited similar tensile strength but significantly higher yield strength than the 15 vol% SiCp(3.5 µm)/Al microcomposite. The creep resistance of the 1 vol% Si_3N_4 (15 nm)/Al nanocomposite was about two orders of magnitude higher than that of the 15 vol% SiCp (3.5 µm)/Al microcomposite. Cao *et al.* (2008a) reported that by adding SiC nanoparticles in Mg–4Al–1Si melt, the yield strength of the matrix was improved by 33%, UTS by 23%, and the ductility of the matrix was not reduced. In Cao's experiment, microsized SiC clusters could still be observed on the grain boundary without impairing the overall mechanical properties. It was explained that the negative effect caused by some SiC microclusters was compensated by the significantly positive effects of the grain refining and strengthening of the well-dispersed SiC nanoparticles. The effects of adding reinforcements to the mechanical properties of metal matrix are listed in Table 6.3.

Even with a limited number of encouraging results, most of the research shows that adding nanoparticles into the metal matrix could improve yield strength and hardness, but also lead to a negative impact on ductility. The results of different experiments are not constant and most tested properties

Table 6.3 Physical and mechanical properties of MMNCs compared with monolithic metal

Forming method	Properties of MMNCs compared to monolithic metal matrix	Reference
Disintegrated melt deposition + hot extrusion	UTS and work of fracture of MMNCs increased by 17% and 156% respectively Hardness decreased by 17%	Paramsothy *et al.* (2011a, 2011b)
Ball mill + hot pressing	Elastic modulus of MMNCs increased by 8.5%	Nouri *et al.* (2012)
Ball mill + plasma spray forming	Elastic modulus increased by 78%	Laha *et al.* (2009)
Electroless copper deposition + spark plasma sintering	Yield strength higher Young's modulus two times higher	Daoush *et al.* (2009)
In-situ produced particle + cold pressing	Improved compression strength, hardness and wear resistance	Fathy *et al.* (2011)
Disintegrated melt deposition + hot extrusion	Improved hardness, yield strength, UTS, elastic modulus, and ductility	Hassan *et al.* (2008)
Mechanical milling + hot pressing	Improved hardness and wear resistance	Hosseini *et al.* (2010)
Mechanical mixing + hot pressing + hot extrusion	1 vol.%Si_3N_4/Al has comparable tensile strength to that of the 15 vol.%SiCp/Al composite, and higher yield strength and ductility	Ma *et al.* (1996)
Ultrasonic-aided casting	Yield strength increased by 51%, UTS increased by 14%, ductility decreased by 18%	Cao *et al.* (2008a)
Ball milling + BP-ECAP (back pressure, equal-channel angular pressing)	Compressive yield strength increased from 50 MPa to 260 MPa with adding 5% C	Goussous *et al.* (2009)
Co-deposition + hot pressing	Microhardness increased from 86 to 89.9 HV. Electricity conductivity increased from 86.5% IACS to 134.6% IACS	He *et al.* (2008)

are inferior to predicted values. The major reason for this is that, although nanosized reinforcement is predicted to greatly improve the properties of MMNCs, controlling other process parameters can be very challenging, yet extremely important. For instance, the porosity ratio is an important factor that affects the properties of the final product. Also, the porosity ratio is affected by the consolidation approach. Hot isostatic pressing, hot extrusion, equal channel angular pressing, or laser sintering will make a difference to the density of bulk material, which eventually affects the mechanical properties of MMNCs. The dislocation is strongly influenced by the degree of deformation that occurs during consolidation. The grain size can also influence properties via the Hall–Petch relationship. The grain size can be refined by ultrasonic vibration and large plastic deformation. The role of clean interfaces, which will effectively permit load transfer, can also greatly affect the properties.

6.3.2 Strengthening mechanisms for particulate strength MMNCs

There are two strengthening mechanisms that are typically associated with conventional MMCs – direct and indirect strengthening. Direct strengthening is the result of load transfer from the matrix to the reinforcement. Indirect strengthening is caused by the effect of reinforcement on the matrix microstructure or deformation mode, such as finer grain size and dislocation strengthening induced by mismatch between the reinforcement and the matrix (Sanaty-Zadeh, 2011). The strengthening mechanisms of MMNCs are similar to traditional MMCs. However, due to the dimensional difference, some mechanisms have a more significant effect than others. The model proposed by Zhang and Chen (2006) was the first to predict the yield strength of particulate reinforced MMNCs. Zhang and Chen's model added the Orowan strengthening mechanism to the Ramakrishnan model, a MMCs reinforcing model. The model agrees with some published data. Sanaty-Zadeh (2011) quantitatively analyzed the Zhang and Chen model and the Clyne model, added Hall–Petch, elastic modulus mismatch, and work hardening into the model, and proposed a modified Clyne model.

Strengthening mechanisms

The different strengthening mechanisms of MMNCs analyzed in this section are the Hall–Petch relationship, Orowan strengthening, the Taylor relationship and the load-bearing effect.

Hall–Petch relationship
The Hall–Petch relationship is used to predict the effect of grain size on the strength of materials.

$$\sigma_y = \sigma_0 + \frac{k_y}{d^{1/2}} \tag{6.1}$$

where σ_y is the strength with grain refinement, σ_0 is the initial strength, k_y is a constant, and d is the grain size.

For nanocomposite materials, the grain size could be affected by process parameters, reinforcement size, and volume ratio. The most significant factor is the process used to produce the nanocomposite. For some processes, the influence of nanosize reinforcement has proven to be able to refine the grain size in some fabrication methods. The existence of nanoparticles can also lead to a smaller grain size during the post-treatment process. Li et al. (2009) pointed out that ceramic particles enclosed by a thin nanocrystalline layer produced an average grain size of 35 nm and this was estimated to be the dominant strengthening mechanism, followed by dislocation strengthening and then secondary-phase strengthening. Goussous et al. (2009) observed a nanoscaled grain size due to large deformation during the back pressure equal-channel angular pressing (BP-ECAP) process; nanograin was the main factor that contributed to the great increase in hardness. Cao et al. (2008a) used ultrasonic vibration to disperse nanoparticles into liquid metal and grain refining was observed. Choi et al. (2011) produced nanocomposite powder of different grain sizes by changing the milling time. It has also been shown that grain refinement occurs when a large amount of Y_2O_3 is added to the matrix. MMNCs should have a higher strength due to the grain refining effect of nanoparticles during solidification and also the heat treatment process.

Based on the analysis above, the Hall–Petch mechanism can be a major strengthening mechanism in routes where the grain refining effect is achieved either by adding nanoreinforcements or by altering the manufacturing process.

Orowan and dispersion strengthening

Orowan strengthening is used to express the strengthening effect of dimension and interparticulate spacing of secondary-phase dispersoids. Orowan strengthening caused by the resistance of closely spaced hard particles to the passing of dislocations is important for MMNCs, but is not a significant strengthening factor for MMCs. Zhang and Chen's model takes it into consideration as a major strengthening factor. The general equation is (Sanaty-Zadeh, 2011):

$$\sigma_{Or} = \frac{M0.4Gb}{\pi(1-v)^{1/2}} \frac{\ln(\bar{d}/b)}{\lambda} \qquad [6.2]$$

where $\bar{d} = (2/3)^{1/2}d$, M is the mean orientation factor, G is the shear modulus of the matrix (Pa), b is the Burgers vector (m), v is the Poisson ratio, d is the particle diameter, and λ is the interparticle spacing. For nanosize reinforcement, Orowan bowing, the Orowan dislocation bowing mechanism, and the Orowan loop mechanism are used to explain the effect of Orowan strengthening for MMNCs.

Taylor relationship
The Taylor relationship describes the contribution of dislocation density to the strength of the material. In the case of composite materials, especially ceramic reinforced metal composites, the dislocation caused by the mismatch between reinforcement and matrix in thermal and elastic moduli can be one of the most significant strengthening factors. The following equation expresses the strengthening caused by thermal mismatch:

$$\sigma_d = M\beta Gb(\rho^{CTE})^{1/2} \qquad [6.3]$$

where M is the Taylor factor, β is a constant, ρ^{CTE} is the density of dislocation caused by coefficient of thermal expansion (CTE mismatch), G is the shear modulus of the matrix, and b is the magnitude of the Burgers vector.

The elastic modulus mismatch between reinforcement and matrix can also be significant. For nanosize reinforcement, due to the large surface to volume ratio, the interfacial area between the reinforcement and the matrix increases significantly. In this case, dislocation strengthening caused by thermal and elastic modulus mismatch could be a significant strengthening mechanism.

Load-bearing effect
The load-bearing effect describes transferring of the load between matrix and particles. In the Zhang and Chen (2006) model the load-bearing transfer strengthening factor to yield strength is described as

$$\sigma_l = 0.5V_p\sigma_{ym} \qquad [6.4]$$

where V_p is the volume fraction of the particles and σ_{ym} is the yield strength of the matrix.

It is important that the load-bearing effect is based on a strong bond between matrix and reinforcement. Also, for most MMNCs, only a small amount of reinforcement, volume ratio from 0.2% to 10%, will be added due to agglomeration of nanoparticles; the load-bearing effect of MMNCs is small.

Current models

To successfully predict the mechanical properties of MMNCs, the strengthening mechanism should be decided upon based on the process. The most commonly used methods are the Zhang and Chen model and the Sanaty-Zadeh model.

Published by Woodhead Publishing Limited, 2013

Zhang and Chen model
This is a modified version of the shear-lag model and is mostly commonly
used in recent works (Zhang and Chen, 2008). However, in this model, only
dislocation density, the Orowan strengthening effect, and the load-bearing
improvement factor are considered:

$$\sigma_{yc} = \sigma_{ym}(1 + f_l)(1 + f_d)(1 + f_{Or}) \qquad [6.5]$$

where σ_{yc} is the yield strength of the MMNC, σ_{ym} is the yield strength of the
monolithic matrix with the same process, and f_l, f_d, f_{Or} are improvement
factors from the load-bearing effect, enhanced dislocation density by
thermal mismatch, and the Orowan strengthening effect.

Sanaty-Zadeh model
Sanaty-Zadeh (2011) proposed a modified Clyne method as follows:

$$\sigma_{yc} = \sigma_{ym} + [(\Delta\sigma_l)^2 + (\Delta\sigma_{Or})^2 + (\Delta\sigma_{H-P})^2 + (\Delta\sigma_{EM})^2 + (\Delta\sigma_{CTE})^2$$
$$+ (\Delta\sigma_{WH})^2]^{1/2} \qquad [6.6]$$

The Sanaty-Zadeh model takes into account the Hall–Petch strengthening
mechanism, which was not considered before. The Sanaty-Zadeh model is in
good agreement with experimental values.

6.4   Different manufacturing methods for MMNCs

Based on the sources of nanoreinforcement, MMNC fabrication methods
can be categorized as *ex-situ* and *in-situ*.

For the *ex-situ* route, nanoreinforcements are externally added to the
matrix. Due to the existence of electrostatic and van der Waals forces
between nanoreinforcements, especially nanoparticles, the added nanoreinforcements tend to agglomerate. Uniform dispersal of nanoreinforcement into a metal matrix is the major concern for *ex-situ* methods. Different
processes can be used to homogeneously disperse the reinforcement. These
ex-situ processes can be classified, based on the process temperature, as
liquid-phase processes, solid-state processes, and two-phase processes.

For *in-situ* routes, nanoreinforcements are produced by chemical
reactions within the materials during the fabrication process. *In-situ* routes
are used to produce both MMNC powders, which are consolidated to form
MMNC bulk material, and MMNCs bulk material directly. Reinforcements
created *in-situ* are usually fine and well distributed. However, *in-situ*
reinforcements have fewer options in terms of material selection than *ex-situ*
reinforcements because of the complex reactions involved in the *in-situ*
fabrication process.

Published by Woodhead Publishing Limited, 2013

6.4.1 *Ex-situ* routes

The major purpose of the different *ex-situ* processes is to disperse nanoscale reinforcement homogeneously into the metal matrix material at different process temperatures.

Liquid-phase MMNC ex-situ *fabrication processes*

In liquid-phase processes, ceramic nanoreinforcements are dispersed into a molten metallic matrix. Many stirring techniques have been used in the casting process to fabricate MMCs. However, when the size of the reinforcements reaches nanolevel, the reinforcements tend to agglomerate and are hard to separate using traditional stirring methods (Li *et al.*, 2004; Yang *et al.*, 2004). A larger energy input is then needed to separate clusters and homogenously disperse the reinforcements into the melt. Ultrasonic cavitation aided casting is a modified method of mechanical stir casting and has been proven to be a successful method to fabricate MMNCs. In ultrasonic cavitation based solidification processing, nanoreinforcements are placed on the surface of molten metal. Ultrasonic vibration is then applied to disperse particulates in the melt matrix. In the process, transient micro 'hot spots' with a temperature of about 5000°C and pressure above 1000 atm can be formed by ultrasonic cavitations. The strong heating and cooling rates during the formation process can break the nanoparticle clusters and also clean particle surfaces (Lan *et al.*, 2004; Li *et al.*, 2007). After applying ultrasonic vibration for a certain amount of time, the melt with nanoparticles is cast into a mould to make bulk MMNC samples.

The advantages of this technology are significant (Yang and Li, 2007; Cao *et al.*, 2008a).

• Cast MMNCs with complex shapes can be formed directly.
• It is one of the most cost-effective MMNC fabrication processes.
• Application of this technology into fabricating Al- and Mg-based nanocomposites has shown improved mechanical properties.

The challenges for casting MMNCs are obvious.

• Methods are needed to improve the wettability of liquid metal and solid ceramic particles.
• Even though this technology has proven effective, the effect of nanoparticles on the solidification of structures is unknown. Zhou and Xu (1997), in an experiment on casting SiC reinforced MMC, reported that SiC particles acted as substrates for heterogeneous nucleation of Si crystals. More research is needed to reveal the mechanism of the process

and to describe the effect of ultrasonic vibration on the movement of particles.

• Other casting defects such as porosity and clusters caused by nanoparticle agglomeration, can be significant. For example, gas trapped between nanoparticles will be released in ultrasonic vibration. Without an effective de-gassing process, this would be deleterious to the mechanical properties of bulk material (Cao *et al.*, 2008b).

Solid-phase MMNC ex-situ fabrication processes

In solid-phase MMNC *ex-situ* fabrication, the nanoreinforcement and metal matrix are mixed in a solid state. The basic process is to mix ceramic and metal particles to fabricate nanocomposite particles and then the particles are consolidated by different methods such as hot pressing, hot extrusion, cold pressing and sintering. Solid-phase fabrication processes always involve a number of steps prior to final consolidation. Goussous *et al.* (2009) also reported the use of back pressure equal-channel angular pressing to consolidate milled Al and carbon black powders.

Mechanical milling is widely used to produce well-mixed ceramic and metal powders. The process means that powders are pre-treated to their proper size. Treated powders are mixed in the appropriate proportion and loaded into a mill along with the grinding medium (generally steel balls). The mix is milled until a steady state is reached. In the mechanical fraction process, the ceramic particles are continually fractured until the size of the milled particles reaches a critical value (Zhang, 2004). Mechanical milling can produce a uniform dispersion of reinforcement particles in the matrix through a repeated process of cold welding, fracturing, and rewelding, until the reinforcement particles are well embedded into the matrix particles. The technology is also widely used in the reinforcement of oxide dispersion strengthened alloys (ODS) through uniform dispersion of very fine oxide particles in the metal matrix. The advantages of milling are simplicity, versatility, and economic viability (Suryanarayana *et al.*, 2001). Besides, in most milling processes, not only are the nanoparticles normally distributed but also the grain size is usually at nanoscale due to severe deformation. Results have shown that even after hot isostatic pressing, the grain size can still be kept between 100 and 400 nm (Ferkel and Mordike, 2001). Besides mechanical milling, Nouri *et al.* (2012) used a combined mechanical milling and solution method to produce well-mixed MWNTs and Al powders, which were consolidated to be MMNCs. The friction stir process is another reported successful method to produce Mg-based nanocomposites (Lee *et al.*, 2006) in which the nanoparticles were uniformly dispersed.

Hot press, hot extrusion, cold press, and sintering methods are widely used to consolidate mixed nanocomposite powders. Some modified processes improve the compaction process to produce high-performance MMNCs. A microwave-assisted sintering process has shown improvements to the mechanical properties of MMCs (Gupta and Wong, 2005). Tun and Gupta (2007) use a microwave-assisted rapid sintering technique coupled with hot extrusion to fabricate magnesium nanocomposites with nano-yttria as reinforcement. The results show that even though the microwave assisted the compaction process, the mechanical properties of the MMNCs did not improve significantly (Tun and Gupta, 2007).

Even though the solid-phase methods are the most frequently used to fabricate nanoparticulate-enhanced MMNCs, there are still some unsolved problems.

- Powder contamination results from the fracture and reaction process of powders during the milling process. Some measures have been taken to avoid contamination of particles, for example using coated milling balls, but the problem is still not thoroughly solved.
- Procedures are complex.
- There is a lack of scientific theory, with no model to explain the whole process (Sherif and Eskandarany, 1998)
- The process is harder to control due to the many process parameters involved.

Two-phase MMNC ex-situ *fabrication processes*

Two-phase processes mostly involve producing MMNC powders by a liquid method and then consolidating them in a solid phase. Paramsothy *et al.* (2011a, 2011b) reported a novel method of disintegrated melt deposition (DMD) + hot extrusion. In that research, mechanical stirring was used to mix CNT powder and superheated ZK60A melt. Two jets of argon were used to disintegrate the melt when the mixed melt was pouring from a small orifice. The disintegrated melt slurry was subsequently deposited on a substrate. Hot extrusion was then used to consolidate the deposited ingot. Hassan (2011) compared the mechanical properties of Y_2O_3/Mg nanocomposite fabricated by DMD and powder metallurgy. Metal matrix hardness, UTS, and ductility were all improved by adding Y_2O_3. DMD-processed material exhibited superior overall mechanical properties while material processed by powder metallurgy exhibited excellent ductility and work of fracture.

Thermal spray methods are used to form MMNCs. Laha *et al.* (2004) used plasma spray forming for an aluminum nanocomposite and the mechanical properties were tested. The nanocomposite had higher hardness and elastic modulus but lower ductility than the monolithic metal. Mazaheri *et al.* (2011) used high-velocity oxygen fuel spraying to deposit particles on a substrate and friction stir processing to consolidate the coating. The drawbacks for thermal spraying are that it produces a relative low density, a softened heat-affected zone, and is a difficult process for bulk material.

6.4.2 *In-situ* MMNC fabrication

In-situ techniques involve a chemical reaction resulting in the formation of a very fine and thermodynamically stable reinforcing ceramic phase within a metal matrix (Hsu *et al.*, 2006). To produce reinforcement in the *in-situ* fabrication process, either elevated temperature or mechanical energy is needed to accelerate the reaction rate. Most of the technologies can be categorized as mechanochemical, i.e. combined process of chemical reaction and mechanical compaction.

Friction stir processing (FSP) can be an *in-situ* production process. For the production of intermetallic reinforced *in-situ* composites, FSP can provide: severe deformation to promote mixing and refining of the constituent phases in the material; elevated temperature to facilitate the formation of intermetallic phase *in-situ*; hot consolidation to form a fully dense solid (Hsu *et al.*, 2006).

High-energy ball milling and mechanical stirring processes are also used to contribute energy for the *in-situ* chemical reaction process. Due to the mechanical energy released in the process, chemical reactions and phase transformations take place (Ying and Zhang, 2000). Ying and Zhang successfully used this technology to fabricate Al_2O_3/Cu metal matrix nanocomposite. In Ying and Zhang's experiment, mechanical alloying was used to produce Cu(Al) solid solution powder. The powder was then milled with CuO powder, with a reaction between Cu(Al) and CuO occurring. Al_2O_3 nanoparticles were observed. Yamasaki *et al.* produced TiN/Fe by milling Fe–Ti in a nitrogen gas atmosphere (Yamasaki *et al.*, 2003).

Besides mechanical alloying, molecular level mixing, electroless deposition, and other thermal–chemical methods have been reported as *in-situ* nanocomposite powder routes. Daoush *et al.* (2009) reported on CNT/Cu nanocomposite powders fabricated by electroless deposition copper on CNT and the powders were then sintered to make MMNCs. Shehata *et al.* (2009) used a reaction process with a mechanical milling process and successfully distributed Al_2O_3 particles uniformly in a Cu matrix. The abrasive wear resistance of the Cu matrix was improved by the addition of Al_2O_3 nanoparticles. In a liquid reaction process, Tu *et al.* (2002) managed to fabricate TiB_2 in a melting Cu matrix by the reaction [Ti] + B_2O_3 + 3C → TiB_2 + 3CO. After the reaction process, the molten metal was rapidly solidified and the particulates uniformly distributed.

Advantages of the *in-situ* process are that the reinforced surfaces generated tend to remain free of contamination. The particles are usually more homogenously distributed. Therefore, an improved reinforcement–matrix interface bond can be achieved. *In-situ* processing is also more cost effective compared to most *ex-situ* routes. The shortcoming of this technique is the complex reaction process. Most of the structure of reinforcement is hard to control. Also, the selection of the reinforcement material is limited. For example, composites with nanosized hard phases with high energy of formation (e.g. SiC, WC, B_4C) are difficult to produce *in-situ*.

6.5   Future trends

Different kinds of methods have been used to fabricate MMNCs. Some of the most exciting attributes of MMNCs are:

- potentially improved performance in terms of strength, hardness, and wear resistance
- elimination of some expensive heat-treatment steps
- large range of material compositions available
- consolidation route based on commercially affordable processes
- fundamental mechanistic behavior similar to conventional engineering metallics.

However, the fabrication of MMNCs is much more complex than the fabrication of MMCs. When particles scale down from microlevel to the nanolevel, the major challenges are as follows.

- The reaction process between the bonding interface is still unclear.
- There is a lack of understanding about the strengthening mechanisms. When the reinforcements scale to nanolevel, new strengthening mechanisms are needed to explain the enhancement theory.
- Different processes have been applied but modeling of these processes is needed.
- Agglomeration and clustering in bulk materials can still be observed. Dispersion during processing needs to be optimized.
- Currently, only low-volume fabrication methods are observed. A transition to high-volume and high-rate fabrication is pivotal to applying the technology to real industry fabrication.
- Cost effectiveness is another factor that hinders the fabrication of nanocomposites.

6.6   References

Ajayan P M, Schadler L S, and Braun P (2003), *Nanocomposite science and technology*, London, Wiley-VCH.

Cao G, Konishi H, and Li X (2008a), Mechanical properties and microstructure of Mg/SiC nanocomposites fabricated by ultrasonic cavitation based nanomanufacturing, *Journal of Manufacturing Science Engineering*, 130(3), 031105-1–031105-6.

Cao G, Choi H, Oportus J, Konishi H, and Li X (2008b), Study on tensile properties and microstructure of cast AZ91D/AlN nanocomposites, *Materials Science and Engineering: A*, 494(25), 127–131.

Choi H J, Shin J H, and Bae D H (2011), Grain size effect on the strengthening behavior of aluminum-based composites containing multi-walled carbon nanotubes, *Composites Science and Technology*, 71, 1699–1705.

Daoush W M, Lim B K, Mo C B, Nam D H, and Hong S H (2009), Electrical and mechanical properties of carbon nanotube reinforced copper nanocomposites fabricated by electroless deposition process, *Materials Science and Engineering: A*, 513–514, 247–253.

Davis E A, and Ward I M (1993), *An introduction to metal matrix composites*, London, Cambridge University Press.

Fathy A, Shehata F, Abdelhameed M, and Elmahdy M (2011), Compressive and wear resistance of nanometric alumina reinforced copper matrix composites, *Materials & Design*, DOI: 10.1016/j.matdes.2011.10.021.

Ferkel H, and Mordike B L (2001), Magnesium strengthened by SiC nanoparticles, *Materials Science and Engineering A*, 298, 193–199.

Goussous S, Xu W, Wu X, and Xia K (2009), Al–C nanocomposites consolidated by back pressure equal channel angular pressing, *Composites Science and Technology*, 69(11–12), 1997–2001.

Gupta M, and Wong W L E (2005), Enhancing overall mechanical performance of metallic materials using two-directional microwave assisted rapid sintering, *Scripta Materialia*, 52, 479–485.

Hashim J, Looney L, and Hashmi M S J (2001a), The wettability of SiC particles by molten aluminium alloy, *Journal of Materials Processing Technology*, 119(20), 324–328.

Hashim J, Looney L, and Hashmi M S J (2001b), The enhancement of wettability of SiC particles in cast aluminium matrix composites, *Journal of Materials Processing Technology*, 119(20), 329–335.

Hassan S F (2011), Effect of primary processing techniques on the microstructure and mechanical properties of nano-Y2O3 reinforced magnesium nanocomposites, *Materials Science and Engineering: A*, 528(16–17), 5484–5490.

Hassan S F, Tana M J, and Gupta M (2008), High-temperature tensile properties of Mg/Al$_2$O$_3$ nanocomposite, *Materials Science and Engineering: A*, 486(1–2), 56–62.

He J, Zhao N, Shi C, Du X, Li J, and Nash P (2008), Reinforcing copper matrix composites through molecular-level mixing of functionalized nanodiamond by co-deposition route, *Materials Science and Engineering: A*, 490(1–2), 293–299.

Hosseini N, Karimzadeh F, Abbasi M H, and Enayati M H (2010), Tribological properties of Al6061–Al2O3 nanocomposite prepared by milling and hot pressing, *Materials & Design*, 31(10), 4777–4785.

Hsu C J, Chang C Y, Kao P W, Ho N J, and Chang C P (2006), Al-Al$_3$Ti nanocomposites produced in situ by friction stir processing, *Acta Materialia*, 54, 5241–5249.

Published by Woodhead Publishing Limited, 2013

Ibrahim I A, Mohamed F A, and Lavernia E J (1991), Particulate reinforced metal matrix composites—a review, *Journal of Materials Science*, 26, 1137–1156.

Kelly A (1985), Composites in context, *Composites Science and Technology*, 23, 171–199.

Kim K T, Cha S I, and Hong S H (2007), Hardness and wear resistance of carbon nanotube reinforced Cu matrix nanocomposites, *Materials Science and Engineering: A*, 449–451(25), 46–50.

Laha T, Agarwal A, McKechnie T, and Seal S (2004), Synthesis and characterization of plasma spray formed carbon nanotube reinforced aluminum composite, *Materials Science and Engineering: A*, 381, 249–258.

Laha T, Chen Y, Lahiri D, and Agarwal A (2009), Tensile properties of carbon nanotube reinforced aluminum nanocomposite fabricated by plasma spray forming, *Composites Part A: Applied Science and Manufacturing*, 40(5), 589–594.

Lan J, Yang Y, and Li X C (2004), Microstructure and microhardness of SiC nanoparticles reinforced magnesium composites fabricated by ultrasonic method, *Materials Science and Engineering A*, 386, 284–290.

Lee C J, Huang J C, and Hsieh P J (2006), Mg based nano-composites fabricated by friction stir processing, *Scripta Materialia*, 54(7), 1415–1420.

Li X, Duan Z, Cao G, and Roure A (2007), *Ultrasonic cavitation based solidification processing of bulk Mg matrix nanocomposite*, Schaumburg, IL, American Foundry Society.

Li X C, Yang Y, and Cheng X D (2004), Ultrasonic-assisted fabrication of metal matrix nanocomposites, *Journal of Materials Science*, 39, 3211–3212.

Li Y, Zhao Y H, Ortalan V, Liu W, Zhang Z H, Vogt R G, Browning N D, Lavernia E J, and Schoenung J M (2009), Investigation of aluminum-based nanocomposites with ultra-high strength, *Materials Science and Engineering: A*, 527(1–2), 305–316.

Ma Z Y, Li Y L, Liang Y, Zheng F, Bi J, and Tjong S C (1996), Nanometric Si_3N_4 particulate-reinforced aluminum composite, *Materials Science and Engineering: A*, 219(1–2), 229–231.

Mazaheri Y, Karimzadeh F, and Enayati M H (2011), A novel technique for development of $A356/Al_2O_3$ surface nanocomposite by friction stir processing, *Journal of Materials Processing Technology*, 211(10), 1614–1619.

Motta M S, Brocchi E A, Solorzano I G, and Jena P K (2004), Complementary Microscopy Techniques Applied to the Characterization of Cu-Al2O3 Nanocomposites, Spain: Formatex Research Centre. Available from: http://www.formatex.org/microscopy2/papers/215-223.pdf [Accessed Feb 10 2012].

Neubauer E, Kitzmantel M, Hulman M, and Angerer P (2010), Potential and challenges of metal-matrix-composites reinforced with carbon nanofibers and carbon nanotubes, *Composites Science and Technology*, 70(16), 2228–2236.

Nie K B, Wang X J, Xu L, Wu K, Hu X S, and Zheng M Y (2011), Microstructure and mechanical properties of SiC nanoparticles reinforced magnesium matrix composites fabricated by ultrasonic vibration, *Materials Science and Engineering: A*, 528(15), 5278–5282.

Nouri N, Ziaei-Rad S, Adibi S, and Karimzadeh F (2012), Fabrication and mechanical property prediction of carbon nanotube reinforced aluminum nanocomposites, *Materials and Design*, 34, 1–14.

Paramsothy M, Chan J, Kwok R, and Gupta M (2011a), Enhanced mechanical response of magnesium alloy ZK60A containing Si_3N_4 nanoparticles, *Composites Part A: Applied Science and Manufacturing*, 42(12), 2093–2100.

Paramsothy M, Chan J, Kwok R, and Gupta M (2011b), Addition of CNTs to enhance tensile/compressive response of magnesium alloy ZK60A, *Composites Part A: Applied Science and Manufacturing*, 42(2), 180–188.

Park Y and Lee S M (2009), Effects of particle size on the thermal stability of lithiated graphite anode, *Electrochimica Acta*, 54(12), 3339–3343.

Prasad Y V R K, Rao K P, and Gupta M (2009), Hot workability and deformation mechanisms in Mg/nano-Al2O3 composite, *Composites Science and Technology*, 69(7–8), 1070–1076.

Ramakrishna G and Ghosh H N (2003), Effect of particle size on the reactivity of quantum size ZnO nanoparticles and charge-transfer dynamics with adsorbed catechols, *Langmuir*, 19(7), 3006–3012.

Salvetat J P, Bonard J M, Thomson N H, Kulik A J, Forró L, Benoit W, and Zu L (1999), Mechanical properties of carbon nanotubes, *Applied Physics A: Materials Science & Processing*, 69(3), 255–260, DOI: 10.1007/s003390050999.

Sanaty-Zadeh A (2011), Comparison between current models for the strength of particulate-reinforced metal matrix nanocomposites with emphasis on consideration of Hall–Petch effect, *Materials Science and Engineering A*, DOI: 10.1016/j.msea.2011.10.043.

Shehata F, Fathy A, Abdelhameed M, and Moustafa S F (2009), Preparation and properties of Al2O3 nanoparticle reinforced copper matrix composites by in situ processing, *Materials & Design*, 30(7), 2756–2762.

Sherif M and Eskandarany E I (1998), Mechanical solid state mixing for synthesizing of SiCp/Al nanocomposites, *Journal of Alloys and Compounds*, 279, 263–271.

Suryanarayana C, Ivanov E, and Boldyrev V V (2001), The science and technology of mechanical alloying, *Materials Science and Engineering A*, 304–306, 151–158.

Tjong S C (2007), Novel nanoparticle-reinforced metal matrix composites with enhanced mechanical properties, *Advanced Engineering Materials*, 9(8), 639–652.

Tu J P, Wang N Y, Yang Y Z, Qi W X, Liu F, Zhang X B, Lu H M, and Liu M S (2002), Preparation and properties of TiB_2 nanoparticle reinforced copper matrix composites by in situ processing, *Materials Letters*, 52(6), 448–452.

Tun K S and Gupta M (2007), Improving mechanical properties of magnesium using nano-yttria reinforcement and microwave assisted powder metallurgy method, *Composites Science and Technology*, 67, 2657–2664.

Varadan V K, Pillai A S, and Mukherji D (2010), *Nanoscience and nanotechnology in engineering*, USA, World Scientific.

Yamasaki T, Zheng Y J, Ogino Y, Terasawa M, Mitamura T, and Fukami T (2003), Formation of metal-TiN/TiC nanocomposite powders by mechanical alloying and their consolidation, *Materials Science and Engineering A*, 350, 168–172.

Yang Y and Li X C (2007), Ultrasonic cavitation based nanomanufacturing of bulk aluminum matrix nanocomposites, *Journal of Manufacturing Science and Engineering*, 129, 497–501.

Yang Y, Lan J, and Li X (2004), Study on bulk aluminum matrix nano-composite fabricated by ultrasonic dispersion of nano-sized SiC particles inmolten aluminum alloy, *Materials Science and Engineering: A*, 380(1–2), 378–383.

Ying D Y and Zhang D L (2000), Processing of Cu-Al$_2$O$_3$ metal matrix nanocomposite materials by using high energy ball milling, *Materials Science and Engineering A*, 286, 152–156.

Zhang D L (2004), Processing of advanced materials using high-energy mechanical milling, *Progress in Materials Science*, 49, 537–560.

Zhang Z and Chen D L (2006), Consideration of Orowan strengthening effect in particulate-reinforced metal matrix nanocomposites: A model for predicting their yield strength, *Scripta Materialia*, 54, 1321–1326.

Zhang Z and Chen D L (2008), Contribution of Orowan strengthening effect in particulate-reinforced metal matrix nanocomposites, *Materials Science and Engineering A*, 483–484, 148–152.

Zhou W and Xu Z M (1997), Casting of SiC reinforced metal matrix composite, *Journal of Materials Processing Technology*, 63, 358–363.

7

Carbon nanotube (CNT) reinforced glass and glass-ceramic matrix composites

T. SUBHANI and M. S. P. SHAFFER,
Imperial College London, UK;
A. R. BOCCACCINI, Imperial College London, UK and
University of Erlangen-Nuremberg, Germany

DOI: 10.1533/9780857093493.2.208

Abstract: In this chapter, glass and glass-ceramic matrix composites containing carbon nanotubes (CNTs) are discussed with an emphasis on their production, properties, microstructures and applications. Composite manufacturing routes require both CNT/matrix powder preparation techniques and their densification by suitable sintering processes. Physical, mechanical, functional and technological properties of the composites are evaluated, including density, hardness, elastic modulus, fracture strength and toughness, electrical and thermal conductivity, wear and friction resistance, and thermal shock, cycling and ageing resistance. Microstructural features are typically characterized by X-ray diffraction and scanning and transmission electron microscopy. Based on the characteristics obtained, potential applications of the composites are considered, together with a discussion of the unresolved manufacturing challenges and desirable, but still unattained, properties.

Key words: carbon nanotubes, heterocoagulation, colloidal mixing, sol–gel techniques, hot-press sintering, spark plasma sintering, pressureless sintering.

7.1   Introduction

Ideal carbon nanotubes (CNTs) possess extraordinary physical and mechanical properties in combination with unique geometrical features, i.e. high aspect ratio and diameter in the nanoscale. Such outstanding characteristics along with their low density make them an attractive

208

reinforcement for advanced composites. This new class of materials, CNT reinforced composites, aims at exploiting the increased strength of CNTs in the absence of macroscopic defects, due to their small size (although atomic scale defects exist), which should lead to high-quality structural materials. Moreover, the tremendously increased surface area of CNTs due to their nanosize produces novel effects, which are still being elucidated. CNTs have been extensively used in polymeric, metallic and ceramic matrices to exploit their remarkable features. However, comparatively little attention has been paid to the incorporation of CNTs into amorphous glasses and partially crystalline glass-ceramics to produce silicate-based composites. The available literature data on these brittle matrix systems indicate that the primary aim of incorporating CNTs in glass and glass-ceramic matrices has been to achieve an increase in fracture toughness. However, the interest in these composites has also led to the investigation of other mechanical and functional properties; very recently, some of the more technological properties have been investigated. Although the field of CNT–glass/glass-ceramic matrix composites is still in its infancy, it is opportune to review current knowledge in this field and to highlight the potential applications of these novel composites relevant to future research. These considerations motivated the preparation of this chapter.

7.2   Carbon nanotubes

The classic allotropes of carbon, diamond, graphite and amorphous carbon, known for centuries, have recently been augmented by new families of pure carbon structure. Both spherical (Kroto *et al.*, 1985) and tubular (Iijima, 1991, Radushkevich and Lukyanovich, 1952) versions of fullerenes have attracted great interest, especially the latter. The unique structure and properties of CNTs have attracted more attention than any other nanomaterial in the last two decades and extensive research has explored the potential for their use in a wide range of scientific and engineering fields.

7.2.1   Structure of carbon nanotubes

Carbon atoms are held together by sp^3 hybrid bonds in diamond. The presence of four strong sigma covalent bonds constructs a tetrahedral structure producing hard but electrically insulating diamond. In contrast, graphite has a layered atomic structure, wherein carbon atoms in each graphene layer are linked through three sp^2 hybrid sigma bonds (Poole and Owens, 2003). The out-of-plane pi bond (Meyyappan, 2005) is delocalized and distributed over the entire graphene layer, making graphite conductive and highly anisotropic. As a result, strong covalent bonding exists along the graphene layers and weak Van der Waals forces are present between the layers. CNTs are related to graphite because they are broadly sp^2 hybridized. However, graphene layers in CNTs are rolled up to form seamless hollow tubes; an individual atomic carbon layer forms a single-walled carbon nanotube (SWCNT), while multiple layers rolled up concentrically constitute a multi-walled carbon nanotube (MWCNT). A structure containing pentagons or half fullerenes closes the tube ends. The interlayer spacing of MWCNTs is close to that of graphite layers ($3.35\,\text{Å}$) (Terrones, 2003). The diameter of the innermost tube can be as small as $0.4\,\text{nm}$ and up to a large fraction of the outer diameter, which possibly can reach more than $100\,\text{nm}$ (Meyyappan, 2005). SWCNTs are synthesized in bundles or ropes of up to roughly 100 tubes arranged in a hexagonal array. SWCNTs are either metallic or semiconducting depending on the chirality and diameter of nanotubes; armchair nanotubes are metallic, while chiral and zigzag nanotubes can be metallic or semiconducting (Poole and Owens, 2003). Generally, a ratio of 1:2 is found between metallic and semiconducting SWCNTs.

7.2.2 Production of carbon nanotubes

CNTs are commonly produced by three methods:

- electric arc discharge (EAD)
- laser ablation or evaporation (LAE)
- chemical vapor deposition (CVD).

An even simpler classification (Shaffer and Sandler, 2007) is into high-temperature (EAD and LAE) and low-temperature (CVD) synthesis techniques. The chosen synthesis route and specific parameters control the quality and yield of CNTs produced. Techniques using high temperature produce highly crystalline CNTs but are costly and provide low yield due to the presence of undesirable carbonaceous by-products including fullerenes and graphitic nanoparticles. Low-temperature synthesis techniques produce cheap CNTs in large quantities but these CNTs are structurally defective and often coated with amorphous carbon.

Electric arc discharge

Electric arc discharge was the technique used by Iijima (1991), when CNTs were characterized for the first time. Typically, a current of 50–100 A, at a voltage of 20–40 V, is passed through a pair of graphite electrodes of diameter 6–12 mm, separated by a distance of 1–4 mm in an inert atmosphere, usually helium. Carbon vaporizes away from the anode and arranges itself into MWCNTs, which deposit on the cathode as a soft and dark black fibrous material (Wilson *et al.*, 2002). Introducing transition metal catalysts into the electrodes, such as Fe, Co, Ni, Cu, Ag, Al, Pd, Pt or a combination thereof, produces SWCNTs (Daenen *et al.*, 2003), which deposit on the walls of the reaction chamber. SWCNTs of diameter 1–5 nm and length of up to 1 μm are produced as well as MWCNTs with large variation in their length, and outer and inner diameter (Dresselhaus, 2001). These CNTs are highly graphitized with few structural defects, and possess good electrical, thermal and mechanical properties. The yield, however, is low due to the presence of other graphitic products, particularly nanoparticles (Ajayan *et al.*, 1997, Gamaly and Ebbesen, 1995).

Laser ablation or evaporation

The laser ablation or evaporation process to produce CNTs was first introduced in 1995 (Guo *et al.*, 1995). This method is quite similar to electric arc discharge except using a different heating mechanism, i.e. a high power laser to evaporate graphite (Terrones, 2003) and, usually, a more controlled heating regime. The flow of an inert gas such as argon or helium drives the vaporized carbon atoms away from the high-temperature zone on a cold copper collector, where CNTs deposit (Poole and Owens, 2003). MWCNTs can be synthesized by this technique (Meyyappan, 2005), but it is more often applied for the synthesis of SWCNTs after impregnating the graphite with transition metal catalysts (1–2%) (Terrones, 2003). The processing parameters, such as the type of inert gas, hydrocarbon source, flow of inert gas, intensity of laser and the furnace temperature, influence the morphology and properties of the CNTs produced. The use of a high-power laser and high-purity graphite makes this technique very expensive (Terrones, 2003) but it produces good-quality SWCNTs at higher yields (> 70%), with good diameter control.

Chemical vapor deposition or catalytic growth process

Chemical vapor deposition has been known for a long time as a means to deposit carbon nanofibrils (Radushkevich and Lukyanovich, 1952); however, from 1998, the method began to be optimized for the synthesis of CNTs (Kong *et al.*, 1998, Ren *et al.*, 1998). In this technique, a source of carbon (usually a hydrocarbon or CO) is heated inside a quartz tube at an intermediate temperature range (500–1100°C) in the presence of a catalyst under an inert atmosphere of argon or helium gas. Hydrocarbon molecules catalytically decompose into hydrogen and carbon; subsequently, carbon atoms rearrange themselves into hexagonal networks on the metal catalyst to grow CNTs (Dresselhaus, 2001). The hydrocarbon source may be a solid (camphor, naphthalene), a liquid (benzene, alcohol, hexane) or a gas (acetylene, methane, ethylene); metal catalysts including Fe, Co, Ni, Mo and

7.1 MWCNTs grown in a CVD process by injecting ferrocene and toluene in a horizontal tube furnace.

metal supports such as MgO, $CaCO_3$, Al_2O_3 or Si are used. Higher yields can be achieved by using a mixture of two catalytic metal particles (Seraphin and Zhou, 1994, Seraphin *et al.*, 1994). Metallic catalysts can be pre-deposited as particles or films on a ceramic substrate or injected as colloids or organometallic vapor along the carbon feedstock. The diameter of the resulting CNTs roughly matches that of the catalyst particle while lengths up to millimeters can be obtained with increasing synthesis time (Pan *et al.*, 1998). CNTs grow by two different methods: (a) tip growth and (b) base or root growth. A weak interaction between metal and substrate allows the growth of CNTs by the tip growth mechanism while a strong interaction favors base growth (Dervishi *et al.*, 2009). Both horizontal and vertical furnaces are used as reaction chambers; the former is more common in a laboratory setup and allows control of the CNTs length by controlling hydrocarbon deposition time, whereas the latter is more often used for continuous mass production of CNTs (Teo *et al.*, 2003). Figure 7.1 shows an image of relatively aligned MWCNTs produced by CVD process by injecting ferrocene and toluene into a horizontal furnace (Subhani *et al.*, 2011).

The advantages of the CVD-based fabrication route include large-scale production at relatively low cost together with a control of CNTs growth by adjusting reaction parameters such as temperature, catalyst, hydrocarbon source and the flow rate of gases. However, CNTs tend to be highly entangled or aligned depending upon the parameters chosen. The application of plasma during growth is reported to enhance the alignment and straightness of the nanotubes at low temperatures (Hofmann *et al.*, 2003).

7.2.3 Properties of carbon nanotubes

Modulus of elasticity

In early measurements, the moduli of elasticity of MWCNTs and SWCNTs were found to be 1.8 TPa and 1.25 TPa, respectively, by measuring the amplitude of their thermal vibrations in a transmission electron microscope (Krishnan *et al.*, 1998, Treacy *et al.*, 1996). Highly crystalline CNTs have an axial stiffness of 1.1 TPa, which is close to theoretical prediction (Belytschko *et al.*, 2002, Lu, 1997, Yao and Lordi, 1998), while the bending stiffness of CNTs determined by atomic force microscopy was found also to be around 1 TPa (Salvetat *et al.*, 1999, Wong *et al.*, 1997). CNTs grown by CVD have comparatively lower stiffness due to the presence of a higher concentration of structural defects than those produced by high-temperature techniques. CNTs tend to bend to high deformations without fracture due to buckling mechanisms involving reversible collapse of their cross-sections (Shaffer and Sandler, 2007). CNTs may, therefore, exhibit different values of stiffness in tension, compression and bending. The modulus of elasticity of CNTs is independent of their chirality, but depends on diameter and on deviation from a parallel alignment of graphitic layers (Shaffer and Sandler, 2007).

Tensile strength

Theoretically, the predicted tensile strength of CNTs is relatively high, 75–135 GPa (Zhao and Zhu, 2011). However, a wide range of values has been observed experimentally; for example, individual SWCNTs and high-temperature grown MWCNT shells showed values of 3.6–100 GPa (Li *et al.*, 2000, Wang *et al.*, 2010, Yu *et al.*, 2000a) and 10–110 GPa (Ding *et al.*, 2007, Peng *et al.*, 2008, Yu *et al.*, 2000b), respectively. The ropes/strands of SWCNTs and low-temperature grown MWCNTs showed very low values of only 1–1.7 GPa (Li *et al.*, 2005, Pan *et al.*, 1999, Zhu *et al.*, 2002). CVD-grown MWCNTs also demonstrated low-compression strength of 2 GPa (Zhao *et al.*, 2011). Nanoloading devices attached with AFM tips are generally operated inside a SEM or TEM to measure tensile strength. A brittle fracture mode has always been observed during tensile loadings of

SWCNTs while MWCNTs fail in tension via a sword-in-sheath mechanism; stepwise or brittle fractures are generally observed during the bending of MWCNTs (Zhao and Zhu, 2011). The low experimental strength values and brittle fracture have been related to defects present in the as-synthesized CNTs (Zhao and Zhu, 2011), as the strength of CNTs is affected by the presence of defects as well as the interactions of nanotubes in a MWCNT and in the bundles of SWCNTs (Shaffer and Sandler, 2007). A recent *in-situ* TEM test verified that CNTs were broken at the defective locations and CNTs with no defects showed strength approaching the theoretically determined value (Wang *et al.*, 2010).

Electrical conductivity

The room-temperature electrical conductivity of metallic SWCNTs was found to be 10^5–10^6 S/m but only 10 S/m for semiconducting SWCNTs (Tans *et al.*, 1997). Generally, the electrical conductivity of SWCNT bundles has been found to swing between 10^4 S/m and 3×10^6 S/m (Bozhko *et al.*, 1998, Fischer *et al.*, 1997, Kim *et al.*, 1998), approaching the in-plane conductivity of graphite 2.5×10^6 S/m (Charlier and Issi, 1995). The electrical conductivities of individual MWCNTs have been found to be in the range of 20 to 2×10^7 S/m (Ebbesen *et al.*, 1996). The helicity of the outermost tubes (Lee *et al.*, 2000), ballistic effects, and the presence of defects (Dai *et al.*, 1996) affect the measured electrical conductivity of CNTs significantly.

Thermal conductivity

High thermal conductivity values (i.e. 6600 W/m.K) have been theoretically predicted for individual SWCNTs (Berber *et al.*, 2000, Che *et al.*, 1999, Hone *et al.*, 1998). However, bulk specimens of SWCNTs demonstrated considerably lower values, i.e. $\leqslant 36$ W/m.K for disordered SWCNTs (Hone *et al.*, 1998) and > 200 W/m.K for aligned SWCNTs (Hone *et al.*, 2000, Pan, 2011). Similarly, the experimental thermal conductivities of individual MWCNTs are very high, e.g. 3000 W/m.K (Kim *et al.*, 2001), while MWCNTs in bulk form showed a value of only 25 W/m.K (Yi *et al.*, 1999). Moreover, values as high as 200 W/m.K have been determined for MWCNTs (Yang *et al.*, 2002) but randomly oriented MWCNTs consolidated by spark plasma sintering (SPS) showed a very low value of only 4.2 W/m.K (Zhang *et al.*, 2005). As well as the effects of anisotropy, this sharp decrease in thermal conductivity of bulk CNT specimens may be associated with the interfacial resistance; bundles, ropes, nests and mats of CNTs have shown considerably lower values, sometimes as low as 0.7 W/m.K (Hone *et al.*, 1998).

7.2.4 Applications of carbon nanotubes

As already mentioned, the notable physical and mechanical properties of CNTs suggest numerous scientific and engineering applications such as supercapacitors, transparent electrodes for organic light emitting diodes, lithium ion batteries, nanowires, field-effect transistors, molecular switches, sensors and filters. The conductivity, robustness, high surface area, accessible network formation and facile functionalization of CNTs are promising features for use as electrochemical electrodes and catalyst supports. Catalytically active sites can be used in fuel cells, immobilization of biomacromolecules and for organic reaction catalysis. In addition, CNTs can be used in the fields of catalysis, biomedical and mechanical applications, and as nanoscale reinforcing elements in composites to achieve improved properties (Schnorr and Swager, 2011).

For mechanical and structural applications, which are of relevance in the context of the present chapter, pure CNT arrays can be shaped into bundles, ropes and strands several centimetres in length. Two-ply yarns of CNTs can also be formed, which can be assembled into sheets of a meter length and a width of a few centimeters. Such arrays can be impregnated with other materials to make composites, a particular example of a route to CNT-based structural composites.

7.2.5 CNTs in polymer, metal, and ceramic composite matrices

CNTs have emerged as potentially outstanding reinforcing agents for use in metals (Bakshi *et al.*, 2010), polymers (Shaffer and Sandler, 2007), and glasses and ceramics (Cho *et al.*, 2009) to overcome intrinsic limitations of these materials and to impart new properties and functionalities. For example, CNTs are usually introduced in polymeric and metallic matrices to increase strength and stiffness along with other functional properties. Glasses and ceramics incorporated with CNTs can exhibit reduced brittleness to become structurally more reliable materials (Cho *et al.*, 2009). In addition, CNTs can offer improvements in the thermal and electrical properties of inorganic matrices.

7.3   Glass and glass-ceramic matrix composites

The versatility of glasses and glass-ceramics in compositions, properties and processing makes them an attractive choice to be used as matrix materials in composites for applications including aerospace, automotive, electronics, biomedical and other specialized fields (Boccaccini, 2002). Silicate glasses can be produced with a broad range of chemistries to control the fiber/

matrix chemical interactions. This chemical composition flexibility also enables the development of glasses with a wide range of thermal expansion coefficients (TECs) in order to match the TEC of the reinforcement. In addition, by controlling the viscosity of glasses, high composite densities can be achieved by promoting viscous flow during sintering. Finally, the low elastic modulus of most glasses means that high-modulus fibers can provide a true reinforcement, unlike the case of polycrystalline ceramic matrix composites where the ratio of moduli is close to one (Chawla, 2003). However, the main drawback of glasses is the low-temperature capability compared with polycrystalline ceramics, which limits applications to modest temperatures.

Carbon, silicon carbide, alumina, boron and metallic reinforcements have been used in glass and glass-ceramic matrices in a variety of shapes, including continuous and chopped fibers, particles, platelets and whiskers. Continuous carbon fibers were the first fibers applied to silica reinforcement (Crivelli-Visconti and Cooper, 1969), producing excellent toughness and crack growth resistance; the composites exhibited non-catastrophic failure. Borosilicate glasses and calcium-silicate glass-ceramic matrices have also been reinforced with aligned carbon fibers for improved composite toughness (Sambell *et al.*, 1972), which was attributed to low fiber/matrix interface strength promoting crack deflection. Prewo used continuous carbon fibers to obtain exceptionally high elastic modulus glass matrix composites (Prewo, 1982). The combination of carbon fibers with a glass matrix offers other benefits since carbon fibers impart lubricity to the composite surface and the glass matrix provides high hardness and wear resistance, as well as dimensional stability (Prewo, 1981).

A wealth of literature on silicon carbide (SiC) fiber reinforced glass and glass-ceramic composites is available, showing improved fracture strength and toughness together with the potential for high-temperature oxidation resistance. Monofilaments and large-diameter SiC fibers have been used as reinforcements in different glass matrices (Aveston, 1971, Bansal, 1997, Cho *et al.*, 1995, Prewo and Brennan, 1980). Moreover, SiC yarns have been used to develop high-strength composites (Prewo and Brennan, 1980, 1982); the highest strength was achieved by using a lithium aluminosilicate matrix (Brennan and Prewo, 1982). The addition of SiC whiskers also substantially increased the strength of lithium aluminosilicate (LAS) matrices (Gadkaree and Chyung, 1986).

Glass and glass-ceramic matrix composites containing nano-reinforcements such as CNTs represent an emerging family of composites that potentially exploits the small size of fillers of high-strength values. Furthermore, the enormous interfacial area generated due to the incorporation of nano-fillers in composites may give rise to new kinds of composites, possessing improved properties.

7.4   Glass/glass-ceramic matrix composites containing carbon nanotubes: manufacturing process

Manufacturing effective composites containing CNTs involves several critical requirements: (a) the availability of high-quality CNTs with intrinsically good mechanical properties; (b) a uniform dispersion of CNTs in the matrix; (c) the development of a suitable interfacial bonding. In the case of CNT–glass/glass-ceramic matrix composites, additional critical issues apply, such as (d) the prevention of CNTs oxidation during high-temperature sintering and (e) the consolidation of composites to near theoretical densities. All these challenges must be resolved to develop reproducible materials that allow a true investigation of the intrinsic effect of CNTs on the properties of glasses and glass-ceramics. In this context, the selection of a suitable composite preparation technique is key. The uniformity of the CNT distribution is usually established in the green body, whilst an appropriate consolidation process with optimum sintering conditions can avoid the degradation of CNTs and ensure high composite densities. Consideration must also be given to the effect of CNT orientation on property (an)isotropy and filler packing.

   The manufacturing of CNT–glass/glass-ceramic composites comprises two stages: composite powder preparation using a suitable CNTs dispersion process and their densification by an appropriate sintering technique. These two steps of composite processing are mutually dependent; an inhomogeneous dispersion of CNTs without full densification is as deleterious to the properties of composites as complete consolidation with poorly dispersed CNTs. A combination of the optimum outcome of these two processes can provide high-quality composites for reliable property characterization.

7.4.1 Composite powder preparation

The techniques used to prepare composite powders (mixtures) of CNTs and glasses/glass-ceramics include conventional powder mixing (Boccaccini et al., 2007), sol–gel techniques (Thomas et al., 2009), colloidal mixing processes (Cho et al., 2011) and in-situ CNT synthesis in the matrix material. Conventional powder mixing processes usually yield a poor dispersion of CNTs due to the lack of driving force to distribute CNTs. As a result, agglomerates are readily formed in the sintered composites due to the high aspect ratio, high surface area and poor interaction of the CNTs with the matrix material. On the other hand, sol–gel (Mukhopadhyay et al., 2010) and colloidal mixing (Cho et al., 2011) techniques have successfully produced homogeneous dispersions of CNTs up to 10wt% and 15wt%, respectively, and offer better densification ability than powder mixing techniques due to the absence of CNTs agglomerates and a fine matrix particle size. *In-situ* synthesis of CNTs is another convenient technique that has been utilized predominantly for CNT reinforced crystalline ceramic matrix composite (CMC) powders (Peigney *et al.*, 1998) but it can also be applied to obtain CNT–glass/glass-ceramic matrix composite systems.

Conventional powder mixing

One of the earliest techniques used to prepare composite powders was powder mixing. In this process, as-synthesized CNTs are mixed with glass or glass-ceramic powders and a composite slurry or suspension is prepared, which is ultrasonicated and/or ball-milled to disperse the CNTs. The composite slurry/suspension is then dried, ground and sieved before the final compaction and sintering to obtain composite bodies. Composite powders containing both glasses (Boccaccini *et al.*, 2005, 2007, Ning *et al.*, 2003a) and glass-ceramics (Giovanardi *et al.*, 2010, Wang *et al.*, 2007, Ye *et al.*, 2006) have been prepared by this technique.

In the first investigation on powder mixing process, CNTs ranging from 3.75wt% to 22.5wt% were incorporated in silica (SiO_2) glass powder (Ning *et al.*, 2003a) to prepare CNT–SiO_2 composites. However, the CNTs were not dispersed well in the composite, even at low loadings (3.75wt%), as easily identified in SEM micrographs. Another study was performed using borosilicate glass powder but a similarly unsatisfactory dispersion of CNTs was found (Boccaccini *et al.*, 2005). Attempts to produce CNT–glass-ceramic composite powders, have also proved less successful; examples include 1wt% CNTs in vanadium-doped silicates (Giovanardi *et al.*, 2010), 3.75wt% CNTs in mullite (Wang *et al.*, 2007) and 11.25wt% CNTs in barium aluminosilicate (Ye *et al.*, 2006). The lack of dispersion is not surprising in these systems, given that the matrix particles are typically large compared to the desired CNT–CNT separation, and the fact that the CNTs tend to be forced into mutual contact around the perimeter of the matrix particles. There is no effective mechanism, during consolidation, to disagglomerate these CNTs or to distribute them within the original matrix particles. On the other hand, if an electrically conducting network is required at low loadings, this type of longer length scale structuring can be beneficial.

Sol–gel processing

Sol–gel processing has shown potential to provide good dispersion of CNTs in glass/glass-ceramic matrix composites up to 10wt% (Mukhopadhyay *et al.*, 2010). In this process, (functionalized or surfactant-stabilized) CNTs are mixed in a (usually aqueous) solution (sol) of molecular precursor of the matrix, which is subsequently gelled (gel) and dried to obtain the inorganic composite. The obtained composite body is either used as the final composite after heat treatment (de Andrade *et al.*, 2007) or crushed, ground, sieved and then sintered to obtain composite samples of the desired shape (Thomas *et al.*, 2009). Although earlier studies showed a degree of CNT agglomeration (Ning *et al.*, 2003b, 2004), later investigations showed a better dispersion of CNTs in glass/glass-ceramic matrices, after functionalization, at 3wt% (Thomas *et al.*, 2009) and then 10wt% loadings (Otieno *et al.*, 2010). However, higher loadings of CNTs (15wt%) still produce agglomerates (Mukhopadhyay *et al.*, 2010). Different types of CNTs, i.e. SWCNTs (Babooram and Narain, 2009, Chu *et al.*, 2008, de Andrade *et al.*, 2007, 2008), DWCNTs (de Andrade *et al.*, 2009) and MWCNTs (Mukhopadhyay *et al.*, 2010, Thomas *et al.*, 2009) have been incorporated in glass/glass-ceramic matrices by the sol–gel process including silica (Seeger *et al.*, 2001), borosilicate (Thomas *et al.*, 2009) and aluminoborosilicate (Chu *et al.*, 2008).

In one of the earliest studies on sol–gel processing (Seeger *et al.*, 2001), tetraethoxysilane (TEOS) was used as the precursor in acidified water to produce a silica glass matrix. CNTs were mixed with the solution and, after condensation, produced a composite body containing 2.5wt% MWCNTs, which was milled to powder and densified by pressureless sintering (Seeger *et al.*, 2002) and laser treatment (Seeger *et al.*, 2003). In earlier attempts, CNTs without functionalization were incorporated in the precursor of the desired matrix (Gavalas *et al.*, 2001); however, later investigations focused on using appropriate surfactants or coupling agents for improved dispersion (Hernadi *et al.*, 2003). Most subsequent studies have used functionalized CNTs in sol–gel processing to produce composites (Babooram and Narain, 2009, Berguiga *et al.*, 2006, DiMaio *et al.*, 2001, Zhan *et al.*, 2003, Zhang *et al.*, 2006, Zheng *et al.*, 2007, 2008).

Transparent silica glass matrix composites were also prepared by reinforcing SWCNTs (DiMaio *et al.*, 2001) and MWCNTs (Xu *et al.*, 2009) by sol–gel processing. CNT–SiO_2 glass composites for non-linear optical properties (Xu *et al.*, 2009, Zhan *et al.*, 2005), microwave attenuation (Xiang *et al.*, 2005) and electromagnetic interference shielding (Xiang *et al.*, 2007) provide further examples of sol–gel processing. In addition to bulk composites, thin composite films of CNT–SiO_2 were also produced by this technique and deposited by spin coating (Loo *et al.*, 2007) and dip coating techniques (Berguiga *et al.*, 2006, de Andrade *et al.*, 2007, Lopez *et al.*, 2010). CNT–mesoporous silica composites were also produced by the sol–gel technique to develop novel biomaterials (Vila *et al.*, 2009).

Colloidal mixing

Colloidal mixing is an emerging technique to homogeneously disperse CNTs in glass and glass-ceramic matrices up to high loadings of > 19vol% (15wt %) (Cho *et al.*, 2011). In this process, the surface chemistry of CNTs and the glass powder suspensions is adjusted to encourage the coating of glass particles on CNTs or vice versa, depending upon their size. The manipulation of the surface chemistry of the two composite constituents results in the development of opposite surface charges on them; as a result, similarly charged particles repel each other but attract oppositely charged particles during the mixing of the two suspensions, a process called heterocoagulation. Organic surfactants and dispersants can be used to develop positive or negative surface charges on CNTs and glass powders (Cho *et al.*, 2009).

The earliest works on colloidal mixing techniques find their roots in sol–gel processes, where surface charges were developed on CNTs by using surfactants and functionalized CNTs were then mixed with the precursor of the desired matrix. In one of these studies, CNTs were dispersed in aqueous solution of cetyltrimethyl ammonium bromide (CTAB), which served as the cationic surfactant before mixing it with the alcoholic TEOS solution to form the silica glass matrix (Ning *et al.*, 2004). CNT–SiO_2 composite powders showed better dispersion quality of CNTs than obtained by direct mechanical mixing processes. In another study, CNTs were dispersed in silica glass using different surfactants before mixing with the sol–gel solution to prepare composite films (Loo *et al.*, 2007).

In the second stage of the development of colloidal processing, surfactant-assisted dispersions of CNTs were mixed with commercially available colloidal glass particle suspensions to prepare CNT–glass composite powders. For example, CNTs were mixed with colloidal silica to obtain composite powders (Guo *et al.*, 2007a, 2007b, Sivakumar *et al.*, 2007) leading to a high dispersion quality, which was found to be better than that obtained by direct mixing and sol–gel processes.

More recently, colloidal mixing processes have been developed that provide high-quality CNT dispersions in glass matrices (Arvanitelis *et al.*, 2008, Cho *et al.*, 2011, Subhani *et al.*, 2011). Aqueous CNT dispersions are obtained by treating CNTs with a mixture of sulphuric and nitric acids (Shaffer *et al.*, 1998), which not only purify them from catalyst particles and amorphous carbon produced during their synthesis but also shorten their lengths and decorate them with acidic functional groups (i.e. carboxylic acid and other oxygen-containing groups). These surface functionalities stabilize CNTs electrostatically in aqueous suspensions and develop a negative surface charge. The development of negative surface charge on CNTs requires positive surface charge on the glass particles to encourage

7.2 MWCNTs treated with a mixture of sulphuric and nitric acid to obtain stable aqueous dispersion.

heterocoagulation, which is produced using cationic surfactants, and finally suitable composite powder suspensions are obtained by colloidal mixing. The composite powder suspensions are dried, ground and sieved to obtain powders for subsequent sintering into solid compacts. A calcination process is usually performed on dried composite powders before sintering to remove organic surfactants and CNT oxidation debris at temperatures less than 400°C. Figure 7.2 shows an image of MWCNTs after treating with sulphuric and nitric acid to develop surface charges for stable aqueous dispersion.

In the earliest of these studies, 1wt% CNTs were dispersed in silica glass (Arvanitelis *et al.*, 2008) but high CNT contents were loaded successfully (19vol%) in subsequent investigations (Cho *et al.*, 2011). Both pressureless sintering (PLS) (Subhani *et al.*, 2011) and spark plasma sintering (SPS) (Cho *et al.*, 2011) have been used to prepare CNT–SiO_2 composites at near theoretical densities. A uniform distribution of CNTs was obtained in the silica glass powder, which was retained after sintering. SEM images of 5wt% MWCNTs composite powder before and after sintering are shown in Fig. 7.3 and Fig. 7.4, respectively. The sintered sample exhibits homogeneously dispersed CNTs. Individual CNTs are well separated from the others without showing agglomerates.

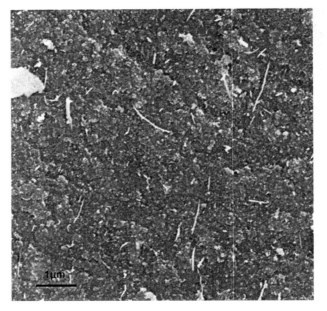

7.3 Green body of 5wt% MWCNTs reinforced silica glass composite powders. Uniform distribution of MWCNTs in silica glass was achieved by heterocoagulation in a colloidal mixing process.

7.4 5wt% MWCNTs reinforced silica glass composite densified by pressureless sintering. Homogeneous dispersion of MWCNTs is evident in the amorphous glass matrix.

In-situ synthesis

In-situ synthesis involves the direct growth of CNTs on matrix powders by CVD processes. The matrix particles clearly have to be stable at the synthesis conditions (elevated temperatures in hydrocarbon atmospheres) and have to be compatible with the CNT catalyst system used. To date, several crystalline ceramic matrices have been directly grafted with CNTs, including Al_2O_3, SiC, TiN, Fe_2N and $BaTiO_3$ (Cho *et al.*, 2009) but glass/glass-ceramic matrix composite powders have not yet been prepared by this method. Possible problems associated with this technique include the retention of metal catalyst and amorphous carbon produced during CNT growth. The non-uniform dispersion of CNTs and the poor density of composites after consolidation are additional drawbacks.

7.4.2   Composite densification techniques

Three densification techniques are commonly used to consolidate CNT–glass/glass-ceramic matrix composites – hot-press sintering (HPS) (Boccaccini *et al.*, 2005), spark plasma sintering (SPS) (Cho *et al.*, 2011) and pressureless sintering (PLS) (Subhani *et al.*, 2011), although laser treatment (Seeger *et al.*, 2003) and high-pressure (de Andrade *et al.*, 2008) techniques have also been occasionally used. In SPS, composite powders are internally heated by passing pulsed DC current through a graphite die, while in HPS and PLS, composite powders are consolidated by an external heat source (Dobedoe *et al.*, 2005). The rapid heating in SPS results in lower sintering temperatures and shorter durations compared with HPS and PLS. 100% relative densities have sometimes been achieved by HPS (Ye *et al.*, 2006) and SPS (Guo *et al.*, 2007a). Vacuum or protective atmospheres of nitrogen or argon are used during sintering to avoid the oxidation of CNTs. In addition, alignment of CNTs has been observed in densified composites sintered by SPS and HPS (Cho, 2010), as these techniques involve uniaxial pressure. In contrast, hot isostatic pressing (HIP) and PLS provide randomly oriented CNTs in the final composites, provided composite powders are pre-compacted by cold isostatic pressing (CIP) (Subhani *et al.*, 2011). In short, SPS is a time efficient and effective route to good densities, but PLS is cheaper and more flexible in terms of composite size and shape, but at the cost of a comparatively lower density.

The following examples highlight the maximum CNT loadings successfully incorporated in glass/glass-ceramic matrix composites densified using the three consolidation techniques. 10vol% MWCNT–barium aluminosilicate composites were densified by HPS in graphite dies at 1600°C for 1 h at a pressure of 20 MPa in nitrogen atmosphere (Ye *et al.*, 2006). SPS was used to densify up to 19vol% (15wt%) MWCNT–SiO_2 composites in graphite dies at 1200°C for 20 min at a pressure of 75 MPa in an argon atmosphere (Cho *et al.*, 2011). 13vol% (10wt%) MWCNT–SiO$_2$ composites were consolidated by PLS at a temperature of 1200°C for 3 h in an argon atmosphere; the composite powders were compacted by uniaxial and isostatic pressing before PLS (Subhani *et al.*, 2011). The densities of the composites are discussed in Section 7.6.1.

7.5   Microstructural characterization

The microstructure of CNT–glass/glass-ceramic matrix composites is usually characterized using SEM, TEM and XRD. High-resolution TEM (HRTEM) is used to observe CNT/matrix interfacial features. A variety of CNT composites containing glasses and glass-ceramics have been microstructurally characterized including silica (Ning *et al.*, 2004, Xu *et al.*, 2009, Zhan *et al.*, 2005, Zheng *et al.*, 2008), borosilicate (Thomas *et al.*, 2009), barium aluminosilicate (Ye *et al.*, 2006), aluminoborosilicate (Mukhopadhyay *et al.*, 2010, Otieno *et al.*, 2010) and vanadium-doped silicates (Giovanardi *et al.*, 2010).

7.5.1   X-ray diffraction (XRD)

XRD is used to confirm the amorphous structure of glass matrices after sintering and to determine the extent of glass devitrification (Arvanitelis *et al.*, 2008, Babooram and Narain, 2009, Boccaccini *et al.*, 2005, 2007, Cho *et al.*, 2011, de Andrade *et al.*, 2009, Ning *et al.*, 2003a, 2003b, Zhang *et al.*, 2006). In the case of glass-ceramic matrices, it identifies the presence of crystalline phases in the glassy phase (Chu *et al.*, 2008, Giovanardi *et al.*, 2010, Mukhopadhyay *et al.*, 2010, Wang *et al.*, 2007). Cristobalite usually forms in silica and borosilicate glass matrices when SPS (de Andrade *et al.*, 2009, Guo *et al.*, 2007a), HPS (Boccaccini *et al.*, 2005, Ning *et al.*, 2003a) or PLS (Subhani *et al.*, 2011) are used to sinter the composites. However, careful control of sintering conditions can avoid crystallization that can considerably affect the composite microstructure and properties, confounding attempts to understand the underlying science (Cho *et al.*, 2011, Ning *et al.*, 2003a, Subhani *et al.*, 2011). Cristobalite can also be suppressed by the addition of Al$_2$O$_3$ but causes the crystallization of mullite and transforms an amorphous glass into a glass-ceramic matrix, as observed in CNT–aluminoborosilicate composites (Chu *et al.*, 2008, Mukhopadhyay *et al.*, 2010). It can be argued that a large interfacial surface area in CNT–glass/glass-ceramic matrix composites should accelerate matrix crystallization through heteronucleation, but studies on CNT–SiO$_2$ composites have instead identified a suppression of matrix devitrification, possibly due to the dissolution of carbon (Subhani *et al.*, 2011).

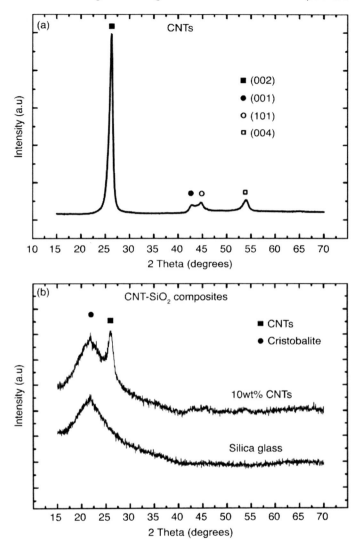

7.5 XRD spectra of (a) as-synthesized CNTs (b) silica glass and CNT–SiO₂ composites containing 10wt% CNTs; an amorphous matrix is observed with little crystallization in both silica glass and composites.

XRD is frequently used to detect the retention of CNTs in composites after sintering involving high temperatures, i.e. up to 1200°C, as in CNT–SiO$_2$ composites (Subhani *et al.*, 2011). A characteristic (002) XRD peak of as-synthesized CNTs (Fig. 7.5(a)) is clearly identified in composite specimens compared to pure silica glass densified by PLS (Fig. 7.5(b)).

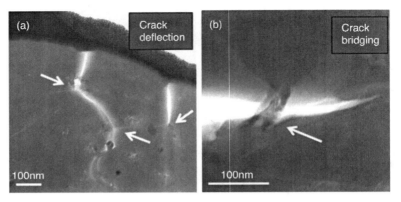

7.6 Toughening mechanisms in CNT–SiO$_2$ composites: (a) crack deflection and (b) CNT bridging. CNT pull-out is shown in Fig. 7.4.

7.5.2 Scanning electron microscopy (SEM)

SEM is primarily used to observe the dispersion quality of CNTs in composites and to verify their presence after sintering. SEM images of composites prepared by mechanical mixing generally show agglomerates associated with inhomogeneous CNT dispersion (Boccaccini *et al.*, 2005). However, micrographs of the composites produced by sol–gel and especially colloidal mixing techniques show well-dispersed CNTs in glass/glass-ceramic matrices (Cho *et al.*, 2011, Mukhopadhyay *et al.*, 2010, Subhani *et al.*, 2011, Thomas *et al.*, 2009). Fig. 7.4 shows extensive pull-out of CNTs from the freshly fractured surface of 5wt% CNT–SiO$_2$ composite without any indication of apparent porosity; an amorphous matrix is also visible without any appreciable indication of crystallization.

SEM also characterizes as-synthesized CNTs (Fig. 7.1), CNT suspensions (Fig. 7.2), and the incorporation of CNTs in glass/glass-ceramic matrices as composite green bodies (Fig. 7.3). The location of CNTs in glass-ceramic matrix composites is also evidenced by SEM, as CNTs generally occupy one of the four locations in glass-ceramic matrices: within grains, along grain boundaries, bridging the grains and within intergranular pores. Conversely, in composites containing amorphous glasses, CNTs are only seen dispersed in a continuous bulk matrix without any effect of grains or grain boundaries (Fig. 7.4). Potential toughening mechanisms in CNT–glass/glass-ceramic matrix composites, including crack deflection (Fig. 7.6(a)), CNTs bridging (Fig. 7.6(b)) and CNTs pull-out (Fig. 7.4), are also revealed by SEM.

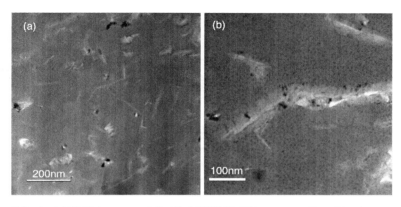

7.7 (a, b) TEM images of 5wt% MWCNT–SiO$_2$ composites. Random alignment of MWCNTs can be seen in the amorphous glass matrix.

7.5.3 Transmission electron microscopy (TEM)

TEM provides information on the actual state of embedded CNTs in composites (i.e. their dispersion and orientation), as well as the degree of crystallinity after processing and the state of the fiber–matrix interface. Figure 7.7 shows TEM images of CNT–SiO$_2$ composites revealing homogeneous dispersion and random orientation of CNTs in a silica glass matrix.

TEM also reveals the structural characteristics of CNTs before and after their incorporation in matrices, i.e. possible CNT curvature and damage. Due to their flexible nature and the intrinsic curvature often associated with CVD growth, images of bent CNTs are common in composites with glass (Ning *et al.*, 2004, Thomas *et al.*, 2009, Xu *et al.*, 2009, Zhan *et al.*, 2005, Zheng *et al.*, 2008) and glass-ceramic (Mukhopadhyay *et al.*, 2010, Otieno *et al.*, 2010, Ye *et al.*, 2006) matrices. Localized curvatures in the CNT walls are also observed, which may be due to structural defects produced during synthesis (Berguiga *et al.*, 2006) or due to the effect of pressure applied during sintering to densify composites. In addition, folds and kinks have been noted in the walls of CNTs (Otieno *et al.*, 2010) and curled and imperfect CNT layered structures are commonly observed (Zhan *et al.*, 2005).

Another aspect of considerable importance in the characterization of composites is related to the interface, as the interface characteristics can significantly affect the composite properties. HRTEM is used to reveal the type of contact between CNTs and the glass/glass-ceramic matrices. Mechanical bonding is reported to form predominantly in CNT–glass/glass-ceramic matrix composites without the existence of an intermediate phase (Subhani, 2012a). However, chemical bonding at the CNT/matrix

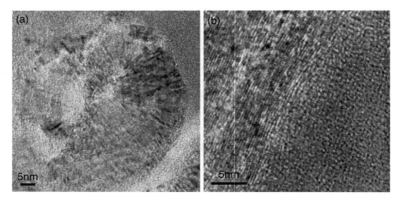

7.8 HRTEM images of (a) transverse and (b) longitudinal cross-sections of 5wt% MWCNT–SiO$_2$ composite showing mechanical bonding between MWCNTs and silica glass.

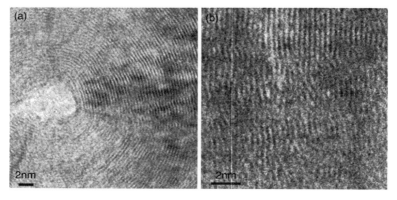

7.9 HRTEM images showing (a) transverse cross-section of a MWCNT reinforced in silica glass matrix composite and (b) parallel layers of MWCNTs also evident in longitudinal cross-section.

interface due to carbothermal reduction has also been reported (Seeger *et al.*, 2002). Figure 7.8 shows HRTEM images of the interface in CNT–SiO$_2$ composites revealing physical contact of CNTs with the surrounding glass matrix without the formation of any new phase due to chemical reaction between CNTs and silica (Subhani, 2012b). Similar observations of the absence of an intermediate phase have been reported in other CNT composites with glass/glass-ceramic matrices such as SiO$_2$ (Ning *et al.*, 2004), borosilicate (Thomas *et al.*, 2009), aluminoborosilicate (Mukhopadhyay *et al.*, 2010, Otieno *et al.*, 2010) and barium aluminosilicate (Ye *et al.*, 2006). HRTEM has also been used to observe longitudinal and transverse cross-sections of CNTs embedded in composites (Fig. 7.9).

Acid-treated CNTs possess corroded surface layers and it can be speculated that these rough CNT surfaces produce mechanical keying with the glass matrix. Such a mechanical interlocking is the likely cause of a mechanical bonding between CNTs and the matrix. When pristine CNTs are used to reinforce glasses/glass-ceramics without functionalization, the presence of surface defects produced during their growth serves to develop mechanical keying. One recent study suggested that CNTs are present in a stressed state in glass/glass-ceramic matrices due to the application of compressive load during sintering of the composites, or simply differential thermal contraction; in response, CNTs apply a tensile stress on the surrounding bulk matrix leading to intimate contact at the interface (Subhani, 2012a).

7.6   Properties

7.6.1   Physical properties

The density of CNT–glass/glass-ceramic matrix composites directly affects their properties; indeed, achievement of near complete densification is a prerequisite for enhanced mechanical properties. SPS and HPS have provided dense CNT–glass/glass-ceramic composites with silica (Cho, 2010), aluminoborosilicate (Mukhopadhyay *et al.*, 2010) and barium aluminosilicate (Ye *et al.*, 2006) as matrices. In some cases, both SPS and HPS techniques are reported to have produced 100% densification of CNT–glass/glass-ceramic matrix composites containing up to 7.5wt% CNTs (Guo *et al.*, 2007a) compared to conventional PLS which yielded ~98% relative density in composites containing the same CNTs fraction (Subhani *et al.*, 2011). Further increases in CNT content can lead to a gradual decrease in the density of composites, although this effect was comparatively more pronounced in the case of PLS than SPS or HPS (Subhani *et al.*, 2011). Figure 7.10 summarizes the densification results available in literature for a variety of CNT-glass/glass-ceramic matrix composites produced by HPS, SPS and PLS.

Good composite density is not only related to the densification technique employed but it is also associated with the uniformity of CNT dispersion in the green body and with the degree of powder packing of the matrix material. A suitable composite mixing process homogeneously disperses CNTs in a well-packed particulate matrix, avoiding CNT agglomeration that otherwise produces unfilled glass areas and can induce the development of porosity.

Interest in conventional PLS is driven by its cost-effectiveness and its adaptability to parts with different shapes and dimensions. Initial attempts, however, did not show good densification results for 10wt% CNT–

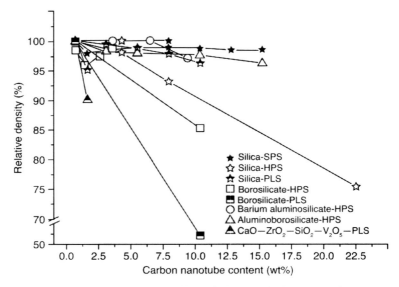

7.10 Percentage relative densities of glass and glass-ceramic matrix composites containing CNTs available in the literature.

borosilicate composites; relative densities of only 51% were obtained (Boccaccini *et al.*, 2005). A change in the composite powder processing technique from direct powder mixing to colloidal mixing provided good density of silica glass composites, i.e. 95% of relative density for composites with (only) 1wt% CNTs (Arvanitelis *et al.*, 2008). Recently, the use of PLS was revived and high CNT loadings (10wt%) were incorporated in silica glass to achieve densities greater than 96% (Subhani *et al.*, 2011); comparatively lower CNTs contents (2.5–7.5wt%) showed even higher densities of 98–99%. Glass-ceramic matrix composites have been also densified by PLS, for example, CNT–CaO–ZrO$_2$–SiO$_2$ and CaO–ZrO$_2$–SiO$_2$–V$_2$O$_5$ matrix composites containing 1wt% CNTs, but these showed only 90% relative density (Giovanardi *et al.*, 2010).

7.6.2 Mechanical properties

Four mechanical properties of CNT–glass/glass-ceramic matrix composites have been investigated in most studies – hardness, elastic modulus, fracture strength and fracture toughness. The incorporation of CNTs in glasses/glass-ceramic matrices has led to different effects on these properties, as is now explained.

Hardness

The hardness of CNT–glass/glass-ceramic matrix composites has been investigated several times, but the actual effect of CNTs on hardness is still unclear. In the absence of a hardness value of individual CNTs, their effect on the hardness of composites is difficult to predict. As a result, three different hardness trends have been shown in the literature and can be explained as follows.

(a) An increase in hardness has been associated with dense matrices and good bonding between the CNTs and the glass matrix, although no evidence of a strong interface is usually available other than the existence of a simple mechanical bond.

(b) A continuous decrease in hardness has been associated with inhomogeneous dispersion of CNTs in matrices and poor densities of sintered composites.

(c) An initial increase followed by a decrease in hardness has been explained by a combination of (a) and (b), i.e. good interfacial bonding and/or uniform dispersion at low CNTs loading and inhomogeneous distribution of CNTs at higher contents, leading to the formation of CNT agglomerates and poor composite densities.

An early investigation on CNT–SiO_2 glass composites containing 4wt% CNTs showed an increase in hardness of 28% in comparison to that of unreinforced silica glass (de Andrade *et al.*, 2007). Another study on CNT–SiO_2 glass composites used a high-pressure technique to densify composites and showed 44% improvement in hardness of composites by adding only 0.075wt% CNTs (de Andrade *et al.*, 2008). In contrast, borosilicate glass matrix composites containing 10wt% CNTs demonstrated a 22% decrease in hardness, which was attributed to inhomogeneous dispersion of CNTs as a direct mixing technique was used to prepare the composite powder (Boccaccini *et al.*, 2007). Another study on borosilicate glass matrix composites, prepared by the sol–gel technique, demonstrated an increase in hardness of 13% by adding 2wt% CNTs but further addition of CNTs (3wt%) was reported to produce agglomerates and a decreased hardness to a value 7% less than that of pure borosilicate glass (Thomas *et al.*, 2009). CNT composites containing glass-ceramic matrices have also shown a mixed trend in hardness with CNT content. In a study on CNT–aluminoborosilicate glass-ceramic composites, an increase in hardness of 10% and 23% was observed on adding MWCNTs and SWCNTs, respectively (Chu *et al.*, 2008). However, another study by the same group of researches showed a 31% decrease in hardness in 15wt% CNT–aluminoborosilicate composites (Mukhopadhyay *et al.*, 2010).

A model composite system has recently been developed by uniformly

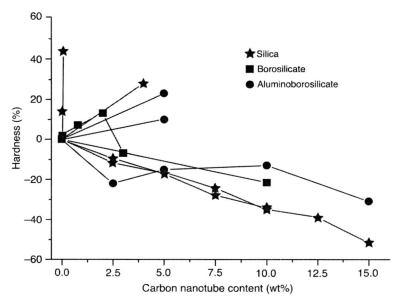

7.11 Percentage variation of the hardness of glass and glass-ceramic matrix composites containing CNTs from the literature.

dispersing CNTs in silica glass by a colloidal mixing technique and consolidating the composite mixtures by PLS and SPS to near theoretical densities (Cho *et al.*, 2011, Subhani *et al.*, 2011). The results of Subhani (2012b) showed a continuous decrease of hardness with CNTs addition. The effect of inhomogeneous CNT dispersion and insufficient densities was avoided in the model silica matrix system investigated. Figure 7.11 shows published hardness data of CNT–glass/glass-ceramic matrix composites, obtained by Vickers macro/microhardness and Berkovich nanohardness techniques. None of the CNT–glass/glass-ceramic matrix composites showed a continuous increase in hardness to high CNT loadings, although improvements up to 5wt% are reported in several systems. In those experiments that probe higher loadings, a systematic decrease in hardness was observed at up to 15wt% CNTs. Sharp increases at very low filler contents are most likely related to changes in matrix morphology or crystallinity.

Elastic modulus

Like hardness, published data for the elastic modulus of CNT–glass/glass-ceramic matrix composites show different trends, as a direct result of different composite microstructures. Considering the high stiffness of CNTs, which should be greater than that of glasses/glass-ceramics, an increase in

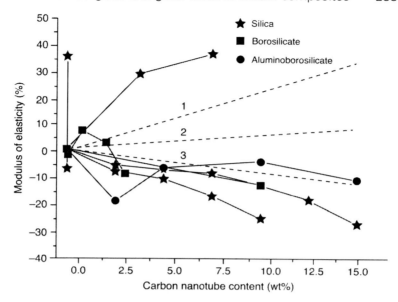

7.12 Percentage variation in modulus of elasticity of glass and glass-ceramic matrix composites containing CNTs from the literature. Dotted lines 1, 2 and 3 show the calculated elastic modulus of three glass matrix composites with three-dimensional randomly oriented CNTs of elastic moduli of 1000 GPa, 500 GPa and 100 GPa; an elastic modulus value of 70 GPa was chosen for glass.

elastic modulus of the composites is expected. However, since the focus of inorganic material reinforcement is usually toughness, relatively few data are directly available (Fig. 7.12) and they are generated by disparate techniques depending on sample size (three- and four-point bending tests, ultrasonic techniques and instrumented hardness). There are some reports of an increase in composite stiffness, for example a 38% enhancement in silica stiffness at 7.5wt% CNTs, though matrix crystallization may play a role, particularly where large increases are observed at low filler loadings (Guo *et al.*, 2007a, Loo *et al.*, 2007). Improvements in stiffness of CNT–borosilicate glass matrix composites at low CNT loadings (2wt%) have been noted in the absence of crystallization effects; however, at higher CNT loadings (3wt%), agglomeration and other consolidation issues lead to a net reduction in performance (Thomas *et al.*, 2009). In other cases, monotonic reductions in stiffness have been attributed to inhomogeneous dispersion of CNTs, insufficient densification or degradation of CNTs during processing, in both glass (Boccaccini *et al.*, 2007, Thomas *et al.*, 2009) and glass-ceramic matrix composites (Mukhopadhyay *et al.*, 2010).

For three-dimensional randomly aligned short fiber composites, a modified rule of mixtures (Krenchel's rule (Krenchel, 1964)) incorporating an orientation efficiency factor (η_0) and length efficiency parameter (η_L) is often used:

$$E_c = \eta_0 \eta_L E_f V_f + E_m(1 - V_f) \qquad [7.1]$$

in which E_c, E_f and E_m are the elastic moduli of the composite, fibers and matrix, respectively, and V_f is the fiber volume fraction. The value of η_0 for three-dimensional random orientation of fibers is 0.2 (Matthews and Rawlings, 1999) and the value of η_L depends upon the fiber length (l), diameter (D), interfiber spacing ($2R$) and shear modulus of the matrix (G_m):

$$\eta_L = 1 - \left(\frac{\tanh[(1/2)\beta l]}{(1/2)\beta l} \right) \qquad [7.2]$$

$$\beta = \left(\frac{8G_m}{E_f D^2 \log_e (2R/D)} \right)^{1/2} \qquad [7.3]$$

Applying the modified rule of mixtures on CNT–glass composites by taking three different values of CNTs (1000, 500 and 100 GPa), and the elastic modulus of glass (70 GPa) provides the composite elastic moduli shown in Fig. 7.12 (Subhani, 2012b). This wide range of values reflect the expected differences between arc-grown and CVD-grown CNTs (Yu *et al.*, 2000a, Zhao and Zhu, 2011); note that the hollow core is generally excluded when reporting the elastic modulus of CNTs but its effect should be considered in the context of composite reinforcement. Higher composite values should be obtained (dotted line 1 in Fig. 7.12) if CNTs possess the theoretically estimated elastic modulus of 1000 GPa; at 500 GPa there is a modest improvement (dotted line 2) while a value of 100 GPa decreases the elastic modulus of the composites (dotted line 3) to a value less than the glass matrix; many CNT–glass/glass-ceramic composites in Fig. 7.12 indeed demonstrate an elastic modulus less than that of the matrix glass. This simple analysis assumes that the transverse properties of CNTs are neglible, which may be reasonable given relatively weak interfacial bonding and the hollow/collapsible nature of the structure. In addition, the length of the CNTs should be considered in more detail; it is worth noting that CNTs near the critical length may be favored for toughening but are sub-optimal for stiffening. For an increased composite modulus in three-dimensional randomly oriented CNT–glass/glass-ceramic matrix composites, a high intrinsic elastic modulus (500–1000 GPa) is required; the value is higher than most CVD-grown CNTs, but the graphitizing effects of high-temperature sintering on the carbon lattice may be beneficial. The porosity generated by insufficient densification also contributes to reducing the elastic modulus.

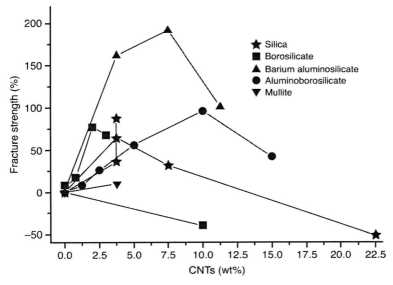

7.13 Percentage variation in fracture strength of glass and glass-ceramic matrix composites containing CNTs available in the literature.

Fracture strength

Significant increases in the fracture strength of CNT–glass/glass-ceramic matrix composites have been observed at low to moderate CNT loadings (<10wt%), using three- and four-point flexural and compression strength tests (Fig. 7.13). However, further increases in CNT content usually lead to composites with reduced mechanical strength (Boccaccini *et al.*, 2007, Ning *et al.*, 2003a, Thomas *et al.*, 2009), although the optimum loading varies with the study, reflecting differences in materials and processing. Microstructure, and the degree of agglomeration, are again key issues, as illustrated by the extra enhancement obtained in 3.75wt% CNT–SiO_2 composites (Ning *et al.*, 2003a), from 65% to 88% on introducing a better surfactant system (Ning *et al.*, 2004).

Fracture toughness

Fracture toughness is the property that has received most attention, since brittleness is the key limitation of many glass-based systems; the objective has been to introduce additional toughening mechanisms through the incorporation of CNTs. Due to the small sample volumes available, a large number of studies have employed Vickers indentation fracture (VIF) toughness techniques to assess fracture toughness. VIF is a simple technique that enables a comparison between toughness values of the composites and

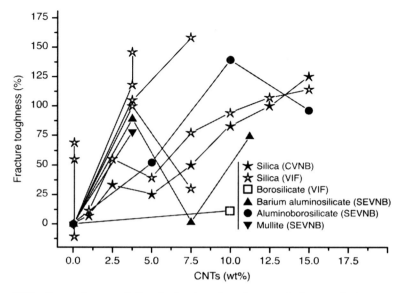

7.14 Percentage variation in fracture toughness of glass and glass-ceramic matrix composites containing CNTs available from the literature.

those of unreinforced glass/glass-ceramic matrices. However, there has been some discussion in the literature about the validity of such measurements, particularly in poorly consolidated materials. Single-edge V-notch beam (SEVNB) and chevron V-notch beam (CVNB) techniques are a better guide to absolute fracture toughness values. Nevertheless, comparative studies have shown that both approaches can offer semi-quantitatively similar trends for well-processed systems, although VIF tends to overestimate CVNB values (Cho *et al.*, 2011).

An increase in fracture toughness has been observed in all composites developed (Fig. 7.14), although in some cases, a drop again occurs at high CNT loadings due to inhomogeneous CNT dispersion. However, the extent of improvement of fracture toughness, although significant, is limited; a maximum increase of 150% has been observed using the indentation technique (Guo *et al.*, 2007a, Ning *et al.*, 2004). This level of improvement in fracture toughness, is not, in itself, expected to deliver structurally reliable composite materials; conventional fiber reinforced composites (discussed earlier) usually exhibit higher fracture toughness. Nevertheless, the improvements observed using CNTs may provide useful benefits in combination with other properties, or indeed in combination with conventional fibers, to create hierarchical composites, as has proved promising for polymer matrix systems (Qian *et al.*, 2010). CNT–glass/glass-ceramic matrix composites tested for fracture toughness include those with silica (Cho *et al.*, 2011, de Andrade *et al.*, 2008, Guo *et al.*, 2007a, Ning *et al.*, 2003a, 2004), borosilicate (Boccaccini *et al.*, 2007), aluminoborosilicate (Mukhopadhyay *et al.*, 2010), barium aluminosilicate (Ye *et al.*, 2006) and mullite (Wang *et al.*, 2007) matrices. Improvements in inorganic matrix CNT composite toughness are usually attributed to conventional fiber mechanisms, such as crack deflection, CNT bridging and CNT pull-out, as characteristic features are often observed by fractography (Figs 7.4 and 7.6). However, scaling considerations, discussed in a recent study (Cho *et al.*, 2011), highlight the lower absolute performance expected for nanofibers compared to microfibers, if only these conventional mechanisms operate. It may, therefore, be most useful to consider the possibility of enabling fundamentally new toughening mechanisms such as shear banding of hollow nanostructures or pull-out of flexible SWNTs over convoluted contour lengths.

7.6.3 Functional properties

The incorporation of CNTs in glass/glass-ceramic matrices also influences their functional properties, including both electrical and thermal conductivities. For completeness, a brief account of these properties is given below.

Electrical conductivity

The incorporation of CNTs in glass/glass-ceramic matrices tremendously increases the electrical conductivity of otherwise insulating glasses and glass-ceramics (Fig. 7.15). At very low CNT contents, electrical conductivity is not increased significantly (Gavalas *et al.*, 2001) but the formation of a conducting percolating network of CNTs rapidly increases electrical conductivity by several orders of magnitude following the scaling law according to percolation theory (Kovacs *et al.*, 2007):

$$\sigma_c = \sigma_{CNT}(\phi - \phi_c)^t \qquad [7.4]$$

where σ_c and σ_{CNT} are the conductivities of composites and CNTs, respectively, Φ is the volume fraction of CNTs in composites, Φ_c is the critical volume fraction (percolation threshold) and the exponent t is the dimensionality of the system. Most of the systems in Fig. 7.15 had a percolation threshold between 1wt% and 3wt% and dimensionality constant between 1 and 3 (Cho, 2010, Mukhopadhyay *et al.*, 2011, Thomas *et al.*, 2009, Xiang *et al.*, 2007). These values of percolation threshold are consistent with the excluded volume theory for high aspect ratio particles (i.e. CNTs of typical aspect ratio of 100–1000). Low percolating thresholds represent agglomeration of CNTs (de Andrade

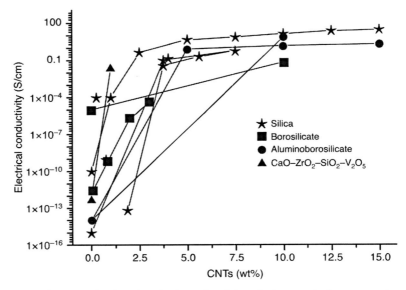

7.15 Increase in the electrical conductivity of glass and glass-ceramic matrix composites containing CNTs available in the literature.

et al., 2009) or the structuring of CNTs around large micron-size grains in glass-ceramic matrices (Subhani, 2012a), while a value > 1.0wt% shows a homogenous dispersion of CNTs with individual CNTs well-separated from others (Bauhofer and Kovacs, 2009). Higher aspect ratio CNTs give lower percolation threshold and higher absolute conductivity of the composite system due to fewer junctions, but are susceptible to alignment effects (i.e. anisotropy in electrical conductivity) (Otieno *et al.*, 2010). In principle, the dimensionality constant should provide geometrical information about the CNT network, whether controlled by agglomeration or excluded volume effects, as also observed in CNT–polymer composites (Bauhofer and Kovacs, 2009).

Thermal conductivity

The thermal diffusivity (α) and specific heat capacity (Cp) of CNT–glass/glass-ceramic matrix composites are generally measured by laser flash technique and differential scanning calorimetry, respectively. Thermal conductivity (Kc) is then calculated by computing the two measured values along with sintered density (ρ) of composites using the following equation:

$$K_c = \alpha C_p \rho \qquad [7.5]$$

Since CNTs exhibit high individual thermal conductivities greater than

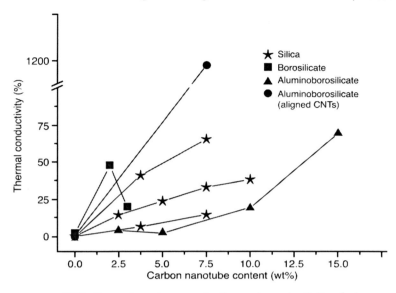

7.16 Percentage increase in thermal conductivity of glass and glass-ceramic matrix composites containing CNTs from the literature.

graphite (2000 W/m.K) and diamond (900 W/m.K) (Lee, 1994), it was hoped that composites of extremely high thermal conductivity might be produced. However, the extent of increment (Fig. 7.16) is typically rather low (<100%), due to the high interfacial density associated with nanoscale particles. Similar levels of enhancement have been observed at elevated temperature (e.g. 650°C) (Ning *et al.*, 2003b). It is noteworthy that the characteristic percolation behavior observed for electrical conductivity is not mirrored in the thermal response. The fundamental reason for this is that there are far fewer orders of magnitude of variation between the thermal conductivities of fiber and matrix than in the electrical case. Only modest improvements in thermal conductivity are possible on incorporating discontinuous CNTs in glass/glass-ceramic matrices. Many reasons may limit the enhancement, including the presence of structural defects in CNTs, non-uniform dispersion of CNTs producing agglomerates, insufficient densification of composites leading to porosity, preferential alignment of CNTs producing anisotropic composite properties and the presence of new phases at the CNT/glass interfaces; however, even apparently good-quality composites disappoint. Simple rules of mixture expectations have never been met, even once the random orientation of the CNTs is taken into account (conductivity is only high axially), due to the strong influence of interfaces. On the basis of the Maxwell–Garnett effective medium approach, the thermal conductivity of a composite can be calculated (Nan *et al.*, 2003):

$$K_c = K_m + K_f \cos^2\theta V_f \qquad [7.6]$$

where K_m and K_f are the thermal conductivities of the matrix and fibers, respectively, V_f is the fiber volume and θ is the angle between fiber axis and the given direction. For aligned and random orientation of fibers, $\cos^2\theta$ is 1 and 1/3, respectively (Nan *et al.*, 2003). The calculated value of the thermal conductivity of CNT–glass/glass-ceramic matrix composites obtained from equation 7.6 is generally two orders of magnitude higher than the experimental value. The thermal resistance due to large interfacial surface area between CNTs and glass matrix is the most likely reason for the reduced increment of thermal conductivity of composites. An expression to calculate thermal conductivity of composites incorporating interfacial thermal resistance is available and has proven quite widely applicable (Nan *et al.*, 2004):

$$K_c = K_m + \frac{K_f L}{2R_i K_f + L} \cos^2\theta V_f \qquad [7.7]$$

where L is the length of fibers and R_i is the interfacial thermal resistance, also known as Kapitza resistance.

7.6.4 Technological properties

A variety of technological properties of CNT–glass/glass-ceramic matrix composites are particularly relevant to potential industrial applications. Recently, a model system based on CNT–SiO$_2$ composites has been used to explore such properties, including wear and friction resistance and thermal shock, cycling and ageing resistance, as now discussed (Subhani, 2012b).

Wear and friction resistance

No published data on the wear and friction properties of CNT–glass/glass-ceramic composites are available, although related studies on CNT–ceramic composites have been performed (Ahmad *et al.*, 2010, An *et al.*, 2003) showing a decrease in wear rate at low CNT loadings (<5wt%) but an increase at higher CNT contents, as observed with many primary mechanical properties (Ahmad *et al.*, 2010). Wear and friction experiments on CNT–glass/glass-ceramic composites can be performed by different methods; commonly available techniques are the ball-on-disk and pin-on-disk tests, in which a ball or a pin of a hard material slides under pressure against the flat surface of the composite specimen. A variety of materials ranging from hardened steels to ceramics and different sizes (diameters) of pins and balls are

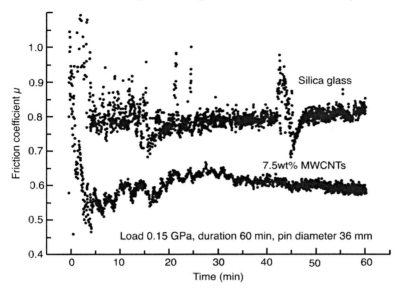

7.17 Friction coefficients of silica glass and 7.5wt% MWCNT–SiO$_2$ composite. A decrease in friction coefficient can be seen in the composite (from 0.8 to 0.6).

used with varying applied loads and sliding speeds depending on the composite specimens to be tested. The frictional force is recorded by a load cell attached to the wear testing machine to find the friction coefficient (μ), which is the ratio of the frictional force (F_f) and the applied force (F) between the sliding bodies, i.e. ball/pin and the composite specimen:

$$\mu = \frac{F_f}{F} \qquad [7.8]$$

The volume loss (wear volume) of the specimen is calculated using a profilometer or measurement of the decrease in weight of the specimen. Wear rate (W) is calculated by inserting the wear volume (V), sliding distance (L) and the applied load (F), as follows:

$$W = \frac{V}{LF} \qquad [7.9]$$

In an unpublished study on model CNT–SiO$_2$ composites (Subhani, 2012d), the pin-on-disk technique was used to measure wear volumes and the friction coefficients of silica glass and 7.5wt% CNT–SiO$_2$ composites. A hardened steel (CPM-10V) pin of diameter 36 mm slid against the specimen in a reciprocating motion at a load of 0.15 GPa for 60 min. A decrease in friction coefficient from 0.8 to 0.6 was observed on incorporating CNTs in silica glass, as shown in Fig. 7.17. However, the wear volume of the composite was found

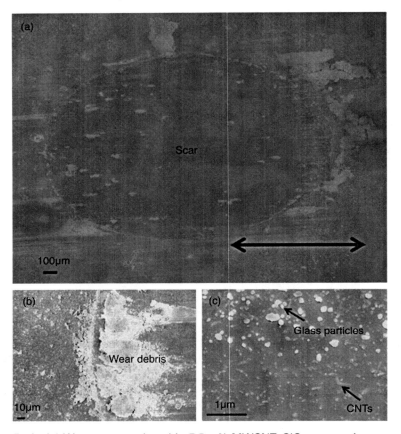

7.18 (a) Wear scar produced in 7.5wt% MWCNT–SiO$_2$ composite specimen after test. (b) Wear debris containing glass particles is evident. (c) Pulled out glass particles are also shown at the edge of the scar; CNTs have been exposed on the scar surface.

to be more than that of the glass specimen. This result suggests that the incorporation of CNTs in glasses/glass-ceramics decreases the friction coefficient by acting as a solid lubricant between the contacting surface of the sliding pin and the specimen. A possible reason for the unexpected increase in wear volume may be the decrease in hardness of composites by adding CNTs. Figure 7.18(a) shows a scar produced on the 7.5wt% CNT–SiO$_2$ composite specimen after the wear test. Wear debris produced is shown at the edge of the scar (Fig. 7.18(b)) along with the pulled out glass particles (Fig. 7.18(c)). In the process, the reinforced CNTs have been exposed due to the ploughing of composite surface by the sliding pin (Fig. 7.18(c)). A detailed study is required to fully explore the wear and friction properties of CNT–glass/glass-ceramic matrix composites and the wear mechanisms involved.

Thermal shock, cycling, and ageing behavior

Glasses and glass-ceramics are sensitive to sudden temperature variations because they generally have poor thermal conductivity and low tensile strength (Arnold *et al.*, 1996). The brittle nature and internal stresses produced due to thermal gradients usually result in the catastrophic failure of glasses and glass-ceramics. CNTs possess high strength and thermal conductivity and their incorporation in glasses/glass-ceramics can produce thermally conductive composites with enhanced resistance to thermal shock.

Thermal shock resistance is experimentally found by subjecting a material to a sudden change (e.g. quenching) in temperature from an initial temperature (T_i) to final temperature (T_f) and measuring the change in fracture strength. T_f is generally kept constant (room temperature) while T_i is gradually increased until the critical temperature (T_c) is achieved, which is the temperature at which a sudden drop in fracture strength is noticed due to the formation of cracks in the specimen. The difference between two temperatures ($\Delta T = T_c - T_f$) is the safe range wherein a specimen remains unaffected by a thermal shock. The resistance to thermal shock is characterized by a thermal shock resistance parameter (R) (Boccaccini *et al.*, 1999), which is related to the fracture strength (σ), Poisson's ratio (v), thermal expansion coefficient (α) and modulus of elasticity (E) of a material:

$$R = \frac{\sigma(1 - v)}{\alpha E}$$
[7.10]

A high value of R shows better resistance of a material to thermal shock. Other factors, including specific heat capacity, fracture toughness, dispersion quality of reinforcing phase and the geometry of the composite, may also affect thermal shock resistance.

No reports are available describing thermal shock, cycling and ageing behaviors of CNT–glass/glass-ceramic matrix composites. However, unpublished data (Subhani, 2012c) of a model CNT–SiO_2 composite showed that the incorporation of CNTs in silica glass did not reduce the thermal shock resistance of otherwise highly resistant silica glass after quenching from 500°C and 1000°C to room temperature (20°C); a further increase in temperature to 1200°C initiated devitrification of silica glass, which reduced the strength of CNT–SiO_2 composites. Figure 7.19 shows SEM images of the surfaces of 2.5wt% CNT–SiO_2 composite specimens that had been thermally shocked from 1000°C and 1200°C. Figure 7.19(a) shows no evidence of cracks while Fig. 7.19(b) shows surface cracks due to cristobalite formation, which was also witnessed in specimens heated to 1200°C and slowly cooled to room temperature without any thermal shock.

Repeating the thermal shock cycles (thermal cycling) on CNT–SiO_2 composites, i.e. heating up to 1000°C and quenching in water at 20°C up to

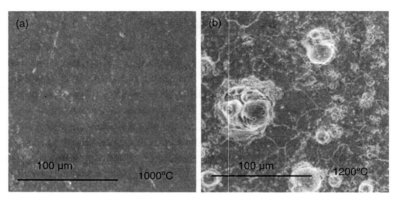

7.19 2.5wt% MWCNT–SiO$_2$ composites after thermal shock test from (a) 1000°C and (b) 1200°C. No cracks can be seen in specimen quenched from 1000°C while the specimen quenched from 1200°C showed cracks due to the formation of cristobalite.

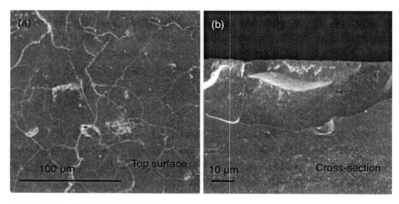

7.20 2.5wt% MWCNT–SiO$_2$ composites after thermal shock test comprising 20 cycles from 1000°C. Cracks can be seen on (a) surface and (b) cross-section of the specimen due to cristobalite formation because of thermal cycling.

20 times, did not lead to failure of the composites due to thermal shock but it encouraged devitrification in silica glass. Figure 7.20 shows images of the surface and cross-section of CNT–SiO$_2$ composites thermally cycled up to 1000°C, 20 times; surface cracks only formed due to the crystallization of cristobalite at 1000°C on repeated heating of the specimens. It can be inferred from the results on CNT–SiO$_2$ composites that the incorporation of CNTs is likely to increase thermal shock resistance when reinforced in glass systems susceptible to thermal shock.

Thermal ageing of CNT–glass/glass-ceramic composites is another important area of research in order to find the maximum working

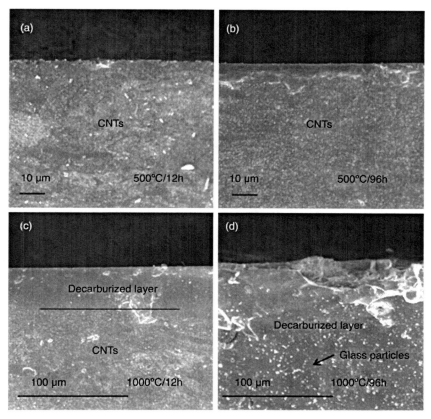

7.21 2.5wt% MWCNT–SiO$_2$ composites thermally aged at 500°C and 1000°C for 12 h and 96 h. MWCNTs did not oxidize at 500°C ((a) and (b)) but at 1000°C, MWCNTs started oxidizing ((c) and (d)), producing a decarburized surface layer.

temperature at which such composites can be used for long periods of time without oxidizing the embedded CNTs. The effect of ageing temperature and time was studied in the same unpublished investigation on CNT–SiO$_2$ composites (Subhani, 2012c), which showed that CNTs did not oxidize at a temperature of 500°C but started oxidizing at 750°C and the rate of CNTs oxidation increased at 1000°C. Figure 7.21 shows cross-sectional views of 2.5wt% CNT–SiO$_2$ composites thermally aged at 500°C and 1000°C for 12 h and 96 h; an increased decarburized depth in CNT–SiO$_2$ composite specimens is evident with the increase in time and temperature. Finally, the effect of CNT concentration in composites was also assessed by thermally ageing CNT–SiO$_2$ composites with CNT loadings from 2.5wt% to 10wt%. A completely CNT-free, porous silica glass was obtained by heating the 10wt% CNT–SiO$_2$ composite, as shown in Fig. 7.22.

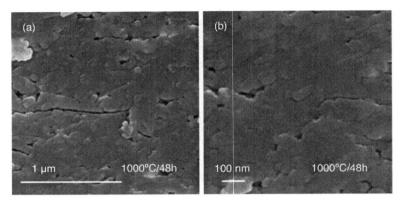

7.22 (a, b) 10wt% MWCNT–SiO$_2$ composites thermally aged at 1000°C for 48 h. MWCNTs oxidized completely, leaving behind a porous silica glass matrix.

7.7 Applications

Low density, improved fracture toughness, enhanced electrical and thermal conductivity, resistance to oxidation at moderate to high temperatures, and higher hardness and stiffness than polymers and some metals make CNT–glass/glass-ceramic matrix composites an attractive choice for potential applications where structural and thermal properties are required. In aerospace, electronic and engine components, enhanced thermal conductivity is required for fast heat dissipation. CNT–glass/glass-ceramic composites, due to their higher thermal conductivities than the pure matrices, may be used for space structural applications and as heat sink materials to dissipate heat from electronic components. Applications requiring a sudden temperature change may also be considered for CNT composites due to their resistance to thermal shock and cycling, for example in the handling of low-melting metals and glasses. CNT composites can also be used for applications demanding intermediate temperatures up to 500°C and normal air environments. Glass matrix composites containing CNTs can be developed for the thermal protection of C/C composites and also as thermal barrier coatings.

The tremendous increase in electrical conductivity of insulating glasses by incorporating CNTs, dispersed or aligned, has added a new material family for the electroceramics industry, in which electrical conductivity can be tailored to the required application. The anisotropy in electrical and thermal conductivity of CNTs can be used to fabricate glass matrix composites with aligned CNTs, which can provide the desired thermal/electrical conductivity in specified directions only. Due to their (modest) increased fracture toughness, CNT–glass/glass-ceramic matrix composites are not (yet) attractive for structural applications but they may be used in applications under low or moderate loads. Moreover, the low friction resistance of CNT–glass matrix composites can be exploited for applications demanding anti-friction materials such as in pump manufacturing and components for the automotive industry. Like carbon fiber–glass/glass-ceramic composites, components such as bearings, seals, brake and gear systems can be manufactured using CNT–glass/glass-ceramic composites.

Finally, the process of developing a porous glass after burning incorporated CNTs may also be used for certain applications where controlled (nano)porosity is required, such as in filters, catalyst supports or sorbents.

7.8   Conclusions and scope

A comprehensive analysis of published data in the field of CNT–glass/glass-ceramic matrix composite reveals the novelty of this new class of nanocomposites and the significant scope that they have for technological applications. During the last ten years, considerable progress has been made and consolidated glass matrix composites with uniformly dispersed CNTs at relatively high loadings (15wt%) are now a reality. Optimization of composite powder preparation techniques to disperse CNTs homogeneously in glass/glass-ceramic matrices and their densification with different consolidation techniques is a significant achievement that delivers a variety of manufacturing routes to suit the varying dimensional requirements of composite products.

Although the effect of randomly dispersed CNTs on the hardness and stiffness of glass/glass-ceramic matrices is still not clearly understood, moderate improvements in fracture strength and toughness and thermal conductivity have been conclusively observed; the significant increase in electrical conductivity is particularly promising. Incorporating highly aligned CNTs of high aspect ratio in one or more well-specified directions remains to be achieved. Although challenging, such a configuration of CNTs in composites will increase their functional and structural properties tremendously, especially thermal conductivity (Otieno et al., 2010).

The interfaces in nanocomposites occupy a large area; it might be possible in future to develop in-situ interfaces formed by the chemical reaction between CNTs and the surrounding glass phase resulting in better composite properties. Uniformly dispersed CNTs may also be utilized as a carbon source for in-situ development of carbides in glass/glass-ceramic matrices.

The optimum quality and nature of CNTs remains an open question, and the answer is likely to vary with application. CNTs produced by different techniques offer varying characteristics in their purity, crystallinity, straightness, diameter and aspect ratio. The selection of SWCNTs, DWCNTs or MWCNTs for reinforcement in glass/glass-ceramic matrices may have a strong influence on phenomenology, particularly on toughening mechanisms. The lateral flexibility of CNTs and their high axial stiffness are contrasting characteristics. It remains to be seen whether thin flexible SWNTs or stiff, straight MWNTs are most effective; most likely this diameter range will span a transition from conventional fiber theory to new mechanisms requiring new understanding. The development of *in-situ* CNT composite synthesis methods is another area for further research; *in-situ* CNT processing could lead to defect-free CNTs without catalysts and carbon by-products and without affecting the surface properties of CNTs.

A useful increase in functional properties, moderate increase in mechanical characteristics and improved technological properties suggest the potential use of CNT–glass/glass-ceramic matrix composites in numerous industrial sectors. Although the extent of improvement in these composites (especially fracture toughness) is not as significant as that obtained with continuous/chopped carbon fiber reinforced glass/glass-ceramic matrix composites, the possibility of tailoring the functional properties and the simplicity of production (in comparison with continuous fiber reinforced composites) justifies continued R&D efforts in this field. Combinations of CNT fillers with conventional fiber reinforcements may prove particularly beneficial in the near term, as has begun to be demonstrated in polymer matrix systems (Qian *et al.*, 2010).

7.9   References

Ahmad, I., Kennedy, A. and Zhu, Y. Q. 2010. Wear resistant properties of multi-walled carbon nanotubes reinforced Al_2O_3 nanocomposites. *Wear*, 269, 71–78.

Ajayan, P. M., Redlich, P. and Ruhle, M. 1997. Balance of graphite deposition and multishell carbon nanotube growth in the carbon arc discharge. *Journal of Materials Research*, 12, 244–252.

An, J. W., You, D. H. and Lim, D. S. 2003. Tribological properties of hot-pressed alumina-CNT composites. *Wear*, 255, 677–681.

Arnold, M., Boccaccini, A. R. and Ondracek, G. 1996. Theoretical and experimental considerations on the thermal shock resistance of sintered glasses and ceramics using modelled microstructure-property correlations. *Journal of Materials Science*, 31, 463–469.

Arvanitelis, C., Jayaseelan, D. D., Cho, J. and Boccaccini, A. R. 2008. Carbon nanotube-SiO2 composites by colloidal processing. *Advances in Applied Ceramics*, 107, 155–158.

Aveston, J. 1971. Strength and toughness in fiber reinforced ceramics. In: *Conference on Properties of Fiber Composites*, 1971. IPC Science and Technology Press.

Babooram, K. and Narain, R. 2009. Fabrication of SWNT/silica composites by the sol-gel process. *ACS Applied Materials & Interfaces*, 1, 181–186.

Bakshi, S. R., Lahiri, D. and Agarwal, A. 2010. Carbon nanotube reinforced metal matrix composites – a review. *International Materials Reviews*, 55, 41–64.

Bansal, N. P. 1997. Mechanical behavior of silicon carbide fiber-reinforced strontium aluminosilicate glass-ceramic composites. *Materials Science and Engineering A – Structural Materials Properties Microstructure and Processing*, 231, 117–127.

Bauhofer, W. and Kovacs, J. Z. 2009. A review and analysis of electrical percolation in carbon nanotube polymer composites. *Composites Science and Technology*, 69, 1486–1498.

Belytschko, T., Xiao, S. P., Schatz, G. C. and Ruoff, R. S. 2002. Atomistic simulations of nanotube fracture. *Physical Review B*, 65, 8.

Berber, S., Kwon, Y. K. and Tomanek, D. 2000. Unusually high thermal conductivity of carbon nanotubes. *Physical Review Letters*, 84, 4613–4616.

Berguiga, L., Bellessa, J., Vocanson, F., Bernstein, E. and Plenet, J. C. 2006. Carbon nanotube silica glass composites in thin films by the sol-gel technique. *Optical Materials*, 28, 167–171.

Boccaccini, A. R. 2002. Glass and glass ceramic matrix composite materials: a review. *Interceram*, 51, 24–34.

Boccaccini, A. R., Pfeiffer, K. and Kern, H. 1999. Thermal shock resistant Al2TiO5–glass matrix composite. *Journal of Materials Science Letters*, 18, 1907–1909.

Boccaccini, A. R., Acevedo, D. R., Brusatin, G. and Colombo, P. 2005. Borosilicate glass matrix composites containing multi-wall carbon nanotubes. *Journal of the European Ceramic Society*, 25, 1515–1523.

Boccaccini, A. R., Thomas, B. J. C., Brusatin, G. and Colombo, P. 2007. Mechanical and electrical properties of hot-pressed borosilicate glass matrix composites containing multi-wall carbon nanotubes. *Journal of Materials Science*, 42, 2030–2036.

Bozhko, A. D., Sklovsky, D. E., Nalimova, V. A., Rinzler, A. G., Smalley, R. E. and Fischer, J. E. 1998. Resistance vs. pressure of single-wall carbon nanotubes. *Applied Physics A – Materials Science & Processing*, 67, 75–77.

Brennan, J. J. and Prewo, K. M. 1982. Silicon-carbide fiber reinforced glass-ceramic matrix composites exhibiting high-strength and toughness. *Journal of Materials Science*, 17, 2371–2383.

Charlier, J. C. and Issi, J. P. 1995. Electrical conductivity of novel forms of carbon. In: *8th International Symposium on Intercalation Compounds, May 28–Jun 01 1995 Vancouver, Canada*. Pergamon-Elsevier Science Ltd, pp. 957–965.

Chawla, K. K. 2003. *Ceramic Matrix Composites*. Kluwer Academic Publishers.

Che, J. W., Cagin, T. and Goddard, W. A. 1999. Thermal conductivity of carbon nanotubes. In: *7th Annual Foresight Conference on Molecular Nanotechnology, Oct 15–17 1999 Santa Clara, California*. IoP Publishing Ltd, pp. 65–69.

Cho, J. 2010. *Processing and Characterisation of Inorganic Matrix Composites Containing Carbon Nanotubes*. PhD Thesis, Imperial College London.

Cho, J., Boccaccini, A. R. and Shaffer, M. S. P. 2009. Ceramic matrix composites containing carbon nanotubes. *Journal of Materials Science*, 44, 1934–1951.

Cho, J., Inam, F., Reece, M. J., Chlup, Z., Dlouhy, I., Shaffer, M. S. P. and Boccaccini, A. R. 2011. Carbon nanotubes: do they toughen brittle matrices? *Journal of Materials Science*, 46, 4770–4779.

Cho, K. Kerans, R. J. and Jepsen, K. A. 1995. Selection, fabrication and failure behaviour of SiC monofilament-reinforced glass composites. *Ceramic Engineering and Science Proceedings*, 15, 815–823.

Chu, B. T. T., Tobias, G., Salzmann, C. G., Ballesteros, B., Grobert, N., Todd, R. I. and Green, M. L. H. 2008. Fabrication of carbon-nanotube-reinforced glass-ceramic nanocomposites by ultrasonic in situ sol-gel processing. *Journal of Materials Chemistry*, 18, 5344–5349.

Crivelli-Visconti, I. and Cooper, G. A. 1969. Mechanical properties of a new carbon fiber material. *Nature*, 221, 754–755.

Daenen, M. J., de Fouw, R. D., Hamers B., Janssen P. G. A., Schouteden K. and Veld M. A. J. 2003. *The Wondrous World of Carbon Nanotubes: A Review on Current Carbon Nanotubes Technologies*. Eindhoven University of Technology.

Dai, H. J., Wong, E. W. and Lieber, C. M. 1996. Probing electrical transport in nanomaterials: Conductivity of individual carbon nanotubes. *Science*, 272, 523–526.

de Andrade, M. J., Lima, M. D., Stein, L., Bergmann, C. P. and Roth, S. 2007. Single-walled carbon nanotube silica composites obtained by an inorganic sol-gel route. *Physica Status Solidi B – Basic Solid State Physics*, 244, 4218–4222.

de Andrade, M. J., Lima, M. D., Bergmann, C. P., Ramminger, G. D., Balzaretti, N. M., Costa, T. M. H. and Gallas, M. R. 2008. Carbon nanotube/silica composites obtained by sol-gel and high-pressure techniques. *Nanotechnology*, 19, 7.

de Andrade, M. J., Weibel, A., Laurent, C., Roth, S., Bergmann, C. P., Estournes, C. and Peigney, A. 2009. Electrical conductive double-walled carbon nanotubes: Silica glass nanocomposites prepared by the sol-gel process and spark plasma sintering. *Scripta Materialia*, 61, 988–991.

Dervishi, E., Li, Z. R., Xu, Y., Saini, V., Biris, A. R., Lupu, D. and Biris, A. S. 2009. Carbon nanotubes: synthesis, properties, and applications. *Particulate Science and Technology*, 27, 107–125.

DiMaio, J., Rhyne, S., Ballato, J., Czerw, R., Xu, J., Webster, S., Carroll, D. L., Fu, K. and Sun, Y. P. Year. Transparent silica glasses containing single walled carbon nanotubes. In: Marker, A. J. and Davis, M. J., eds. *Conference on Inorganic Optical Materials III, Aug 2 2001 San Diego, California*. International Society of Optical Engineering, pp. 48–53.

Ding, W., Calabri, L., Kohlhaas, K. M., Chen, X., Dikin, D. A. and Ruoff, R. S. 2007. Modulus, fracture strength, and brittle vs. plastic response of the outer shell of arc-grown multi-walled carbon nanotubes. *Experimental Mechanics*, 47, 25–36.

Dobedoe, R. S., West, G. D. and Lewis, M. H. 2005. Spark plasma sintering of ceramics: understanding temperature distribution enables more realistic comparison with conventional processing. *Advances in Applied Ceramics*, 104, 110–116.

Dresselhaus, M. S. 2001. Nanotubes – burn and interrogate. *Science*, 292, 650–651.

Ebbesen, T. W., Lezec, H. J., Hiura, H., Bennett, J. W., Ghaemi, H. F. and Thio, T. 1996. Electrical conductivity of individual carbon nanotubes. *Nature*, 382, 54–56.

Fischer, J. E., Dai, H., Thess, A., Lee, R., Hanjani, N. M., Dehaas, D. L. and Smalley, R. E. 1997. Metallic resistivity in crystalline ropes of single-wall carbon nanotubes. *Physical Review B*, 55, R4921–R4924.

Gadkaree, K.P. and Chyung, K. 1986. Silicon carbide whisker-reinforced glass and glass-ceramic matrix. *American Ceramic Society Bulletin*, 65, 370–376.

Gamaly, E. G. and Ebbesen, T. W. 1995. Mechanism of carbon nanotube formation in the arc-discharge. *Physical Review B*, 52, 2083–2089.

Gavalas, V. G., Andrews, R., Bhattacharyya, D. and Bachas, L. G. 2001. Carbon nanotube sol-gel composite materials. *Nano Letters*, 1, 719–721.

Giovanardi, R., Montorsi, M., Ori, G., Cho, J., Subhani, T., Boccaccini, A. R. and Siligardi, C. 2010. Microstructural characterisation and electrical properties of multiwalled carbon nanotubes/glass-ceramic nanocomposites. *Journal of Materials Chemistry*, 20, 308–313.

Guo, S. Q., Sivakumar, R. and Kagawa, Y. 2007a. Multiwall carbon nanotube-SiO2 nanocomposites: Sintering, elastic properties, and fracture toughness. *Advanced Engineering Materials*, 9, 84–87.

Guo, S. Q., Sivakumar, R., Kitazawa, H. and Kagawa, Y. 2007b. Electrical properties of silica-based nanocomposites with multiwall carbon nanotubes. *Journal of the American Ceramic Society*, 90, 1667–1670.

Guo, T., Nikolaev, P., Thess, A., Colbert, D. T. and Smalley, R. E. 1995. Catalytic growth of single-walled nanotubes by laser vaporization. *Chemical Physics Letters*, 243, 49–54.

Hernadi, K., Ljubovic, E., Seo, J. W. and Forro, L. 2003. Synthesis of MWNT-based composite materials with inorganic coating. *Acta Materialia*, 51, 1447–1452.

Hofmann, S., Ducati, C., Robertson, J. and Kleinsorge, B. 2003. Low-temperature growth of carbon nanotubes by plasma-enhanced chemical vapor deposition. *Applied Physics Letters*, 83, 135–137.

Hone, J., Whitney, M. and Zettl, A. 1998. Thermal conductivity of single-walled carbon nanotubes. In: *International Conference on Science and Technology of Synthetic Metals (ICSM 98), Jul 12–18 1998 Montpellier, France*. Elsevier Science, pp. 2498–2499.

Hone, J., Llaguno, M. C., Nemes, N. M., Johnson, A. T., Fischer, J. E., Walters, D. A., Casavant, M. J., Schmidt, J. and Smalley, R. E. 2000. Electrical and thermal transport properties of magnetically aligned single walt carbon nanotube films. *Applied Physics Letters*, 77, 666–668.

Iijima, S. 1991. Helical microtubules of graphitic carbon. *Nature*, 354, 56–58.

Kim, G. T., Choi, E. S., Kim, D. C., Suh, D. S., Park, Y. W., Liu, K., Duesberg, G. and Roth, S. 1998. Magnetoresistance of an entangled single-wall carbon-nanotube network. *Physical Review B*, 58, 16064–16069.

Kim, P., Shi, L., Majumdar, A. and Mceuen, P. L. 2001. Thermal transport measurements of individual multiwalled nanotubes. *Physical Review Letters*, 87, 215502.

Kong, J., Soh, H. T., Cassell, A. M., Quate, C. F. and Dai, H. J. 1998. Synthesis of individual single-walled carbon nanotubes on patterned silicon wafers. *Nature*, 395, 878–881.

Kovacs, J. Z., Velagala, B. S., Schulte, K. and Bauhofer, W. 2007. Two percolation thresholds in carbon nanotube epoxy composites. *Composites Science and Technology*, 67, 922–928.

Krenchel, H. 1964. *Fiber Reinforcement*. Akademisk Forlag.

Krishnan, A., Dujardin, E., Ebbesen, T. W., Yianilos, P. N. and Treacy, M. M. J.

1998. Young's modulus of single-walled nanotubes. *Physical Review B – Condensed Matter*, 58, 14013–14019.

Kroto, H. W., Heath, J. R., Obrien, S. C., Curl, R. F. and Smalley, R. E. 1985. C-60 – Buckminsterfullerene. *Nature*, 318, 162–163.

Lee, J. O., Park, C., Kim, J. J., Kim, J., Park, J. W. and Yoo, K. H. 2000. Formation of low-resistance ohmic contacts between carbon nanotube and metal electrodes by a rapid thermal annealing method. *Journal of Physics D – Applied Physics*, 33, 1953–1956.

Lee, W. E. 1994. *Ceramic Microstuctures: Property Control by Processing*. Kluwer Academic Publishers.

Li, F., Cheng, H. M., Bai, S., Su, G. and Dresselhaus, M. S. 2000. Tensile strength of single-walled carbon nanotubes directly measured from their macroscopic ropes. *Applied Physics Letters*, 77, 3161–3163.

Li, Y. J., Wang, K. L., Wei, J. Q., Gu, Z. Y., Wang, Z. C., Luo, J. B. and Wu, D. H. 2005. Tensile properties of long aligned double-walled carbon nanotube strands. *Carbon*, 43, 31–35.

Loo, S., Idapalapati, S., Wang, S., Shen, L. and Mhaisalkar, S. G. 2007. Effect of surfactants on MWCNT-reinforced sol-gel silica dielectric composites. *Scripta Materialia*, 57, 1157–1160.

Lopez, A. J., Urena, A. and Rams, J. 2010. Fabrication of novel sol-gel silica coatings reinforced with multi-walled carbon nanotubes. *Materials Letters*, 64, 924–927.

Lu, J. P. 1997. Elastic properties of carbon nanotubes and nanoropes. *Physical Review Letters*, 79, 1297–1300.

Matthews, F. L. and Rawlings, R. D. 1999. *Composite Materials: Engineering and Science*. Woodhead Publishing.

Meyyappan, M. 2005. *Carbon Nanotubes: Science and Applications*. CRC Press.

Mukhopadhyay, A., Chu, B. T. T., Green, M. L. H. and Todd, R. I. 2010. Understanding the mechanical reinforcement of uniformly dispersed multiwalled carbon nanotubes in alumino-borosilicate glass ceramic. *Acta Materialia*, 58, 2685–2697.

Mukhopadhyay, A., Otieno, G., Chu, B. T. T., Wallwork, A., Green, M. L. H. and Todd, R. I. 2011. Thermal and electrical properties of aluminoborosilicate glass-ceramics containing multiwalled carbon nanotubes. *Scripta Materialia*, 65, 408–411.

Nan, C. W., Shi, Z. and Lin, Y. 2003. A simple model for thermal conductivity of carbon nanotube-based composites. *Chemical Physics Letters*, 375, 666–669.

Nan, C. W., Liu, G., Lin, Y. H. and Li, M. 2004. Interface effect on thermal conductivity of carbon nanotube composites. *Applied Physics Letters*, 85, 3549–3551.

Ning, J. W., Zhang, J. J., Pan, Y. B. and Guo, J. K. 2003a. Fabrication and mechanical properties of SiO_2 matrix composites reinforced by carbon nanotube. *Materials Science and Engineering A – Structural Materials Properties Microstructure and Processing*, 357, 392–396.

Ning, J. W., Zhang, J. J., Pan, Y. B. and Guo, J. K. 2003b. Fabrication and thermal property of carbon nanotube/SiO_2 composites. *Journal of Materials Science Letters*, 22, 1019–1021.

Ning, J. W., Zhang, J. J., Pan, Y. B. and Guo, J. K. 2004. Surfactants assisted processing of carbon nanotube-reinforced SiO2 matrix composites. *Ceramics International*, 30, 63–67.

Otieno, G., Koos, A. A., Dillon, F., Wallwork, A., Grobert, N. and Todd, R. I. 2010. Processing and properties of aligned multi-walled carbon nanotube/ aluminoborosilicate glass composites made by sol-gel processing. *Carbon*, 48, 2212–2217.

Pan, R. Q. 2011. Diameter and temperature dependence of thermal conductivity of single-walled carbon nanotubes. *Chinese Physics Letters*, 28, 066104-1–066104-4.

Pan, Z. W., Xie, S. S., Chang, B. H., Wang, C. Y., Lu, L., Liu, W., Zhou, M. Y. and Li, W. Z. 1998. Very long carbon nanotubes. *Nature*, 394, 631–632.

Pan, Z. W., Xie, S. S., Lu, L., Chang, B. H., Sun, L. F., Zhou, W. Y., Wang, G. and Zhang, D. L. 1999. Tensile tests of ropes of very long aligned multiwall carbon nanotubes. *Applied Physics Letters*, 74, 3152–3154.

Peigney, A., Laurent, C., Dumortier, O. and Rousset, A. 1998. Carbon nanotubes Fe alumina nanocomposites. Part I: Influence of the Fe content on the synthesis of powders. *Journal of the European Ceramic Society*, 18, 1995–2004.

Peng, B., Locascio, M., Zapol, P., Li, S. Y., Mielke, S. L., Schatz, G. C. and Espinosa, H. D. 2008. Measurements of near-ultimate strength for multiwalled carbon nanotubes and irradiation-induced crosslinking improvements. *Nature Nanotechnology*, 3, 626–631.

Poole, C. P. and Owens, F. J. 2003. *Introduction to Nanotechnology*. John Wiley.

Prewo, K. M. 1981. *Research on Graphite Reinforced Glass Matrix Composites*. NASA.

Prewo, K. M. 1982. A compliant, high failure strain, fiber-reinforced glass-matrix composite. *Journal of Materials Science*, 17, 3549–3563.

Prewo, K. M. and Brennan, J. J. 1980. High-strength silicon-carbide fiber-reinforced glass-matrix composites. *Journal of Materials Science*, 15, 463–468.

Prewo, K. M. and Brennan, J. J. 1982. Silicon-carbide yarn reinforced glass matrix composites. *Journal of Materials Science*, 17, 1201–1206.

Qian, H., Greenhalgh, E. S., Shaffer, M. S. P. and Bismarck, A. 2010. Carbon nanotube-based hierarchical composites: a review. *Journal of Materials Chemistry*, 20, 4751–4762.

Radushkevich, L. V. and Lukyanovich, V. M. 1952. O strukture ugleroda obrazujucegosja pri termiceskom razlozenii okisi ugleroda na zeleznom kontakte. *Zhurnal Fizicheskoi Khimii*, 26, 8.

Ren, Z. F., Huang, Z. P., Xu, J. W., Wang, J. H., Bush, P., Siegal, M. P. and Provencio, P. N. 1998. Synthesis of large arrays of well-aligned carbon nanotubes on glass. *Science*, 282, 1105–1107.

Salvetat, J. P., Kulik, A. J., Bonard, J. M., Briggs, G. A. D., Stockli, T., Metenier, K., Bonnamy, S., Beguin, F., Burnham, N. A. and Forro, L. 1999. Elastic modulus of ordered and disordered multiwalled carbon nanotubes. *Advanced Materials*, 11, 161–165.

Sambell, R. A. J., Phillips, D. C. and Bowen, D. H. 1972. Carbon fiber composites with ceramic and glass matrices 1. Discontinuous fibers. *Journal of Materials Science*, 7, 663–675.

Schnorr, J. M. and Swager, T. M. 2011. Emerging applications of carbon nanotubes. *Chemistry of Materials*, 23, 646–657.

Seeger, T., Redlich, P., Grobert, N., Terrones, M., Walton, D. R. M., Kroto, H. W. and Ruhle, M. 2001. SiOx-coating of carbon nanotubes at room temperature. *Chemical Physics Letters*, 339, 41–46.

Seeger, T., Kohler, T., Frauenheim, T., Grobert, N., Ruhle, M., Terrones, M. and Seifert, G. 2002. Nanotube composites: novel SiO2 coated carbon nanotubes. *Chemical Communications*, 34–35.

Seeger, T., De La Fuente, G., Maser, W. K., Benito, A. M., Callejas, M. A. and Martinez, M. T. 2003. Evolution of multiwalled carbon-nanotube/SiO2 composites via laser treatment. *Nanotechnology*, 14, 184–187.

Seraphin, S. and Zhou, D. 1994. Single-walled carbon nanotubes produced at high-yield by mixed catalysts. *Applied Physics Letters*, 64, 2087–2089.

Seraphin, S., Zhou, D., Jiao, J., Minke, M. A., Wang, S., Yadav, T. and Withers, J. C. 1994. Catalytic role of nickel, palladium, and platinum in the formation of carbon nanoclusters. *Chemical Physics Letters*, 217, 191–198.

Shaffer, M. and Sandler, J. 2007. Carbon nanotube/nanofiber polymer composites. In: *Processing and Properties of Nanocomposites*. World Scientific Publishing.

Shaffer, M. S. P., Fan, X. and Windle, A. H. 1998. Dispersion and packing of carbon nanotubes. *Carbon*, 36, 1603–1612.

Sivakumar, R., Guo, S. Q., Nishimura, T. and Kagawa, Y. 2007. Thermal conductivity in multi-wall carbon nanotube/silica-based nanocomposites. *Scripta Materialia*, 56, 265–268.

Subhani, T. 2012a. *Microstructural and Interfacial Charaterization of Silica Glass Matrix Composites Reinforced with Carbon Nanotubes*. Imperial College London, unpublished report.

Subhani, T. 2012b. *Silica and Borosilicate Glass Matrix Composites Containing Carbon Nanotubes*. PhD Thesis, Imperial College London.

Subhani, T. 2012c. *Thermal Shock, Cyling and Ageing Behaviour of Carbon Nanotube Reinforced Glass Matrix Composites*. Imperial College London, unpublished report.

Subhani, T. 2012d. *Wear and Friction Properties of Carbon Nanotube Reinforced Glass Matrix Composites*. Imperial College London, unpublished report.

Subhani, T., Shaffer, M. S. P., Boccaccini, A. R. and Lee, W. E. 2011. Densification of carbon nanotubes reinforced glass matrix composites using pressureless sintering. Unpublished report.

Tans, S. J., Devoret, M. H., Dai, H. J., Thess, A., Smalley, R. E., Geerligs, L. J. and Dekker, C. 1997. Individual single-wall carbon nanotubes as quantum wires. *Nature*, 386, 474–477.

Teo K., S. C., Chhowalla M., Milne W. 2003. *Encyclopedia of Nanoscience and Nanotechnology*. American Scientific Publishers.

Terrones, M. 2003. Science and technology of the twenty-first century: Synthesis, properties and applications of carbon nanotubes. *Annual Review of Materials Research*, 33, 419–501.

Thomas, B. J. C., Shaffer, M. S. P. and Boccaccini, A. R. 2009. Sol-gel route to carbon nanotube borosilicate glass composites. *Composites Part A – Applied Science and Manufacturing*, 40, 837–845.

Treacy, M. M. J., Ebbesen, T. W. and Gibson, J. M. 1996. Exceptionally high Young's modulus observed for individual carbon nanotubes. *Nature*, 381, 678–680.

Vila, M., Hueso, J. L., Manzano, M., Izquierdo-Barba, I., De Andres, A., Sanchez-Marcos, J., Prieto, C. and Vallet-Regi, M. 2009. Carbon nanotubes-mesoporous silica composites as controllable biomaterials. *Journal of Materials Chemistry*, 19, 7745–7752.

Wang, J., Kou, H. M., Liu, X. J., Pan, Y. B. and Guo, J. K. 2007. Reinforcement of mullite matrix with multi-walled carbon nanotubes. *Ceramics International*, 33, 719–722.

Wang, M. S., Golberg, D. and Bando, Y. 2010. Tensile tests on individual single-walled carbon nanotubes: linking nanotube strength with its defects. *Advanced Materials*, 22, 4071–4075.

Wilson, M., Kannangara, K. K., Smith, G., Simmons, M. and Raguse, B. 2002. *Nanotechnology: Basic Science and Emerging Technologies*. Chapman & Hall/ CRC Press.

Wong, E. W., Sheehan, P. E. and Lieber, C. M. 1997. Nanobeam mechanics: elasticity, strength, and toughness of nanorods and nanotubes. *Science*, 277, 1971–1975.

Xiang, C. S., Pan, Y. B., Liu, X. J., Sun, X. W., Shi, X. M. and Guo, J. K. 2005. Microwave attenuation of multiwalled carbon nanotube-fused silica composites. *Applied Physics Letters*, 87, 123103-1–123103-3.

Xiang, C. S., Pan, Y. and Guo, J. K. 2007. Electromagnetic interference shielding effectiveness of multiwalled carbon nanotube reinforced fused silica composites. *Ceramics International*, 33, 1293–1297.

Xu, H., Pan, Y. B., Zhu, Y., Kou, H. M. and Guo, J. K. 2009. Transparent multi-walled carbon nanotube-silica composite prepared by hot-pressed sintering and its nonlinear optical properties. *Journal of Alloys and Compounds*, 481, L4–L7.

Yang, D. J., Zhang, Q., Chen, G., Yoon, S. F., Ahn, J., Wang, S. G., Zhou, Q., Wang, Q. and Li, J. Q. 2002. Thermal conductivity of multiwalled carbon nanotubes. *Physical Review B*, 66, 165440-1–165440-6.

Yao, N. and Lordi, V. 1998. Young's modulus of single-walled carbon nanotubes. *Journal of Applied Physics*, 84, 1939–1943.

Ye, F., Liu, L. M., Wang, Y. J., Zhou, Y., Peng, B. and Meng, Q. C. 2006. Preparation and mechanical properties of carbon nanotube reinforced barium aluminosilicate glass-ceramic composites. *Scripta Materialia*, 55, 911–914.

Yi, W., Lu, L., Zhang, D. L., Pan, Z. W. and Xie, S. S. 1999. Linear specific heat of carbon nanotubes. *Physical Review B*, 59, R9015–R9018.

Yu, M. F., Files, B. S., Arepalli, S. and Ruoff, R. S. 2000a. Tensile loading of ropes of single wall carbon nanotubes and their mechanical properties. *Physical Review Letters*, 84, 5552–5555.

Yu, M. F., Lourie, O., Dyer, M. J., Moloni, K., Kelly, T. F. and Ruoff, R. S. 2000b. Strength and breaking mechanism of multiwalled carbon nanotubes under tensile load. *Science*, 287, 637–640.

Zhan, H. B., Chen, W. Z., Wang, M. Q., Zheng, C. and Zou, C. L. 2003. Optical limiting effects of multi-walled carbon nanotubes suspension and silica xerogel composite. *Chemical Physics Letters*, 382, 313–317.

Zhan, H. B., Zheng, C., Chen, W. Z. and Wang, M. Q. 2005. Characterization and nonlinear optical property of a multi-walled carbon nanotube/silica xerogel composite. *Chemical Physics Letters*, 411, 373–377.

Zhang, H. L., Li, J. F., Yao, K. F. and Chen, L. D. 2005. Spark plasma sintering and thermal conductivity of carbon nanotube bulk materials. *Journal of Applied Physics*, 97, 114310-1–114310-5.

Zhang, Y. J., Shen, Y. F., Han, D. X., Wang, Z. J., Song, J. X. and Niu, L. 2006. Reinforcement of silica with single-walled carbon nanotubes through covalent functionalization. *Journal of Materials Chemistry*, 16, 4592–4597.

Zhao, J. and Zhu, J. 2011. Electron microscopy and in situ testing of mechanical deformation of carbon nanotubes. *Micron*, 42, 663–679.

Zhao, J. O., He, M. R., Dai, S., Huang, J. Q., Wei, F. and Zhu, J. 2011. TEM observations of buckling and fracture modes for compressed thick multiwall carbon nanotubes. *Carbon*, 49, 206–213.

Zheng, C., Feng, M., Zhen, X., Huang, J. and Zhan, H. B. 2007. Effect of doping levels on the pore structure of carbon nanotube/silica xerogel composites. *Materials Letters*, 61, 644–647.

Zheng, C., Feng, M., Zhen, X., Huang, J. and Zhan, H. B. 2008. Materials investigation of multi-walled carbon nanotubes doped silica gel glass composites. *Journal of Non-Crystalline Solids*, 354, 1327–1330.

Zhu, H. W., Xu, C. L., Wu, D. H., Wei, B. Q., Vajtai, R. and Ajayan, P. M. 2002. Direct synthesis of long single-walled carbon nanotube strands. *Science*, 296, 884–886.

8

Ceramic ultra-thin coatings using atomic layer deposition

X. LIANG, D. M. KING and A. W. WEIMER,
University of Colorado, USA

DOI: 10.1533/9780857093493.2.257

Abstract: Ultra-thin films can be coated on primary fine particles without significant aggregation by atomic layer deposition (ALD) in a fluidized bed reactor. Precursor doses can be delivered to the bed of particles sequentially and, in most cases, can be utilized at nearly 100% efficiency without precursor breakthrough and loss, with the assistance of an in-line downstream mass spectrometer. A multitude of applications can be addressed in a competitive fashion using fine particles that have been surface-modified using ALD in scalable, high-throughput unit operations. Several examples of the applications of conformal ALD coatings have been discussed, including oxidation-resistant metals or ceramics, coatings that enable biomedical applications including tissue engineering, and corrosion-resistant particles for next-generation batteries, capacitors or fuel cells. It is expected that the technology of thin film coating by particle ALD will play a major role in the field of advanced materials.

Key words: atomic layer deposition (ALD), particles, ultra-thin coating, ceramic, fluidized bed reactor.

8.1   Introduction

A thin film is defined as a layer of material of thickness ranging from fractions of a nanometer to several micrometers. The thickness of an ultra-thin film is in the range of several nanometers. Ceramic materials (e.g. alumina, zirconia, and silicon carbide) are well known for their refractory characteristics and outstanding mechanical properties. Ceramic thin films can be deposited onto bulk materials (substrates) to achieve properties not easily attainable or unattainable by modifying the bulk substrates alone.

257

The functions of thin film coatings can be divided into two categories – surface protection and surface functionalization. These types of thin coatings have numerous applications in many different fields. For example, alumina coatings are highly resistant to mechanical wear and corrosion, and the use of such coatings on cutting tools can extend the life of these items by many times and can be used under more aggressive conditions (e.g. higher cutting speeds) to increase productivity. On the other hand, alumina has excellent dielectric properties and adheres to many surfaces quite well. These properties make alumina attractive in the silicon microelectronics and thin film device industries as an insulator, diffusion barrier, and protective coating.

Thin film deposition can either be carried out in liquid phase or vapor phase. Liquid methods are relatively simple, but less than ideal, since there is oftentimes little ability to control film growth, which makes the conformal coating of films onto particle substrates difficult. In addition, washing, drying, separation, and other additional handling processes are typically required, which can make them energy and time intensive and impurities may still remain on the substrates if not carried out properly. Vapor-phase thin film techniques have significant advantages over liquid-phase techniques, including being more applicable to a wider array of substrate and coating materials, increased flexibility and control over process conditions, and better access to surfaces during deposition due to the efficient nature of gas–solid contacting reactor systems. Traditional vapor-phase deposition includes physical vapor deposition (PVD) and chemical vapor deposition (CVD). Both PVD and CVD are well-known techniques for the deposition of solids from gaseous precursors in the production of solid-state devices, the formation of protective coatings, and the structural design of ceramics. CVD processes are heavily used in major markets in the cutting tool industry and the semiconductor industry to produce thin films, to name a few. However, they do not offer the best control or the highest material quality of the thin film techniques, as they are typically used for their speed rather than precision. During a CVD reaction, the chemical reactants are allowed to coincide in the gas phase making this technology dependent upon reaction time and temperature. CVD techniques are not able to deposit ultra-thin films or thin films on primary particles or to inherently control the location and the thickness of the film. Often, CVD films must be 'overbuilt' to achieve a certain level of barrier performance, and uniformity remains a crucial problem in traditional thin film production in many industries.

Atomic layer deposition (ALD) (Suntola, 1992; George et al., 1996; Leskela and Ritala, 2002) can overcome many of the typical drawbacks of traditional deposition techniques and can be used to coat particles with ultra-thin and conformal coatings. ALD is a multi-step gas-phase thin film deposition method that forms chemical bonds with the surface of the substrate. ALD is an analogue of the CVD technique, and is most appropriate for binary compounds because a binary CVD reaction can easily be separated into two half-reactions. In ALD, the film growth takes place in a cyclic manner. The simplest case includes four steps: (1) exposure of the first precursor; (2) purge or evacuation of the reactor chamber; (3) exposure of the second precursor; (4) purge or evacuation of the reactor chamber. The film thickness is therefore simply determined by the growth rate per cycle and the number of cycles completed.

The unique self-limiting film growth mechanism allows a number of advantageous features, such as excellent conformality and uniformity, pinhole-free films, low impurity content, independence of line of sight, simple and accurate film thickness control, and low processing temperature (Suntola, 1992; George et al., 1996; Leskela and Ritala, 2002). This self-limiting growth mechanism also ensures good reproducibility and relatively straightforward scale-up. Depending on the coating material and process conditions, one ALD cycle deposits ~0.01–0.20 nm film thicknesses. This slow growth rate provides for the control of the atomic-scale deposition. This is also the major limitation of ALD processing, but this limitation is losing its significance when weighed against the numerous benefits of ALD relative to other techniques.

ALD processes are typically operated under some degree of vacuum. The vacuum transport medium has the important advantages of a decreased chance of contamination and clear access to the deposition surface. A large number of materials can be prepared by ALD, including, but not limited to, Al_2O_3 (Ott et al., 1996; Yun et al., 1998; Jeon et al., 2002), ZnO (Ferguson et al., 2005; Hamann et al., 2008; King et al., 2008a), TiO_2 (Ritala et al., 1993; Kumagai et al., 1995; Ferguson et al., 2004), SiO_2 (Klaus et al., 1997a, 1997b; McCool and DeSisto, 2004), ZnS (Suntola and Hyvarinen, 1985; Suntola, 1992), HfO_2 (Ding et al., 2003; Chang et al., 2004; Kukli et al., 2004), AlN (Kim et al., 2009; Bosund et al., 2011), TiN (Ritala et al., 1995; Satta et al., 2002; Elam et al., 2003), BN (Ferguson et al., 2002), Fe_2O_3 (Scheffe et al., 2009; Martinson et al., 2011), CoO (Scheffe et al., 2010; De Santis et al., 2011), MnO (Nilsen et al., 2003; Burton et al., 2009), W (Wilson et al., 2008), Pt (Aaltonen et al., 2003; Zhu et al., 2007; King et al., 2008b; Li et al., 2010) and Pd (Senkevich et al., 2003; Ten Eyck et al., 2005; Elam et al., 2006; Lu and Stair, 2010). This vacuum environment also allows operating thin film deposition in plasma. Plasma is a partially ionized gas and contains a great deal of energy, which can activate film deposition processes at lower temperatures. Plasma-enhanced CVD (Suchaneck et al., 2001) and ALD (Yun et al., 2004; Ten Eyck et al., 2007) have been carried out at low temperatures for film deposition on thermally sensitive substrates. Reviews on the surface chemistry of ALD are given by George et al. (1996), George (2010) and Puurunen (2005). For other aspects of ALD than its surface chemistry, the reader is referred to different views on the development of ALD and applications over the years by Goodman and Pessa (1986), Suntola (1989, 1992), George *et al.* (1996), Malygin *et al.* (1996), Leskela and Ritala (1999, 2002), Niinisto *et al.* (1996) and George (2010). It is not the intent of this chapter to give an extensive review of the subject of thin film coating, but the focus will be on the topic of ceramic ultra-thin film coatings on particles by ALD in gas-phase and related applications.

8.2   Ultra-thin ceramic films coated on ceramic particles by atomic layer deposition (ALD)

8.2.1   Particle ALD in a fluidized bed reactor

Fine particles have gained increased interest in a variety of fields for different applications (Bruchez *et al.*, 1998; Panyam and Labhasetwar, 2003; Daniel and Astruc, 2004). The ability to modify the surface of particles with thin films, which can add new surface functionalities and keep the bulk properties of the particles largely the same, has an increasing number of new applications as new systems are studied. Additional functionality in thin films can be achieved by depositing multiple layers of different materials. For example, nanocoatings on fine capsules can modify the drug release characteristics and provide physical and chemical protection for the drugs (Theobald *et al.*, 2003; Bose and Bogner, 2007); the coating of magnetic iron nanoparticles with photoactive TiO_2 films can be used for the degradation of aqueous contaminants, while the magnetic moment of the substrates keeps the composite recoverable (Zhou *et al.*, 2010).

ALD is an ideal technique to deposit conformal, pinhole-free films on particle substrate surfaces with precision thickness control (Leskela and Ritala, 2002). CVD processes cannot be adequately performed on submicron-sized particles due to the tendency for these particles to aggregate. The typical process windows for ALD vs. CVD processes for particle coating are shown in Fig. 8.1. Film deposition on particles using ALD was first reported in 2000 for the deposition of Al_2O_3 on BN particles (Ferguson *et al.*, 2000). High surface area (~$40\,m^2/g$) BN particles were pressed into a tungsten grid using a polished stainless steel die and a manual press, and were coated in a viscous flow reactor using the alumina ALD chemistry (Dillon *et al.*, 1995; Ott *et al.*, 1996). Conformal ALD coating on a whole particle surface by such methods is difficult to impossible, since there are inherently some contact points between the particle and the tungsten grid. The contact points will not be coated, though ALD is independent of line of sight. Equipment that has good gas–solid contacting properties is needed to apply conformal ALD coating on a whole particle.

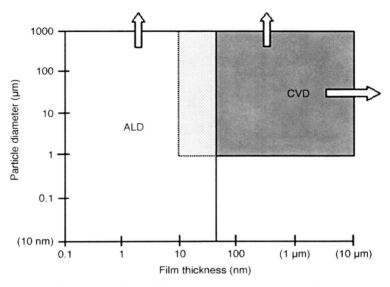

8.1 Depiction of ALD vs. CVD process windows for particle coating.

The cost effectiveness of the process also requires that both exposure and purge steps are rapidly completed.

A flow-type reactor with moving substrate particles is a good solution for such requirements. One major effort for ALD coating on particles consisted of using fluidized bed reactors (FBRs). The first success with this process was demonstrated by Wank *et al.* (2004a, 2004b). This unit operation was also used to deposit Al_2O_3 on micron-sized BN (~10 μm), as well as Ni (~150 μm) particles. In the design of an FBR as shown in Fig. 8.2 (King *et al.*, 2007), a porous metal disc is used as a gas distributor and a porous metal filter is used at the inside top of the reactor column to retain the particles in the reactor at all times. Mechanical vibration has proven to be an effective means for improving the fluidization of cohesive particles (Wank *et al.*, 2001; Hakim *et al.*, 2005c; Xu and Zhu, 2006). Vibration can break up the large agglomerates generated by cohesive interparticle forces, thereby achieving stable fluidization. The reactor itself is maintained at reduced pressure using a vacuum pump. After these successes, the size scales of the substrate particles were pushed further down to nanoscale using an FBR (Hakim *et al.*, 2005a, 2005b, 2006). For example, ultra-thin Al_2O_3 films were deposited on TiO_2 nanoparticles with a primary particle size of 21 nm (Hakim *et al.*, 2006). A rotary reactor is another kind of reactor that has been demonstrated for ALD coating on particles. Al_2O_3 ALD on nanoparticles in a rotary reactor was reported by McCormick *et al.* (2007). In the design of a rotary reactor, a cylindrical drum with porous

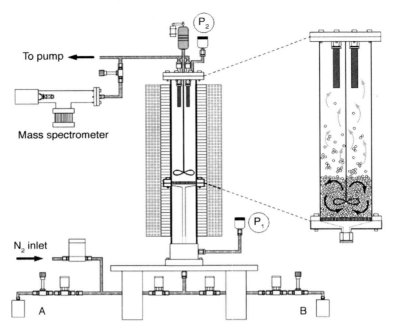

8.2 Schematic illustration of ALD fluidized bed reactor (King *et al.*, 2007). Reproduced by permission of Elsevier.

metal walls was positioned inside a vacuum chamber. The porous cylindrical drum was rotated by a magnetically coupled rotary feedthrough.

The FBR has many advantages, including good mixing, large gas–solid contact area, high efficiency of mass and heat transfer, and large-batch processing capability. With appropriate expansion of a bed of particles, granular materials are transformed into a fluid-like state through contact with a gas or a liquid. The rigorous mixing during fluidization provides excellent fluid–solid contacting. In an FBR system, gas precursors can be introduced into the bed for reaction with the surface of the particles, while the particles are circulating in a liquid-like state. Ultra-thin films can be coated on primary nanoparticles without significant aggregation during the ALD coating process. This phenomenon can be explained by a behavior called dynamic aggregation (Hakim *et al.*, 2005c), which is observed during the fluidization of nanoparticles. The native cohesive properties of the nanoparticles will form agglomerates that are several times larger than the primary particles. Dynamic agglomerates partially break apart and reform due to constant solids recirculation and gas flow through the bed of particles. External forces, such as mechanical vibration, can improve fluidization quality by helping to partially overcome interparticle forces. This serves to reduce the average size and increase the disengagement rate of particles within agglomerates in the bed. Even though this breakage is partial and temporary during a single collision event, due to recirculation and frequent impacts, the entire surface area of each of the primary particles will be eventually exposed and the individual nanoparticles can be coated (Hakim *et al.*, 2005b).

8.2.2   *In situ* mass spectrometry

Since all precursors and products are gaseous at operating conditions, residual gas analysis (RGA) is an efficient technique to observe the termination of each half-reaction. *In situ* mass spectrometry has been successfully used for real-time monitoring of gaseous products and reactants throughout the ALD reaction in an FBR (King *et al.*, 2007). Here, alumina ALD is used as an example to demonstrate this powerful technique. Al_2O_3 ALD films have been deposited using repeated exposures of trimethylaluminum (TMA) and H_2O in an ABAB... sequence (Dillon *et al.*, 1995; Ott *et al.*, 1996). Al_2O_3 ALD is derived from the following binary CVD reaction:

$$2Al(CH_3)_3 + 3H_2O \rightarrow Al_2O_3 + 6CH_4 \qquad [8.1]$$

This binary reaction can be divided into two half-reactions:

$$A : 2AlOH\ ^* + 2Al(CH_3)_3 \rightarrow 2\left[AlOAl(CH_3)_2\right]\ ^* + 2CH_4 \qquad [8.2]$$

$$B : 2\left[AlOAl(CH_3)_2\right]\ ^* + 3H_2O \rightarrow Al_2O_3 + 2AlOH\ ^* + 4CH_4 \quad [8.3]$$

where the asterisks indicate the surface species (Dillon *et al.*, 1995; Ott *et al.*, 1996). The expected reaction product is CH_4, which, however, is also formed directly from the fragmentation of TMA in the mass spectrometer (Juppo *et al.*, 2000). To distinguish the amount of CH_4 formed as a reaction product formed directly from TMA, D_2O was used instead of H_2O. The reaction product from TMA and D_2O was expected to be CH_3D, instead of CH_4. The m/z peaks of interest (i.e. primary and fragmentation peaks) were 28 for N_2, 42 ($-AlCH_3$), 57 ($-Al(CH_3)_2$) and 72 ($Al(CH_3)_3$) for TMA, 18 ($-OD$) and 20 (D_2O) for D_2O, and 17 for CH_3D.

The *in situ* mass spectrometry results of Al_2O_3 ALD on 16 μm high-density polyethylene (HDPE) particles at 77°C, shown in Fig. 8.3 (King *et al.*, 2009), indicate that the chemistry of TMA and D_2O ALD is self-limiting. When TMA was dosed into the reactor (half reaction A), an instantaneous increase of the signal of CH_3D byproduct was not seen. This is expected, since there was no deuterium element in the system before the very first dose of D_2O into the reaction system. TMA is an extremely reactive precursor and, as such, very high surface conversions (in some

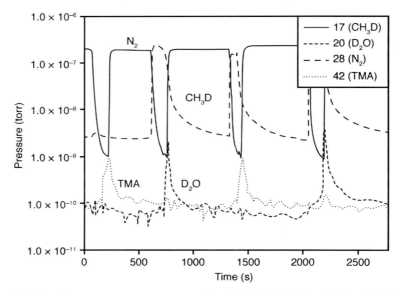

8.3 *In situ* mass spectrometry results of Al₂O₃ ALD using TMA and D₂O
(King *et al.*, 2009). Reproduced by permission of the Electrochemical
Society.

instances greater than 98%) have been observed at the point of breakthrough (King *et al.*, 2007). It is clear that all TMA entering the reactor was
completely utilized until the time at which the TMA signal increased, which
is called the 'breakthrough' time. Just after the TMA breakthrough time, the
TMA dose was stopped, and N₂ was fed into the reactor to purge any
residual reaction product or unused TMA from the system. After the N₂
purge, D₂O was dosed (half reaction B) and the CH₃D signal increased
instantaneously. As the D₂O dose proceeded, the reaction product began to
decrease, and the signal of D₂O ($m/z = 20$) appeared. At this point, the
surface reaction was nearly complete and N₂ was fed into the reactor to
purge any residual reaction product or unreacted D₂O from the system.
After N₂ was purged from the system, the TMA was again dosed into the
reactor to begin another A half reaction, and there was an instantaneous
increase of the CH₃D byproduct. This is reasonable since the particle
surfaces had been saturated with −OD groups. During both precursor
doses, it is apparent that the reactions were self-limiting and self-
terminating, because the reaction product increased and then decreased
while the reactants were still being dosed. If these reactions had not been
self-limiting, the product would continue to be generated as long as the
reactants were dosed. This real-time monitoring strategy allows for
optimizing the dose time of precursors to prevent process overruns and

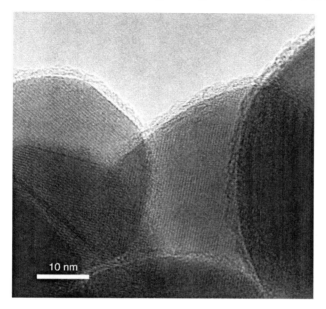

8.4   HRTEM image of 2.5 nm BN films deposited on ZrO_2 nanoparticles (Ferguson *et al.*, 2002). Reproduced by permission of Elsevier.

excess precursor waste. The process is also commercially significant since it can operate in a simple process for practical large-scale applications.

8.2.3   Characterization of ALD thin films

Different analytical techniques are needed to verify the composition and the uniformity of ALD films deposited on particle substrates. X-ray photoelectron spectroscopy (XPS), Fourier-transform infrared (FTIR), and inductively coupled plasma atomic emission spectroscopy (ICP-AES) can analyze the composition of the ALD films. Powder X-ray diffraction (XRD) can study the crystallinity of ALD films and measure the crystallite size propagation of some ALD films with the number of cycles. Transmission electron microscopy (TEM) is a useful technique to observe film/particle interfaces. Normally, the particles are put directly onto Cu grids with a holey carbon overlay film. ALD films thinner than ~5 nm may be difficult to observe using standard TEM technique alone. High-resolution transmission electron microscopy (HRTEM) is needed for these thin films. One example of an ultra-thin BN ALD film on ZrO_2 nanoparticles observed by HRTEM is shown in Fig. 8.4 (Ferguson *et al.*, 2002). The deposited films on micron-sized particles can be better observed by cross-sectional HRTEM. Cross-sectional TEM samples can be prepared by cutting thin slices of a cured

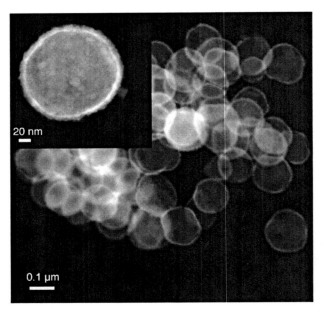

8.5 STEM images of ZnO films deposited on 100 nm SiO$_2$ spherical particles.

epoxy resin loaded with the particles of interest. The thickness of the polymer slices is normally ~50 nm. The thickness of the ALD films can be directly observed by TEM. Scanning transmission electron microscopy (STEM) utilizes similar equipment and provides a z-contrast image, such that materials with some difference in molecular weight can be readily observed. Figure 8.5 is one example of an STEM image, which shows discrete ZnO ALD films deposited on individual 100 nm SiO$_2$ particles. Energy dispersive spectroscopy (EDS) detector unit equipped with TEM can be used for elemental analysis while imaging.

For ALD thin film coating on particles, the key challenge in trying to individually coat primary nanoparticles is to overcome their natural tendency to form soft agglomerates. One method to verify that particles are coated individually is to check the morphology of the particles before and after ALD coating, which can be directly observed by SEM. The particle size distribution can also be measured by a particle size analyzer. These two methods are very useful for micro-sized particle substrates. For nanoparticles, one effective method is the determination of the surface area. If nanoparticles are coated as aggregates, the surface area will dramatically decrease. Nevertheless, there is an expected change in the surface area of the nanoparticles as the density and the particle size will change as the thickness of the ALD films increases. With an increase in the thickness of ALD films, the particle size is larger, which would result in a smaller surface area. However, with thicker ALD films, the density of the coated particles could decrease if the density of the ALD films is smaller than that of particle substrate. This could result in a higher surface area. Therefore, the surface area of the particles could increase or decrease with an increase in the number of ALD coating cycles, but the expected change would be calculable prior to running an experiment. The specific surface area of the particles can be calculated by the Brunauer–Emmett–Teller (BET) method from N_2 adsorption isotherms obtained at $-196°C$. The thickness of an ALD film on a particle surface can also be estimated using the measured content of ALD film on the particle surface, the density of ALD film, and the surface area of the uncoated particles (Wank *et al.*, 2004b; Liang *et al.*, 2008).

8.3   Using ultra-thin ceramic films as a protective layer

Magnetic metal nanoparticles are used in many applications, such as high-density data storage, magneto-optical switches, novel photoluminescent materials, inks, biomedical diagnosis, catalysis, environmental remediation, and others. One of the biggest limitations for extending the use of metal nanoparticles is their instability at ambient conditions. Owing to their very high surface area, most metal nanoparticles spontaneously oxidize when exposed to air. One approach to prevent oxidation is to individually coat metal nanoparticles with a uniform film that prevents the diffusion of oxygen into the bare metal surface (Hakim *et al.*, 2007). Al_2O_3 is one of the most frequently studied ALD barrier films. Extensive work has been done for the preparation of oxidation-resistant iron nanoparticles via the decomposition of iron oxalate particles followed by *in situ* passivation with ultra-thin alumina films deposited by ALD (Hakim *et al.*, 2007).

The pyrophoric nature of metal nanopowders makes them extremely hazardous, which has limited their use in practical applications. Rather than pursuing metal nanoparticles directly, the *in situ* decomposition of metal salts followed by ALD coating within an FBR to produce passivated metal nanopowders can be extremely beneficial to a variety of industries. Several metal oxalate powders have been studied, including those of iron, nickel, cobalt, and copper; when the salts were decomposed the remaining materials were shown to retain their metallic properties. The resulting particles are micron-sized, although highly porous due to the macroscopic release of CO or CO_2 upon decomposition. Several size reduction processes have been employed, including standard ball milling and cryogenic milling. The latter allowed particles to be as small as 30–50 nm, which provided a suitable test vehicle for a proof of concept that even metal nanopowders could effectively be passivated. Alumina ALD films have been conformally coated on the surface of iron nanoparticles (Fig. 8.6) (Hakim *et al.*, 2007). The

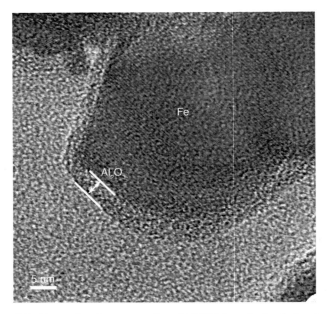

8.6 Example of cross-sectional HRTEM analysis of alumina-coated iron nanoparticles (Hakim *et al.*, 2007). Reproduced by permission of IOP Publishing.

temperature-dependent oxidation behavior of uncoated Fe nanopowder and the oxidation resistance of Fe nanopowder coated by 30 and 50 cycles of Al_2O_3 ALD is shown in Fig. 8.7. It is clear that the Fe nanopowder began to oxidize almost instantaneously in air, and resulted in a mass gain of 20% at a temperature of 250°C. For the 30 cycle film (~6 nm film thickness), the onset of oxidation was delayed until a temperature of approximately 350°C. The 10 nm film (50 cycles) showed no signs of oxidation when held at a temperature of 400°C for over three hours. The resulting elemental analyses from these experiments showed that the oxygen content of the 50 cycle material matched the anticipated content for Al_2O_3 coated on iron metal, not an oxidized form. Similar results have been obtained for Al_2O_3 ALD films coated on the other core metals produced from the decomposition of their respective oxalates. However, at much higher temperatures, the mismatch between thermal expansion coefficients of the core and shell is too great and the passivating layers may be fractured. This problem can be overcome by depositing alternating multilayers of materials to perform thermal expansion matching between the composite film and the core particle. The magnetic moment of the micron-sized iron spheres coated by various thicknesses of Al_2O_3 remained unchanged, which was anticipated due to the extreme thinness of each shell (Hakim *et al.*, 2007).

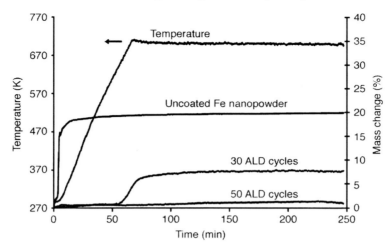

8.7 TGA plot of the thermal stability of uncoated and alumina-coated Fe nanoparticles at elevated temperatures in the presence of air (Hakim *et al.*, 2007). Reproduced by permission of IOP Publishing.

8.4   Enhanced lithium-ion batteries using ultra-thin ceramic films

Lithium ion (Li-ion) batteries are believed to be quite promising to store energy for vehicles because of their high energy-to-weight ratios (Sun *et al.*, 2005, 2009; Carpenter *et al.*, 2008). In order to employ Li-ion battery technologies in next-generation hybrid electric and/or plug-in hybrid electric vehicles (HEVs and PHEVs), these batteries must satisfy many requirements. Examples of the near-term needs that will help bring Li-ion battery technologies to market include electrodes fabricated from inexpensive environmentally benign materials with long lifetimes (near-term targets are 5000 charge-depleting cycles, 15-year calendar life), stability over a wide temperature range (from $-46°C$ to $+66°C$), a high energy density, and a high rate capability (Jung *et al.*, 2010b). Current efforts include the development of low-cost electrode materials, better electrolytes, and low-cost packaging for such batteries. One effective path to cost reduction is to improve power capability and extend battery life, thereby reducing the overbuilding of Li-ion batteries while maintaining an energy density suitable for transportation applications (Chen *et al.*, 2010). Thin film coatings on Li-ion battery electrode powders have proven to be an effective way to improve the capacity retention, rate capability, and even the thermal stability of cathode materials. Different coating materials have been studied, including carbon (Chen and Dahn, 2002; Belharouak *et al.*, 2005; Cao *et al.*, 2007), metal oxides (e.g. Al_2O_3, ZrO_2, ZnO, SiO_2, and TiO_2) (Cho *et al.*, 2001; Li *et al.*, 2006) and metal phosphates (e.g. $AlPO_4$) (Cho *et al.*, 2003).

Based on the morphology of the coating, the surface coating can be divided into three subgroups – rough coatings, a core–shell structure, and ultra-thin film coating (Chen *et al.*, 2010). For rough coatings, the majority of coating strategies have been based on solution techniques such as the sol–gel method (Cho *et al.*, 2000; Sun *et al.*, 2009). These coating technologies can result in an incomplete coating or a thick coating, which brings a trade-off between Li^+ diffusion (power and/or energy density) and protection (lifetime) controlled by the thickness (nm) of the film (Kim *et al.*, 2002). The core–shell structure takes advantage of both the high electrochemical performance of the core material and the excellent stability of the shell material. However, core–shell structured materials generally have a thick coating shell, sometimes up to 1 μm, which brings problems of crystal structure mismatch and slow transport of electrons and Li-ions. Normally, relatively thick films deposited onto substrates exhibit mechanical stresses, including intrinsic and thermal stresses. In battery systems, repeated charge/discharge cycling will create additional mechanical stresses between the core and the shell due to volumetric changes, which can lead to fatigue and the formation of cracks between the two types of materials.

By contrast, an ultra-thin film coating of thickness down to sub-nanometer levels has the potential to reduce the mechanical stress and to alleviate the requirements on the electron and Li-ion conductivity for the coating materials. ALD can prepare pinhole-free surface coatings on battery electrode powders with precise control of the coating layer thickness down to 0.1 nm. Since ALD is a layer-by-layer process, the slow growth rate can reduce the intrinsic stress induced during the growth of the films. Unlike physical coating methods, strong chemical bonds are created to maintain the physical integrity between the substrate and coating layer prepared by ALD. Ultra-thin ALD films are much more flexible than thick coatings, and the thermal stress can be greatly reduced.

Jung *et al.* (2010a) demonstrated that conformal Al_2O_3 ALD films greatly increased the performance of $LiCoO_2$ powders, as shown in Fig. 8.8. Alumina ALD films were deposited on commercially available micron-sized $LiCoO_2$ powders at 180°C using alternating reactions of TMA and water. The coated $LiCoO_2$ powders with thicknesses of only 0.3 to 0.4 nm using two ALD cycles exhibited a capacity retention of 89% after 120 charge–discharge cycles in the 3.3–4.5 V (vs. Li/Li^+) range, while the bare $LiCoO_2$ powders displayed only a 45% capacity retention. However, with thicker alumina ALD films, the capacity started to decrease significantly and showed a negligible value of $20 \, mA \, h \, g^{-1}$ after ten ALD cycles (~2 nm thick). The loss of capacity resulted mainly from the electronically insulating character of the Al_2O_3 ALD film. The electronic conductivity of $LiCoO_2$ powders was significantly reduced from 2×10^{-4} to $5 \times 10^{-5} \, S \, cm^{-1}$ after only two ALD cycles. The conductivity continuously decreased with

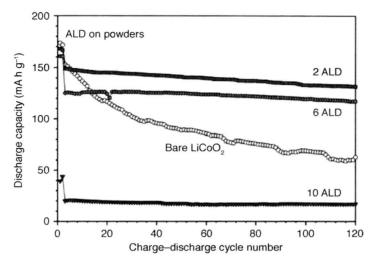

8.8 Charge–discharge cycle performance of electrodes fabricated
using bare LiCoO₂ powders and the Al₂O₃ ALD coated LiCoO₂ powders
using 2, 6, and 10 ALD cycles (Jung *et al.*, 2010a). Reproduced by
permission of the Electrochemical Society.

increasing number of ALD cycles. The reduction in electron conductivity
could result in slower charge–discharge kinetics. The Al_2O_3 ALD film could
also reduce Li-ion conductivity. It is important to note that the charging to
4.5 V is atypically high for a $LiCoO_2$ battery system. This is another
important feature of the coating, that the safety aspect of the battery system
is greatly improved during non-standard operation such as overcharging.

One set of experiments by Jung *et al.* (2010b) supported the conclusion of
the hindrance of electron transport in the Al_2O_3 film resulting from the
complete ALD coverage on the active material powder. Five cycles of ultra-
thin alumina ALD films were deposited directly on natural graphite (NG)
composite electrodes. For comparison, NG electrodes prepared with Al_2O_3
ALD coatings on powder and bare NG were tested using charge–discharge
cycling at an elevated temperature of 50°C. The bare NG displayed a
relatively rapid decay in reversible capacity versus the number of charge–
discharge cycles. In contrast, the capacity retention was dramatically
improved by performing only five cycles of Al_2O_3 ALD directly on the
electrode. The charge–discharge capacity retention for the electrode coated
with five Al_2O_3 ALD cycles was 98% for 200 charge–discharge cycles,
normalized to the reversible capacity at the third charge–discharge cycle,
with negligible kinetic hindrance. Clearly, direct deposition of ALD films on
finished electrodes is another viable technique to produce ALD-enabled
battery materials, which is another route to protecting the surface of the

active powders in electrodes while maintaining an interparticle electronic pathway.

Recently, Scott *et al.* (2011) demonstrated that ultra-thin alumina ALD coatings on nano-$LiCoO_2$ cathode materials can be cycled at a remarkable rate performance with durable high energy. The nano cathode particles allow for an increase in electrode–electrolyte contact area and shorten the Li-ion insertion distances, thus enhancing the power density compared to micrometer-sized particles. For example, the two ALD cycle coated nano-$LiCoO_2$ particles (~400 nm) delivered a discharge capacity of 133 mA h g^{-1} with currents of 1400 mA g^{-1} (7.8 C), corresponding to a 250% improvement in reversible capacity compared to bare nanoparticles when cycled at this high rate. High rate capability is another requirement for electric vehicle technologies in order to make them suitable for more rapid acceleration and higher efficiency recovery during regenerative braking.

The results with ultra-thin Al_2O_3 ALD films are exceptional. To determine if the effect of Al_2O_3 ALD is unique to Al_2O_3, Jung *et al.* (2010a) coated commercially available micron-sized $LiCoO_2$ powders with four cycles of ZnO ALD films at 180°C using alternating reactions of diethylzinc and water. The results of the similar experiments showed that the ultra-thin ZnO ALD films did not display enhanced performance, but the atomic fraction of Zn [Zn/(Zn + Co)] on cathode particles dramatically decreased from 0.49 before any charge–discharge cycling to 0.01 after ten charge–discharge cycles. In contrast, the initial atomic fraction of Al [Al/(Al + Co)] was 0.55 before cycling and 0.53 after ten charge–discharge cycles, which is within experimental error. Therefore, the main reason may be the instability of the ZnO ALD layer on the $LiCoO_2$ particles. This is expected because, with each battery system, the composition of the electrolyte must be paired with the appropriate set of coating materials that are minimally soluble in the electrolyte. For example, coatings for lead acid batteries must be stable in H_2SO_4, coatings for alkaline batteries must be stable in KOH, and coatings for Li-ion batteries must be stable in $LiPF_6$.

For some cases, the main limitation of ALD processing is the slow growth rate. However, for Li-ion battery powders, only a few atomic layers of an ALD coating significantly improved the capacity retention of $LiCoO_2$ and thicker Al_2O_3 coating was not necessary and deteriorated the electrochemical performance of the $LiCoO_2$. The application of thin ALD films for battery electrode powders can take full advantage of the ALD technique without the typical disadvantage of growth rate. Therefore, this simple ALD process is broadly applicable and provides new opportunities for the electrochemical industry to design novel nanostructured electrodes that are highly durable even while cycling at high rates. ALD is believed to be an important technique in the research of high-performance materials for batteries, but research is still in the incubation stage with very few reports in the literature. It is still not entirely understood why such a small number of atomic layers of oxide can sustain continuous corrosion reaction for extended period of time, while only a few more create such a dramatic reduction in performance. Major research efforts are needed in the near future to fundamentally understand the mechanism for the enhanced performance.

8.5   Using ultra-thin ceramic films in tissue engineering

Porous polymers have gained increased interest in the field of tissue engineering. This is attributed to their unique physicochemical properties, such as an interconnected pore structure, large surface area, and small pore size. A good scaffold should give structural support, provide the physicochemical signals to control cellular interactions, and also provide sites for cell attachment, migration, and tissue in-growth (Hutmacher, 2000; Liu and Ma, 2004). Bone graft substitutes should consist of highly porous structures, which can induce high bioactivity and permit tissue in-growth and thus anchor the prosthesis with the surrounding bone to prevent the loosening of implants (Hutmacher, 2000; Liu and Ma, 2004). Owing to the poor bioactivity and mechanical properties of pure porous polymers, many strategies have been developed to improve the bioactivity and mechanical properties of these low-cost materials. Ceramics such as alumina, titania, and zirconia have excellent biocompatibility and bone bonding (Piconi and Maccauro, 1999; Warashina et al., 2003). The inclusion of a bioactive ceramic phase in polymer substrates can reinforce the porous structures of the polymer and enhance the bioactivity and subsequent tissue interaction.

   Most porous polymer/ceramic composites are produced via incipient wetting methods. Solvent-based methods, however, have the risk of leaving potentially toxic organic solvent residues and, as such, regulations often inhibit their usage commercially. Also, a potential negative effect of nanoparticle-containing scaffolds is the possibility of the migration of nanoparticles within the body and their distribution via the blood stream, leading to pathologies with unknown consequences (nanopathologies) (Gerhardt et al., 2007). A novel process to produce reinforced porous structures with enhanced biocompatibility and improved tissue interaction is to coat inside the pores as well as on the surface of the porous polymers with ultra-thin ceramic films, while maintaining the original porous structure of the substrates. The ceramic films can be coated on some dense substrates by traditional methods such as CVD, but it poses a great challenge in coating inner surfaces of pores as the coating area is partially enclosed in the material's matrix and has a high ratio of curvature. Films grown by CVD will block the pores of the porous polymers. In addition, typical CVD processes require high operating temperatures, which are much higher than the softening and melting temperatures of the polymers. Therefore, it is difficult to impossible to deposit ultra-thin ceramic films inside the pores as well as on the surface of the porous polymers by CVD methods. ALD is an ideal method for this kind of coating.

A process to synthesize a porous polymer ceramic composite material has been demonstrated and the properties of the composite materials have been investigated for a variety of applications (Liang *et al.*, 2007, 2009). Highly porous polystyrene divinyl benzene (PS-DVB) particles (Fig. 8.9(a)) were successfully coated with alumina ALD films at the atomic level in an FBR at 33°C. This fabrication technique was considered to be novel since it allowed for the ceramic layers to form on any surface site the gaseous molecules could reach, meaning that Al_2O_3 was successfully deposited on the external surface and along all of the internal pore walls of the particles. A cross-sectional TEM image of the Al_2O_3 ALD films deposited along the walls of the porous polymer is shown in Fig. 8.9(b). The dark lines along the edges are Al_2O_3 layers, and the lightest gray areas are the pores. This functionalized nanocomposite system can be utilized for a variety of applications, many of which can be enabling in the field of biotechnology.

Ultra-thin alumina films can increase the bioactivity of the polymer substrate. The ability of a biomaterial to induce hydroxyapatite formation on its surface is one criterion for designing a scaffold for bone regeneration (Neo *et al.*, 1992; Suchanek and Yoshimura, 1998). To examine the ability

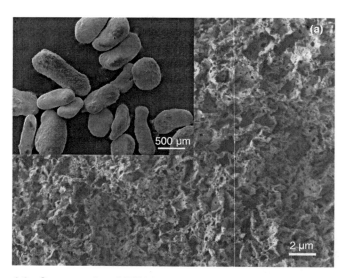

8.9 Cross-sectional SEM image of one uncoated highly porous PS-DVB particle (a) and TEM image of one PS-DVB particle coated with 50 cycles of alumina by ALD (b) (Liang *et al.*, 2007). Reproduced by permission of American Chemical Society.

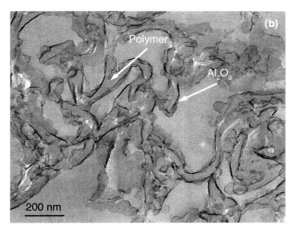

8.9   (continued)

of polymer/ceramic composites to nucleate and deposit hydroxyapatite on
the material surface *in vitro*, uncoated and ceramic-coated samples were
soaked in simulated body fluid (SBF) for different periods of time. This
process is known in the literature as a 'biomimetic process' associated with
bone (Kokubo, 1996). The samples were characterized after immersion in
SBF for one to two weeks. Field emission scanning electron microscopy
(FESEM) examination of all tested samples indicated changes in the
appearance and morphology (microstructure) of the materials after
incubation in SBF. Figure 8.10 shows the FESEM images of the Al_2O_3-
coated porous polymer samples after incubation in SBF at 37°C. Compared
to Fig. 8.10(a), after one week, the surface of the polymer substrate was
roughened and several small, individual hydroxyapatite particles (Fig. 8.10
(b)) were formed. After two weeks (Fig. 8.10(c)), the amount of
hydroxyapatite present on the substrate surface was greatly increased.
Moreover, nanoscale hydroxyapatite particles were homogeneously dispersed and formed a fairly dense coating on the substrate surface. The
substrate surface appeared considerably rougher in comparison to the
surface before hydroxyapatite formation. In order to verify the composition
of the hydroxyapatite, EDS was performed to analyze the composition of
the particular matter formed on the substrate surface. The EDS spectrum
(Fig. 8.10(d)) confirmed that a considerable amount of calcium (Ca) and
phosphor (P) was present on the roughened surface, and in relative amounts
that are typical of hydroxyapatite compositions.

The impetus behind this study was to demonstrate a convenient and
universal surface treatment strategy for modifying the surface properties of
porous polymer materials and thus a method to promote cell adhesion for
tissue engineering. Surface modification of porous polymers with ultra-thin

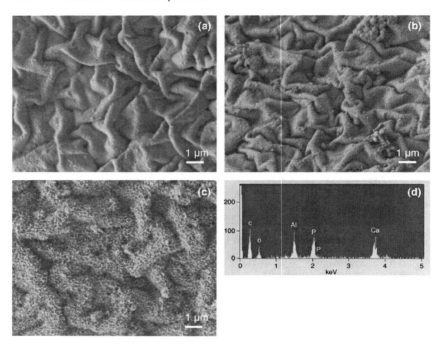

8.10 FESEM images of the polymer/alumina composite before (a) and after incubation in SBF at 37°C for one week (b) and two weeks (c), (d) EDS spectrum of the polymer/alumina composite after incubation in SBF for two weeks (Liang *et al.*, 2009). Reproduced by permission of American Chemical Society.

ceramic ALD films provides a new class of polymer/ceramic composite materials with potentially greater material strength and enhanced bioactivity for tissue regeneration. Future work should focus on the deposition of ceramic films on more appropriate tissue engineering porous scaffolds that are being used in bone tissue engineering, such as poly (lactic acid), poly (lactide-co-glycolide) and poly(2-hydroxyethyl methacrylate) (Liang *et al.*, 2009).

8.6   Conclusions and future trends

This chapter has reviewed the novel particle ALD process. It has been shown that a fluidized bed reactor is a robust unit operation for the batch functionalization of fine and ultra-fine particles using the ALD technique. *In situ* mass spectrometry is a useful tool that allows for optimization of the process variables at any scale. 100% precursor utilization was shown to be attainable, and the surface conversion at the point of unreacted precursor breakthrough is a function of the reactivity of the precursor. Many materials can be readily deposited using this flexible reactor configuration and virtually any particle substrate with the appropriate surface functionality can be coated using this gas-phase adsorption process. Internal surfaces of porous particle substrates can also be easily coated. Particle ALD has been used to deposit many types of ceramic metal oxides on different kinds of particle substrates. In addition, particle ALD provides an opportunity to use unconventional precursors that normally would have been cost prohibitive. The quality of the deposited layers is unparalleled from a particle functionalization perspective.

Particle ALD has been successfully demonstrated at lab scale and pilot scale. Scale-up of the reactors, reproducibility, and uniformality of the film properties in large-scale production and carrying out the thin film deposition process under atmospheric pressure are other important deposition technology related issues that need further attention and work. If integrated into the function of a system, the lifetime and technical performance of the resulting components (cutting tools, batteries, phosphors, flexible polymers for organic photovoltaics and light emitting diodes, etc.) can be significantly increased. For many applications, the benefits obtained through thin films far exceed their manufacturing costs. Even if these costs are high, the resulting increase in performance of the components for their applications is justified by highly competitive improvements. Ideally, however, ALD processes can also be used to reduce production costs for manufacturers of existing products without sacrificing performance. Specifically, particle ALD processes are anticipated to be a cost-effective alternative to replace some conventional CVD and sol–gel liquid solution processes on the basis of reduction in raw materials usage to achieve a given level of performance (i.e. improved cost/performance benefit). A multitude of applications can be addressed in a competitive fashion using ultra-fine particles that have been surface-modified using the ALD technique in scalable, high-throughput unit operations. There is no doubt that the emerging technology of thin film coating by particle ALD will play a major role in the field of advanced materials.

8.7   References

Aaltonen T, Ritala M, Sajavaara T, Keinonen J, and Leskela M (2003), 'Atomic layer deposition of platinum thin films,' *Chemistry of Materials* **15**[9] 1924–1928.

Belharouak I, Johnson C, and Amine K (2005), 'Synthesis and electrochemical analysis of vapor-deposited carbon-coated LiFePO$_4$,' *Electrochemistry Communications* **7**[10] 983–988.

Bose S and Bogner R H (2007), 'Solventless pharmaceutical coating processes: A review,' *Pharmaceutical Development and Technology* **12**[2] 115–131.

Bosund M, Sajavaara T, Laitinen M, Huhtio T, Putkonen M, Airaksinen V M, and

Lipsanen H (2011), 'Properties of AlN grown by plasma enhanced atomic layer deposition,' *Applied Surface Science* **257|17|** 7827–7830.

Bruchez M, Moronne M, Gin P, Weiss S, and Alivisatos A P (1998), 'Semiconductor nanocrystals as fluorescent biological labels,' *Science* **281|5385|** 2013–2016.

Burton B B, Fabreguette F H, and George S M (2009), 'Atomic layer deposition of MnO using Bis(ethylcyclopentadienyl) manganese and H_2O,' *Thin Solid Films* **517|19|** 5658–5665.

Cao Q, Zhang H P, Wang G J, Xia Q, Wu H Q, and Wu Y P (2007), 'A novel carbon-coated $LiCoO_2$ as cathode material for lithium ion battery,' *Electrochemistry Communications* **9|5|** 1228–1232.

Carpenter J A, Gibbs J, Pesaran A A, Marlino L D, and Kelly K (2008), 'Road transportation vehicles,' *MRS Bulletin* **33|4|** 439–444.

Chang H S, Baek S K, Park H, Hwang H, Oh J H, Shin W S, Yeo J H, Hwang K H, Nam S W, Lee H D, Song C L, Moon D W, and Cho M H (2004), 'Electrical and physical properties of HfO_2 deposited via ALD using $Hf(OtBu)_4$ and ozone atop Al_2O_3,' *Electrochemical and Solid State Letters* **7|6|** F42–F44.

Chen Z H and Dahn J R (2002), 'Reducing carbon in $LiFePO_4$/C composite electrodes to maximize specific energy, volumetric energy, and tap density,' *Journal of the Electrochemical Society* **149|9|** A1184–A1189.

Chen Z H, Qin Y, Sun Y K, and Amine K (2010), 'Role of surface coating on cathode materials for lithium-ion batteries,' *Journal of Materials Chemistry* **20 |36|** 7606–7612.

Cho J, Kim Y J, and Park B (2000), 'Novel $LiCoO_2$ cathode material with Al_2O_3 coating for a Li ion cell,' *Chemistry of Materials* **12|12|** 3788–3791.

Cho J, Kim Y J, Kim T J, and Park B (2001), 'Zero-strain intercalation cathode for rechargeable Li-ion cell,' *Angewandte Chemie – International Edition* **40|18|** 3367–3369.

Cho J, Kim Y W, Kim B, Lee J G, and Park B (2003), 'A breakthrough in the safety of lithium secondary batteries by coating the cathode material with $AlPO_4$ nanoparticles,' *Angewandte Chemie – International Edition* **42|14|** 1618–1621.

Daniel M C and Astruc D (2004), 'Gold nanoparticles: Assembly, supramolecular chemistry, quantum-size-related properties, and applications toward biology, catalysis, and nanotechnology,' *Chemical Reviews* **104|1|** 293–346.

De Santis M, Buchsbaum A, Varga P, and Schmid M (2011), 'Growth of ultrathin cobalt oxide films on Pt(111),' *Physical Review B* **84|12|** 125430.

Dillon A C, Ott A W, Way J D, and George S M (1995), 'Surface-chemistry of Al_2O_3 deposition using $Al(CH_3)_3$ and H_2O in a binary reaction Sequence,' *Surface Science* **322|1–3|** 230–242.

Ding S J, Hu H, Lim H F, Kim S J, Yu X F, Zhu C X, Li M F, Cho B J, Chan D S H, Rustagi S C, Yu M B, Chin A, and Kwong D L (2003), 'High-performance MIM capacitor using ALD high-k HfO_2-Al_2O_3 laminate dielectrics,' *IEEE Electron Device Letters* **24|12|** 730–732.

Elam J W, Schuisky M, Ferguson J D, and George S M (2003), 'Surface chemistry and film growth during TiN atomic layer deposition using TDMAT and NH_3,' *Thin Solid Films* **436|2|** 145–156.

Elam J W, Zinovev A, Han C Y, Wang H H, Welp U, Hryn J N, and Pellin M J (2006), 'Atomic layer deposition of palladium films on Al_2O_3 surfaces,' *Thin Solid Films* **515|4|** 1664–1673.

Ferguson J D, Weimer A W, and George S M (2000), 'Atomic layer deposition of ultrathin and conformal Al_2O_3 films on BN particles,' *Thin Solid Films* **371[1–2]** 95–104.

Ferguson J D, Weimer A W, and George S M (2002), 'Atomic layer deposition of boron nitride using sequential exposures of BCl_3 and NH_3,' *Thin Solid Films* **413 [1–2]** 16–25.

Ferguson J D, Yoder A R, Weimer A W, and George S M (2004), 'TiO_2 atomic layer deposition on ZrO_2 particles using alternating exposures of $TiCl_4$ and H_2O,' *Applied Surface Science* **226[4]** 393–404.

Ferguson J D, Weimer A W, and George S M (2005), 'Surface chemistry and infrared absorbance changes during ZnO atomic layer deposition on ZrO_2 and $BaTiO_3$ particles,' *Journal of Vacuum Science & Technology A* **23[1]** 118–125.

George S M (2010), 'Atomic layer deposition: an overview,' *Chemical Reviews* **110[1]** 111–131.

George S M, Ott A W, and Klaus J W (1996), 'Surface chemistry for atomic layer growth,' *Journal of Physical Chemistry* **100[31]** 13121–13131.

Gerhardt L C, Jell G M R, and Boccaccini A R (2007), 'Titanium dioxide (TiO_2) nanoparticles filled poly(D,L lactid acid) (PDLLA) matrix composites for bone tissue engineering,' *Journal of Materials Science – Materials in Medicine* **18[7]** 1287–1298.

Goodman C H L and Pessa M V (1986), 'Atomic layer epitaxy,' *Journal of Applied Physics* **60[3]** R65–R81.

Hakim L F, Blackson J, George S M, and Weimer A W (2005a), 'Nanocoating individual silica nanoparticles by atomic layer deposition in a fluidized bed reactor,' *Chemical Vapor Deposition* **11[10]** 420–425.

Hakim L F, George S M, and Weimer A W (2005b), 'Conformal nanocoating of zirconia nanoparticles by atomic layer deposition in a fluidized bed reactor,' *Nanotechnology* **16[7]** S375–S381.

Hakim L F, Portman J L, Casper M D, and Weimer A W (2005c), 'Aggregation behavior of nanoparticles in fluidized beds,' *Powder Technology* **160[3]** 149–160.

Hakim L F, McCormick J A, Zhan G D, Li P, George S M, and Weimer A W (2006), 'Surface modification of titania nanoparticles using ultrathin ceramic films,' *Journal of the American Ceramic Society* **89[10]** 3070–3075.

Hakim L F, Vaughn C L, Dunsheath H J, Carney C S, Liang X H, Li P, and Weimer A W (2007), 'Synthesis of oxidation-resistant metal nanoparticles via atomic layer deposition,' *Nanotechnology* **18[34]** 345603.

Hamann T W, Martinson A B E, Elam J W, Pellin M J, and Hupp J T (2008), 'Aerogel templated ZnO dye-sensitized solar cells,' *Advanced Materials* **20[8]** 1560–1564.

Hutmacher D W (2000), 'Scaffolds in tissue engineering bone and cartilage,' *Biomaterials* **21[24]** 2529–2543.

Jeon W S, Yang S, Lee C S, and Kang S W (2002), 'Atomic layer deposition of Al_2O_3 thin films using trimethylaluminum and isopropyl alcohol,' *Journal of the Electrochemical Society* **149[6]** C306–C310.

Jung Y S, Cavanagh A S, Dillon A C, Groner M D, George S M, and Lee S H (2010a), 'Enhanced stability of $LiCoO_2$ cathodes in lithium-ion batteries using surface modification by atomic layer deposition,' *Journal of the Electrochemical Society* **157[1]** A75–A81.

Jung Y S, Cavanagh A S, Riley L A, Kang S H, Dillon A C, Groner M D, George S M, and Lee S H (2010b), 'Ultrathin direct atomic layer deposition on composite electrodes for highly durable and safe Li-ion batteries,' *Advanced Materials* **22** [19] 2172–2176.

Juppo M, Rahtu A, Ritala M, and Leskela M (2000), 'In situ mass spectrometry study on, surface reactions in atomic layer deposition of Al_2O_3 thin films from trimethylaluminum and water,' *Langmuir* **16**[8] 4034–4039.

Kim J S, Johnson C S, and Thackeray M M (2002), 'Layered $xLiMO_2 \cdot (1\text{-}x)Li_2M'O_3$ electrodes for lithium batteries: a study of $0.95LiMn_{0.5}Ni_{0.5}O_2 \cdot 0.05Li_2TiO_3$,' *Electrochemistry Communications* **4**[3] 205–209.

Kim K H, Kwak N W, and Lee S H (2009), 'Fabrication and properties of AlN film on GaN substrate by using remote plasma atomic layer deposition method,' *Electronic Materials Letters* **5**[2] 83–86.

King D M, Spencer J A, Liang X H, Hakim L F, and Weimer A W (2007), 'Atomic layer deposition on particles using a fluidized bed reactor with in situ mass spectrometry,' *Surface & Coatings Technology* **201**[22–23] 9163–9171.

King D M, Liang X H, Carney C S, Hakim L F, Li P, and Weimer A W (2008a), 'Atomic layer deposition of UV-absorbing ZnO films on SiO_2 and TiO_2 nanoparticles using a fluidized bed reactor,' *Advanced Functional Materials* **18**[4] 607–615.

King J S, Wittstock A, Biener J, Kucheyev S O, Wang Y M, Baumann T F, Giri S K, Hamza A V, Baeumer M, and Bent S F (2008b), 'Ultralow loading Pt nanocatalysts prepared by atomic layer deposition on carbon aerogels,' *Nano Letters* **8**[8] 2405–2409.

King D M, Liang X H, and Weimer A W (2009), 'Functionalization of fine particles using atomic and molecular layer deposition,' *ECS Transactions* **25**[4] 163–190.

Klaus J W, Ott A W, Johnson J M, and George S M (1997a), 'Atomic layer controlled growth of SiO_2 films using binary reaction sequence chemistry,' *Applied Physics Letters* **70**[9] 1092–1094.

Klaus J W, Sneh O, and George S M (1997b), 'Growth of SiO_2 at room temperature with the use of catalyzed sequential half-reactions,' *Science* **278**[5345] 1934–1936.

Kokubo T (1996), 'Formation of biologically active bone-like apatite on metals and polymers by a biomimetic process,' *Thermochimica Acta* **280** 479–490.

Kukli K, Ritala M, Lu J, Harsta A, and Leskela M (2004), 'Properties of HfO_2 thin films grown by ALD from hafnium tetrakis(ethylmethylamide) and water,' *Journal of the Electrochemical Society* **151**[8] F189–F193.

Kumagai H, Matsumoto M, Toyoda K, Obara M, and Suzuki M (1995), 'Fabrication of titanium-oxide thin-films by controlled growth with sequential surface chemical-reactions,' *Thin Solid Films* **263**[1] 47–53.

Leskela M and Ritala M (1999), 'ALD precursor chemistry: Evolution and future challenges,' *Journal de Physique Iv* **9**[P8] 837–852.

Leskela M and Ritala M (2002), 'Atomic layer deposition (ALD): from precursors to thin film structures,' *Thin Solid Films* **409**[1] 138–146.

Li C, Zhang H P, Fu L J, Liu H, Wu Y P, Ram E, Holze R, and Wu H Q (2006), 'Cathode materials modified by surface coating for lithium ion batteries,' *Electrochimica Acta* **51**[19] 3872–3883.

Li J H, Liang X H, King D M, Jiang Y B, and Weimer A W (2010), 'Highly dispersed Pt nanoparticle catalyst prepared by atomic layer deposition,' *Applied Catalysis B –Environmental* **97|1–2|** 220–226.

Liang X H, George S M, Weimer A W, Li N H, Blackson J H, Harris J D, and Li P (2007), 'Synthesis of a novel porous polymer/ceramic composite material by low-temperature atomic layer deposition,' *Chemistry of Materials* **19|22|** 5388–5394.

Liang X H, Zhan G D, King D M, McCormick J A, Zhang J, George S M, and Weimer A W (2008), 'Alumina atomic layer deposition nanocoatings on primary diamond particles using a fluidized bed reactor,' *Diamond and Related Materials* **17|2|** 185–189.

Liang X H, Lynn A D, King D M, Bryant S J, and Weimer A W (2009), 'Biocompatible interface films deposited within porous polymers by atomic layer deposition (ALD),' *ACS Applied Materials & Interfaces* **1|9|** 1988–1995.

Liu X H and Ma P X (2004), 'Polymeric scaffolds for bone tissue engineering,' *Annals of Biomedical Engineering* **32|3|** 477–486.

Lu J L and Stair P C (2010), 'Low-temperature ABC-type atomic layer deposition: synthesis of highly uniform ultrafine supported metal nanoparticles,' *Angewandte Chemie – International Edition* **49|14|** 2547–2551.

Malygin A A, Malkov A A, and Dubrovenskii S D (1996), 'The chemical basis of surface modification technology of silica and alumina by molecular layering method,' *Adsorption on New and Modified Inorganic Sorbents* **99** 213–236.

Martinson A B F, DeVries M J, Libera J A, Christensen S T, Hupp J T, Pellin M J, and Elam J W (2011), 'Atomic layer deposition of Fe_2O_3 using ferrocene and ozone,' *Journal of Physical Chemistry C* **115|10|** 4333–4339.

McCool B A and DeSisto W J (2004), 'Self-limited pore size reduction of mesoporous silica membranes via pyridine-catalyzed silicon dioxide ALD,' *Chemical Vapor Deposition* **10|4|** 190–194.

McCormick J A, Cloutier B L, Weimer A W, and George S M (2007), 'Rotary reactor for atomic layer deposition on large quantities of nanoparticles,' *Journal of Vacuum Science & Technology A* **25|1|** 67–74.

Neo M, Kotani S, Nakamura T, Yamamuro T, Ohtsuki C, Kokubo T, and Bando Y (1992), 'A comparative-study of ultrastructures of the interfaces between 4 kinds of surface-active ceramic and bone,' *Journal of Biomedical Materials Research* **26|11|** 1419–1432.

Niinisto L, Ritala M, and Leskela M (1996), 'Synthesis of oxide thin films and overlayers by atomic layer epitaxy for advanced applications,' *Materials Science and Engineering B – Solid State Materials for Advanced Technology* **41|1|** 23–29.

Nilsen O, Fjellvag H, and Kjekshus A (2003), 'Growth of manganese oxide thin films by atomic layer deposition,' *Thin Solid Films* **444|1–2|** 44–51.

Ott A W, McCarley K C, Klaus J W, Way J D, and George S M (1996), 'Atomic layer controlled deposition of Al_2O_3 films using binary reaction sequence chemistry,' *Applied Surface Science* **107** 128–136.

Panyam J and Labhasetwar V (2003), 'Biodegradable nanoparticles for drug and gene delivery to cells and tissue,' *Advanced Drug Delivery Reviews* **55|3|** 329–347.

Piconi C and Maccauro G (1999), 'Zirconia as a ceramic biomaterial,' *Biomaterials* **20|1|** 1–25.

Puurunen R L (2005), 'Surface chemistry of atomic layer deposition: A case study for the trimethylaluminum/water process,' *Journal of Applied Physics* **97**[12] 121301.

Ritala M, Leskela M, Niinisto L, and Haussalo P (1993), 'Titanium isopropoxide as a precursor in atomic layer epitaxy of titanium-dioxide thin-films,' *Chemistry of Materials* **5**[8] 1174–1181.

Ritala M, Leskela M, Rauhala E, and Haussalo P (1995), 'Atomic layer epitaxy growth of TiN thin-films,' *Journal of the Electrochemical Society* **142**[8] 2731–2737.

Satta A, Schuhmacher J, Whelan C M, Vandervorst W, Brongersma S H, Beyer G P, Maex K, Vantomme A, Viitanen M M, Brongersma H H, and Besling W F A (2002), 'Growth mechanism and continuity of atomic layer deposited TiN films on thermal SiO_2,' *Journal of Applied Physics* **92**[12] 7641–7646.

Scheffe J R, Frances A, King D M, Liang X H, Branch B A, Cavanagh A S, George S M, and Weimer A W (2009), 'Atomic layer deposition of iron(III) oxide on zirconia nanoparticles in a fluidized bed reactor using ferrocene and oxygen,' *Thin Solid Films* **517**[6] 1874–1879.

Scheffe J R, Li J H, and Weimer A W (2010), 'A spinel ferrite/hercynite water-splitting redox cycle,' *International Journal of Hydrogen Energy* **35**[8] 3333–3340.

Scott I D, Jung Y S, Cavanagh A S, An Y F, Dillon A C, George S M, and Lee S H (2011), 'Ultrathin coatings on nano-$LiCoO_2$ for Li-ion vehicular applications,' *Nano Letters* **11**[2] 414–418.

Senkevich J J, Tang F, Rogers D, Drotar J T, Jezewski C, Lanford W A, Wang G C, and Lu T M (2003), 'Substrate-independent palladium atomic layer deposition,' *Chemical Vapor Deposition* **9**[5] 258–264.

Suchaneck G, Norkus V, and Gerlach G (2001), 'Low-temperature PECVD-deposited silicon nitride thin films for sensor applications,' *Surface & Coatings Technology* **142** 808–812.

Suchanek W and Yoshimura M (1998), 'Processing and properties of hydroxyapatite-based biomaterials for use as hard tissue replacement implants,' *Journal of Materials Research* **13**[1] 94–117.

Sun Y K, Myung S T, Kim M H, Prakash J, and Amine K (2005), 'Synthesis and characterization of $Li[(Ni_{0.8}Co_{0.1}Mn_{0.1})_{0.8}(Ni_{0.5}Mn_{0.5})_{0.2}]O_2$ with the microscale core-shell structure as the positive electrode material for lithium batteries,' *Journal of the American Chemical Society* **127**[38] 13411–13418.

Sun Y K, Myung S T, Park B C, Prakash J, Belharouak I, and Amine K (2009), 'High-energy cathode material for long-life and safe lithium batteries,' *Nature Materials* **8**[4] 320–324.

Suntola T (1989), 'Atomic layer epitaxy,' *Acta Polytechnica Scandinavica – Electrical Engineering Series* [64] 242–270.

Suntola T (1992), 'Atomic layer epitaxy,' *Thin Solid Films* **216**[1] 84–89.

Suntola T and Hyvarinen J (1985), 'Atomic layer epitaxy,' *Annual Review of Materials Science* **15** 177–195.

Ten Eyck G A, Senkevich J J, Tang F, Liu D L, Pimanpang S, Karaback T, Wang G C, Lu T M, Jezewski C, and Lanford W A (2005), 'Plasma-assisted atomic layer deposition of palladium,' *Chemical Vapor Deposition* **11**[1] 60–66.

Ten Eyck G A, Pimanpang S, Juneja J S, Bakhru H, Lu T M, and Wang G C (2007),

'Plasma-enhanced atomic layer deposition of palladium on a polymer substrate,' *Chemical Vapor Deposition* **13**[6–7] 307–311.

Theobald J A, Oxtoby N S, Phillips M A, Champness N R, and Beton P H (2003), 'Controlling molecular deposition and layer structure with supramolecular surface assemblies,' *Nature* **424**[6952] 1029–1031.

Wank J R, George S M, and Weimer A W (2001), 'Vibro-fluidization of fine boron nitride powder at low pressure,' *Powder Technology* **121**[2–3] 195–204.

Wank J R, George S M, and Weimer A W (2004a), 'Coating fine nickel particles with Al_2O_3 utilizing an atomic layer deposition-fluidized bed reactor (ALD-FBR),' *Journal of the American Ceramic Society* **87**[4] 762–765.

Wank J R, George S M, and Weimer A W (2004b), 'Nanocoating individual cohesive boron nitride particles in a fluidized bed by ALD,' *Powder Technology* **142**[1] 59–69.

Warashina H, Sakano S, Kitamura S, Yamauchi K I, Yamaguchi J, Ishiguro N, and Hasegawa Y (2003), 'Biological reaction to alumina, zirconia, titanium and polyethylene particles implanted onto murine calvaria,' *Biomaterials* **24**[21] 3655–3661.

Wilson C A, McCormick J A, Cavanagh A S, Goldstein D N, Weimer A W, and George S M (2008), 'Tungsten atomic layer deposition on polymers,' *Thin Solid Films* **516**[18] 6175–6185.

Xu C B and Zhu J (2006), 'Parametric study of fine particle fluidization under mechanical vibration,' *Powder Technology* **161**[2] 135–144.

Yun S J, Kang J S, Paek M C, and Nam K S (1998), 'Large-area atomic layer deposition and characterization of Al_2O_3 film grown using $AlCl_3$ and H_2O,' *Journal of the Korean Physical Society* **33** S170–S174.

Yun S J, Lim J W, and Lee J H (2004), 'Low-temperature deposition of aluminum oxide on polyethersulfone substrate using plasma-enhanced atomic layer deposition,' *Electrochemical and Solid State Letters* **7**[1] C13–C15.

Zhou Y, King D M, Li J H, Barrett K S, Goldfarb R B, and Weimer A W (2010), 'Synthesis of photoactive magnetic nanoparticles with atomic layer deposition,' *Industrial & Engineering Chemistry Research* **49**[15] 6964–6971.

Zhu Y, Dunn K A, and Kaloyeros A E (2007), 'Properties of ultrathin platinum deposited by atomic layer deposition for nanoscale copper-metallization schemes,' *Journal of Materials Research* **22**[5] 1292–1298.

9

High-temperature superconducting ceramic nanocomposites

A . O. TONOYAN and S. P. DAVTYAN,
State Engineering University of Armenia, Armenia

DOI: 10.1533/9780857093493.2.284

Abstract: This chapter reviews the preparation and study of high-temperature superconducting (SC) nanocomposites based on SC ceramics and various polymeric binders. Regardless of the size of the ceramic grains, any increase in their quantity results in an increase of resistance to rupture and modulus and a decrease in limiting deformation, whilst a similar increase in the average ceramic grain size worsens resistance properties. Investigation of the SC, thermo-chemical, mechanical and dynamic-mechanical properties of the samples are discussed. Superconducting properties of the polymer–ceramic nanocomposites are explained by intercalation of macromolecule fragments into the interstitial layer of the ceramic grains, a phenomenon that leads to a change in the morphological structure of SC nanocomposites.

Key words: polymer–ceramic nanocomposites, superconducting, physical–mechanical properties, dynamic–mechanical properties, morphological structure.

9.1 Introduction

This chapter reviews recent research on the effect of various processing techniques and parameters on the superconducting (SC) characteristics of high-temperature SC polymer–ceramic composites. Following the discovery of high-temperature superconductivity in perovskite systems containing oxygen by Bednorz and Muller [1, 2], the literature on this topic has become a veritable avalanche, as both fundamental and applied research on the fabrication of metal [3–14] and polymer [15–26] binder composites of various geometries possessing SC properties has been undertaken. Polymer–

284

ceramic composites using thermoplastic and reactoplastic polymeric matrices are described in various sources [15–26]. Haupt *et al.* [15], prepared products by incineration of a polymeric matrix. Incineration of the organic part, as a rule, is accompanied by oxidation and thermal destruction, which consumes oxygen irreversibly, thus making the ceramic grains amorphous. Loss of SC properties of materials produced via incineration can be explained in this way. This demands a full circle restoration. It must be noted that the coke formed after incineration also has a negative impact on SC properties. This technique can be used to produce polymer–ceramic composites designed to protect high-temperature superconductors against humidity [13, 14].

It is known that particulate polymers, as a rule, improve a number of the features of composites (hardness, impact strength and heat resistance, for example). This is mainly related to the formation of a special interfacial layer between the filler and polymeric binder. High-temperature SC polymer–ceramic composites can be obtained both by the conventional hot pressing of a ceramic mixture with some ready-made, highmolecular weight binder and by the polymerization filler method [21]. Hot pressing [18–20, 22–28] of the $Y_1Ba_2Cu_3O_{7-x}$ oxide ceramic and superhigh molecular polyethylene mixture at $200°C$ destroys the SC properties, which are restored only after treatment of samples in a dry oxygen stream [22].

With regard to conventional fillers, several specific properties of perovskite high-temperature superconductors (such as their layered structure, developed surface of the ceramic grains, catalytic properties, and free oxygen dislocated on the surface of the ceramic grains) have a significant impact not only on the formation of a phase boundary and, consequently, on the physical–mechanical properties, but on the SC properties of the polymer–ceramic composites as well. Regardless of the nature of the binder, the critical transition temperature (T_c) into the SC state of polymer–ceramic composites increases by 1–3 K. This increase in the transition temperature is due to the interaction of the polymer chains with the surface of the ceramic grains.

It could be expected that such an interaction should change the packing and structure of the polymer chains as well as the conformation at the interphase. Interphase phenomena at the ceramic–polymer boundary have been investigated, for example, for superhigh molecular weight polyethylene $+ Y_1Ba_2Cu_3O_{7-x}$ ceramic. The influence of crystalline binders on the valence state of Cu^{2+}(I) in the ceramic has also been investigated.

It is interesting to study the influence of the environment on the SC properties of SC ceramics and polymer–ceramic composites at ambient temperatures. It is known that the bulk oxygen content determines the properties of high-temperature oxide ceramics of the Y–Ba–Cu–O system and the pattern in which oxygen fills the crystal structure. The presence of an unstable phase is a feature of current-generation SC ceramics. Non-stability of the lattice structure is linked with the deficiency of the atoms of oxygen in the elementary cell, accompanied by the accumulation of regular patterns of vacancies in the Cu–O atom chains. As a result, the increase of non-stability in the lattice structure of the $Y_1Ba_2Cu_3O_{7-x}$ ceramic over time can change its SC characteristics, for example its critical temperature of transition into the SC state (T_i, K) or the width of transition (ΔT_c, K).

9.2   Material preparation, characterization and testing

In order to study the effects of different preparation factors, powders of the corresponding polymer and $Y_1Ba_2Cu_3O_{7-x}$ ceramic were preliminarily blended in an agate mill to give a homogeneous mixture for the formation of items (plate, rod, tube, ring, etc.). These pre-made mixtures were then filled into previously heated (130, 150, 160, 200°C) forms and pressed at 100 mPa for 5, 10, 20, or 30 min.

In another series of experiments focused on the formation of polymer–ceramic SC items from $Y_1Ba_2Cu_3O_{7-x}$ ceramic powder at ambient temperature, specimens were pressed with follow-up imbibing of methyl methacrylate, both with initiator (azobisisobutyronitrile (AIBN)) and without. To counterbalance vaporization of the monomer, the specimens were placed in sealed glass forms and polymerized at 60–80°C. As a control for the reaction end, other specimens were tested concomitantly in a DAK-11 microcalorimeter.

For the gas-phase polymerization of ethylene influenced by $Y_1Ba_2Cu_3O_{6.97}$ (particle size <50 μm), the surface was activated at 197°C over the course of 4 hours, then cooled down to ambient temperature in a dry air environment. Part of the thus-obtained ceramic was used for the determination of the critical transition temperature to the SC state, whilst the other part was placed in the polymerization reactor. At room temperature, under a pressure of 20 bar, ethylene was introduced into the reactor filled with hexane, and alkyl aluminum was injected under vigorous agitation (rotation speed of 100 rpm). Ethylene was consumed for 3 hours, which is a major sign of ethylene polymerization.

Structures of the high-temperature SC ceramic and its composites were determined using X-ray analysis on a DRON 2.0 instrument (λCuKr) in the $15° \leq 20° \leq 130°$ range of angles at room temperature. The critical temperatures of the SC transitions were measured by the αC-magnetic susceptibility method at a frequency of 1 kHz, when the amplitude of the magnetic field was 10 mE. The physico-mechanical properties were determined on an INSTRON rupture machine, and thermo-oxidation destruction was monitored via a derivatographic method using a MOM brand Q-1500 instrument.

Superconductive composite samples based on superhigh molecular polyethylene (SHMPE) take the form of $3 \times 1 \times 0.1\,cm^3$ plates (matrix:filler = 100:0; 85:15; 50:50; 15:85 mass ratios). These samples were used for studying dynamic mechanical properties. The mechanical relaxation properties of the SHMPE composites were measured using a Du Pont DMA instrument under 0.1 and 0.2 mm oscillation amplitudes. Structural peculiarities of SC polymer–ceramic composites have been investigated using a 'Tesla' electron-microscope, and the kinetics of polymerization were investigated using a micro-calorimeter (DAK-11) in the presence of $Y_1Ba_2Cu_3O_{6.97}$ ceramic. Mn, Zn, Co and Ni containing metal monomers were synthesized according to methods present in the literature [27].

The frontal polymerization [28] of metal monomers was carried out in vertically placed glass ampoules. For this purpose, ceramic and metal monomer powders were combined into a homogenous mixture using an agate mill, and used to fill a glass ampoule with an inner diameter of 0.8 cm. The samples of SC ceramic and polymer–ceramic composites were kept in air at ambient temperatures, and the enthalpy of the filled crystalline polyethylene was determined on a DSM-3A differential scanning calorimeter.

Quinol ether [O,O-(bis-(1,3,5-tri-tert-butyl-4-oxocyclohexadiene-2,5-yl)-2-methyl-5-propyl-benzo-quinone dioxime, 1% by mass of divinyl rubber (DR)] was added to stitched DR to obtain the corresponding composite specimens. The blend was well mixed with the ceramic and the obtained mixture was polymerized at 200°C under the impact of a specific pressure (100 mPa) over the course of 30 min. This produced 1.5 mm thick plates, which were subsequently cut into blades and tested on an INSTRON rupture machine. Five measurements were taken for each specimen composition, which were later averaged as physical–mechanical characteristics.

In another series of experiments, the fractionation of $Y_1Ba_2Cu_3O_{7-x}$ ceramic on diffuser–confuser sieves elaborated at the Institute of Chemical Physics of the Russian Science Academy was examined. The density of various fractions of powder was determined on hydrostatic scales, while the BET method was used for the determination of the specific surface area. Initial DR had a molecular mass of 10^6 and a polydispersity distribution of 2.5. An average particle size fractionation in between 5 and 30 μm was used.

The used $Y_1Ba_2Cu_3O_{7-x}$ ceramic was made up of two composites, $Y_1Ba_2Cu_3O_{6.97}$ and $Y_1Ba_2Cu_3O_{6.92}$, with SC critical transition initiation temperatures (T_c) of 93 K and 91.5 K and width (ΔT) of 6.5 and 6.0 K. For the formation of polymer–ceramic composites via the hot pressing method, the following polymeric binders were used: SHMPE $T_{melt} = 128$–135°C and ramified polyethylene (RPE) with $T_{melt} = 105$–108°C; isotactic polypropylene (iPP) with $T_{melt} = 167$–171°C; polybutene (PB) with $T_{melt} = 135$°C; copolymer of ethylene with tetrafluoroethylene (brand name, F-40) with

T_{melt} = 265–278°C; polyvinilidenefluoride (PVF) with T_{melt} = 171–180°C; polyvinyl alcohol (PVA) with T_g = 85°C; polyformaldehyde (PFA), polymethylmetacrylate (PMA) with T_g = 100–105°C; polystyrene (PS) with T_g = 98–102°C. In addition, copolymers of styrene (ST) with methyl metacrylate (MMA) (ST content 80, 60 and 40 mol/%), DR with average molecular mass equal to 106 and 2.5 polydispersity were used. All of the polymers used in this work were fine powders and, as an antioxidant, Irganox® 1010 and NG-2246 in 5% weight ratio versus the polymeric binder were used.

9.3   Superconducting (SC) properties of polymer–ceramic nanocomposites manufactured by hot pressing

Polymer–ceramic composites with the various binders used exhibit no SC properties immediately after hot pressing [20, 22, 24] at 200°C for 30 min. The Meissner effect, for example, is missing. The absence of SC properties can be explained by the depletion of oxygen from the orthorhombic SC phase of the ceramic $Y_1Ba_2Cu_3O_{7-x}$ after pressing at 200°C. The released oxygen interacts irreversibly with the polymeric binder, causing its thermo-oxidative destruction. Investigations carried out on composites with superhigh molecular polyethylene using a MOM 1500 derivatographic instrument show that, at 160°C and higher temperatures, weight loss took place, which indicates that macromolecules of the binder decompose under thermo-oxidative conditions. Oxygen participates in the thermo-oxidative destruction of the binder by desorbing and diffusing into the polymeric phase from the nucleus of the ceramic grains. It seems that free oxygen dislocated on the surface of the grains of the oxide ceramic reacts with the polymeric phase. Indirect poof of this assumption comes from restoration of the SC properties of the composites (see Table 9.1) under an atmosphere of dry oxygen at the α transition temperatures of the polymeric binders. The characteristic curve of the SC transition ($Y_1Ba_2Cu_3O_{7-x}$ + superhigh molecular polyethylene) obtained for the samples after restoration is shown in Fig. 9.1. Further proof can be seen in the results of the experiments carried out when Irganox® 1010 (a polymeric antioxidant) was added to the initial mixture. It is known that antioxidant additives in a polymeric matrix substantially reduce the rate of oxidative destruction of the polymers [29–31].

Consequently, it can be deduced that the introduction of minute quantities of antioxidant (0.5m.Kt% of binder) in the polymer–ceramic composite decreases the rate of oxidation reactions. This then retards the depletion of oxygen from the surface of the ceramic grains. Directly after

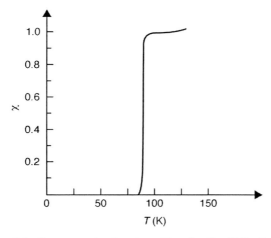

9.1 Superconducting transition for the $Y_1Ba_2Cu_3O_{7-x}$ composite with SHMPE. Ceramic/binder weight ratio, 80:20; pressing temperature, 200°C [18].

pressing and without implementation of the restoration stage, composites with small amounts of antioxidant have actually been shown to possess SC properties, as demonstrated by the data given in Table 9.1.

In both composites containing antioxidants and those without, when the nature and quantity of the polymer remain intact, the SC transition critical temperature (T_c) stays practically unchanged whilst the transition end temperature (T_f) changes. In samples without antioxidants, a wider (T_c-T_f) transition temperature span is seen, which seems to be linked to a non-homogeneous distribution of the oxygen deposited on the surface of the ceramic grain following the restoration stage.

The interaction of composites with various binders shows that both the beginning and end of the transition depend considerably upon the nature and chemical composition of the polymeric binder. In the case of SC polymer–ceramic composites, the SC critical temperature increases by 2–3 K, except for PVA and PFA, as seen in Table 9.1. It is supposed that this T_c temperature elevation can be ascribed to the physical interaction of separate fragments of the macromolecules of the binder with the surface of the $Y_1Ba_2Cu_3O_{7-x}$ ceramic. The elevation increases up to the intercalation of these fragments into the interstitial layer of the ceramic grains, or the anchoring of the fragments in free oxygen positions.

For composites with PVA, the observed decrease of T_c by five degrees suggests the possible interaction of some polymeric chain fragments with the ceramic surface. In this case, reactionary OH groups, such as those in water, alcohol or weak acids [32], distort the orthorhombic phase in nearby surface layers of the ceramic, thus decreasing the critical temperature of the

Table 9.1 Superconducting properties of polymer–ceramic nanocomposites after restoration

Composite	Ceramic:binder weight ratio	T_c (K)	T_f (K)	Notes
SHMPE + $Y_1Ba_2Cu_3O_{7-x}$	80:20	96	84	Obtained after the
	85:15	96	84	restoration stage
	85:15	96	84	
RPE + $Y_1Ba_2Cu_3O_{7-x}$	80:20	94	80	Obtained after the restoration stage
Ф-40 + $Y_1Ba_2Cu_3O_{6.97}$	75:25	96	77	Obtained after the restoration stage
PB + $Y_1Ba_2Cu_3O_{7-x}$	80:20	96	83	Obtained after the restoration stage
PVF + $Y_1Ba_2Cu_3O_{7-x}$	85:15	90	75	Obtained after the restoration stage
PFA + $Y_1Ba_2Cu_3O_{7-x}$	80:20	88	76	Obtained after the restoration stage
SHMPE + Irganox + $Y_1Ba_2Cu_3O_{7-x}$	80:20	96	89	Obtained without restoration
RPE + Irganox + $Y_1Ba_2Cu_3O_{7-x}$	80:20	94	85	Obtained without restoration
PVA + Irganox + $Y_1Ba_2Cu_3O_{7-x}$	85:15	90	80	Obtained without restoration

superconducting transition. For the polyformaldehyde composition, the considerable T_c decrease can be explained by the high inclination of polyformaldehyde towards thermo-oxidative destruction, which enhances the consumption of oxygen in the ceramic. Moreover, even at the comparatively lower temperature [30] of polyformaldehyde thermo-oxidation, this was accompanied by a de-polymerization reaction; the resultant release of gaseous formaldehyde could, in turn, adversely affect the SC properties of the nanocomposites.

9.3.1   Influence of thermal regimes on the SC properties of polymer–ceramic nanocomposites

To ensure steady contact between the polymeric binder and the ceramic grains, hot pressing of the mixture should be performed at temperatures slightly exceeding the melting or vitrification temperature of the polymeric matrix. The temperature of vitrification for acryl and vinyl homo- and copolymers does not exceed 100°C, whilst the melting temperature of the polyolefins is 120–125°C (apart from polypropylene (PP), for which $T_{melt} = 170–175°C$). This is why most samples studied were pressed at 130°C.

After pressing at 130°C, samples sustain their SC properties. Nevertheless, the samples based on polyethylene possess the same values of T_c and ΔT_c as

Table 9.2 Superconducting characteristics of nanocomposites based on $Y_1Ba_2Cu_3O_{6.97}$ ceramic (t_{press} = 130°C, τ_{press} = 4 min)

Binder	Content of binder (wt%)	Antioxidant (5% by weight of binder)	T_c (K)	ΔT_c (K)	Difference in parameters of the elementary cell (Å)	Index of oxygen
HDPE	10		92.1	7.0	0.0732	6.98
	15		91.8	7.0	0.0719	6.97
SHMPE	10		91.2	7.0	0.0660	6.93
	15		91.5	~5.0	0.0703	6.96
	20		91.8	6.0	0.0719	6.97
	15	Irganox 1010	91.7	6.0	0.0719	6.97
	15	NG-2246	91.2	6.0	0.0708	6.96
PMMA	10		92.3	8.0	0.0707	6.96
	15		93.7	7.0	0.0710	6.96
	20		91.7	7.0	0.0689	6.95
	15	NG-2246	93.2	6.0	0.0747	6.98
PS	10		92.0	6.0	0.0719	6.97
	15		93.1	7.0	0.0752	6.99
	20		92.3	8.0	0.0703	6.96
	15	NG-2246	93.0	7.0	0.0751	6.99
SPL Styrene–MMA (60:40 mole%)	10		91.7	7.0	0.0689	6.95
	15		92.3	7.0	0.0659	6.93
	20		92.1	7.0	0.0666	6.93
	15	NG-2246	92.0	7.0	0.0659	6.93
SPL styrene–MMA (80:20 mole%)	10		92.1	8.0	0.0689	6.95
	15		91.9	7.0	0.0707	6.96
	20		92.6	8.0	0.0739	6.98
	15	NG-2246	93.4	~9.0	0.0767	6.998

the initial ceramic, whilst for the acryl and vinyl homo- and copolymers, the initial temperature of transition to the SC properties is 1–2 K higher (Table 9.2). There is no distinct relationship between the amount of ceramic in the composite, the values of T_c and ΔT_c for the interval of concern, and the degree of filling of the SC polymer–ceramic materials (80–90 wt%). An increase of the press load of the composites, up to 200 MPa, does not change the T_c and ΔT_c values. Furthermore, the addition of antioxidants (5% of the mass of binder) at the given pressing temperature of the composite does not notably affect either the ceramic's crystalline lattice structure or the critical temperatures of transition to the SC state.

As previously indicated, in pressing the polymer–ceramic composites at 200°C for 30 min, the SC properties of the materials are lost. Cutting the pressing duration down to 4 min at the same temperature for composites based on SHMPE (content of ceramic is 85% by weight) retains the same critical SC transition parameters as the initial ceramic ($T_c = 92$ K, and $\Delta T_c = 8$ K). For PP under the same conditions at 180°C, T_c remains constant (91.6 K) while the transition width increases sharply ($\Delta T_c \geq 8$ K). It is known that the presence of tertiary atoms of carbon in the macro chain weakens carbon–carbon bonds in that polymer, making it less thermo-stable than polyethylene [31]. It is assumed that this inclination of the PP to decompose under thermo-oxidation conditions, combined with the participation of oxygen from the SC orthorhombic phase in this process, are the main causes of the observed phenomena.

To obtain confirmation of this supposition, antioxidants (Irganox® 1010 and NG-2246) were additionally introduced into the matrix, which should decrease the thermo-oxidative destruction rate of the polymer substantially. However, the addition of Irganox® 1010 contracted the width of the temperature transition into the SC state up to the characteristic values of the initial ceramic ($\Delta T_c \approx 8$ K), whereas NG-2246 did not affect the ΔT_c of the composites based on PP. This is because NG-2246 is not an effective antioxidant for polyolefins.

Elevation of the pressing temperature of the composites based on PMMA (content of ceramic 85% by weight) up to 160°C and a longer duration (15 min, cooling down to 40°C) increases ΔT_c appreciably (≥ 8 K), while T_c remains steady (91.9–92.3 K). When NG-2246 antioxidant was added or the formation duration was cut down to the conventional time of 4 min, the characteristics ($T_c = 91.5$ K and $\Delta T_c = 8$ K) were restored.

Thus, all of the results outlined in Table 9.2 can be explained by the competing action of two processes taking place in parallel:

• interaction of separate elements or fragments of the macromolecule of polymeric binder with the surface of the ceramic grain before their intercalation into the interstitial layer of the ceramic
• thermo-oxidative destruction of the polymeric binder.

In this case, factors that promote the interaction of the binder macromolecule elements with the surface of the ceramic (flexibility of the macro chains, elevation of the temperature, etc.), as well as the decrease of the rate of thermo-oxidative destruction (reducing the temperature of pressing, introduction of antioxidants), promote, or at least maintain, the critical SC properties that they possess in the initial state. It is supposed that oxygen contained in the ceramic grains plays an active role in the process of thermo-oxidative destruction of the polymeric matrix. Nevertheless, to determine whether thermo-oxidative destruction of the polymeric binder is a governing factor responsible for the widening of ΔT_c, the thermo-chemical properties of the non-filled polymeric binder, as well as those of the composites based on them, was investigated.

Table 9.3 Thermal stability of polymeric binders and SC nanocomposites based on them ($t_{press} = 130°C$, $\tau_{press} = 4$ min)

Polymeric binder	Binder content (wt%)	Type of antioxidant used (5% of the mass of the binder)	Thermo-oxidation destruction initiation temperature (°C)	Weight loss at 300 C (calc. on the polymeric mass) (mass%)	Notes
SHMPE	100		195	2.3	The used ceramic
	10		185	~33.0	has a low oxygen
	15		190	~21.0	index and a wide
	20		195	~20.0	transition into the
					SC state
					temperature
					interval
PMMA	100		170	21.0	Conventional
	15	NG-2246	155	33.0	ceramic is used,
	15		235	21.0	$Y_1Ba_2Cu_3O_{6.97}$
SPL	100		165	20	—
Styrene–	15	NG-2246	125	—	—
MMA (60:40	15		220	~5	—
mole%)					

9.3.2 Influence of thermo-oxidative destruction on the SC properties of polymer–ceramic nanocomposites

Results of derivatographic analysis of the samples under investigation are presented in Table 9.3 and Figs 9.2 and 9.3. The results obtained prove that the $Y_1Ba_2Cu_3O_{6.97}$ oxide ceramic is stable up to 300°C, as revealed by thermo-gravimetric curves (TG) and differential thermal analysis (DTA). No displacement of apexes was observed. Study of derivatograms of non-filled SHMPE (Fig. 9.2) and the composites based on it (Fig. 9.3) allows us to conclude that between 150 and 195°C, the exothermic peaks on the DTA curves are conditioned by oxidation of the matrix at these temperatures. Furthermore, this conclusion is substantiated by the weight increase of the samples, as observed on the TG curves.

The initiation temperature for the oxidation of SHMPE in the composites (Fig. 9.3) is lowered by 10–15°C compared with the non-filled polymer. Beginning between 185 and 195°C, thermo-oxidative destruction of the composite is initiated, accompanied by a loss in sample weight. This means that the formation of composites with SHMPE binder at 200°C is accompanied by thermo-oxidative destruction of the matrix. PMMA–SC ceramic composites are thermally less stable. Introduction of the SC ceramic (filling rate 85% by weight) decreases the initiation temperature for the thermo-oxidation destruction of PMMA from 170°C down to 155°C.

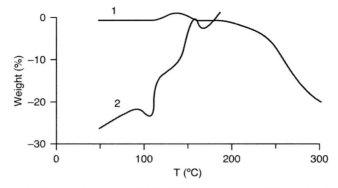

9.2 Weight loss (% of SHMPE) (curve 1) and differential thermal assay (curve 2) via thermo-oxidative destruction of the SHMPE. Heating rate is 3.2 °C/min [19].

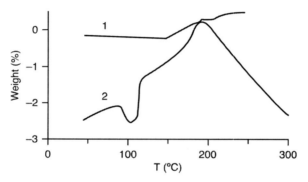

9.3 Weight loss (% of binder) (curve 1) and differential thermal assay (curve 2) of SHMPE + $Y_1Ba_2Cu_3O_{6.97}$. SHMP:$Y_1Ba_2Cu_3O_{6.97}$ = 20:80 (in mass ratio of the monomer). Heating rate is 3.2°C/min [19].

The addition of NG-2246 as an antioxidant decreases the rate of thermo-oxidative processes occurring in the matrix, thus increasing the initiation temperature of binder destruction up to 235°C. These results fully support the assumption made above, and correlate well with observed peculiarities in the SC characteristics of PMMA–ceramic composites. Composites based on styrene with methyl methacrylate (Table 9.3) are similarly effective in the presence of NG-2246 antioxidant.

9.3.3 Polymerization of methyl methacrylate in the presence of $Y_1Ba_2Cu_3O_{7-x}$ ceramics

For composites obtained by the polymerization of methyl methacrylate (MMA) in pressed samples of $Y_1Ba_2Cu_3O_{7-x}$ ceramic, the measurement of

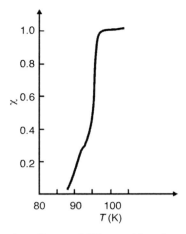

9.4 Curve of SC transition for composite obtained during polymerization of MMA in pressed samples of the $Y_1Ba_2Cu_3O_{7-x}$ ceramic at 80°C. Ceramic:binder weight ratio is 88:12 [21].

SC characteristics shows that, in all samples, the Meissner effect was observed without a restoration stage. The character of transition to the SC state for composites with MMA binder, obtained through polymerization filling, is given in Fig. 9.4.

Unexpected results were obtained while studying the influence of the $Y_1Ba_2Cu_3O_{7-x}$ oxide ceramic on the kinetics of MMA polymerization in the presence and absence of the initiator AIBN. The results revealed that for long periods (around 4–5 hours) no polymerization takes place. However, in the absence of an initiator, polymerization occurs quite quickly when the filling rate was higher ($Y_1Ba_2Cu_3O_{7-x}$ MMA + AIBN = 90:10; 85:15, % by weight). For elucidation of such anomalies, the influence of various quantities of $Y_1Ba_2Cu_3O_{7-x}$ ceramic on the polymerization kinetics of MMA, both in the presence and absence of AIBN, was specifically investigated.

A series of kinetic curves showing this influence of $Y_1Ba_2Cu_3O_{7-x}$ ceramic on the polymerization of MMA is presented in Fig. 9.5 and Fig. 9.6(a), for the presence and absence of the ceramic respectively. From kinetic curve 1 (Fig. 9.5) it can be seen that in the absence of ceramic, the kinetics of the initiated polymerization of MMA follows a conventional trend. The maximum heat evolution is about 130 kcal/degree, which corresponds to the heat effect of MMA polymerization. Addition of $Y_1Ba_2Cu_3O_{7-x}$ (Fig. 9.6, curves 2–5) decreases not only the rate of polymerization, but the marginal rate of conversion as well.

This is in contrast to the effects observed when polymerization of MMA is carried out in the absence of AIBN. In this case, as demonstrated by

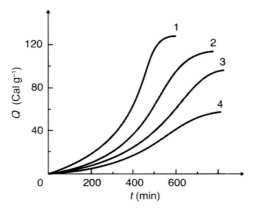

9.5 Influence of added $Y_1Ba_2Cu_3O_{7-x}$ on the kinetics of polymerization of MMA under the impact of 3×10^{-2} m × l^{-1} at 60°C. $Y_1Ba_2Cu_3O_{7-x}$ content (in g): 0 (curve 1); 0.1 (curve 2); 0.2 (curve 3); 0.3 (curve 4) [19].

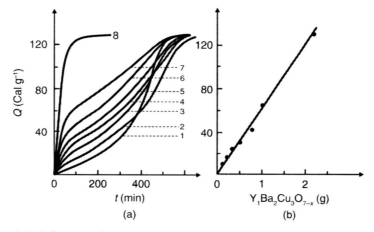

9.6 Influence of the added $Y_1Ba_2Cu_3O_{7-x}$ on the kinetics of polymerization of MMA without initiator at 60°C (a) and the dependence of Q on added ceramic (b). $Y_1Ba_2Cu_3O_{7-x}$ content (in g): 0 (curve 1), 0.1 (curve 2); 0.2 (curve 3); 0.3 (curve 4); 0.5 (curve 5); 0.7 (curve 6); 1 (curve 7); 3.8 (curve 8). Concentration of AIBN is 3×10^{-2} m·l^{-1} without ceramic [19].

Fig. 9.6(a), the rate of polymerization somehow increases non-conventionally. For this particular case (i.e. addition of the ceramic) a considerable increase in the rate of polymerization at the initial stages of the process takes place and a corresponding decrease in the gel effect in the later stages of conversion is noted. These indicated anomalies become conspicuous when the amount of ceramic added is increased in the initial reaction mixture.

The observed change in the kinetic character of MMA polymerization is understandable if it is considered that some specific interaction takes place for the ceramic, both with the monomer and with the initiator. It is known [33] that perovskites of the $Y_1Ba_2Cu_3O_{7-x}$ type possess catalytic properties for numerous chemical reactions. In this case, it could be suggested that some separate localities of the ceramic grains can interact with MMA and AIBN. In this instance, if the rate of interaction between the ceramic and the initiator is far higher than with the monomer, then some part of the initiator interacts with the active centers of the $Y_1Ba_2Cu_3O_{7-x}$ grain surface, filling them fully and thus excluding the possibility of interaction with MMA. Thus, some part of the initiator is 'blocked' by the ceramic's surface and takes no part in the initiation reaction.

In this particular case, increasing the amount of ceramic added enhances the share of the 'blocked' initiator and, consequently, reduces the rate of polymerization (Fig. 9.6(a), curves 2–5). As to the kinetics of polymerization obtained in the absence of AIBN, it seems that the monomer interacts with the active localities of the ceramic grains. This leads to the formation of primary active centers of polymerization, which are fixed on the solid $Y_1Ba_2Cu_3O_{7-x}$ surface.

It is hard to apply the conventional mechanisms of chain growth and rupture for this example because the macromolecule end groups, which would otherwise by capable of participating in reactions, are non-mobile in this particular case. This excludes the possibility of chain quenching. From this knowledge, the sharp increase in the rate of polymerization at the initial, fast stages of the reaction becomes understandable. As to the mechanism of the growth of the chains, the monomer reacts with an active center while approaching it. It is apparent that elongation of the macromolecule captures these centers and consequently diminishes the accessibility of the centers to monomers. Finally, the kinetics of the chain growth reactions on the ceramic surface turn into diffusion control, leading to a decrease in the rate of polymerization.

The second region of the kinetic curves (Fig. 9.6(a), curves 2–7) shows a sharp decrease in the polymerization rate, and a stepwise increase is observed. This is linked with polymerization that occurs not on the surface, but in the bulk monomer – the layer of macromolecules that, upon reaching a definite dimension, encompasses the active chain and partially desorbs from the surface of the ceramic. The active end of the chain is thus uncovered and when these uncovered ends accumulate further, the rate of homogeneous polymerization increases up to the point of complete utilization of the monomer. It should therefore be noted that the rate of bimolecular chain termination in the bulk is low because the furled and uncovered macro chains have little mobility.

For this mechanism of chain initiation and growth, it is obvious that an increase in the amount of added ceramic will enhance conversion in the first region, where high rates of polymerization occur. A subsequent decrease in polymerization in the second region is observed (Fig. 9.6(a), curves 2–8). It is interesting to note that the extent of conversion, which is reached in the first region of the kinetic curves, is linearly dependent upon the amount of added ceramic (Fig. 9.6(b)). From the data shown in this figure, the quantity of ceramic needed to achieve complete polymer conversion in the region where the high rate of polymerization occurs was estimated. As expected (Fig. 9.6(a), curve 8), addition of 90% by weight of $Y_1Ba_2Cu_3O_{7-x}$ assures complete polymerization in the active polymerization centers on the ceramic surface.

The mechanism and topochemistry of the process can alternatively be explained if the active polymerization centers formed and fixed on the surface of the SC ceramic grains are of the radical type. Initiation of polymerization (in the absence of AIBN) begins from the surface of the ceramic when the filler content in the reaction mixture is too high (90% by weight). In such cases, the polymerization process is characterized by a high rate and, presumably, is localized at the ceramic–monomer interstitial boundary.

It is known that fixation of the macromolecule ends on the surface of the filler sharply decreases their mobility and, correspondingly, changes the kinetic parameters of polymerization. In particular, it substantially decreases the constant rate of the bimolecular chain rupture. This is the main reason for the steadiness of the process. In the reaction mixture, a second region appears in the kinetic curves of polymerization with an increase in monomer concentration. This suggests a decrease in the rate of the process. This is possibly the result of accumulated polymer on the surface of the ceramic, hindering the accessibility of the monomer molecules to the active polymerization centers. Occlusion of the active centers by the macromolecules occurs and the growth rate is controlled by diffusion.

Furthermore, because of transference of the chain to the monomer (analogous to blocked polymerization) the kinetics of the process become salient at a certain MMA concentration (20–25% by weight). This facilitates the possible transition of radicals from the surface of the filler into the bulk. Presumably, in this case, the rate of blocked polymerization is lower than in the initial stages of implanted polymerization. This is because the diffusion retardation imposed by the filler surface is less conspicuous. Although the present qualitative model of the mechanism and topochemistry of the surface MMA satisfactorily describes the observed peculiarities, this topic remains open for further study and discussion.

9.4   Mechanical properties of SC polymer–ceramic nanocomposites

As a rule, particulate polymers improve a number of composite qualities (including hardness, impact strength, work of disintegration and heat resistance) [34]. This is mainly linked with the formation of a special interfacial layer, as between the filler and polymeric binder. With regard to conventional fillers, several specific properties of perovskite high-temperature superconductors (such as the layered structure, developed surface of the ceramic grains, catalytic properties and free oxygen dislocated on the surface of ceramic grains) will have an outstanding impact not only on the formation of the phase boundary and consequently on the physical–mechanical properties, but also on the superconducting (SC) properties of the polymer–ceramic composites. This section is devoted to the discovery of these issues.

Determination of the physical–mechanical properties of SC polymer–ceramic nanocomposites is of interest not only at ambient temperatures, but also in lower temperature ranges. Particularly interesting is their behavior at the critical temperature of transition into the SC state. For SC polymer–ceramic nanocomposites, the limiting strength of rupture (σ), elasticity modulus (E) and elongation (ε) were found to be at ambient temperatures [20–27, 35], whereas with SHMPE binder, temperatures were found to be close to 193 K. The SC state temperature of the composite was 77 K. The magnitudes of σ, E and ε were determined via distention of the samples at ambient and 77 K temperatures, and under compression conditions at 193 K. In the latter case, the ceramic:binder ratio in the composites was also varied.

The values of σ, E and ε are presented in Table 9.4. Juxtaposition of the durability data for different binders shows that the highest rupture resistance and elasticity modulus are observed for polyvinyl alcohol binder, ethylene and tetrafluoroethylene copolymer, for practically the same filling index. It is noteworthy that polymeric materials and filling systems with lower usage temperatures produced enhanced durability and elasticity, but simultaneously made deformity capabilities very inadequate. Deformation ability is used as a criterion for the workability of polymeric materials, especially at lower temperatures, which is why the measurements for composites based on SHMPE were conducted at lower temperatures. Compression durability measurements for composites based on SHMPE (conducted at 193 K) show that, for materials containing 90, 85 and 80% weight of ceramic, the index is 34, 61 and 60 MPa, respectively. Comparison of the rupture properties at ambient and cryogen temperatures is also interesting. Typical diagrams of elasticity for $Y_1Ba_2Cu_3O_{6.97}$ nanocomposites with SHMPE are presented in Fig. 9.7. It is apparent that for the same

Table 9.4 Physical–mechanical properties of SC polymer–ceramic nanocomposites

Composite	Weight ratio of ceramic and binder	Deformation method	Temperature of deform- ation (K)	σ (MPa)	E (MPa)	ε (%)
SHMPE + Y$_1$Ba$_2$ Cu$_3$O$_{6.92}$	85:15	Elongation	300	30	100	10.0
	85:15		193	60	80	1.0
	85:15		77	100	—	0.1
RPE + Y$_1$Ba$_2$Cu$_3$ O$_{6.92}$	80:20	Elongation	300	15	75–80	9.0–10.0
F-40 + Y$_1$Ba$_2$Cu$_3$ O$_{6.92}$	75:25	Elongation	300	32	150	7.2
i-PB + Y$_1$Ba$_2$Cu$_3$ O$_{6.92}$	80:20	Elongation	300	28	100–110	8.3
PVA + Irganox + Y$_1$Ba$_2$Cu$_3$O$_{6.92}$	85:20	Elongation	300	34	130	7.5

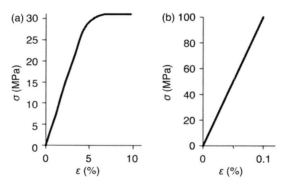

9.7 Elongation curves for Y$_1$Ba$_2$Cu$_3$O$_{6.97}$ nanocomposites with SHMPE obtained at ambient (a) and liquid nitrogen (b) temperatures. Ceramic: binder weight ratio is 85:15 [26].

filling ratio at ambient temperatures, the samples have a region of high elasticity (Fig. 9.7(a)). In contrast, distention of the samples at 77 K brings about brittle decomposition of the samples in the Hooke's deformity region, where the elongation of SC polymer–ceramic nanocomposites only reaches 0.1%.

9.4.1 Influence of particle size and degree of filling on the mechanical and SC properties of polymer–ceramic nanocomposites

In order to understand the process of formation and properties of filled polymeric materials, it is important to investigate their resistance characteristics, as dependent on the average size of particles, the nature of the filler and the binder. It is noteworthy that the issue of stability of the properties of a polymeric binder is disputable if it has a three-dimensional structure. Nevertheless, the influence of filler particle size and the nature of the binder on the physical–mechanical properties of composites based on epoxy binders has been demonstrated [20, 24, 26].

The results of investigations into the influence of filler particle size and concentration on the physical–mechanical properties of polymer–ceramic nanocomposites, obtained on the basis of $Y_1Ba_2Cu_3O_{6.97}$ and divinyl rubber (DR), are outlined in Table 9.5. The table shows the influence of the size and filling degree of particles on ε, σ and E and also on the beginning (T_c) and end (T_f) temperatures of the transition into a superconductive state for nanocomposites with cross-linked DR. The data indicate that the higher the ceramic content, the higher the rupture strength and modulus of elasticity and the lower the limiting deformation, which are all independent of the particle size of the $Y_1Ba_2Cu_3O_{6.97}$.

The considerable increase in σ and E with an increase in the rate of filling evinces the presence of full contact in the polymer–filler interface and the absence of scaling of the binder from the filler during the deformation of the tested samples. The absence of scaling is also indicated by the stress–deformation curves (Fig. 9.7), where no yield point (point of fluidity) is observed. These results confirm the earlier conclusion [25, 26] about the existence of sufficiently strong interactions between the binder and the surface of the $Y_1Ba_2Cu_3O_{6.97}$ ceramic grains.

This is additionally confirmed by data on the influence of ceramic grain size on σ and E for the same filling rate. Furthermore, as outlined in Table 9.5, an increase of the mean filler size decreases the values of σ and E. This fact can be explained by the diminution of the general contacting surface of the binder with the filler, causing the general energy of their interaction to reduce, and consequently decreasing the limiting strength and modulus of elasticity.

It is interesting that, with an increase in the average particle size, a tendency for increased composite deformation capacity is observed. It is probable that the increase of particle size has a positive effect, providing enhancing efficiency in preventing the propagation of cracks at high degrees of deforming capacity. Table 9.5 also provides data on the influence of average ceramic grain size on critical temperature, both at the beginning of the SC transition (T_c) and at its end (T_f).

The initial non-fractioned $Y_1Ba_2Cu_3O_{6.97}$ ceramic possesses the characteristics of: $T_c = 93\,K$ and $T_f = 87\,K$. Juxtaposition of the initial ceramic composite with DR, plus different ceramic particle size, results in dissimilar SC properties. For nanocomposites with a mean particle size of 5–10 μm, the value of T_c is lower by 5–10 degrees than in the original ceramic. In addition, the transition width (T_c-T_f) widens. When the mean particle size

Table 9.5 Influence of particle size and filling rate on σ, E, ε, T_c and T_f

Mean particle size (µm)	Specific surface area (cm²/g)	Filing rate (mass %)	σ (MPa)	E (MPa)	ε (%)	T_c (°C)	T_f (°C)
5	1132	10	1.85	65	280		
		20	2.60	85	260		
		30	5.50	115	210		
		40	10.00	151	160		
		50	17.50	190	100	87	3
15	755	10	1.10	55	277		
		20	2.00	74	263		
		30	4.50	100	230		
		40	9.00	132	183		
		50	17.0	165	130	91	3
25	453	10	0.80	50	275		
		20	1.80	68	264		
		30	4.00	90	225		
		40	7.20	120	200		
		50	15.0	150	157	95	8
35	323	10	0.80	48	272		
		20	1.60	60	270		
		30	3.50	80	252		
		40	6.20	100	120		
		50	15.50	130	180	95	9

becomes greater than 20–25 µm, the critical parameters are sustained and even improved compared with those of the initial ceramic sample. The decrease in critical temperature and increase in transition width for nanocomposites with a mean particle size of 5–10 µm, are related to two factors:

- lower content of small, particle-sized fractions of the ceramic during the orthorhombic phase
- enhanced utilization of oxygen in the reactions of thermo-oxidation destruction of the binder compared with the bigger fractions.

The critical temperature of transition into the SC state for nanocomposites with ceramic of mean particle size 20–25 µm and higher is elevated. This can be explained [19, 23, 25] by the intercalation of macromolecule elements of the binder into the interstitial layers of the ceramic grains, as noted earlier.

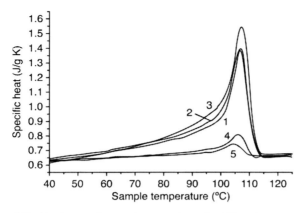

9.8 Heat capacity vs. temperature for SC polymer–ceramic samples obtained at different initial temperatures (T_0) and with different ratios of ceramic:RPE. T_0 (°C) = 130 (curves 1, 4, 5), 140 (curve 2), 160 (curve 3). $Y_1Ba_2Cu_3O_{6.97}$:RPE = 90:10 (curves 1–3), 97:3 (curve 4), 99:1 (cruve 5) [25].

Table 9.6 Melting temperatures (T_m) and enthalpies (ΔH_m) of nanocomposites depending on content

$Y_1Ba_2Cu_3O_{6.97}$: i-PP (mass%)	T_m (°C)	ΔH_m calc. for 1 g sample (J g^{-1})	ΔH_m calc. for 1 g RPE (J g^{-1})	Crystallinity (%)
90:10	107	8.4	84	28.6
97:3	107	2.9	97	33
99:1	105	1.3	133	45

9.4.2 Thermo-physical properties and morphological characteristics of SC polymer–ceramic nanocomposites

The thermal capacity dependence of polymer–ceramic nanocomposites with ramified polyethylene (RPE) is shown in Fig. 9.8. From curves 1–3 it can be seen that the initial formation temperature of polymer–ceramic SC samples has practically no influence on the heat of fusion. As expected, the heat of fusion changes dramatically depending on the samples' content. Melting points and enthalpies for SC polymer–ceramic nanocomposites with RPE binders are presented in Table 9.6, using values calculated from experimentally obtained data.

The strong increase of ΔH_m value (calculated for RPE) could be the result of two reasons. One is that the increase in quantity of SC ceramics leads to an increase of crystallinity. The second and more possible cause is that RPE macromolecule fragments intercalate [25, 26] into the interstitial layers between the ceramic grains, leading to changes in the crystalline RPE

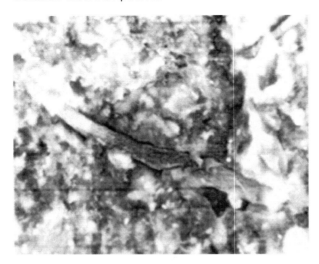

9.9 Y$_1$Ba$_2$Cu$_3$O$_{6.97}$:RPE = 90:10 SC polymer–ceramic nanocomposite [25].

morphology. SC polymer–ceramic nanocomposite samples were investigated using electron microscopy (Fig. 9.9). It is seen that in this case, as for RPE binders, there are fibrils present, which are not typical for RPE. It is possible that they are the result of the intercalation of RPE macromolecule fragments into the layered structure of the ceramics. Such a binding presumably influences the mobility of some RPE macro chains, and it is logical to assume that crystallization of such macromolecules occurs by cooperative interaction between them. In the case of SC polymer–ceramic nanocomposites with i-PP isotactic polypropylene binders, the phenomena observed above are presented more clearly. As can be seen from the dependence of heat capacity on temperature (Fig. 9.10; curves 1, 2 and 3), in this case the heat of fusion splits into two components. One can assume that this split is connected with the presence of two different types of structure within the polymer–ceramic nanocomposites. Here, the content of the fibrils is appreciably higher than that of the RPE binder.

9.5   Interphase phenomena in SC polymer–ceramic nanocomposites

Figures 9.11 and 9.12 show the temperature dependence of the elastic modulus (E) and of the loss tangent (tanδ) for pure SHMPE and a polymer–ceramic nanocomposite with 15% filler. Both E and tanδ, increase in line with an increase in the amount of ceramic filler. In both curves, two transitions are seen. The step in E and the peak in tanδ around $-100°C$ are

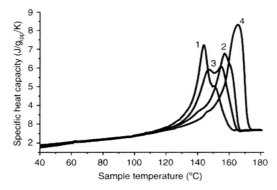

9.10 Heat capacity vs. temperature for SC polymer–ceramic samples obtained with different ratios of ceramic:i-PP. $Y_1Ba_2Cu_3O_{6.97}$:i-PP = 85:15 (curve 1), 70:30 (curve 2), 50:50 (curve 3); pure i-PP (curve 4) [25].

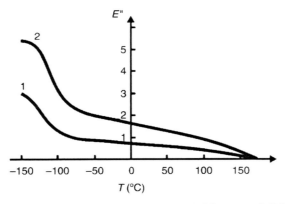

9.11 Temperature dependence of E for pure SHMPE (curve 1) and SHMPE–ceramic nanocomposite (curve 2); ceramics content (wt%) = 15% [25].

related to a relaxation process. The broad melting region of SHMPE from about 50°C up to 170°C yields a further softening of the samples and a large peak in tanδ. With an increasing amount of ceramic filler, the peaks in tanδ are increased and shifted to higher temperatures (see Fig. 9.12).

It should be noted that the observed change in the loss tangent curves is a rare feature for conventional, non-nanosized ceramic polymer nanocomposites [36, 37]. Increasing mechanical loss peaks have been linked with the plate-like structure of the filler [38]. It seems that, in the present case, we have the reverse situation. Some parts of the macromolecules penetrate by intercalation mechanisms into the sandwich structure of the filler, thus creating effects similar to that shown in curve 2 of Fig. 9.12.

The increase in mechanical loss cannot be explained by the agglomeration

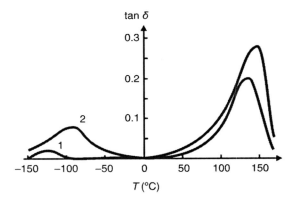

9.12 Temperature dependence of tanδ for pure SHMPE (curve 1) and SHMPE–ceramic nanocomposite (curve 2); ceramics content (wt%) = 15% [25].

of small particles of the filler because the formation of such aggregates can only take place at higher degrees of filling [39]. It seems that the observed increase in mechanical loss is a result of adsorption of the binder on the surface of the filler and intercalation of fragments of the macromolecules into the interlayer space of the ceramic grains. Such an interaction can change the structure of the polymeric matrix near the boundary of the particles and, as a consequence, yield an increase in mechanical losses.

It is known [40] that, in some cases, the filler shifts the maximum mechanical losses and T_g towards higher temperatures. It is assumed that the magnitude of the shift is proportional to the surface area of the filler, which explains the polymer–filler interaction. The non-additive contribution of the added ceramics on the T_g shift (Table 9.7) points not only to adsorption interactions, but also to the intercalation of the fragments of SHMPE macromolecules into the interlayer space of the ceramic grains, as indicated earlier.

It is obvious that such an interaction limits the mobility of the macromolecules, thus changing the packing density of the polymeric chains. As a result, a morphology change around the phase boundary may occur. To prove these considerations, the temperatures and enthalpies of melting for a range of SHMPE–$Y_1Ba_2Cu_3O_{6.97}$ nanocomposites were measured directly, using differential scanning calorimetry (Table 9.8). With an increase in filler content, the enthalpy of melting recalculated for the polymer fraction (excluding the filler) increases, as shown in Table 9.8. Again, there is some discrepancy with the curves shown in Fig. 9.11 and 9.12. In the figures, the melting peak shifts to higher temperatures with increasing filler content. This increase in enthalpy is linked with either the overall degree of crystallization or with a change in morphology of the

Table 9.7 Dynamic–mechanical characteristics of nanocomposites of $Y_1Ba_2Cu_3O_{6.97}$ with SHMPE

Weight ratio SHMPE: ceramic filler	E ($T=-150°C$)	E ($T=-100°C$)	E ($T=25°C$)	T_c (°C)	tan δ (first transition)	T_α (°C)	tan δ (second transition)
100:0	2.95	1.5	1.1	−127	0.01	134	0.2
85:15	5.1	3.4	1.6	−95	0.06	143	0.25
50:50	10.1	6.5	3.1	−94	0.065	155	—
15:85	—	—	4.5	—	—	157	0.27

Table 9.8 Influence of filling rate on the temperature and enthalpy of melting of the binder in SHMPE–$Y_1Ba_2Cu_3O_{6.97}$ nanocomposites

Weight ratio SHMPE: $Y_1Ba_2Cu_3O_{6.97}$	T_{melt} (°C) (extrapolated onset of DSC peak)	ΔH_{melt} per gram of SHMPE (J/g)	Crystallinity (%)
100:0	134	115.0	39.1
85:15	143	116.5	39.7
50:50	155	122.5	41.7
15:85	157	123.5	42.0

binder at the interphase. However, based on the obtained results, the dominant role of any of the factors mentioned cannot be definitely proven. For this purpose, it is necessary to conduct thorough electron microscope investigations of the nanocomposite samples.

Investigation of the structures of SC polymer–ceramic nanocomposites by scanning electron microscopy has shown that, for both the amorphous and the crystalline polymer matrices, a complete and uniform covering of the ceramic grains by the binder occurs (Fig. 9.13 and Fig. 9.14). This indicates a strong interaction at the polymer–ceramic interface. Electron microscope investigations of polymer–ceramic nanocomposite samples obtained under conditions of variable filling (Fig. 9.14) provide support for such an assertion. Furthermore, fibril structures are observed in the nanocomposites based on crystalline polymers (SHMPE, PP), independent of the SC ceramic filler content (Fig. 9.13 and 9.14).

As seen in Fig. 9.14(a), in $Y_1Ba_2Cu_3O_{6.97}$:i-PP = 85:15, singular fibrils appear in the field of vision of the microscope. With an increase in ceramic additive (Fig. 9.14(b) and (c)) the number of fibrils becomes significantly higher. It should be noted that there are no such fibrilar structures in the initial, unfilled semi-crystalline polymers (Fig. 9.15(a) and (b)). This indicates that, during the formation of polymer–ceramic nanocomposites, SC ceramics play a special role in the formation of the crystalline structure of the polymeric matrix.

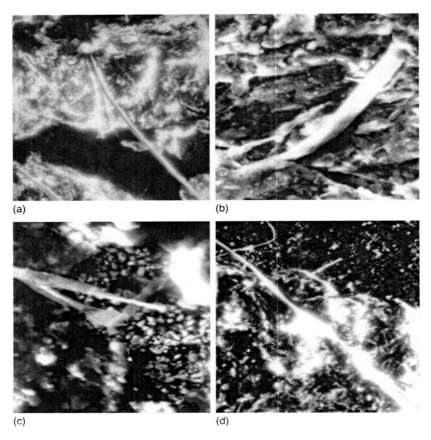

9.13 Morphology of polymer–ceramic nanocomposites with SHMPE (a) and (b) and i-PP (c) and (d) as binders. $Y_1Ba_2Cu_3O_{6.97}$:binder (wt%) = 85:15 (a) and (c), 80:20 (b) and (d) [25].

9.6   Influences on the magnetic properties of SC polymer–ceramic nanocomposites

It is known that high-temperature superconducting (sc) oxide ceramics possess their own localized magnetic moments, producing a Cu^{2+} EPR signal. Nevertheless, it should be noted that there are two types of copper atoms in $Y_1Ba_2Cu_3O_{7-x}$ ceramics: Cu^{2+} (I) and Cu^{2+} (II). The first is in the CuO chain along the axis direction, while the second is in the CuO_2 planes along the ab plane. For a long time, the nature of the $Y_1Ba_2Cu_3O_{7-x}$ ceramics' EPR response was unclear [41].

Investigations [42, 43] have been conducted into the dependence of Cu^{2+} EPR signal intensity with simultaneous registration of X-ray absorption near edges structure (XANES) at the Cu^{2+} k edges of the same signals [44],

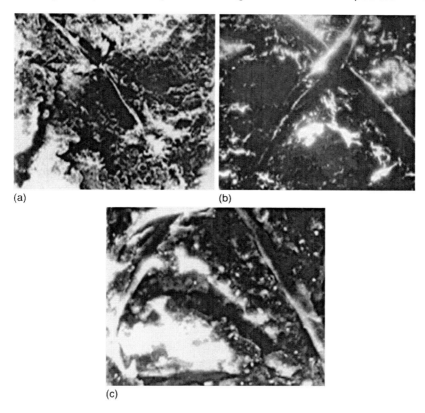

(a)   (b)

(c)

9.14 Polymer–ceramic nanocomposites with different ceramic:i-PP ratio. $Y_1Ba_2Cu_3O_{6.97}$:i-PP = 85:15 (a), 70:30 (b), 50:50 (c) [25].

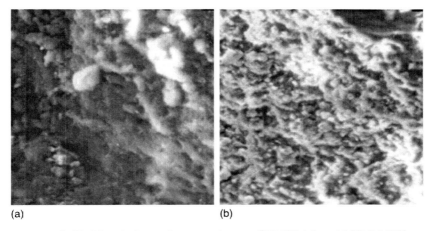

(a)   (b)

9.15 Morphology of pure polymers SHMPE (a) and i-PP (b) [23].

as well as Cu^{2+} EPR signal intensity dependence on the substitution degree of $Cu^{2+}(I)$ in $Y_1Ba_2Cu_3O_{7-x}$ by Fe atoms. The results showed [45] that EPR signals correspond to $Cu^{2+}(I)$ in the chains and not to the $Cu^{2+}(II)$ in the CuO_2 planes.

Analysis of EPR signals for polymer–ceramic nanocomposites showed that $Cu^{2+}(I)$ EPR signals depend on the binder. Figure 9.16 shows $Cu^{2+}(I)$ EPR spectra for $Y_1Ba_2Cu_3O_{6.97}$ and polymer–ceramic nanocomposites with various polymeric binders: polystyrene (ps), polymethylmetacrylate (PMMA) and polyethylene (PE). The results demonstrate that the particles of the layered ceramic feature nanocomposite structures, where the ceramic grains are the precursors of the macromolecules. The addition of a polymer changes the valence state of $Cu^{2+}(I)$, as demonstrated by curves 1–4 of Fig. 9.16. This indicates intermolecular interaction between the $Y_1Ba_2Cu_3O_{0.67}$ ceramic grains and elements of the polymer chains. Such an interaction can be explained by the intercalation of some elements or fragments of the macromolecules into the layered structure of the ceramic. During such an intercalation, superposition of the unpaired electron of the $Cu^{2+}(I)$ $3d_{x2-y2}$ orbital with the orbital of corresponding elements in the polymer chains occurs. As a result, the $Cu^{2+}(I)$ EPR response is altered because of the change of the valence state of $Cu^{2+}(I)$.

It is interesting to elucidate whether the intensity of the $Cu^{2+}(I)$ EPR signal is dependent on the filling rate (on the fraction of binder in the ceramic, for example). To answer this question, polymer–ceramic nano-composites were investigated with various $Y_1Ba_2Cu_3O_{6.97}$:SHMPE ratios: 100:0; 99:1; 97:3; 95:5; 93:7; 90:10 and 80:20 (in accordance to their mass %). The obtained data are presented in Fig. 9.17. Curves 1–6 of the figure demonstrate that the intensity of the EPR output signal depends on the binder content. It is interesting to note that a bigger shift is observed for smaller quantities of additive (SHMPE) (curves 2 and 3) and compared with the pure ceramic (curve 1). Further increase of the binder content reduces the signal intensity (curves 4–6).

9.7   The use of metal-complex polymer binders to enhance the SC properties of polymer–ceramic nanocomposites

Doping of some atoms into a ceramic's lattice structure is one technique presently used in the search for new superconducting (SC) ceramic nanocomposites in order to enhance the onset of the SC transition temperature. As such, the use of metal-complex polymers as binders has been explored as a possible method for regulating both the critical transition temperature and the width of the SC state.

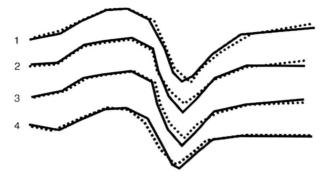

9.16 Cu^{2+}(I) EPR spectra of SC $Y_1Ba_2Cu_3O_{7-x}$ ceramics ($T_i = 92.0\,K$); reference, dotted curves; nanocomposites, solid line curves. Curve 1, 15% PS and 85% ceramic. Curve 2, 15% PMMA and 85% ceramic. Curve 3, 20% PE and 80% ceramic. Curve 4, 15% copolymer of ST and MMA and 85% ceramic [24].

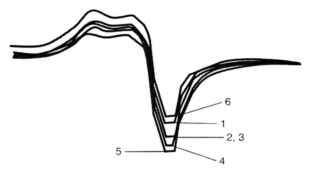

9.17 Cu^{2+}(I) of SC $Y_1Ba_2Cu_3O_{6.97}$ ceramic (curve 1) and nanocomposites with SHMPE: curve 2, 1% SHMPE; curve 3, 3% SHMPE; curve 4, 5% SHMPE; curve 5, 10% SHMPE; curve 6, 20% SHMPE [24].

It is known that acrylamide (AAm) complexes of metal nitrates of the first transition group are able to polymerize at frontal regimes. The essence of frontal polymerization is in localized heating of the sample edge, initiating polymerization [28]. The heat evolved is transmitted to neighboring layers by a heat-conductance mechanism, where, in turn, polymerization begins. Thus, the heat wave front propagates over the entire volume. As metal-containing monomers are able to polymerize frontally, complexes like $(AAm)_4(H_2O)_2(MO_3)_2$ (M = Mn, Co, Y, Cu, etc.) could be used. A previous investigation showed that frontal polymerization of AAm complexes in the presence of SC ceramic is possible only within a limited temperature range. It was shown experimentally that the lower temperature limit for carrying out the reaction (100°C) is given by the stability of frontal polymerization upon the propagation of vertical heat waves from up to down. This sharply

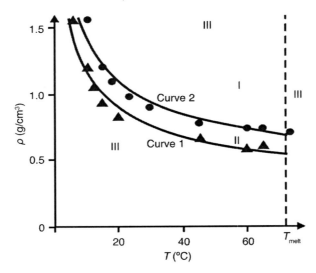

9.18 Density–temperature diagram for three regimes of frontal polymerization [28].

reduces and limits the yield of nanocomposites. The upper temperature limit (200°C) is determined by the thermal degradation of the nitrate groups in the nanocomposites obtained.

The influence of density, temperature and reactor diameter on the structure of heat waves, rate of front propagation and range of existence of steady-state regimes was investigated. Experimental results obtained at different temperatures and densities are summarized in Fig. 9.18. Here, curve 2 is the domain of steady-state heat polymerization waves, which is limited by the straight line corresponding to limiting packing of the reaction media (ρ_{lim}) and full melting of the crystalline monomer. There are no wave regimes of frontal polymerization below curve 1, when $\rho > \rho_{\text{lim}}$ and $T \geq T_{\text{melt}}$.

For mass ratios of the ceramic:metal polymer $\leq$ 80:20, the addition of minute quantities of $Y_1Ba_2Cu_3O_{7-x}$ does not allow frontal waves to travel from up to down, at temperatures up to 100°C. At the same time, the formation of the propagating frontal regimes of heat waves is observed at temperatures above 60°C, when the wave initiates from the bottom and propagates from bottom up. This situation is explained by the gas evolution during frontal polymerization, as well as by the inhibition of some constituents of the gases evolved from the process of polymerization.

The results of an investigation into the SC properties of Mn-, Co-, Ni- and Zn-containing polymer–ceramic nanocomposites are presented in Table 9.9, which shows that the onset of the transition into the SC state (T_c) is shifted towards higher temperatures in comparison to the initial ceramic,

Table 9.9 Superconducting characteristics of polymer–ceramic nanocomposites with Mn, Co, Ni, Zn metals

$Y_1Ba_2Cu_3O_{6.98}$		Metal-monomer complex		Metal	T_c (K)	T_f (K)
Mass (g)	Mass (%)	Mass (g)	Mass (%)			
0.293	43	0.388	57	Mn	95	87
0.396	50	0.396	50	Mn	94	85
0.90	70	0.29	30	Mn	94	84
0.518	73	0.196	27	Mn	93	85
0.90	70	0.39	30	Mn	95	85
0.416	67	0.209	33	Mn	94	83
0.552	78	0.156	22	Mn	95	85
0.325	51	0.318	49	Co	93	83
0.432	60	0.283	40	Co	92	84
0.503	70	0.228	30	Co	92	84
0.486	70	0.208	30	Ni	95	83
0.552	78	0.156	22	Zn	95	86
0.416	67	0.209	33	Zn	94	83

$T_c = 93$ K, $T_f = 78$ K. The SC onset increase reaches 1–3 K for T_c and more than 5 K for T_f.

It is known from the literature that the $Y_1Ba_2Cu_3O_{7-x}$ high-temperature SC ceramic exhibits an anti-ferromagnetic transition, with a transition into the spin glassy (vitreous) state before the desired transition into the SC state. Presumably, anti-ferromagnetic and high-temperature SC states are co-existent. Co, Ni and Mn are known to be anti-ferromagnetic metals which intercalate into the interstitial layers of the ceramic, causing the three-dimensional properties of the ceramic grains to change. This explains why T_c increases and the transition width decreases. This interpretation is supported by our data on the change of valence states of Cu in polymer–ceramic nanocomposites.

9.8   Aging of SC polymer–ceramic nanocomposites

The prevailing orientation of the crystallites in the (110) direction in the $Y_1Ba_2Cu_3O_{6.92}$ sample is practically stable over time, and is strong in absolute value intensity line. The $Y_1Ba_2Cu_3O_{6.97}$ specimen is oriented along (006) and the absolute value of this reflex is not changed with the progression of time. Furthermore, the Y_2BaCuO_5 semiconductor phase is identified as (211) – 2% in content. The texture in both of these specimens is controlled by the relative intensities of the peaks (110), (103), (005), (014), (113), (006), (020) and (200). As a result, the SC properties of $Y_1Ba_2Cu_3O_{6.97}$ specimens are relatively stable over time (Table 9.10), whilst

Table 9.10 Superconducting characteristics of high-temperature SC ceramics and polymer–ceramic nanocomposites based on $Y_1Ba_2Cu_3O_{6.97}$: SC transition temperature (T_c, K), the transition width (ΔT_c, K) and the orthorhombic distortion of the lattice structure ($=(b - a)/a, \eta$)

Time, (months)	SC characteristics of specimens	$Y_1Ba_2Cu_3$ $O_{6.97}$	Type of polymeric binder ($Y_1Ba_2Cu_3O_{7-x}$: polymer component = 85:15%)			
			PE	PS+NG -2246	ST:MMA (40:60 mol%)	SPL (ST:MMA) (80:20 mole%) + NG-2246
0	η	0.0189	0.0185	0.0197	0.0194	0.0202
	T_c	92.0	91.8	93.0	92.6	93.4
	ΔT_c	6.5	6.5	7.0	6.5	9.0
6	η	0.0185	0.020	0.0185	0.0185	0.0185
	T_c	91.7	92.8	91.7	92.1	92.2
	ΔT_c	6.0	6.0	≈ 7	≈ 7	≈ 7
12	η	0.0185	0.0196	0.0181	0.0180	0.0180
	T_c	91.8	92.8	91.7	92.0	92.0
	ΔT_c	≈ 8	≈ 8	≈ 9	≈ 8	≈ 9

some insignificant property changes are reversible and can be ascribed to auto-oscillating processes. Such auto-oscillating processes are typical for structurally unstable electronic systems in solids, and cannot be attributed to the aging of the structure. It seems that the Y_2BaCuO_5 semiconductor phase in $Y_1Ba_2Cu_3O_{6.97}$ is the oxygen transport phase [46], which is present in definite amounts in high-temperature SC ceramics, and supports the stabilization of the SC characteristics on behalf of vacant oxygen positions. Some increase in the T_c value immediately after polymer–ceramic nanocomposite formation is linked with intercalation of the fragments of the macromolecule binders into the interlayer space between the ceramic grains during the hot pressing process [25, 26].

The change in SC properties is dependent upon the chemical composition of the polymeric binder, as kinetic investigations of the aging process in SC polymer–ceramic nanocomposites have shown. The SC characteristics of ST–MMA copolymers are worse, whereas in the case of the polyethylene polymer, improvement is observed. An increase in the critical temperature of transition for the SC state in nanocomposites with polyethylene binders is possibly linked with processes taking place after the formation stages of polymer–ceramic nanocomposites.

Previous work showed the presence of interactive forces between the polymeric binder elements and the surface of the ceramic grains and the intercalation of these elements into the interlayer space of the ceramic particles. As it is evident from the data in Table 9.10, one can assume that interaction of the elements of the polymeric binder with the surface of ceramic grains, although slow, nevertheless proceeds at ambient temperatures through the intercalation mechanism. For this reason, some increase in T_c is observed (1–1.5 degrees).

As already stated, for the SC nanocomposites of ST with MMA copolymers, the aging process results in a decrease in T_c by 1–1.5°C, an increase in ΔT_c and a decrease in η. The presence of an antioxidant (NG-2246) does not influence the stated characteristics. These kinds of characteristic changes can be explained by the enhanced inclination of the ST–MMA copolymer binders towards destruction under the impact of ultraviolet radiation. To obtain more substantial data on the elevation of T_c during the aging of polymer–ceramic nanocomposites based on the polyethylene matrix, similar investigations have also been carried out on the samples obtained by the gas-phase polymerization of ethylene in the presence of $Y_1Ba_2Cu_3O_{6.97}$ ceramics.

9.8.1 Preparation of SC polymer–ceramic nanocomposites by gas-phase polymerization of ethylene

The catalytic properties of oxide ceramic perovskites have been known for a long time [47–49]. As a result, questions have naturally been posed over the issue of catalytic activation of a ceramic surface for polymerization of gas-phase monomers (ethylene, propylene, etc.) without the use of polymerization catalysts on the surface. Presumably, the layered structure of the crystalline orthorhombic phase will allow coordinating ethylene (or other kinds of olefins) on the surface of the $Y_1Ba_2Cu_3O_{7-x}$ ceramic grains and, in this case, the polymerization process may proceed in the presence of polymerization (alkyl-aluminum) co-catalysts.

It should be noted that, in this case, the quantity of forming polymer was regulated by the polymerization time permitted for the preparation of nanocomposites with various binder contents. In this case, there is the Meissner effect on all samples that did not have added antioxidants. This is explained by the decreasing diffusion velocity of oxygen located on the surface of the ceramic grains as a result of the formation of the polymer covering around the oxide ceramic particles. Parallel displacement of curves on the SC transition towards enhancement of T_c from 46 K (for initial ceramics) up to 70 K (Fig. 9.19) with increasing polyethylene content was observed [19–21].

The decrease in T_c in the initial $Y_1Ba_2Cu_3O_{6.97}$ ceramics from 92 K up to 46 K, and accordingly for the obtained nanocomposites, is possibly linked to high-temperature activation of the surface of the ceramic grains. During development of the SC state, three transition temperatures were detected on the curves (Fig. 9.19) of the SC transition for all of the nanocomposites: low

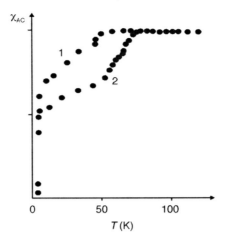

9.19 Dependence of critical temperature of transition into SC state and the width of (T_c–T_f) on the weight ratio of ceramic and polyethylene [21].

Table 9.11 Dependence of critical temperatures of transition into SC state and the width of (T_c–T_f) on the weight ratio of the ceramic and polyethylene

Ceramic:polyethylene weight ratio	$T_{l.c.}$ (K)	$T_{l.f.}$ (K)	$T_{m.c.}$ (K)	$T_{m.f.}$ (K)	$T_{h.c.}$ (K)	$T_{h.f.}$ (K)
100:0	10	4.5	~30	~10	46	~30
80:20	10	4.5	~50	10	70	~50
50:50					79	~63

temperature ($T_{l.c.}$ ~5 K), which does not depend on the formula of the composition; mid-temperature ($T_{m.c.}$); and comparatively high temperature ($T_{h.c.}$), which depends appreciably on the content of polyethylene in the nanocomposites (Table 9.11).

Different degrees of degradation in various fractions of the ceramic cause the transition into a SC state to have a stepwise character and, accordingly, decrease the rate of intercalation in relation to the degree of amorphization of the ceramic grains. It is interesting to note that the beginning of the high-temperature transition ($T_{h.c.}$), as well as its width, substantially depend upon the weight ratio of the composite components. As previously stated, the observed increase in $T_{h.c.}$ and $T_{h.c.}$–$T_{h.f.}$ can be explained by sufficient intercalation of the separate fragments of polyethylene or co-catalyst into the interstitial layer of the $Y_1Ba_2Cu_3O_{6.97}$ during gas-phase polymerization of the ethylene.

As in previous experiments, the SC nanocomposite containing 20m.Kt% PE was kept in air at ambient temperatures over the course of 1 year. The SC characteristics at 6 and 12 months showed an increase in T_c by 5 K and

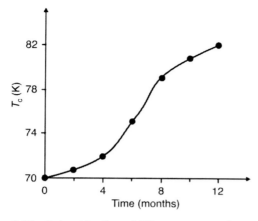

9.20 Aging kinetics of SC nanocomposites with 20% PE content [21].

12 K respectively (Fig. 9.20). These results confirm the supposition that, even at ambient temperatures, the process of intercalation of macromolecule fragments into the layered structure of ceramic grains takes place (albeit slowly), leading to the formation of local nanocomposites.

9.8.2 Aging of the SC characteristics of polymer–ceramic nanocomposites with Mn, Co, Zn and Ni

Samples of metal-containing SC polymer–ceramic nanocomposites with various metal polymers were used to investigate the effects of aging. The results of measurements of SC nanocomposite characteristics before and after aging, in the course of 1 year, are given in Table 9.12. It can be seen that during aging of Mn-, Co-, Zn- and Ni-containing polymer–ceramic nanocomposites, T_c increased by 1–3 K, while the SC transition width narrowed down to 4–6 K. There are, as yet, no reasonable explanations for this.

9.9   Conclusions

The superconducting (SC) characteristics of polymer–ceramic composites are determined by two main competing factors.

1. Interaction of the elements of the binder macromolecule with the surface of the ceramic. Further intercalation of these elements into the interstitial layer of the ceramic causes some elevation of T_c (by 1–3 K) of the composites compared with the initial ceramic.
2. The stability of the polymeric binder against thermo-oxidative destruction via macromolecule disintegration (for example as a result

Table 9.12 The effect of aging on SC characteristics of polymer–ceramic nanocomposites with Mn, Co, Zn and Ni

$Y_1Ba_2Cu_3O_{6.98}$		Metal–monomer complex		Metal	Pressing duration (min)	T_c/T_c (K)	T_f/T_f (K)
Mass (g)	Mass (%)	Mass (g)	Mass (%)				
0.293	43	0.388	57	Mn	10	95/96	87/89
0.396	50	0.396	50	Mn	5	94/96	85/87
0.90	70	0.29	30	Mn	5	94/95	84/88
0.518	73	0.196	27	Mn	5	93/96	85/90
0.90	70	0.39	30	Mn	5	95/96	85/91
0.416	67	0.209	33	Mn	10	94/96	83/86
0.552	78	0.156	22	Mn	5	95/96	85/88
0.325	51	0.318	49	Co	5	93/95	83/85
0.432	60	0.283	40	Co	5	92/9	84/89
0.503	70	0.228	30	Co	2	92/95	84/87
0.486	70	0.208	30	Ni	5	95/96	83/85

of composite products evolving upon destruction of the matrix). The possible participation of oxygen, produced by the ceramic during the orthorhombic phase of thermo-oxidative destruction of matrix conditions, may broaden the temperature interval of the SC transition.

On the basis of data obtained on the nature of the change of physical–mechanical and SC properties of high-temperature SC nanocomposites, dependent on granulometric composition and filling degree, it can be concluded that certain key factors play a significant role. These factors include the peculiarities of forming the boundary of the interstitial layer, the role of the oxide and the highly conductive ceramic–binder, its structure, and the adhesion of the binder with the surface of the ceramic. For the nanocomposites considered here, an increase in melting temperature was found. This can be explained by the intercalation of macromolecule fragments into the interstitial layer of the ceramic grains. This phenomenon leads to a change in the morphological structure of the SC nanocomposites obtained, as has been proven by electron microscopy.

The ceramic–binder boundary plays an important role in the SC and mechanical properties of SC polymer–ceramic nanocomposites based on SHMPE. According to data on dynamic–mechanical properties obtained over a wide temperature interval, it can be concluded that the peculiarities of the formation of the interface within the ceramic–binder boundary are the most important factor. Data on electron microscopy and EPR signals on Cu^{2+} (I) are sound proof of the presence of intercalated fragments of the macromolecules in the interlayer space of the ceramic grains, leading to the formation of nanostructures.

The activation of $Y_1Ba_2Cu_3O_{6.97}$ on the surface of the ceramic grains therefore allows their use as catalysts for olefin gas-phase polymerization. The occurrence of metal monomer frontal polymerization in the presence of $Y_1Ba_2Cu_3O_{6.97}$ ceramics facilitates the preparation of metal-containing polymer–ceramic SC nanocomposites.

As a result of aging of SC polymer–ceramic nanocomposites at room temperature, intercalation of macromolecule fragments takes place, binding the interlayer space of ceramics. This process results in the formation of intercalated nanostructures between the ceramic grains.

9.10   References

1.   Bednorz, J.G. and Muller, K.A. 'Possible high-Tc superconductivity in the BaLaCuO System' *Z. Phys.*, 1986, B64, 189–97.
2.   Bednorz, J.G. and Muller, K.A. 'Perovskite-type oxides—The new approach to high-Tc superconductivity' *Rev. Mod. Phys.*, 1988, 60, 585–600.
3.   Meyer, H.M., Hill, D.M., Wagener, T.J., Gao, Y., Weaver, J.H., Capone, D. W., and Goretta, K.C. 'Electronic structures of the $YBa_2Cu_3O_{7-x}$ surface and its modification by sputtering and adatoms of Ti and Cu' *Phys. Rev. B*, 1988, 38, 6500–12.
4.   Gao, Y., Wagener, T.J., Weaver, J.H., and Capone, D.W. II, 'Interface formation of semiconductors with high-Tc superconductors: Ge/$La_{1.85}Sr_{0.15}CuO_4$' *Phys. Rev. B*, 1988, 37, 515–8.
5.   Romana, L.T. and Wilshaw, P.R. 'High-resolution microchemistry and structure of grain-boundaries in bulk $Y_1Ba_2Cu_3O_{7-x}$' *Supercond. Sci. Tech.*, 1989, 6, 285–90.
6.   Lindbery, P.A., Shen, Z.X., Wells, B.O., Dessau, D.S., Mitzi, D.B., Lindau, I., Spicer, W.E., and Kapitulnik, A. 'Reaction of Rb and oxygen overlayers with single-crystalline $Bi_2Sr_2CaCu_2O_{8+\delta}$ superconductors' *Phys. Rev. B*, 1989, 39, 2890–93.
7.   Golyamina, E.M. and Likov, A.N. 'Method of HTSC protecting from chemical action' *Russ. Supercond.: Phys. Chem. Tech.*, 1989, 2, 51–62.
8.   Asoka-Kumar, P.S., Mahamuni, Sh., Kulkarni, P., Mulla, I.S., Chandrachood, M., Sinha, A.P.B., Nigavekar, A.S., and Kulkarni, S.K. 'Room temperature reaction of Ni/$Bi_2Sr_2CaCu_2O_8$' *Interface J. Appl. Phys.*, 1990, 67, 3184–91.
9.   Hanic, F., Gergel, M.; Chromik, S., Plesch, G., and Strbik, V. 'Preparation of thin epitaxial multilayer $YBa_2Cu_3O_{7-x}$(Ag) films by aerosol deposition' *Bull. Mater. Sci.*, 1991, 14, 485–92.
10. Asoka Kumar, P.S., Mahamuni, Sh., Nigavekar, A.S., and Kulkarni, S.K. 'Reactivity and surface modification at a Bi/$Bi_2Sr_2CaCu_2O_8$ interface' *Physica C*, 1992, 201, 145–50.
11. Kovtun, E.D., Pulyaeva, I.V., Experiandova, L.P., and Mateichenko, P.V. 'Study of surface of Y-Ba-Cu/Ag ceramics obtained from wet powders' *Physica C*, 1994, 235–40, 415–16.
12. Sato, K., Ohkura, K., Hayashi, K., Munetsugu, U., Fujikami, J., and Takeshi, K. 'High field generation using silver-sheathed BSCCO conductor' *Physica B*, 1996, 216, 258–60.
13. Ueyama, M., Ohkura, K., Hayashi, K., Kobayashi, S., Muranaka, K., Hikata,

T., Saga, N., Hahakura, S., and Sato, K. 'Development of Ag-sheathed Bi(2223) superconducting wires and coils' *Physica C*, 1996, 263, 172–5.

14. Docenko, V.I., Braude, V.I., Ivanchenko, L.G., Kislyak, I.F., Puzanova, A.A., and Shevchenko, A.A. 'The Structure and SC parameters YBaCuO-Ag powder composites, obtained by the static (cold and hot) and explosive pressing' *Low Temp. Phys.*, 1996, 22, 1222–5.

15. Haupt, S.G., Lo., R.K., Zhao, J.A., and McDevitt, J.T. 'Conductive polymer/ superconductor thin film assemblies' *Mol. Cryst. Liquid Cryst.*, 1994, 256, 571–6.

16. Lo, R.K., Ritchie, J.E., Zhou, J.P., Zhao, J., McDevitt, J.T., Xu, F., and Mirkin, C.A. 'Polypyrrole growth on $YBa_2Cu_3O_{7-\delta}$ modified with a self-assembled monolayer of N-(3-aminopropyl)pyrrole: Hardwiring the "electroactive hot spots" on a superconductor electrode' *J. Am. Chem. Soc.*, 1996, 118, 11295–6.

17. Chen, K., Xu, F, Chad, Mirkin, A., Lo, R.K., Nanjundaswamy, K.S., Zhou, J. P., and McDevitt, J.T. 'Do alkanethiols adsorb onto the surfaces of Tl−Ba−Ca−Cu−O-based high-temperature superconductors? The critical role of H_2O content on the adsorption process' *Langmuir*, 1996, 12, 2622–4

18. Tonoyan, A.O., Davtyan, S.P., Martirosyan, S.A., and Mamalis, A.G. 'High-temperature superconducting polymer-ceramic composites' *J. Mater. Process. Technol.*, 2001, 108, 201–12.

19. Ayrapetyan, S.M., Tonoyan, A.O., Arakelova, E.R., and Davtyan, S.P. 'High-temperature superconducting polymer-ceramic composites and their properties' *Polym. Sci.*, 2001, A43, 1814–8.

20. Ayrapetyan, S.M., Tonoyan, A.O., Arakelova, E.R., Saakyan, A.A., and Davtyan, S.P. 'Influence of heat regimes, addition of anti-oxidants and the processes of thermo-oxidation destruction on the SC properties of the polymer-ceramic composites' *Chem. J. Armenia*, 2001, 52, 563–72.

21. Tonoyan, A.O., Ayrapetyan, S.M., and Davtyan, S.P. 'Interphase phenomena in high-temperature superconducting polymer-ceramic nanocomposites' *Chem. J. Armenia*, 2001, 52, 76–82.

22. Davtyan, S.P., Tonoyan, A.O., Ayrapetyan, S.M., and Manukyan, L.S. 'A note on the peculiarities of producing high-temperature super conducting polymer-ceramic composites' *J. Mater. Process. Technol.*, 2005, 160, 306–14.

23. Davtyan, S.P., Tonoyan, A.O., Tataryan, A.A., and Schick, Ch. 'Interphase phenomena in superconductive polymer-ceramic nano-composites' *Compos. Interface*, 2006, 13, 535–46.

24. Davtyan, S.P., Tonoyan, A.O., Tataryan, A.A., Schick, Ch., and Sargsyan, A.G. 'Physical-mechanical, thermophysical and superconducting properties of polymer-ceramic nano composites' *J. Mater. Process. Technol.*, 2007, 163, 734–42.

25. Davtyan, S.P., Tonoyan, A.O., and Schick, Ch. 'Intercalated nanocomposites based on high-temperature superconducting ceramics and their properties' *Materials*, 2009, 2, 2154–87.

26. Davtyan, S.P., Tonoyan, A.O., and Schick, Ch. *Encyclopedia of Polymer Composites: Properties, Performance and Applications.* Chapter 6. Polymer science and technology. Editors: Mikhail Lechkov and Sergej Prandzhev, Nova Science, NewYork, 2009, pp. 225–80.

27. Pomogaylo, A.D. and Savotjanov, V.S. *Metal Containing Monomers and Polymers Based on It.* Chemistry, Moscow, 1988.
28. Davtyan, S.P., Hambartsumyan, A.F., Davtyan, D.S., Tonoyan, A.O., Hayrapetyan, S.M., Bagyan, S.H., and Manukyan, L.S. 'The structure, rate and stability of auto waves during polymerization of Co metal-complexes with acryl amide' *Eur. Polym. J.*, 2002, 38, 2423–34.
29. Torsueva, E.S., Belostotskaya, I.S., Komissarova, N.L., and Shlyapnikov, Yu. A. 'Rules for antioxidant action of 3,6-di-tert-butylpyrocatechol' *Proc. Russ. Acad. Sci. Ser. Chem.*, 1976, 9, 2130–3.
30. Yenikopolyan, N.S. and Volfson, S.A. *Chemistry and Technology of Poly-Formaldehyde.* Chemistry, Moscow, 1968.
31. Madorski, S. *Thermal Decomposition of Organic Polymers.* MIR, Moscow, 1967.
32. Afanasiadi, L.I., Blank, A.B., Kvichko, L.A., Kotok, L.A., Livienko, Yu.G., Moghilko, E.T., Nartova, Z.M., Pavlyuk, V.A., Pirogov, A.S., Pulyayeva, I.V., and Sheshina, S.G. 'Effect of chemical degradation on superconducting properties of the HTSC ceramics' *Bull. Mater. Sci.*, 1991, 14, 335–8.
33. Shapligin, I.S., Kakhan, B.G., and Lazarev, W.B. 'Obtainment and properties of Ln_2CuO_4 (Ln–La, Pr, Nd, Sm, Gd) compounds and their some solid solutions' *J. Inorg. Chem.*, 1979, 24 1478–82.
34. Farris, R.J. 'Ultrasonic assessment of cumulative internal damage in filled polymers' *J. Appl. Polym. Sci.*, 1964, 8, 25–32.
35. Davtyan, S.P. and Tonoyan, A.O. *Fundamentals of Nanotechnology. Nanoparticles and Polymeric Nanocomposites.* Gitutyun, Yerevan, 2011.
36. Davtyan, S.P., Tonoyan, A.O., and Berlin A.A. 'Advances and problems of frontal polymerization processes' *Rev. J. Chem.*, 2011, 1, 56–92.
37. Joong Tark Han and Kilwon Cho. 'Layered silicate-induced enhancement of fracture toughness of epoxy molding compounds over a wide temperature range' *Macromol. Mater. Eng.*, 2005, 290, 1184–91.
38. Kraus, G. and Gruver, J.T. 'Thermal expansion, free volume, and molecular mobility in a carbon black-filled elastomer' *J. Polym. Sci.*, 1970, 43, Part A-2 8, 571–81.
39. Ball, G.L. and Salyer, I.O. 'Development of a viscoelastic composition having superior vibration damping capability' *J. Acoust. Soc. Am.*, 1966, 39, 663–72.
40. Bohme, R.D. 'Aluminum reinforcement of some polyethylenes' *J. Appl. Polym. Sci.*, 1968, 12, 1097–1107.
41. Gorl, U. and Parkhouse, A. 'Investigations on the reaction silica/organosilane and organosilane/polymer Part 4: Studies on the chemistry of the silane sulfur chain' *Kautsch. Gummi Kunst.*, 1999, 52, 588–97.
42. Viglin, N.A., Ustinov V.V., and Osipov V.V. 'Spin injection maser' *JETP Lett.*, 2007, 86, 193–6.
43. Asaturyan, R., Ent, R., Fenker, H., Gaskell, D., Huber, G.M., Jones, M., Mack, D., Mkrtchyan, H., Metzger, B., Novikoff, N., Tadevosyan, V., Vulcan, W., and Wood, S. 'The aerogel threshold Cherenkov detector for the high momentum spectrometer in hall C at Jefferson Lab.' *Nucl. Instrum. Meth. Phys. Res. A*, 2005, 548, 364–74.
44. Asaturian, R.A., Sarkissian, A.G., Ignatian, E.L., and Begoian, K.G.

'Peculiarities of Fe intercalation in $YBa_2Cu_{3-x}Fe_xO_y$ for small values X' *Solid State Commun.*, 1995, 95, 389–91.

45. Nicolais, L. and Narkis, M. 'Stress-strain behavior of styrene-acrylonitrile/glass bead composites in the glassy region' *Polymer Eng. Sci.*, 1971, 11, 194–9.

46. Clarke, R. and Uher, C., 'High pressure properties ol graphite and its intercalation compounds' *Adv. Phys.*, 1984, 33, 469–566.

47. Lazarev, V.B., Gavrichev, K.S., and Greenberg, J.H. 'Thermodynamics of high-Tc superconductors' *Pure Appl. Chem.*, 1991, 63, 1341–6.

48. Goodentough, Y.B., Demazen, G., Pouchard, M., and Hagenmuller, P. 'Sur une nouvelle phase oxigence du cuivze + III: $SrLaCuO_4$' *J. Solid State Chem.*, 1973, 8, 325–30.

49. Ganguly, P. and Rao, C.N.R. 'Electron transport properties of transition metal oxide synthesis with the K_2NiF_4 structure' *Mater Res. Bull.*, 1973, 8, 405–12.

10

Nanofluids including ceramic and other nanoparticles: applications and rheological properties

G. PAUL, Indian Institute of Technology, India and
I. MANNA, CSIR-Central Glass and Ceramic Research
Institute, India

DOI: 10.1533/9780857093493.2.323

Abstract: Fluids with suspended nanoparticles (metallic or ceramic in spherical or non-spherical shapes), forming a stable colloid and maintaining a quasi-single-phase state that can offer an extraordinary level of heat transport property at very low levels of dispersion (< 1 vol%), are called nanofluids. This chapter provides an extensive discussion of the potential application areas where different types of nanofluids can be utilized. Different aspects of nanofluids are now being widely researched across the globe, using both experimental and simulation tools. This chapter highlights the parameters that affect the rheological properties of nanofluids. The results published in the literature suggest that the viscosity of nanofluids is directly related to the amount of nanoparticle dispersion (in comparison to that of the base fluid), except at very low concentrations. Sometimes the increase can be many-fold. Theoretical models that explain the increase in the viscosity of nanofluids are compiled and presented.

Key words: nanofluids, potential applications, viscosity, concentration, temperature, modeling.

10.1 Introduction

Energy is key to all human endeavor and development. Harnessing energy at an affordable price without endangering nature or safety is one of the biggest challenges faced by mankind. Since thermal power continues to provide the largest share of non-renewable forms of energy, efficient heat

323

transfer during power/electricity generation has assumed an added importance. There are two ways to increase heat transport efficiency (Duncan and Peterson, 1994; Eastman *et al.*, 2004):

- designing improved cooling devices, such as increasing the surface by fins, microchannels, integrated spot cooling and miniaturized cryodevices
- improving the heat transfer capability or efficiency of working fluids.

The effectiveness of updating the design of cooling devices as a conventional method to increase the heat transfer rate, however, has reached a limit (Eastman *et al.*, 2004). With increasing demands for machines and devices to operate efficiently, the development of new heat transfer fluids with higher thermal conductivity and greater cooling efficiency is now an absolute necessity. Most modern large-scale energy production systems are reliant on the effective working of heat transfer fluids and any enhancement in their properties would directly benefit current energy production. Conventional heat transfer fluids have very poor thermal properties, and improving their thermal properties is thus a key area of research. Historically, thermal properties of colloidal systems have been of little interest to the scientific community. However, due to recent advances in nanoparticle colloid production, such fluids are being explored for new uses like heat transfer. The use of solid particles as an additive into the base fluid is one technique for thermal property enhancement.

As solids possess very high thermal conductivity in comparison to conventional heat transfer fluids (Fig. 10.1), it is expected that thermal properties should be enhanced by dispersing solids in fluids. Since this idea was introduced by Maxwell (1904), many scientists and researchers have made continued efforts to increase the thermal conductivity of fluids by dispersing millimeter or micrometer sized particles in the fluids. Initially, experiments started by blending milli- or micrometer sized particles in fluids to form suspensions. Maxwell (1773) was a pioneer in this area who presented a theoretical basis for calculating the effective thermal conductivity of a suspension. His efforts were followed by numerous theoretical and experimental studies, such as those by Hamilton and Crosser (1962) and Wasp *et al.* (1977). These models work very well in predicting the thermal conductivity of slurries.

However, these studies were limited to the suspension of micro- to macrosized particles, and such suspensions bear a number of disadvantages, including:

- rapid settling of the coarse particles from suspension when not in use/circulation
- abrasion of the surface of the channels caused by these particles

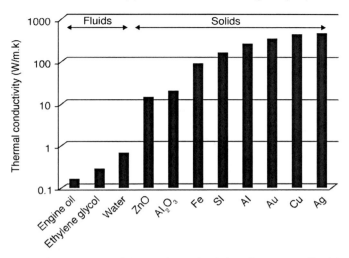

10.1 Comparison of thermal conductivity of common liquids and solids.

- clogging of microchannels
- erosion of pipelines
- increase in pressure drop.

These problems substantially limited the application of milli- or micrometer sized dispersed particles. Thus, the success of enhancing the thermal conductivity of a fluid by dispersing solid particles seems to crucially depend on the ability to develop a stable suspension without the danger of sedimentation, clogging or erosion.

10.2   The development of nanofluids

Recent advances in materials technology have made it possible to produce nanometer-sized particles that can overcome the above stated problems to a large extent. More than a decade ago, it was demonstrated that fluids with suspended nanoparticles, forming a stable colloid and maintaining a quasi-single-phase state (called nanofluids), can offer an extraordinary level of heat transport property for a very small amount of dispersion (<1 vol%). The term nanofluid was coined by Choi (1995) of Argonne National Laboratory, USA, in 1995. Earlier studies of tailored nanofluids showed that there is about 40% enhancement in thermal conductivity with 0.3 vol% of copper nanoparticles of about 10 nm size (Eastman *et al.*, 2001). Nanofluids have now emerged as a promising field in the nanotechnology domain and represent the meeting point of nanoscience, nanotechnology, and thermal engineering.

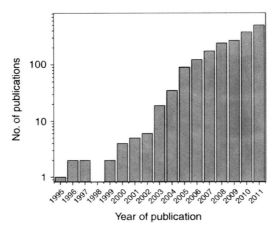

10.2 Nanofluid technology publications rate.

Since the advent of nanofluids, the excellent potential of these fluids for heat transfer applications has led both industry and academia to launch research and development efforts in nanofluid technology. This effort has increased considerably over the years, as corroborated by the near-exponential increase in the number of publications in this field (Fig. 10.2). In a span of 15 years over 1500 articles in the research field of nanofluid technology have been published.

10.3 Potential benefits of nanofluids

Nanoparticles have unique properties such as a large surface area to volume ratio, low density, size-dependent physical properties, and low kinetic energy. The combination of these properties can be advantagous in nanofluid development. Nanofluids have several advantages and benefits over conventional thermal fluids, as follows.

- The original properties of the base fluids are retained. That is, they behave like pure liquids and incur little penalty in pressure drop due to the fact that the dispersed nanoparticles are very tiny, meaning that they can be stably suspended in fluids with or without the help of surfactants (Xuan and Li, 2003).
- Dispersion stability is high with predominant Brownian motion of particles. Since the particles are very small, their weight is small and the chances of sedimentation are less. This reduced sedimentation can overcome one of the major drawbacks of suspensions, the settling of particles, making nanofluids more stable.
- Higher heat conduction is possible by using nanofluids due to the large specific surface area of the nanoparticles. Another advantage is the

mobility of the particles, attributable to their tiny size, which may bring about micro-convection in the fluid and hence increased heat transfer. It has already been found that the thermal conductivity of nanofluids increases significantly with a rise in temperature (Das *et al.*, 2003a), which may be attributed to heat conduction due to micro-convection and particle morphology.

- Suspension of particles of nanometer size reduces the probability of particle clogging as compared to conventional slurries, thus enabling system miniaturization.
- Nanoparticles used as dispersoids in nanofluids are very small and do not carry as much momentum as their milli- or micro-counterparts; thus, the kinetic energy and momentum they impart to solid surfaces is small. Consequently, erosion of components such as pipelines, pumps, and heat exchangers will be greatly reduced.
- Enhanced thermo-physical properties (including thermal conductivity) by varying the particle concentration make nanofluids suitable for various applications. The increased thermal conductivity of the fluids also ensures an increment in heat transfer rate and is expected to save much pumping power.

10.4   Applications of nanofluids

The improved thermo-physical properties and the above stated benefits of nanofluids make them a promising candidate for a wide range of industrial and medical applications (Yu *et al.*, 2007a). Some of the fields of application, including transportation, microelectronic cooling, manufacturing, biomedicine, and defense, are now briefly discussed.

10.4.1 Heat transfer devices

Heat rejection requirements are continually increasing due to trends towards faster speeds (in the multi-GHz range) and smaller features (towards <100 nm) for microelectronic devices, more power output for engines, and brighter beams for optical devices. Nanofluids, due to their projected enhanced heat transfer capability, can be used to combat such ever increasing heat rejection requirements. Since nanofluids utilize particles in suspension that are in the nanometer range, problems due to sedimentation and clogging should not pose a major problem. Hence they can be used as cooling agents even in microchannel heat transfer devices. Nanofluids can also be used as a coolant in heat exchangers, including car radiators, to efficiently extract heat at a much faster rate. In addition, experiments have shown that the critical heat flux is improved by using nanofluids compared to conventional coolants which are expected to improve the boiling heat transfer in devices.

10.4.2 Coolants in manufacturing processes

Most traditional and advanced manufacturing processes require a coolant to absorb the high amount of heat generated during manufacturing operations. Conventional coolants have poor heat absorbing quality, thus increasing the amount of coolant that needs to be circulated. This increased amount of coolant circulating at a higher rate increases the pumping power required, which in turn increases the motor size. Because of their efficient heat transfer characteristics, nanofluids can remove heat at a higher rate than conventional coolants. Nanofluids can thus be thought of as a replacement for conventional coolants, leading to a decrease in the size of the coolant pumping motor.

Oil-based nanofluids containing silver nanoparticles could find application in deep-hole drilling, where they could help improve the drilling penetration rate and clean, lubricate, and cool the drill bit. Thus, nanofluids could significantly improve drilling speeds and hence make it possible to extract more oil (Phuoc and Lyons, 2007). Shen (2008) studied wheel wear and tribological characteristics in wet, dry, and minimum quantity lubrication (MQL) grinding of cast iron. Water-based alumina and diamond nanofluids were applied in the MQL grinding process and the grinding results were compared with those of pure water. Nanofluids demonstrated the benefits of reducing grinding forces, improving surface roughness, and preventing burning of the workpiece. In contrast to dry grinding, MQL grinding could considerably lower the grinding temperature. Further studies are warranted on tribological properties using nanofluids for a wider range of particle loadings as well as on the erosion rate of radiator material in order to help develop predictive models for nanofluid wear and erosion in engine systems.

10.4.3 Coolants in electronic devices

Miniaturization has been a major trend since the middle of the twentieth century in the field of science and technology. This has led to a dramatic increase in the power densities of integrated circuits and microprocessors. This current trend is expected to continue in the future. Existing aircooling system designs have reached their limit and liquid cooling systems are taking over to fulfill the need for higher high heat transfer capacity required for cooling high-performance electronic devices (Yu, 2007a). Water-based nanofluids have been used as the working medium in a circular heat pipe designed as a heat spreader to be used in a central processing unit (CPU) in a notebook or a desktop computer. The results showed a significant reduction in thermal resistance of the heat pipe with the nanofluid as compared with de-ionized water (Tsai et al., 2004). Nanometer sized particles suspended in nanofluids can easily flow in the microchannels without clogging, thus enhancing the cooling of micro-electro-mechanical systems (MEMS) under high heat flux conditions. This has been numerically tested by Chein and Huang (2005) using copper dispersed water nanofluids in silicon micro-channels. The performance of the microchannels was found to improve due to increased thermal conductivity and thermal dispersion effects.

Nguyen et al. (2007a) investigated the heat transfer enhancement and behavior of Al_2O_3–water nanofluid with the intention of using it in a closed cooling system designed for microprocessors or other electronic devices. The experimental results indicate that the inclusion of nanoparticles into distilled water produces a significant increase in convective heat transfer coefficient.

Further research on the use of nanofluids in electronic cooling applications will lead to the development of the next generation of cooling devices that will eventually incorporate nanofluids for ultrahigh heatflux electronic systems.

10.4.4 Coolants in transportation

The transportation industry has strong demands to improve the perfor-mance of vehicle heat transfer fluids and related cooling technologies. Ethylene glycol (EG) and water mixed in a ratio of 50:50 is almost universally used as an automotive coolant. However, the heat transfer properties of this anti-freeze coolant are quite poor, and could be improved by the addition of nanoparticles in suspension. To observe the heat transfer properties of nanofluids as automotive coolants, numerical simulations have been carried out on nanofluids containing CuO dispersed nanoparticles in a mixture of EG and water. By using the same engine and cooling parameters as used for standard coolant, nanofluids have been observed to exhibit a higher heat transfer coefficient, resulting in lower engine and coolant temperatures (Yu et al., 2007b).

Nanofluids have also been studied for the automatic transmission of four-wheel drive vehicles. Tzeng et al. (2005) used CuO and Al_2O_3 dispersed engine oil based nanofluids to study the temperature distribution on the exterior of a rotary-blade-coupling transmission at four engine operating speeds (400, 800, 1200, and 1600 rpm). It was observed that CuO nanofluids produced the lowest transmission temperatures both at high and low rotating speeds.

The drive to employ nanofluids as automotive fluids carries on with several projects exploring various aspects of design, synthesis, and proper-ties of nanofluids. Timofeeva et al. (2009) attempted to utilize boehmite alumina dispersed EG/water based nanofluids in actual truck cooling systems. Recent results from a research project involving industry and a university point to the use of nanoparticles in lubricants to enhance tribological properties such as load carrying capacity, wear resistance, and friction reduction between moving mechanical components. Such results are encouraging for improving heat transfer rates in automotive systems through the use of nanofluids.

10.4.5 Applications in biomedicine

Endeavors concerning the development of integrated nanodrug delivery systems can enable convenient monitoring and controlling target-cell responses to pharmaceutical stimuli, understanding biological cell activities, or creating new drug development processes. While conventional drug delivery is characterized by the 'high-and-low' phenomenon, microdevices facilitate precise drug delivery by both implanted and transdermal techniques. When a drug is dispensed conventionally, the drug concentration in the blood will increase, peak, and then drop as the drug is metabolized, and the cycle is repeated for each drug dose. In case of nanodrug delivery systems, controlled drug release can take place over an extended period of time. Thus, the desired drug concentration will be sustained within the therapeutic window as required. A nanodrug supply system (bio-MEMS) was studied by Kleinstreuer *et al.* (2008) where the principal concern was the conditions for delivering uniform concentrations at the microchannel exit of the supplied nanodrugs. A heat flux, which depends on the levels of nanofluid and purging fluid velocity, was added to ascertain that drug delivery to the living cells occurred at an optimal temperature.

A new initiative takes advantage of several properties of certain nanofluids for use in cancer imaging and drug delivery. This initiative involves the use of iron-based nanoparticles as delivery vehicles for drugs or radiation in cancer patients. Magnetic nanofluids are to be used to guide particles through the bloodstream to a tumor using magnets. This will allow doctors to deliver high local doses of drugs or radiation without damaging nearby healthy tissue, which is a significant side effect of traditional cancer treatment methods. A nanofluid containing magnetic nanoparticles also acts as a super-paramagnetic fluid which, in an alternating electromagnetic field, absorbs energy producing controllable hyperthermia. By enhancing chemotherapeutic efficacy, hyperthermia is able to produce a preferential radiation effect on malignant cells (Chiang *et al.*, 2007).

10.5   The rheology of nanofluids

Although some articles have been published on the viscosity of nanofluids, very limited data have been reported analyzing the rheological behavior of nanofluids; most nanofluid studies focus on heat transfer behavior including thermal conduction, phase change, and convective heat transfer issues. The relationship between thermal conductivity and viscosity and the role of viscosity on the thermal behavior of nanofluids warrant extensive studies to understand the underlying mechanism of heat conduction in nanofluids. Nanofluid viscosity is a critical parameter that governs the amount of particles suspended in the fluid. It is suggested that the viscosity of nanofluids depends on a number of parameters, such as temperature and concentration and size of nanoparticles.

Efforts have been made to analyze the effect of the above-mentioned parameters on the viscosity of nanofluids. An overview of the relevant studies is summarized in Table 10.1 where the variation of relative viscosity is listed as a function of relevant parameters for several nanofluids. The effect of these parameters is discussed in the following sections. Table 10.1 indicates that the most common nanomaterial used as a dispersoid in studies analyzing the viscosity of nanofluids is Al_2O_3, although some studies have utilized TiO_2 and SiO_2 as the dispersed particles.

10.5.1 Effect of particle volume concentration

Similarly to thermal conductivity, viscosity is also strongly dependent on the concentration of particle dispersion in the base fluid. In the literature (Table 10.1), it has been reported that nanofluid viscosity in most cases is higher than that of the base fluid. The ratio of viscosity of the nanofluid to that of the base fluid increases with an increase in particle concentration. The available literature data (Table 10.1) on the relative viscosity (η_{eff}/η_f) of nanofluids (with respect to the base fluid) as a function of particle concentration are summarized in Fig. 10.3. In the figure (and others in this chapter), the references in the legend represent the first three letters of the first author followed by the last two digits of the year of publication. It is evident that relative viscosity of the nanofluid in most cases gradually increases as a function of particle concentration, although Williams *et al.* (2008) report an exponential increase in the relative viscosity at a very low concentration of dispersed particles. It is interesting to note that the degree of increase of viscosity is more than three-fold compared to the viscosity of base fluid, which cannot be explained by the classical equation of Einstein (1906). There is thus an urgent need to investigate the underlying mechanisms for the increase in viscosity of nanofluids with increasing particle concentration.

10.5.2 Effect of temperature

The variation of viscosity of nanofluids (with respect to the base fluid) as a function of temperature as reported in the literature (Table 10.1) is

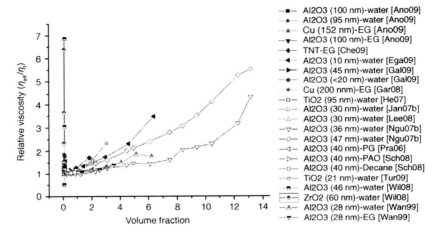

Relative viscosity (η_{eff}/η_f)

Volume fraction

- ─■─ Al2O3 (100 nm)-water [Ano09]
- ⋯♦⋯ Al2O3 (95 nm)-water [Ano09]
- ⋯▲⋯ Cu (152 nm)-EG [Ano09]
- ─▼─ Al2O3 (100 nm)-EG [Ano09]
- ─◆─ TNT-EG [Che09]
- ─◀─ Al2O3 (10 nm)-water [Ega09]
- ─▶─ Al2O3 (45 nm)-water [Gal09]
- ─●─ Al2O3 (<20 nm)-water [Gal09]
- ⋯★⋯ Cu (200 nnm)-EG [Gar08]
- ─□─ TiO2 (95 nm)-water [He07]
- ⋯○⋯ Al2O3 (30 nm)-water [Jan07b]
- ⋯△⋯ Al2O3 (30 nm)-water [Lee08]
- ─▽─ Al2O3 (36 nm)-water [Ngu07b]
- ─◇─ Al2O3 (47 nm)-water [Ngu07b]
- ─◁─ Al2O3 (40 nm)-PG [Pra06]
- ⋯▷⋯ Al2O3 (40 nm)-PAO [Sch08]
- ⋯○⋯ Al2O3 (40 nm)-Decane [Sch08]
- ⋯☆⋯ TiO2 (21 nm)-water [Tur09]
- ─■─ Al2O3 (46 nm)-water [Wil08]
- ─▲─ ZrO2 (60 nm)-water [Wil08]
- ─▲─ Al2O3 (28 nm)-water [Wan99]
- ─▼─ Al2O3 (28 nm)-EG [Wan99]

10.3 Relative viscosity of different nanofluids at different particle concentrations summarized from the available literature (Table 10.1).

Table 10.1 Viscosity details of nanofluids

Authors	Nanofluid	Average particle size (nm)	Temperature (°C)	Shear rate (s^{-1})	Volume fraction (%)	Average viscosity ratio (η_{eff}/η_f)
Wang et al. (1999)	Al$_2$O$_3$–water Al$_2$O$_3$–EG	28	—	—	1.0–6.0 1.25–3.50	1.09–1.77 1.07–1.39
Das et al. (2003b)	Al$_2$O$_3$–water	38	20	50–500	1.0 4.0	1.09 1.36
			60	50–500	1.0	1.08
					4.0	1.38
Prasher et al. (2006)	Al$_2$O$_3$– propelyne glycol	50 40 27	30–60	0.7–100	0.5	1.054 1.058 1.072
		50 40 27	30–60	0.7–100	2.0	1.158 1.186 1.239
		50 27 40	30–60	0.7–100	3.0	1.240 1.286 1.362
Chen et al. (2007b)	TiO$_2$–EG	25	—	0.05– 10 000	0.003–1.8	1.001–1.220
		25	20 30 40	1–200	0.86	1.046 1.054 1.064
		25	20–60	—	0.10 0.21 0.42 0.86 1.80	1.006 1.024 1.046 1.096 1.228

Table 10.1 (cont.)

Authors	Nanofluid	Average particle size (nm)	Temperature (°C)	Shear rate (s^{-1})	Volume fraction (%)	Average viscosity ratio (η_{eff}/η_f)
Chevalier et al. (2007)	SiO$_2$–ethanol	35 94 190	Ambient	5000– 50 000	0.011– 0.070	1.16–1.96 1.14–1.84 1.05–1.45
He et al. (2007)	TiO$_2$–water	95 95–210	25	0.1–1000	0.26–1.18 0.6	1.04–1.11 1.05–1.07
Jang et al. (2007a)	Al$_2$O$_3$–water (tube dia. = 1.773 mm)	30±5	26	—	0.02–0.3	1.010–1.075
	Al$_2$O$_3$–water (tube dia. = 1.023 mm)					1.017–1.095
	Al$_2$O$_3$–water (tube dia. = 0.710 mm)					1.020–1.105
	Al$_2$O$_3$–water (tube dia. = 0.581 mm)					1.024–1.112
	Al$_2$O$_3$–water (tube dia. = 0.351 mm)					1.034–1.150
	Al$_2$O$_3$–water (tube dia. = 0.310 mm)					1.037–1.164
Jang et al. (2007b) and Lee et al. (2008)	Al$_2$O$_3$–water	30±5	21	—	0.01–0.30	1.0008– 1.0290
Namburu et al. (2007a)	SiO$_2$– EG/water (60/40 by weight)	50	−35–50	—	2.0 4.0 6.0 8.0 10.0	1.08 1.21 1.34 1.57 1.96
		20 50 100	−35–50	—	8.0	1.75–2.11 1.63–1.92 1.40–1.75
Namburu et al. (2007b)	CuO– EG/water (60/40 by weight)	29	−35–50	—	1.0–6.12	1.22–4.03

(Continued)

Table 10.1 (cont.)

Authors	Nanofluid	Average particle size (nm)	Temperature (°C)	Shear rate (s⁻¹)	Volume fraction (%)	Average viscosity ratio (η_{eff}/η_f)
Nguyen et al. (2007b)	Al_2O_3–water	36	Ambient	—	2.2–14	1.01–4.31
		47				1.06–5.51
	Al_2O_3–water	36	20–70	—	1.0	1.14
			20–65		4.5	1.51
			20–60		7.0	2.27
			20–55		9.0	4.43–3.62
	Al_2O_3–water	47	20–70	—	1.0	1.06
			20–65		4.0	1.61
			20–60		7.0	2.08
			20–60		9.0	5.36–4.37
	CuO–water	29	20–65	—	1.0	1.14
			20–65		4.0	1.92
			20–55		7.0	4.39–3.72
			20–50		9.0	9.77–7.18
Timofeeva et al. (2007)	Al_2O_3–EG	11	23	—	0.5–10	0.87–55.96
		20				0.87–85.11
		40				0.84–2.04
	Al_2O_3–water	11	23	—	0.5–10	0.87–497.11
		20				0.86–561.79
		40				0.87–2.68
Chen et al. (2008)	TNT–water	260	25	—	0.12–0.60	1.04–1.83
		260	25	50–500	0.12	1.11
					0.24	1.45
					0.60	2.73
Garg et al. (2008)	Cu–EG	200	25	3–3000	0.6	1.05
					1.5	1.12
				3000	0.005–0.020	1.05–1.24
Han et al. (2008)	Indium–PAO	30	30–90	—	8.0	1.31–1.22
Murshed et al. (2008a)	TiO_2–water	15	25–55	—	0.1	1.04–1.15
Murshed et al. (2008b)	TiO_2–water	15	—	—	0.01–0.05	1.25–1.86
	Al_2O_3–water	80	—	—	0.01–0.05	1.03–1.83
Schmidt et al. (2008)	Al_2O_3–PAO	40	22	0.33–3270	0.25–1.00	1.02–1.07
	Al_2O_3–Decane	40	22	0.33–327	0.25–1.00	1.02–1.09
Williams et al. (2008)	Al_2O_3–water	46	21–80	—	0.015–0.060	0.53–6.88
	ZrO_2–water	60	21–80	—	0.007–0.027	1.54–6.38

Table 10.1 (cont.)

Authors	Nanofluid	Average particle size (nm)	Temperature (°C)	Shear rate (s^{-1})	Volume fraction (%)	Average viscosity ratio (η_{eff}/η_f)
Anoop et al. (2009)	Al_2O_3–water	50	20	10–1000	1.0	1.10
					4.0	1.45
					6.0	1.62
	Al_2O_3–EG	50	20	10–1000	1.0	1.12
					4.0	1.25
					6.0	1.36
	Al_2O_3–water	100	20	200	0.005–0.060	1.02–1.57
		95			0.01–0.06	1.20–1.77
	Al_2O_3–EG	100	20	200	0.005–0.060	1.06–1.30
	Cu–EG	152	20	200	0.005–0.060	1.08–1.33
	Al_2O_3–EG	50	20–50	200	2.0	1.13
					4.0	1.21
					6.0	1.31
	Al_2O_3–water	50	20–50	200	0.5	1.06
					1.0	1.10
					2.0	1.21
					4.0	1.38
					6.0	1.58
Chen et al. (2009)	TNT–EG	10 × 100	40	0.1–1100	0.10	1.17
					0.21	1.28
					0.42	1.37
			20	1–1100	0.86	1.92–1.30
			30			2.06–1.42
			40			26.67–1.49
			50			104.12–1.46
			60			217.94–1.49
			20–60	2000	0.10	1.033
					0.21	1.070
					0.42	1.162
					0.86	1.263
					1.80	1.709
Egan et al. (2009)	Al_2O_3–water	10±5	—	—	0.3–6.3	1.08–3.48
Pastoriza–Gallego et al. (2009)	Al_2O_3–water	40–50	10–60	—	0.0013–	1.01–1.37
	Al_2O_3–water	<20			0.029	1.06–1.84
Singh et al. (2009)	SiC–water	130	15–35	—	1.85	1.35
			15–45		3.70	1.88
			15–55		7.40	2.55–2.69
Turgut et al. (2009)	TiO_2–water	21	13–55	—	0.2–3.0	1.04–2.34
Xie et al. (2010)	MgO–EG	20	10–60	—	0.01–0.05	1.03–1.30

(Continued)

Table 10.1 (cont.)

Authors	Nanofluid	Average particle size (nm)	Temperature (°C)	Shear rate (s^{-1})	Volume fraction (%)	Average viscosity ratio (η_{eff}/η_f)
Shima et al. (2010)	CuO–EG	10	5–55	—	0.18	1.15–1.23
					0.54	1.23–1.39
					0.92	1.42–1.54
					1.31	1.52–1.84

summarized in Fig. 10.4. It is observed that the viscosity of the nanofluids is greater than that of the base fluid, the ratio being greater than 1 in most cases. In general, most studies reported a gradual decrease in the viscosity of nanofluids as well as that of base fluid with increasing temperature. However, the slope of the curve remains almost identical in the concerned temperature range. In other words, Fig. 10.4 shows that relative viscosity remains nearly constant as a function of temperature for most of the cases reported in the literature. The majority of studies were conducted from ambient to higher temperatures, but Namburu *et al.* (2007a, 2007b) reported viscosity values of a nano-SiO_2 or CuO dispersed EG and water mixture (60:40 by weight) based nanofluid in the temperature range of $-35°C$ to $50°C$. In some of the studies, the viscosity of the nanofluid was observed to increase by more than twice that of the viscosity of base fluid.

10.5.3 Effect of particle size

Although very limited data are available, some studies have reported the effect of particle size on viscosity of nanofluids. The viscosity ratio of several nanofluids (with respect to the base fluid) at different particle sizes is summarized from the data available in literature (in Fig. 10.5). From Fig. 10.5 it can be observed that, in general, the viscosity ratio of nanofluids decreases with an increase in the size of dispersed particles. However, there are some variations in the literature, where the viscosity ratio has been reported to increase with an increase in particle size (Nguyen *et al.*, 2007b), and, in another study, viscosity first increases and then decreases with an increase in particle size for 3.0 vol% Al_2O_3–propylene glycol nanofluid (Prasher *et al.*, 2006). The mechanism for such contradictory behavior needs to be analyzed, which requires in-depth theoretical study of the phenomenon.

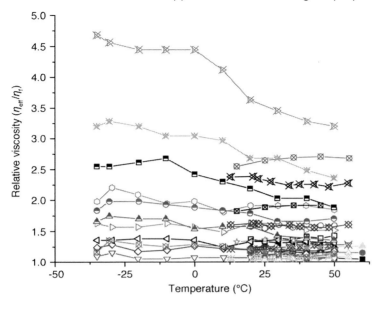

10.4 Relative viscosity of nanofluids as a function of temperature summarized from the available literature (Table 10.1).

10.6 Modeling the viscosity of nanofluids

Investigations into the viscosity of colloidal suspensions started with the analysis of Einstein (1906) in which particles were assumed to be small, rigid, uncharged, without attractive forces. Under such assumptions, particles move at the velocity of the streamline in line with the particle center, as shown in Fig. 10.6(a). This leads to the rotational motion of the particle known as the vorticity of the shear field and hence the dispersion viscosity is calculated as:

$$\frac{\eta}{\eta_0} = 1 + [\eta]\phi + O(\phi)^2 \qquad [10.1]$$

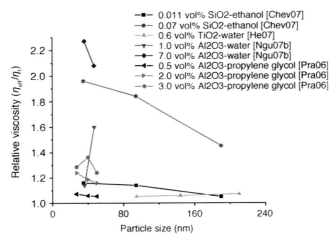

10.5 Variation of relative viscosity of different nanofluids as a function of dispersed particle size summarized from the available literature (Table 10.1).

where $[\eta]$ is the intrinsic viscosity with a value of 2.5 for mono-dispersed hard spheres and equation 10.1 is applicable for concentrations $\phi < 0.01$.

For particle concentrations $\phi > 0.01$, the hydrodynamic interactions between particles are considered as the disturbance of fluid around one particle interacting with that around other particles. The viscosity of colloidal suspensions in such a condition was predicted by Batchelor (1977) as:

$$\frac{\eta}{\eta_0} = 1 + [\eta]\phi + k_{\mathrm{H}}([\eta]\phi)^2 + O(\phi)^3 \qquad [10.2]$$

where k_{H} is the Huggin's coefficient and can be considered as the interaction parameter characterizing the colloidal interactions between particles. Equation 10.2 is valid for $\phi = 0.1$ for flows dominated by particle-pair microstructures (shown in Fig. 10.6(b)).

At higher concentrations ($\phi > 0.01$), multi-particle collisions (Fig. 10.6 (c)) become more prominent and hence the third and higher order terms have to be rigorously analyzed. Considering such factors, Kreiger and Dougherty (1959) derived a semi-empirical relationship for viscosity as a function of the volume fraction of dispersed particles:

$$\frac{\eta}{\eta_0} = \left(1 - \frac{\phi}{\phi_{\mathrm{m}}}\right)^{-[\eta]\phi_{\mathrm{m}}} \qquad [10.3]$$

where ϕ_{m} is the maximum particle packing fraction varying in the range of 0.495 to 0.540 under quiescent conditions and is approximately 0.605 at high

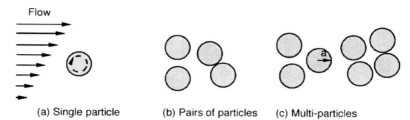

Flow

(a) Single particle   (b) Pairs of particles   (c) Multi-particles

10.6 (a–c) Schematic illustration of interactions of particles with shear flow (Chen *et al.*, 2007a).

shear rates. For $\phi < 0.01$, equation 10.3 reduces to Einstein's equation (equation 10.1) if a monomial expansion is performed and to Batchelor's equation (equation 10.2) if a binomial expansion is performed.

Cheng and Law (2003) utilized equation (equation 10.1) and used a numerical expansion to predict the viscosity ratio of nanofluids for dilute concentrations empirically as:

$$\frac{\eta}{\eta_0} = 1 + \frac{5}{2}\phi + \frac{35}{8}\phi^2 + \frac{105}{16}\phi^3 + \frac{1155}{128}\phi^4 + \frac{3003}{256}\phi^5 + \cdots \qquad [10.4]$$

As the dynamic effects of particles and fluid are not considered in the above, the model is applicable for only dilute concentrations of nanofluids. Using the above stated theories, Chen *et al.* (2007a) studied the effect of high shear viscosity, shear thinning behavior, and temperature effect on the behavior of viscosity and inferred that the rheological behavior of nanofluids can be categorized into four categories: (a) dilute nanofluids where Einstein's equation (equation 10.1) is applicable; (b) semi-dilute nanofluids with aggregation of nanoparticles; (c) semi-concentrated nanofluids where aggregation of nanoparticles and shear thinning behavior can be observed; and (d) concentrated nanofluids with interpenetration of aggregation.

Lu and Fan (2008) used a molecular dynamics simulation to predict the effect of particle concentration and size on the viscosity of nanofluids assuming an *NVT* ensemble. The shear viscosity simulated using numerical data quite accurately fitted with the experimental data for Al_2O_3 dispersed in water and EG based nanofluids (Fig. 10.7). Anoop *et al.* (2009) predicted the viscosity ratio of electrostatically stabilized nanofluids by modifying equation 10.1. It was considered that the electrical double layer introduces an additional increase in viscosity brought about by electroviscous forces. The intrinsic viscosity value was modified as:

$$[\eta]_{EV} = [\eta](1 + p) \qquad [10.5]$$

where $[\eta]_{EV}$ and $[\eta]$ are the intrinsic viscosity values in the presence of electroviscous forces and with uncharged particles respectively and p is the

(a)

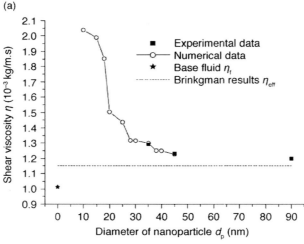

(b)

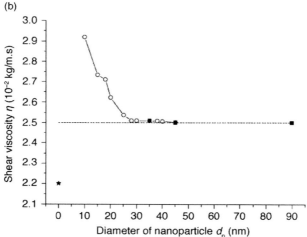

10.7 Shear viscosity as a function of diameter of Al_2O_3 in water (a) and ethylene glycol (b) (volume fraction = 5%) (Lu and Fan, 2008).

primary viscous coefficient. According to Adachi *et al.* (1998), p is proportional to κ^{-1}, which is the double layer thickness (also known as the Debye length) and is expressed as:

$$p = \frac{(r + \kappa^{-1})^3 - r^3}{r^3} \qquad [10.6]$$

Computing the value of the Debye length, the average value of p has been assumed to be 3, which gives the value of $[\eta]_{EV}$ as 10. Thus, the viscosity ratio was modified as:

$$\frac{\eta}{\eta_0} = 1 + 10\phi \qquad\qquad [10.7]$$

Masoumi *et al.* (2009), considering the effect of Brownian motion of the nanoparticles, showed a negligible contribution of Reynolds number for 10 nm Al_2O_3 particles dispersed in water and assumed no interparticle interaction to predict the effective viscosity of nanofluids as:

$$\frac{\eta}{\eta_0} = 1 + \frac{\rho_p V_B d_p^2}{72 C \delta \eta_f} \qquad\qquad [10.8]$$

where δ is the distance between nanoparticles, V_B is the Brownian velocity, d_p is the particle diameter, and c is the correction factor. Experimental results from Namburu *et al.* (2007b), Chen *et al.* (2007a), and Nguyen *et al.* (2007b) have been compared using this model and the data quite accurately fit the values predicted by the model, suggesting that Brownian motion has a major contribution in the phenomenal increase in viscosity of nanofluids.

10.7   Summary and future trends

Since the advent of nanofluids in the late 1990s, extensive investigations have attempted to characterize the different properties of nanofluids. This chapter has provided an insight into the study of the synthesis and thermophysical properties of nanofluids, including a discussion of the potential fields for application of nanofluids. Although an amount of experimental research has been carried out by several research groups across the globe, there is a severe void regarding explanation of the experimental results. As surveyed from the literature, the viscosity of nanofluids has been observed to increase in comparison to the base fluid, sometimes the increase being manyfold; theories that can explain such level of enhancement still remain elusive. Several mechanisms have been proposed to explain the anomalous enhancement in viscosity of nanofluids, but one integrated theory has still not been predicted that can singlehandedly explain the enhancement in viscosity irrespective of the nature and type of nanofluid. Further detailed research in theoretical explanation of the experimental results observed in nanofluids is thus a major area for future focus.

10.8   References

Adachi, Y., Nakaishi, K. and Tamaki, M. (1998), Viscosity of a dilute suspension of sodium montmorillonite in an electrostatically stable condition, *J. Colloid Interf. Sci.*, Vol. 198, pp. 100–105.

Anoop, K. B., Kabelac, S., Sundararajan, T. and Das, S. K. (2009), Rheological and flow characteristics of nanofluids: Influence of electroviscous effects and particle agglomeration, *J. Appl. Phys.*, Vol. 106, p. 034909.

Batchelor, G. K. (1977), Effect of Brownian-motion on bulk stress in a suspension of spherical-particles, *J. Fluid Mech.*, Vol. 83, pp. 97–117.

Chein, R. and Huang, G. (2005), Analysis of microchannel heat sink performance using nanofluids, *Appl. Therm. Eng.*, Vol. 25, No. 17, pp. 3104–3114.

Chen, H., Ding, Y. He, Y. and Tan, C. (2007a), Rheological behaviour of ethylene glycol based titania nanofluids, *Chem. Phys. Lett.*, Vol. 444, pp. 333–337.

Chen, H., Ding, Y. and Tan, C. (2007b), Rheological behaviour of nanofluids, *New J. Phys.*, Vol. 9, pp. 367(1–25).

Chen, H., Yang, W., He, Y., Ding, Y., Zhang, L., Tan, C., Lapkin, A. A. and Bavykin, D. V. (2008), Heat transfer and flow behaviour of aqueous suspensions of titanate nanotubes (nanofluids), *Powder Technol.*, Vol. 183, pp. 63–72.

Chen, H., Ding, Y. and Lapkin, A. (2009), Rheological behaviour of nanofluids containing tube/rod-like nanoparticles, *Powder Technol.*, Vol. 194, pp. 132–141.

Cheng, N.-S. and Law, A.W.-K. (2003), Exponential formula for computing effective viscosity, *Powder Technol.*, Vol. 129, pp. 156–160.

Chevalier, J., Tillement, O. and Ayela, F. (2007), Rheological properties of nanofluids flowing through microchannels, *Appl. Phys. Lett.*, Vol. 91, p. 233103.

Chiang, P. C., Hung, D. S., Wang, J. W., Ho, C. S. and Yao, Y. D. (2007), Engineering water-dispersible Fe-Pt nanoparticles for biomedical applications, *IEEE T. Magn.*, Vol. 43, No. 6, pp. 2445–2447.

Choi, S. U. S. (1995), Enhancing thermal conductivity of fluids with nanoparticles. In *Proceedings of the 1995 ASME International Mechanical Engineering Congress and Exposition*, San Francisco, CA, USA.

Das, S. K., Putra, N., Thiesen, P. and Roetzel, W, (2003a), Temperature dependence of thermal conductivity enhancement for nanofluids. *Trans. ASME, J. Heat Trans.*, Vol. 125, pp. 567–574.

Das, S. K., Putra, N. and Roetzel, W. (2003b), Pool boiling characteristics of nanofluids, *Int. J. Heat Mass Tran.*, Vol. 46, pp. 851–862.

Duncan, A. B. and Peterson, G. P. (1994), Review of microscale heat transfer, *J. Appl. Mech. Rev.*, Vol. 47, No. 9, pp. 397–428.

Eastman, J. A., Choi, S. U. S., Li, S., Yu, W. and Thompson, L. J. (2001), Anomalously increased effective thermal conductivity of ethylene glycol-based nanofluids containing copper nanoparticles, *Appl. Phys. Lett.*, Vol. 78, pp. 718–720.

Eastman, J. A., Phillpot, S. R., Choi, S. U. S. and Keblinski, P. (2004), Thermal transport in nanofluids, *Ann. Rev. Mater. Res.*, Vol. 34, No. 1, pp. 219–246.

Egan, V. M., Walsh, P. A. and Walsh, E. J. (2009), On viscosity measurements of nanofluids in micro and mini tube flow, *J. Phys. D: Appl. Phys.*, Vol. 42, p. 165502.

Einstein, A. (1906), Eine neue bestimmung der molekul-dimension (A new determination of the molecular dimensions), *Ann. Phys.*, Vol. 19, pp. 289–306.

Garg, J., Poudel, B., Chiesa, M., Gordon, J. B., Ma, J. J., Wang, J. B., Ren, Z. F., Kang, Y. T., Ohtani, H., Nanda, J., McKinley, G. H. and Chen, G. (2008), Enhanced thermal conductivity and viscosity of copper nanoparticles in ethylene glycol nanofluid, *J. Appl. Phys.*, Vol. 103, p. 074301.

Hamilton, R. L. and Crosser, O. K. (1962), Thermal conductivity of heterogeneous two-component systems, *Ind. Eng. Chem. Fundamen.*, Vol. 1, No. 3, pp. 187–191.

Han, Z. H., Cao, F. Y. and Yang, B. (2008), Synthesis and thermal characterization of phase-changeable indium/polyalphaolefin nanofluids, *Appl. Phys. Lett.*, Vol. 92, p. 243104.

He, Y., Jin, Y., Chen, H., Ding, Y., Cang, D. and Lu, H. (2007), Heat transfer and flow behaviour of aqueous suspensions of TiO2 nanoparticles (nanofluids) flowing upward through a vertical pipe, *Int. J. Heat Mass Tran.*, Vol. 50, pp. 2272–2281.

Jang, S. P., Lee, J.-H. and Hwang, K. S. (2007a), Particle concentration and tube size dependence of viscosities of Al$_2$O$_3$-water nanofluids flowing through micro- and minitubes, *Appl. Phys. Lett.*, Vol. 91, p. 243112.

Jang, S. P., Hwang, K. S., Lee, J.-H., Kim, J. H., Leeand B. H. and Choi, S. U. S. (2007b), Effective thermal conductivities and viscosities of water-based nanofluids containing Al$_2$O$_3$ with low concentration. In *Proceedings of the 7th IEEE International Conference on Nanotechnology*, p. 188.

Kleinstreuer, C., Li, J. and Koo, J. (2008), Microfluidics of nano-drug delivery, *Int. J. Heat Mass Tran.*, Vol. 51, pp. 5590–5597.

Krieger, I. M. and Dougherty, T. J. (1959), A mechanism for non-Newtonian flow in suspensions of rigid spheres, *Trans. Soc. Rheology*, Vol. 3, pp. 137–152.

Lee, J.-H., Hwang, K. S., Jang, S. P., Lee, B. H., Kim, J. H., Choi, S. U. S. and Choi, C. J. (2008), Effective viscosities and thermal conductivities of aqueous nanofluids containing low volume concentrations of Al$_2$O$_3$ nanoparticles, *Int. J. Heat Mass Tran.*, Vol. 51, pp. 2651–2656.

Lu, W.-Q. and Fan, Q.-M. (2008), Study for the particle's scale effect on some thermophysical properties of nanofluids by a simplified molecular dynamics method, *Eng. Anal. Bound. Elem.*, Vol. 32, pp. 282–289.

Masoumi, N., Sohrabi, N. and Behzadmehr, A. (2009), A new model for calculating the effective viscosity of nanofluids, *J. Phys. D: Appl. Phys.*, Vol. 42, p. 055501.

Maxwell, J. C. (1773), *Treatise on Electricity and Magnetism*. Clarendon Press. Oxford, UK.

Maxwell, J. C. (1904), *A Treatise on Electricity and Magnetism*, 2nd edn., Oxford University Press, Oxford, UK.

Murshed, S. M. S., Tan, S. H. and Nguyen, N. T. (2008a), Temperature dependence of interfacial properties and viscosity of nanofluids for droplet-based microfluidics, *J. Phys. D: Appl. Phys.*, Vol. 41, pp. 085502(1-5).

Murshed, S. M. S., Leong, K. C. and Yang, C. (2008b), Investigations of thermal conductivity and viscosity of nanofluids, *Int. J. Therm. Sci.*, Vol. 47, pp. 560–568.

Namburu, P. K., Kulkarni, D. P., Dandekar, A. and Das, D. K. (2007a),

Experimental investigation of viscosity and specific heat of silicon dioxide nanofluids, *Micro Nano Lett.*, Vol. 2, No. 3, pp. 67–71.

Namburu, P. K., Kulkarni, D. P., Misra, D. and Das, D. K. (2007b), Viscosity of copper oxide nanoparticles dispersed in ethylene glycol and water mixture, *Exp. Therm. Fluid Sci.*, Vol. 32, pp. 397–402.

Nguyen, C. T., Roy, G., Gautheir, C. and Galanis, N. (2007a), Heat transfer enhancement using Al_2O_3-water nanofluid for an electronic liquid cooling system, *Appl. Therm. Eng.*, Vol. 27, pp. 1501–1506.

Nguyen, C. T., Desgranges, F., Roy, G., Galanis, N., Maré, T., Boucher, S. and Mintsa, H. A. (2007b), Temperature and particle-size dependent viscosity data for water-based nanofluids – Hysteresis phenomenon, *Int. J. Heat Fluid Flow*, vol. 28, pp. 1492–1506.

Pastoriza-Gallego, M. J., Casanova, C., Páramo, R., Barbés, B., Legido, J. L. and Piñeiro, M. M. (2009), A study on stability and thermophysical properties, density and viscosity of Al_2O_3 in water nanofluid, *J. Appl. Phys.*, Vol. 106, p. 064301.

Phuoc, X. T. and Lyons, D. K. (2007), Nanofluids for use as ultra-deep drilling fluids. U.S. Department of Enegy, Office of Fossil Energy: National Energy Technology Laboratory [Online] Available http://www.netl.doe.gov/publications/factsheets/rd/R&D108.pdf.

Prasher, R., Song, D. and Wang, J. (2006), Measurements of nanofluid viscosity and its implications for thermal applications, *Appl. Phys. Lett.*, Vol. 89, p. 133108.

Schmidt, A. J., Chiesa, M., Torchinsky, D. H., Johnson, J. A., Boustani, A., McKinley, G. H., Nelson, K. A. and Chen, G. (2008), Experimental investigation of nanofluid shear and longitudinal viscosities, *Appl. Phys. Lett.*, Vol. 92, p. 244107.

Shen, B. (2008), Minimum quantity lubrication grinding using nanofluids. [Online] Available http://deepblue.lib.umich.edu/bitstream/2027.42/60683/1/binshen_1.pdf.

Shima, P. D., Philip, J. and Raj, B. (2010), Influence of aggregation on thermal conductivity in stable and unstable nanofluids, *Appl. Phys. Lett.*, Vol. 97, p. 153113.

Singh, D., Timofeeva, E., Yu, W., Routbort, J., France, D., Smith, D. and Lopez-Cepero, J. M. (2009), An investigation of silicon carbide-water nanofluid for heat transfer applications, *J. Appl. Phys.*, Vol. 105, p. 064306.

Timofeeva, E. V., Gavrilov, A. N., McCloskey, J. M. and Tolmachev, Y. V. (2007), Thermal conductivity and particle agglomeration in alumina nanofluids: Experiment and theory, *Phys. Rev. E*, Vol. 76, p. 061203.

Timofeeva, E., Smith, D., Yu, W., Routbort, J. and Singh, D. (2009), Nanofluid development for engine cooling systems. [Online] Available http://www1.eere.energy.gov/vehiclesandfuels/pdfs/merit_review_2009/vehicles_and_systems_simulation/vssp_21_timofeeva.pdf.

Tsai, C. Y., Chien, H. T., Ding, P. P., Chan, B., Luh, T. Y. and Chen, P. H. (2004), Effect of structural character of gold nanoparticles in nanofluid on heat pipe thermal performance, *Mater. Lett.*, Vol. 58, pp. 1461–1465.

Turgut, A., Tavman, I., Chirtoc M., Schuchmann, H. P., Sauter, C. and Tavman, S. (2009), Thermal conductivity and viscosity measurements of water-based TiO2 nanofluids, *Int. J. Thermophys.*, Vol. 30, pp. 1213–1226.

Tzeng, S.-C., Lin, C.-W. and Huang, K. D. (2005), Heat transfer enhancement of nanofluids in rotary blade coupling of four-wheel-drive vehicles, *Acta Mech.*, Vol. 179, pp. 11–23.

Wang, X., Xu, X. and Choi, S. U. S. (1999), Thermal conductivity of nanoparticle-fluid mixture, *J. Thermophys. Heat Tran.*, Vol. 13, pp. 474–480.

Wasp, E. J., Kenny, J. P. and Gandhi, R. L. (1977), *Solid–Liquid Flow Slurry Pipeline Transportation, Series on Bulk Materials Handling*, Vol. 1. Trans. Tech. Publications, Clausthal, Germany.

Williams, W., Buongiorno, J. and Hu, L.-W. (2008), Experimental investigation of turbulent convective heat transfer and pressure loss of alumina/water and zirconia/water nanoparticle colloids (nanofluids) in horizontal tubes, *Trans. ASME, J. Heat Trans.*, Vol. 130, p. 042412.

Xie, H., Yu, W. and Chen, W., (2010), MgO nanofluids: higher thermal conductivity and lower viscosity among ethylene glycol-based nanofluids containing oxide nanoparticles, *J. Exp. Nanosci.*, Vol. 5, No. 5, pp. 463–472.

Xuan, Y. and Li, Q. (2003), Investigation on convective heat transfer and flow features of nanofluids, *Trans. ASME, J. Heat Trans.*, Vol. 125, No. 1, pp. 151–155.

Yu, W., France, D. M., Choi, S. U. S. and Rourborr, J. L. (2007a), Review and assessment of nanofluid technology for transportation and other applications. [Online] Available http://www.ipd.anl.gov/anlpubs/2007/05/59282.pdf.

Yu, W., Saripella, S. K., France, D. M. and Routbort, J. L. (2007b), Effects of nanofluid coolant in a class 8 truck engine, document number 2007-01-2141. [Online] Available http://www.sae.org/technical/papers/2007-01-2141.

Nanofluids including ceramic and other nanoparticles: synthesis and thermal properties

G . P A U L , Indian Institute of Technology, India and
I . M A N N A , CSIR-Central Glass and Ceramic Research
Institute, India

DOI: 10.1533/9780857093493.2.346

Abstract: This chapter reports the detailed study on the synthesis of nanofluids comprising very low concentrations of nanometric metallic or ceramic particles, rods, tubes, etc. The most common ways of preparing nanofluids are the one-step and two-step methods. While the one-step approach usually yields more stable nanofluids, the two-step method is more versatile as it provides the opportunity to disperse a wide variety of nanoparticles in different types of base fluids. However, the main focus of this chapter is on the thermal conductivity of nanofluids, which is the most researched aspect of nanofluids worldwide. An insight into the different parameters that influence the thermal conductivity of nanofluids is presented. In addition to experimental work, the theories used to try to analyze the cause of the anomalous increase in thermal conductivity are also presented.

Key words: nanofluids, thermal conductivity, synthesis, one-step method, two-step method, concentration, temperature, particle size.

11.1 Introduction

Since the pioneering work of Maxwell (1904), there have been several attempts to enhance the thermal properties of fluids by adding solid particles. However, these slurries posed problems of clogging and abrasion of the channels through which the fluid flowed. Recent advances in nanotechnology have made it possible to produce nanometer-sized particles that can overcome these problems. More than a decade ago it was

346

demonstrated that fluids with suspended nanoparticles, forming a stable colloid and maintaining a quasi-single phase state, exhibit high heat transport properties at a very low amount of dispersion (<1 vol%). The term nanofluid was coined by Choi (1995) to describe this new class of fluids. Nanofluids have now emerged as a promising field in nanotechnology research. Among the properties of nanofluids that have been studied, the largest volume of work has focused on characterization of the thermal properties of nanofluids, particularly the high thermal conductivity they display. There has also been research on modeling the anomalous increase in thermal conductivity.

11.2   Synthesis of nanofluids

Nanofluids can be synthesized by adding nanometer-sized particles to a liquid. However, nanofluids are not simply solid–liquid mixtures. Some special requirements are essential, namely an even and stable suspension, adequate durability, negligible agglomeration of particles, and no chemical change of the dispersed particles or fluid. The synthesis procedure of a nanofluid is a key factor on which the thermal properties depend, and the behavior of a nanofluid is highly dependent on the behavior of the base fluids and the dispersed phases, particle concentration, size and morphology, as well as the presence of dispersants or surfactants. The synthesis of stable nanofluids with minimum agglomeration and controlled properties like thermal conductivity and viscosity for heat transfer applications is the main objective of the community dealing with nanofluid technology. The techniques generally utilized for the synthesis of nanofluids are discussed in the following sections.

11.2.1 One-step process

A one-step process is a bottom-up approach that combines the synthesis of nanoparticles with the preparation of nanofluids, as the nanoparticles synthesized (by physical/chemical vapor deposition or chemical methods) are collected in the same fluid/medium. A number of techniques have been reported for the synthesis of nanofluids by a one-step process, including vacuum evaporation onto a running oil substrate (VEROS) (Yatsuya et al., 1978), the direct condensation technique (Eastman et al., 2001), microwave irradiation (Zhu et al., 2004), citrate reduction (Patel et al., 2003; Zhu et al., 2004; Zhang et al., 2006a; Liu et al., 2006; Fuentes et al., 2008; Mishra et al., 2009; Wang and Wei, 2009), submerged arc nanoparticle synthesis system (SANSS) (Lo et al., 2005a, 2005b, 2007), multi-pulse laser ablation (Phuoc et al., 2007), sputtering on running liquid (Tamjid and Guenther, 2010), to name a few. A detailed summary of the different techniques by which nanofluids can be synthesized by a one-step process is given in Table 11.1. Most of the materials used as dispersoids have been metals and their oxides as they are easy to produce and chemically stable in solution. The base fluids include water, ethylene glycol, several types of oils, toluene, dielectric fluids, ethanol, etc.

Early reports show the possibility of synthesis of silver–silicone oil nanofluid synthesis following a VEROS technique (Yatsuya *et al.*, 1978). In this method, metals were evaporated in vacuum onto the surface of running oil, and thus fine particles in the nanometer range were grown on the surface of the oil. Later, Wagener and Gunther (1999) used a modified VERL technique employing high-pressure magnetron sputtering to develop iron– and silver–silicone oil nanofluids. The particles formed by this technique formed agglomerates. To stabilize the fluids, different dispersants were used. The mean sizes of the particles for Fe nanofluids were 15 nm (without surfactant) and 9 nm (with surfactant). The size for Ag nanofluids varied from 5 to 15 nm by varying the pressure of the system. In another method, Eastman *et al.* (2001) employed a direct evaporation condensation (DEC) technique to synthesize copper–ethylene glycol nanofluids. This technique involves the vaporization of a source material to be dispersed into the fluid under vacuum conditions in a chamber containing the base fluid, which is rotated, and a thin film of the fluid is constantly being transported over the top of the chamber. Advantages of this technique are that nanoparticles are produced without oxide layers, the size of the nanoparticles is in a narrow range, and nanofluids without particle agglomeration can be produced. On the contrary, the disadvantages are that the liquid must have a very low vapor pressure and this technique can only produce very limited amounts of nanofluids.

In another widely used one-step process, nanofluids were prepared by the chemical reduction method (Patel *et al.*, 2003; Zhu *et al.*, 2004; Zhang *et al.*, 2006a; Liu *et al.*, 2006; Fuentes *et al.*, 2008; Mishra *et al.*, 2009; Wang and Wei, 2009). In this technique, a precursor solution is chemically reduced by a reducing agent to produce nanoparticles in suspension under boiling conditions. In SANSS, a solid bar of the particle to be dispersed, submerged in the base fluid, is used as an electrode in a vacuum system which melts and vaporizes in the region of high-temperature electrical arc generated by the system. The base fluid medium also vaporizes and rapidly removes the vapor of the solid and cools it to restrain further particle growth. Nanoparticles are then formed from the evaporated solid and are well dispersed in the cooling medium through three transformation stages, namely nucleation, growth, and condensation (Lo *et al.*, 2005a, 2005b). Alternatively, silver–water nanofluids were prepared by multi-pulse laser ablation technique where a silver bar submerged in de-ionized water was ablated using a double-beam approach (Phuoc *et al.*, 2007).

Table 11.1 Different available techniques for the synthesis of nanofluids in a one-step process

Nanofluid	Preparation method
Ag–silicon oil	Vacuum evaporation onto a running oil substrate (VEROS) (Yatsuya *et al.*, 1978)
Fe–silicon oil Ag–silicon oil	Vacuum evaporation on running liquids (VERL) process (Wagener and Gunther, 1999)
Cu–ethylene glycol	Direct condensation technique (Eastman *et al.*, 2001)
Ag–water Au–water	Citrate reduction method (Patel *et al.*, 2003)
Au–toluene	Brust *et al.* procedure (Patel *et al.*, 2003)
Cu–ethylene glycol (EG) Ag–EG Ag–glycerol	Microwave irradiation (Zhu *et al.*, 2004) Microwave dielectric heating (Patel *et al.*, 2005)
CuO–water/EG	Submerged arc nanoparticle synthesis system (SANSS) (Lo *et al.*, 2005a)
Cu–dielectric fluids	SANSS (Lo *et al.*, 2005b)
Ag–ethylene glycol (PVP)	Magnetic stirring at room temperature (Slistan-Grijalva *et al.*, 2005)
Au–toluene	Chemical reaction method (Zhang *et al.*, 2006a)
Cu–water	Chemical reduction method (Liu *et al.*, 2006)
CuO–water	Microwave irradiation (Zhu *et al.*, 2007)
Ag–water	Submerged arc nanoparticle synthesis system (SANSS) (Lo *et al.*, 2007)
SiO_2–water	Sol-precipitation method (Tao *et al.*, 2007)
Ag–water	Multi-pulse laser ablation technique (Phuoc *et al.*, 2007)
Au(core)/Ag(shell)–water	Chemical reduction method (Fuentes *et al.*, 2008)
Indium–PAO oil	Nanoemulsification technique (Han *et al.*, 2008)
Al_2O_3–water	Plasma arc system (Chang and Chang, 2008)
Ag–silicon oil	Magnetron sputtering (Hwang *et al.*, 2008)
Cu–EG	Reduction of copper sulfate using sodium hypophosphite (Kumar *et al.*, 2009)
Spherical Fe_3O_4 in water/various oils	Chemical solution method (Wang and Wei, 2009)
Elliptic Cu nanorods in EG or EG/water mixtures Needlelike CuO nanoparticles in water/various oils	

(Continued)

Table 11.1 (cont.)

Nanofluid	Preparation method
Octahedral Cu_2O nanoparticles in water or water/EG mixtures	
$CePO_4$ nanofibers in water, ethanol, EG, and their mixtures	
Hollow Cu or CuS nanoparticles in water, ethanol, EG, and their mixtures	
Hollow and wrinkled Cu_2O nanoparticles in EG or EG/water mixtures	
Cu_2O(core)/CuS(shell) nanoparticles in water	
Au–water	Pulsed laser ablation in liquids technique (Kim *et al.*, 2009)
Au (PVP)–water	Chemical reaction method (Mishra *et al.*, 2009)
Ag–diethylene glycol	Sputtering on running liquid technique (Tamjid and Guenther, 2010)

11.2.2 Two-step process

The two-step method is extensively used in the synthesis of nanofluids primarily due to the easy availability of several types of nanopowders and the inherent versatility associated with this approach. In this method, nanoparticles are first synthesized by mechanical alloying, chemical reaction, vapor condensation, or decomposition of organic complex and are then dispersed in the base fluid with mechanical agitation (stirring) applying ultrasonic vibration (Hwang *et al.*, 2008; Li *et al.*, 2008) or any such suitable dispersing technique. Generally, ultrasonic equipment is used to intensively disperse the particles and reduce the agglomeration of particles. The stability factor of the nanofluids is taken care of by using surfactants. The main advantage of this two-step synthesis method is that it produces nanoparticles under clean conditions, without undesirable surface coatings and other contaminants (Lee *et al.*, 1999). The major problem is that agglomeration of nanoparticles may still occur. When finely divided solid nanostructures are immersed in liquids, they often do not form a stable dispersion. Many of the particles acquire surface charge and tend to physically aggregate together in clusters. Though these particles can be easily re-dispersed in liquids by mechanical vibration/agitation, they soon cluster together again to form large aggregates that settle down quickly from the suspension. This can be taken care of by adding dispersants during preparation, which form a coating layer around the nanoparticles that is sterically bulky, thus keeping the nanoparticles separated from each other resulting in the formation of stable nanofluids. A detailed summary of nanofluids prepared by a two-step process with and without the use of surfactants is given in the Appendix (Section 11.7).

11.2.3 Other processes

While most nanofluid productions to date have used one of the above (single-step or two-step) techniques, other techniques are also available depending on the particular combination of nanoparticle material and fluid. For example, nanoparticles with specific geometries, densities, porosities, charge, and surface chemistries can be synthesized by templating, electrolytic metal deposition, layer-by-layer assembly, microdroplet drying, and other colloid chemistry techniques. Another process is the shape- and size-controlled synthesis of nanoparticles at room temperature (Cao *et al.*, 2006). The structural characteristics of nanoparticles such as mean particle size, particle size distribution, and shape depend on the synthesis method, which can provide an opportunity to ensure good control over such physical characteristics. These characteristics for nanoparticles in suspensions cannot be easily measured. This fact could account for some of the discrepancies in thermal properties reported in the literature among different experimenters (Yu *et al.*, 2007).

11.3   The thermal conductivity of nanofluids

Thermal conductivity is the most important intrinsic parameter to demonstrate the enhancement potential of heat transfer in nanofluids. It has been shown that the thermal conductivity of a nanofluid is influenced by the heat transfer properties of the base fluid and identity/composition, volume fraction, size, shape of the nanoparticles suspended in the liquid (Xuan and Roetzel, 2000). It is also intuitive to anticipate that spatial and temporal distribution and uniformity of dispersed nanoparticles should affect the thermal conductivity. Until now, the development of a comprehensive theory to predict the thermal conductivity of nanofluids has remained elusive, although some attempts and propositions have been made to calculate the apparent conductivity of a two-phase mixture (Xuan and Li, 2000). Most of the data related to the thermal conductivity of nanofluids are consolidated in the Appendix. The volume of experimental research carried out on different particle material and base fluid combinations, the thermal conductivity enhancement of nanofluids for these particle–fluid combinations, and the techniques utilized for the measurement of thermal conductivity by several researchers are provided in the Appendix as extensively as possible.

11.3.1 Effect of particle concentration

As reported in the Appendix (Section 11.7), it is evident that the most extensive research has concerned the effect of nanoparticle loading concentration on the thermal conductivity of nanofluids. Different research groups have reported thermal conductivity ratio data for different nanofluids measured by different techniques at different temperatures and other conditions such as particle size, pH level of the solution, and with/ without addition of surfactants. Most of the thermal conductivity enhancement ratio data for different nanofluids reported by different research groups as a function of particle concentration have been graphically represented in Fig. 11.1. The references in the legends of the graph represent the first few letters of the first author of the article followed by the last two digits of the year of publication. The same convention has been followed for other figures in this chapter.

As is evident from Fig. 11.1, the general trend shows a gradual increase in the thermal conductivity ratio of nanofluids with an increase in nanoparticle loading concentration. Most of the studies report the investigation of thermal conductivity of nanofluids for a volume fraction less than 1% (shown as the magnified view of Fig. 11.1) since loading the nanofluids with higher concentrations shows an adverse effect on the stability of nanofluids. Many results show the thermal conductivity of the nanofluid to be almost 2.8 or 3 times that of the base fluid. The thermal conductivity ratio for a particular particle–base fluid combination has been observed to vary as a function of concentration for different reports by different research groups. A comparative study of Al_2O_3–water nanofluids (shown in Fig. 11.2) validates the fact that although some results show some identical trends (Wen and Ding, 2005; Yoo et al., 2007; Sundar and Sharma, 2008), large variations can be observed in the results. This variation might be attributed to the different conditions and measurement techniques and differences in particle size, purity, surrounding conditions, and nanoparticle synthesis process.

11.3.2 Effect of nature of dispersed particles

The effect of different nature/type of particles on the thermal conductivity of nanofluids is shown in Fig. 11.3. It can be observed that the metal particles (Au and Cu) dispersed nanofluids show a higher thermal conductivity ratio than the metal oxide particles (Al_2O_3, TiO_2, SiC, CuO) dispersed nanofluids. This is an expected result since metals possess a higher thermal conductivity than metal oxides. It can also be seen that the thermal conductivity ratio achieved by the metal oxide dispersed nanofluids at a much higher concentration can be achieved by the metal dispersed nanofluids at a much

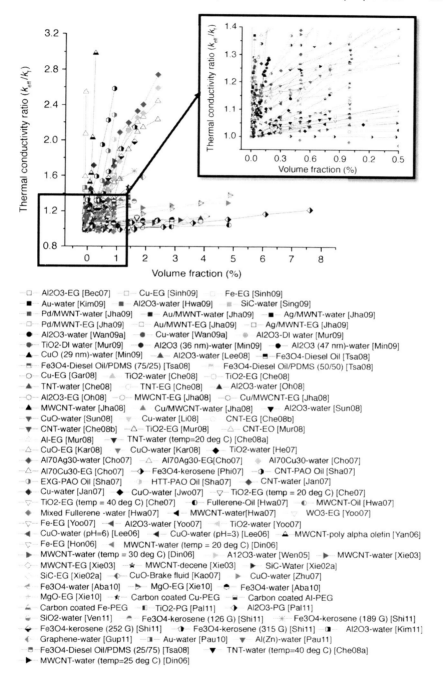

11.1 Thermal conductivity as a function of concentration for different nanofluids.

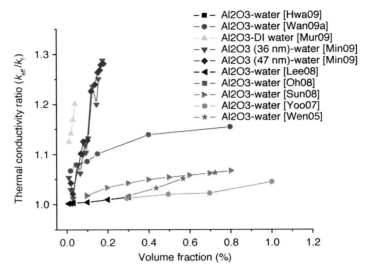

11.2 Thermal conductivity ratio as a function of particle concentration for Al_2O_3–water nanofluids reported by various research groups.

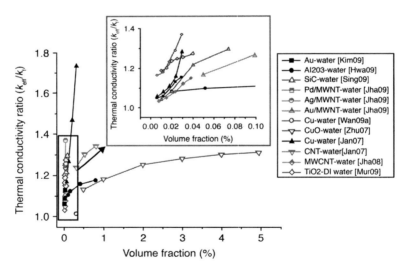

11.3 Thermal conductivity ratio of nanofluids with different types of nanoparticle dispersion.

lower concentration. It may be pointed out that in comparison to metal and metal oxide dispersed nanofluids, carbon nanotube dispersed nanofluids give much higher thermal conductivity enhancement at the same concentration. Thus, it is obvious that the thermal conductivity of the dispersed nanoparticles strongly influences the overall conductivity of the nanofluid.

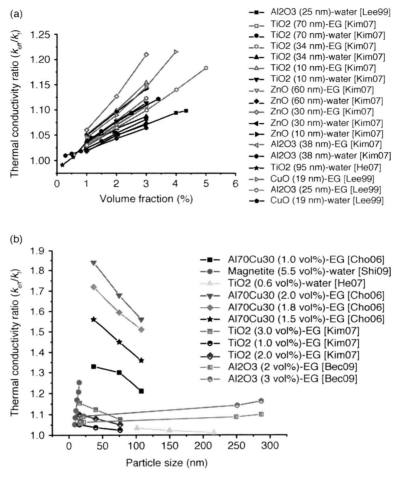

11.4 Thermal conductivity of nanofluids for different sized particles as a function of concentration (a) and as a function of particle size (b).

11.3.3 Effect of particle size

For analyzing the effect of particle size on the thermal conductivity ratio of nanofluids, mainly only spherically shaped particles have been reported. The thermal conductivity ratio as a function of volume concentration for different particle sizes is shown in Fig. 11.4(a). It can be easily seen from the figure that for the same type of particle and base fluid medium, the thermal conductivity ratio for a smaller sized particle is much higher than that for a larger sized particle. This observation is valid for all types of particles and base fluid mediums. This is more clearly demonstrated in Fig. 11.4(b), which graphically represents the thermal conductivity ratio as a function of particle size. The general trend is a decrease in thermal conductivity ratio with an increase in particle size (Chopkar *et al.*, 2006; He *et al*, 2007; Kim *et al*, 2007). But contradicting the trend, Shima *et al.* (2009) and Beck *et al.* (2009) report that for magnetite–water nanofluids, the thermal conductivity ratio increases considerably with an increase in particle size. In most of these cases, particle agglomeration being a key factor that cannot be determined from experiment or theory may contribute largely to the ambiguity of results produced. The accuracy of the particle size reported is also questionable as in many cases the researchers report the data available from the manufacturers.

11.3.4 Effect of particle shape

A few articles concerning the effect of particle shape on the thermal conductivity ratio of nanofluids have been published. The thermal conductivity ratio for different particle shapes as a function of volume concentration is shown in Fig. 11.5. From all of the studies in Fig. 11.5 (a)–(c) it is quite clear that cylindrical particles show the maximum thermal conductivity ratio enhancement irrespective of the kind of particle dispersion. It may be assumed that cylinders form a mesh of elongated particles that conducts heat through the base fluid. This fact is further validated by the study by Jiang *et al.* (2009) which shows that by increasing the aspect ratio of the cylinders, even for the smallest diameter, the thermal conductivity enhancement is the highest. However, cylindrical particles often tend to form agglomerates, are difficult to produce, and are liable to disintegrate into smaller parts during ultrasonic vibration. Hence nanofluids with spherical nanoparticle dispersion are the easiest to prepare and most widely exploited and reported.

11.3.5 Effect of type of base fluid

Studies involving the use of several base fluid media, namely water, ethylene glycol, pump oil, ethanol, refrigerant, and toluene (see Fig. 11.1) for the preparation of nanofluids have been reported. Among these fluids, water and ethylene glycol are the most extensively used. The influence of base fluid medium on the thermal conductivity ratio of nanofluids can be seen in Fig. 11.4(a). It may be pointed out that, in general, the ethylene glycol based nanofluids show higher thermal conductivity ratio compared to water based nanofluids, all other parameters being kept constant. Though water has the highest thermal conductivity among fluids, nanofluids with other base fluid media show a higher thermal conductivity ratio. This result is encouraging because heat transfer enhancement is often most desired when fluids with poorer heat transfer properties are utilized. Ethylene glycol alone is a relatively poor heat transfer fluid compared to water, and mixtures of

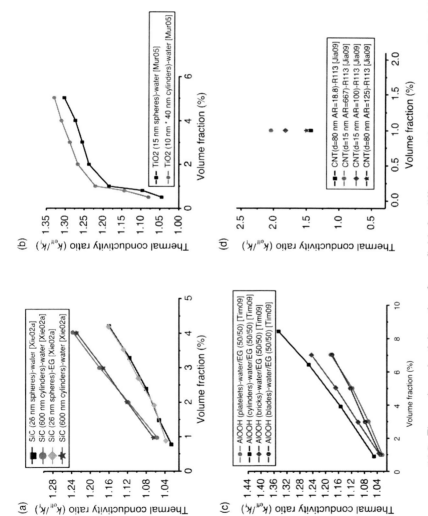

11.5 Thermal conductivity ratio of nanofluids for different particle shapes reported by (a) Xie et al. (2002a), (b) Murshed et al. (2005), (c) Timofeeva et al. (2009), and (d) Jiang et al. (2009).

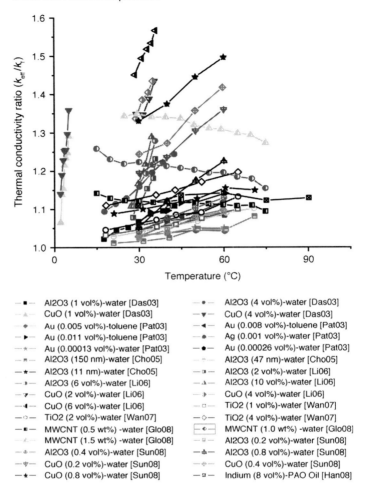

11.6 Thermal conductivity ratio of different nanofluids as a function of temperature.

ethylene glycol and water fall between the two in heat transfer extremes. Though this is the general trend, it is not the overall trend since some articles report the thermal conductivity ratio of water based nanofluids as higher than that of ethylene glycol based nanofluids (Chopkar *et al.*, 2006, 2007).

11.3.6 Effect of temperature

Temperature has a strong influence on the thermal conductivity ratio of nanofluids. The effect of temperature on the thermal conductivity ratio for different nanofluids is graphically represented in Fig. 11.6. Experimental data from several research articles measuring the thermal conductivity as a function of temperature are collated and plotted in a single graph to show a general trend for the nature of variation of thermal conductivity ratio. As is evident from the graph, the general trend follows a linearly increasing thermal conductivity ratio with an increase in temperature. Most researchers reporting thermal conductivity ratio as a function of temperature used Ag, Au, Al_2O_3, and CuO as the dispersoid. The trend of a gradual increase in thermal conductivity ratio with an increase in temperature is encouraging for engine and heat exchanger applications in the transportation industry, where fluids operate at elevated temperatures. However, studies investigating the influence of temperature in the reverse trend of variation (with a decrease in temperature at sub-zero regime) have not been attempted or reported. Such investigations are warranted to probe the possible role of Brownian motion on the enhancement of thermal conductivity of nanofluid.

11.4   Modeling of thermal conductivity

Experimental investigations of the thermal conductivity of nanofluids indicate a substantial increment that needs to be understood and explained through suitable theoretical models. It has also been pointed out that the thermal conductivity of nanofluids depends on the concentration, size, shape/morphology of particles, the base fluid medium, and operating temperature. Originally, the thermal conductivity of nanofluids was attributed to formulations involving the effective thermal conductivity of mixtures from continuum formulations that typically involve only the particle size/shape and volume fraction and assume diffusive heat transfer in both fluid and solid phases. This approach can give a good prediction for micrometer or larger-size solid/fluid systems, but it fails to explain the anomalously high thermal conductivity of nanofluids. Details of previously developed models are given in Table 11.2.

Since most classical models fail to explain the thermal conductivity enhancement of nanofluids, the nanofluid community has intensively investigated this phenomenon and a number of mechanisms have been proposed. Keblinski *et al.* (2002) investigated the possible factors that may enhance the thermal conductivity in nanofluids such as particle size, Brownian motion, the clustering of particles, and the existence of a nanolayer between the nanoparticles and the base fluid. The possible mechanisms are schematically shown in Fig. 11.7. Based on this study several other models have been developed by researchers to validate the thermal conductivity enhancement of nanofluids.

Table 11.2 Conventional models predicting the thermal conductivity of solid–liquid mixtures

Researchers	Classical models	Remarks
Maxwell (1904)	$\frac{k_{eff}}{k_f} = 1 + 3\varphi \frac{k_p-k_f}{2k_f+k_p-\varphi(k_p-k_f)}$	Depends on the thermal conductivities of both phases and volume fraction of solid
Hamilton and Crosser (1962)	$\frac{k_{eff}}{k_f} = \frac{k_p+(n-1)k_f-(n-1)\varphi(k_f-k_p)}{k_p+(n-1)k_f+\varphi(k_f-k_p)}$	Valid for both the spherical and cylindrical particles and $n = 3/\psi$ where ψ is the particle sphericity
Bruggeman (1935)	$(1-\varphi)\frac{k_f-k_{eff}}{2k_{eff}+k_f} + \varphi\frac{k_p-k_{eff}}{2k_{eff}+k_p} = 0$	Valid for spherical particles and considered interaction between particles
Wasp *et al.* (1977)	$\frac{k_{eff}}{k_f} = \frac{k_p+2k_f-2\varphi(k_f-k_p)}{k_p+2k_f+\varphi(k_f-k_p)}$	Special case of Hamilton and Crosser's model with sphericity $\psi = 1$

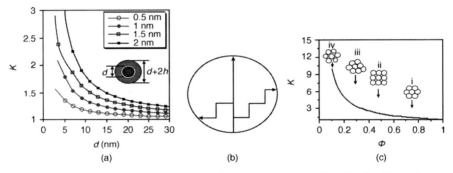

11.7 Possible mechanisms of *k* enhancement (after Keblinski *et al.* (2002)): (a) enhancement of *k* due to the formation of highly conductive layer–liquid structure at liquid/particle interface; (b) ballistic and diffusive phonon transport in a solid particle; (c) enhancement of *k* due to increased effective ϕ of highly conducting clusters.

11.4.1 Effect of liquid layering at the liquid/particle interface

Many researchers have used the concept of a liquid/solid interfacial layer to explain the anomalous improvement of the thermal conductivity in nanofluids. Yu and Choi (2003, 2004) suggested models, based on conventional theory, which consider a liquid molecular layer around the nanoparticles. The theory proposed the existence of a solid-like nanolayer at the interface of the solid particle and bulk liquid which acts as a thermal bridge, thus enhancing thermal conductivity (Fig. 11.8). In these models, the thermal conductivity and volume fraction of the nanoparticles were replaced by those of the equivalent particles, i.e. particles with nanolayers. The models were formulated as follows.

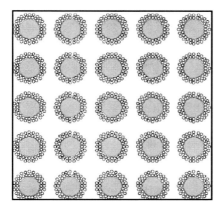

11.8 Schematic cross-section of nanofluid structure consisting of nanoparticles, bulk liquid, and nanolayers at the solid/liquid interface (Yu and Choi, 2003).

Modified Hamilton and Crosser:

$$\frac{k_{eff}}{k_f} = \frac{k_{pe} + 2k_f - 2\phi(k_{pe} - k_f)(1+\beta)^3}{k_{pe} + 2k_f - \phi(k_{pe} - k_f)(1+\beta)^3} \quad [11.1]$$

where

$$k_{pe} = \frac{[2(1-\gamma) + (1+\beta)^3(1+2\gamma)]\gamma}{-(1-\gamma) + (1+\beta)^3(1+2\gamma)} k_p$$

$\beta = h/r$ and $\gamma = k_{layer}/k_p$ for layer thickness of h and particle radius r.
Modified Maxwell:

$$\frac{k_{eff}}{k_f} = 1 + \frac{n\phi_{eff}A}{1 - \phi_{eff}A} \quad [11.2]$$

where

$$A = \frac{1}{3} \sum_{j=a,b,c} \frac{k_{pj} - k_f}{k_{pj} + (n-1)k_f}$$

Xue and Xu (2005) derived an expression for the effective thermal conductivity of nanofluids taking into consideration the thermal conductivity of the solid and liquid, their relative volume fraction, particle size, and interfacial properties. They calculated the thermal conductivity of the complex particle–liquid structure and adopted the model of Bruggeman

(1935) to predict the thermal conductivity of nanofluids as

$$\left(1 - \frac{\phi}{\alpha}\right)\frac{k_{eff} - k_f}{2k_{eff} + k_f} + \frac{\phi}{\alpha}\frac{(k_{eff} - k_2)(2k_2 + k_p) - \alpha(k_p - k_2)(2k_2 + k_{eff})}{(2k_{eff} + k_2)(2k_2 + k_p) + 2\alpha(k_p - k_2)(k_2 - k_{eff})} \quad [11.3]$$

where

$$\alpha = [r/(r + h)]^3$$

and k_2 is the thermal conductivity of the layer.

Murshed et al. (2008) modeled the thermal conductivity of nanofluids considering the interfacial liquid layering concept with the assumption that temperature fields are continuous in the particle, interfacial layer and liquid, and at the interfacial boundaries and the heat fluxes across the interfaces (particle/layer and layer/fluid) are also continuous:

$$k_{eff} = \frac{(k_p - k_{lr})\phi k_{lr}(2\gamma_1^3 - \gamma^3 + 1) + (k_p + 2k_{lr})\gamma_1^3[\phi\gamma^3(k_{lr} - k_f) + k_f]}{\gamma_1^3(k_p + 2k_{lr}) - (k_p - k_{lr})\phi(\gamma_1^3 + \gamma^3 - 1)} \quad [11.4]$$

$$k_{eff} = \frac{(k_p - k_{lr})\phi k_{lr}(\gamma_1^2 - \gamma^2 + 1) + (k_p + k_{lr})\gamma_1^2[\phi\gamma^2(k_{lr} - k_f) + k_f]}{\gamma_1^2(k_p + 2k_{lr}) - (k_p - k_{lr})\phi(\gamma_1^2 + \gamma^2 - 1)} \quad [11.5]$$

Equation 11.4 is applicable for spherical particles while equation 11.5 is applicable for cylindrical particles and $\gamma = 1 + h/a$ and $\gamma_1 = 1 + h/2a$.

In another study, Tillman and Hill (2006) estimated the interfacial layer thickness by considering a steady-state heat conduction condition and assuming the thermal conductivity of the layer as a function of the distance from the particle center. Since most of the earlier reports assumed the layer thickness, this brought a new dimension to the study. The effective thermal conductivity was calculated as:

$$\frac{k_{eff} - k_f}{k_{eff} + 2k_f} = \frac{(k_a - k_f)(2k_a + k_p)\delta + (2k_a + k_f)(k_p - k_a)}{(k_a + 2k_f)(2k_a + k_p)\delta + (2k_a - k_f)(k_p - k_a)}\phi \quad [11.6]$$

where k_a is the nanolayer thermal conductivity and δ is the ratio of the outer and inner interface of the nanolayer.

11.4.2 Effect of Brownian motion

When Keblinski et al. (2002) proposed a heat transfer mechanism in nanofluids, it was summarized that the movement of particles due to Brownian motion was too slow to transport a significant amount of heat through the nanofluid. Later, Jang and Choi (2004) developed a model based on kinetics, thermal diffusion, and Brownian motion of the particles.

The effective thermal conductivity was modeled as:

$$\frac{k_{\text{eff}}}{k_{\text{f}}} = (1 - \phi) + \phi\frac{k_{\text{p}}}{k_{\text{f}}} + 3c\frac{r_{\text{f}}}{r_{\text{p}}}\phi\left(\frac{k_{\text{B}}T}{3\pi\mu_{\text{f}}\phi_{\text{f}}l_{\text{f}}}\right)^2 \text{Pr} \qquad [11.7]$$

The empirical parameter c is a limiting factor for predicting the thermal conductivity of nanofluids. This model was modified by Jang and Choi (2007) considering nano-convection induced by Brownian motion.

Koo and Kleinstreuer (2004) predicted the thermal conductivity of CuO nanofluids modifying the equation of Maxwell (1904) and using the Brownian motion effect. The thermal conductivity was predicted as:

$$k_{\text{eff}} = k_{\text{f}} + 3\phi\frac{k_{\text{p}} - k_{\text{f}}}{2k_{\text{f}} + k_{\text{p}} - \phi(k_{\text{p}} - k_{\text{f}})}k_{\text{f}}$$

$$+ 5 \times 10^4 \beta\rho_{\text{f}}C_{\text{pf}}\phi\left(\frac{k_{\text{B}}T}{2\rho_{\text{p}}r_{\text{p}}}\right)^{1/2}[(-134.63 + 1722.3\phi)$$

$$+ (0.4705 - 6.04\phi)T] \qquad [11.8]$$

where β is related to the Brownian motion of the nanoparticle and empirically determined as:

$$\beta = 0.0137(100\phi)^{-0.8229} \quad \phi < 0.01$$
$$\beta = 0.0011(100\phi)^{-0.7272} \quad \phi > 0.01 \qquad [11.9]$$

On the other hand, Evans *et al.* (2006) used kinetic theory to demonstrate that the hydrodynamic effects associated with Brownian motion of the particles have only a minor effect on the thermal conductivity of nanofluids. This was supported by molecular dynamic simulation studies considering suitable parameters required for the simulation. Prasher *et al.* (2006b) argued that Keblinski *et al.* (2002) and Evans *et al.* (2006) had not considered the energy transport due to convection caused by the Brownian motion of the particles. Accordingly, they analyzed the problem considering convection and the effective thermal conductivity was expressed on the lines of the Maxwell–Garnet model as:

$$\frac{k_{\text{eff}}}{k_{\text{f}}} = \left(1 + \frac{\text{Re}.\text{Pr}}{4}\right)\left(\frac{[k_{\text{p}}(1 + 2\alpha) + 2k_{\text{m}}] + 2\phi[k_{\text{p}}(1 - \alpha) - k_{\text{m}})]}{[k_{\text{p}}(1 + 2\alpha) + 2k_{\text{m}}] - \phi[k_{\text{p}}(1 - \alpha) - k_{\text{m}})]}\right) \qquad [11.10]$$

where $\alpha = 2R_{\text{b}}k_{\text{m}}/d$ is the nanoparticle Biot number, d is the particle diameter, and R_{b} is the interfacial thermal resistance.

Murshed *et al.* (2009) predicted the thermal conductivity of nanofluids considering static and dynamic effects. In considering the dynamic effect on the thermal conductivity of nanofluids, a modified effective diffusion

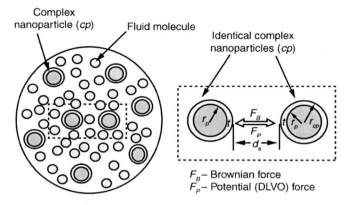

11.9 Dynamic mechanisms of nanoparticles in base fluid (Murshed *et al.*, 2009).

coefficient was used. Also, both Brownian and potential forces were considered on the complex nanoparticles (Fig. 11.9). This led to a very complicated model for determining the thermal conductivity of nanofluids, the details of which have been given elsewhere (Murshed *et al.*, 2009).

11.4.3 Effect of clustering and aggregate formation of nanoparticles

After the proposal of Keblinski *et al.* (2002) regarding the effect of clustering being a mechanism for enhanced thermal heat conduction in nanofluids, Wang *et al.* (2003) used a fractal model to predict the thermal conductivity of nanofluids taking into account the effect of clustered nanoparticles in suspension. The researchers made use of the effective medium theory and the concept of fractal dimensions for nanoparticle clusters to predict the effective thermal conductivity of nanofluids utilizing the model of Bruggeman (1935) (equation 11.3). In the proposed model, the volume concentration ϕ was replaced by the fractal volume fraction $f(r)$:

$$f(r) = \left(\frac{r}{a}\right)^{D_{fl}-3} \qquad [11.11]$$

to get the effective thermal conductivity of the cluster. The effective medium/Maxwell–Garnett theory was modified to predict the thermal conductivity of nanofluids as:

$$\frac{k_{eff}}{k_f} = \left(\frac{(1-\phi) + 3\phi \int_0^\infty \frac{k_{cl}(r)n(r)}{k_{cl}(r)+2k_f}\,dr}{(1-\phi) + 3\phi \int_0^\infty \frac{k_f n(r)}{k_{cl}(r)+2k_f}\,dr}\right) \qquad [11.12]$$

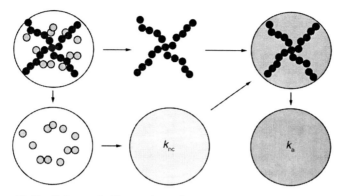

11.10 Schematic illustration of a single aggregate consisting of the backbone (black circles) and dead ends (gray circles). The aggregate is decomposed into dead ends with the fluid and the backbone (Prasher *et al.*, 2006c).

where $n(r)$ is the radius distribution function, which represents the fractal characteristics of the space distribution of clusters.

Prasher *et al.* (2006c) analyzed the effects of aggregation and its kinetics to predict the effective thermal conductivity of nanofluids. They suggested that a fractal cluster is embedded within a sphere of radius equal to r and is composed of a few approximately linear chains called the backbone of the cluster, with other particles called dead ends (Fig. 11.10). The volume fraction of particles belonging to dead ends was calculated as $\phi_{nc} = f$ $(r) - \phi_c$ in which $\phi_c = (r/a)^{d_f - 3}$ is the volume fraction backbone particles.

The thermal conductivity of the aggregate due to dead end particles is calculated from the Bruggeman (1935) equation as:

$$(1 - \phi_{nc}) \frac{k_f - k_{nc}}{k_f + 2k_{nc}} + \phi_{nc} \frac{k_p - k_{nc}}{k_p - 2k_{nc}} = 0 \qquad [11.13]$$

Assuming the backbone particles to form randomly oriented cylindrical chains, the model of Nan *et al.* (2004) can be used to predict the thermal conductivity of an aggregate sphere with both chains and dead ends as:

$$k_a = k_{nc} \frac{3 + \phi_c[2\beta_{11}(1 - L_{11}) + \beta_{33}(1 - L_{33})]}{3 - \phi_c(2\beta_{11}L_{11} + \beta_{33}L_{33})} \qquad [11.14]$$

where

$$L_{11} = 0.5p^2/(p^2 - 1) - 0.5p\cosh^{-1}p/(p^2 - 1)^{1.5}$$
$$L_{33} = 1 - 2L_{11}$$
$$\beta_{11} = \frac{k_{nc} - k_f}{k_f + L_{11}(k_{nc} - k_f)}$$
$$p = r/a$$

The thermal conductivity of the whole nanofluid system has been calculated from the Maxwell–Garnet model as:

$$\frac{k_{eff}}{k_f} = \frac{k_a + 2k_f + 2\phi_a(k_a - k_f)}{k_a + 2k_f - \phi_a(k_a - k_f)} \qquad [11.15]$$

Evans *et al.* (2008) used the theory of Prasher *et al.* (2006c) and compared the effective thermal conductivity predicted from the fractal model using a random walker Monte Carlo algorithm. It was predicted that thermal conductivity enhancement due to aggregation was also strongly dependent on the chemical dimension of the aggregates and the radius of gyration of the aggregate. Wang *et al.* (2009c) proposed a statistical clustering model to determine the macroscopic characteristics of clusters. It has been suggested that the thermal conductivity of a nanofluid can be estimated from the existing effective medium theory without considering the fractal model reported earlier.

11.4.4 Other models

In several other models, molecular dynamic simulation has been utilized to predict the thermal conductivity of nanofluids. For instance, Sarkar and Selvam (2007) used the equilibrium molecular dynamic simulation method utilizing the Green–Kubo formulation to predict the thermal conductivity of copper and argon nanofluids as a function of particle concentration. In another study, Sankar *et al.* (2008) predicted the thermal conductivity enhancement of platinum–water nanofluids using equilibrium molecular dynamic simulation as a function of nanoparticle concentration and showed significant enhancement from a particle concentration of 1 to 7%.

11.5   Summary and future trends

The enhancement in thermal conductivity of nanofluids has been observed to be dependent on a combination of factors such as concentration of nanoparticles dispersed in the base fluid, operating temperature, size of the nanoparticles, and type of surfactant used for preparation of the nanofluid.

The level of enhancement observed for many different kinds of nanofluids can hardly be explained by existing theoretical models in the literature. A single unified theory that may explain the many-fold increase in thermal conductivity still remains elusive. For wide-scale application, the effects of erosion, particle settling, and agglomeration need to be studied in detail. The agglomeration of particles in a nanofluid is aggravated by the two-step process of producing nanofluids where powders are added to liquids. The dispersion and suspension of nanoparticles in a fluid pose a difficult colloidal chemistry problem, and considerable work remains to be done if the two-step process is ever to develop into large-scale production. (The two-step process is currently the most economical way to produce nanofluids and has good potential for scale-up to commercial production levels.) Better characterization of nanofluids is also important for developing engineering designs based on the work of multiple research groups, and fundamental theories to guide this effort should be improved. Important features for commercialization must be addressed, including particle settling, particle agglomeration, surface erosion, and large-scale nanofluid production at acceptable cost.

11.6   References

Abareshi, M., Goharshadi, E. K., Zebarjad, S. M., Fadafan, H. K. and Youssefi, A., (2010), Fabrication, characterization and measurement thermal conductivity of Fe_3O_4 nanofluids, *J. Magn. Mag. Mat.*, Vol. 322, pp. 3895–3901.

Ali, F. M., Yunus, W. M. M., Moksin, M. M. and Talib, Z. A., (2010), The effect of volume fraction concentration on the thermal conductivity and thermal diffusivity of nanofluids: Numerical and experimental, *Rev. Sci. Instrum.*, Vol. 81, p. 074901.

Assael, M. J., Chen, C.-F., Metaxa, I. and Wakeham, W. A. (2004), Thermal conductivity of suspensions of carbon nanotubes in water, *Int. J. Thermophys.*, Vol. 25, pp. 971–985.

Assael, M. J., Metaxa, I. N., Arvanitidis, J., Christofilos, D. and Lioutas, C. (2005), Thermal conductivity enhancement in aqueous suspensions of carbon multi-walled and double-walled nanotubes in the presence of two different dispersants, *Int. J. Thermophys.*, Vol. 26, pp. 647–664.

Assael, M., Metaxa, J. I. N., Kakosimos, K. and Constantinou, D. (2006), Thermal conductivity of nanofluids – experimental and theoretical, *Int. J. Thermophys.*, Vol. 27, No. 4, pp. 999–1017.

Beck, M. P., Sun, T. A. and Teja, S. (2007), The thermal conductivity of alumina nanoparticles dispersed in ethylene glycol, *Fluid Phase Equilibr.*, Vol. 260, pp. 275–278.

Beck, M. P., Yuan, Y., Warrier, P. and Teja, A. S. (2009), The effect of particle size on the thermal conductivity of alumina nanofluids, *J. Nanopart. Res.*, Vol. 11, pp. 1129–1136.

Bruggeman, D. A. G. (1935), Berechnung verschiedener physikalisher Konstanten von heterogenen Substanzen: I. Dielektrizitatskonstanten und Leitfahigkeiten der Mischkorper aus isotropen Substanzen, *Ann. Physik*, Vol. 24, pp. 636–664.

Cao, H. L., Qian, X. F., Gong, Q., Du, W. M., Ma, X. D. and Zhu, Z. K. (2006), Shape- and size- controlled synthesis of nanometer zno from a simple solution route at room temperature, *Nanotechnology*, Vol. 17, pp. 3632–3636.

Chang, H. and Chang, Y. C. (2008), Fabrication of Al$_2$O$_3$ nanofluid by a plasma arc nanoparticles synthesis system, *J. Mater. Process. Tech.*, Vol. 207, pp. 193–199.

Chang, M.-H., Liu, H.-S. and Tai, C. Y. (2011), Preparation of copper oxide nanoparticles and its application in nanofluid, *Powder Technol.*, Vol. 207, pp. 378–386.

Chen, H., Ding, Y. He, Y. and Tan, C. (2007), Rheological behaviour of ethylene glycol based titania nanofluids, *Chem. Phys. Lett.*, Vol. 444, pp. 333–337.

Chen, H., Yang, W., He, Y., Ding, Y., Zhang, L., Tan, C., Lapkin, A. A. and Bavykin, D. V. (2008a), Heat transfer and flow behaviour of aqueous suspensions of titanate nanotubes (nanofluids), *Powder Technol.*, Vol. 183, pp. 63–72.

Chen, L., Xie, H., Li, Y. and Yu, W. (2008b), Nanofluids containing carbon nanotubes treated by mechanochemical reaction, *Thermochim. Acta*, Vol. 477, pp. 21–24.

Choi, C., Yoo, H. S. and Oh, J. M. (2008), Preparation and heat transfer properties of nanoparticle-in-transformer oil dispersions as advanced energy-efficient coolants, *Curr. Appl. Phys.*, Vol. 8, pp. 710–712.

Choi, S. U. S. (1995), Enhancing thermal conductivity of fluids with nanoparticles. In *Proceedings of the 1995 ASME International Mechanical Engineering Congress and Exposition*, San Francisco, CA, USA.

Choi. S. U. S., Zhang, Z. G., Yu, W., Lockwood, F. E. and Grulke, E. A. (2001), Anomalous thermal conductivity enhancement in nanotube suspensions, *Appl. Phys. Lett.*, Vol. 79, pp. 2252–2254.

Chon, C. H., Kihm, K. D., Lee, S. P. and Choi, S. U. S. (2005), Empirical correlation finding the role of temperature and particle size for nanofluid (Al$_2$O$_3$) thermal conductivity enhancement, *Appl. Phys. Lett.*, Vol. 87, p. 153107.

Chopkar, M., Das, P. K. and Manna, I. (2006), Synthesis and characterization of nanofluid for advanced heat transfer applications, *Scripta Mater.*, Vol. 55, pp. 549–552.

Chopkar, M., Kumar, S., Bhandari, D. R., Das, P. K. and Manna, I. (2007), Development and characterization of Al$_2$Cu and Ag$_2$Al nanoparticle dispersed water and ethylene glycol based nanofluid, *Mat. Sci. Eng. B*, Vol. 139, pp. 141–148.

Das, S. K., Putra, N., Thiesen, P. and Roetzel, W. (2003), Temperature dependence of thermal conductivity enhancement for nanofluids. *Trans. ASME, J. Heat Trans.*, Vol. 125, pp. 567–574.

Ding, Y., Alias, H., Wen, D. and Williams, R. A. (2006), Heat transfer of aqueous suspensions of carbon nanotubes (CNT nanofluids), *Int. J. Heat Mass Tran.*, Vol. 49, pp. 240–250.

Eastman, J. A., Choi, S. U. S., Li, S., Yu, W. and Thompson, L. J. (2001), Anomalously increased effective thermal conductivity of ethylene glycol-based nanofluids containing copper nanoparticles, *Appl. Phys. Lett.*, Vol. 78, pp. 718–720.

Evans, W., Fish, J. and Keblinski, P. (2006), Role of Brownian motion hydrodynamics on nanofluid thermal conductivity, *Appl. Phys. Lett.*, Vol. 88, p. 093116.

Evans, W., Prasher, R., Fish, J., Meakin, P., Phelan, P. and Keblinski, P. (2008), Effect of aggregation and interfacial thermal resistance on thermal conductivity of nanocomposites and colloidal nanofluids, *Int. J. Heat Mass Tran.*, Vol. 51, pp. 1431–1438.

Fuentes, R. G., Rojas, J. A. P., Jimé nez-Pérez, J. L., Ramirez, J. F. S., Orea, A. C. and Alvarez, J. G. M. (2008), Study of thermal diffusivity of nanofluids with bimetallic nanoparticles with Au(core)/Ag(shell) structure, *Appl. Surf. Sci.*, Vol. 255, pp. 781–783.

Garg, J., Poudel, B., Chiesa, M., Gordon, J. B., Ma, J. J., Wang, J. B., Ren, Z. F., Kang, Y. T., Ohtani, H., Nanda, J., McKinley, G. H. and Chen, G. (2008), Enhanced thermal conductivity and viscosity of copper nanoparticles in ethylene glycol nanofluid, *J. Appl. Phys.*, Vol. 103, p. 074301.

Glory, J., Bonetti, M., Helezen, M., Mayne-L'Hermite, M. and Reynaud, C. (2008), Thermal and electrical conductivities of water-based nanofluids prepared with long multiwalled carbon nanotubes, *J. Appl. Phys.*, Vol. 103, p. 094309.

Gupta, S. S., Siva, V. M., Krishnan, S., Sreeprasad, T. S., Singh, P. K., Pradeep, T. and Das, S. K. (2011), Thermal conductivity enhancement of nanofluids containing grapheme nanosheets, *J. Appl. Phys.*, Vol. 110, p. 084302.

Hamilton, R. L. and Crosser, O. K. (1962), Thermal conductivity of heterogeneous two-component systems, *Ind. Eng. Chem. Fundamen.*, Vol. 1, No. 3, pp. 187–191.

Han, Z. H., Cao, F. Y. and Yang, B. (2008), Synthesis and thermal characterization of phase-changeable indium/polyalphaolefin nanofluids, *Appl. Phys. Lett.*, Vol. 92, p. 243104.

He, Y., Jin, Y., Chen, H., Ding, Y., Cang, D. and Lu, H. (2007), Heat transfer and flow behaviour of aqueous suspensions of TiO_2 nanoparticles (nanofluids) flowing upward through a vertical pipe, *Int. J. Heat Mass Tran.*, Vol. 50, pp. 2272–2281.

Hong, K. S., Hong, T.-K. and Yang H.-S. (2006), Thermal conductivity of Fe nanofluids depending on the cluster size of nanoparticles, *Appl. Phys. Lett.*, Vol. 88, p. 031901.

Hong, T.-K., Yang, H.-S. and Choi, C. J. (2005), Study of the enhanced thermal conductivity of Fe nanofluids, *J. Appl. Phys.*, Vol. 97, p. 064311.

Hwang, K. S., Jang, S. P. and Choi, S. U. S. 2009), Flow and convective heat transfer characteristics of water-based Al_2O_3 nanofluids in fully developed laminar flow regime, *Int. J. Heat Mass Tran.*, Vol. 52, pp. 193–199.

Hwang, Y., Lee, J.K., Lee, C.H., Jung, Y.M., Cheong, S.I., Lee, C.G., Ku, B.C. and Jang, S.P. (2007), Stability and thermal conductivity characteristics of nanofluids, *Thermochim. Acta*, Vol. 455, pp. 70–74.

Hwang, Y., Lee, J.K., Lee, J.K., Jeong, Y.M., Cheong, S.I., Ahn Y.C. and Kim, S.H. (2008), Production and dispersion stability of nanoparticles in nanofluids, *Powder Technol.*, Vol. 186, pp. 145–153.

Hwang, Y., Park, H. S., Lee, J. K. and Jung, W. H. (2006), Thermal conductivity and lubrication characteristics of nanofluids. *Curr. Appl. Phys.*, Vol. 6s 1, pp. e67–e71.

Jana, S., Salehi-Khojin, A. and Zhong, W.-H. (2007), Enhancement of fluid thermal conductivity by the addition of single and hybrid nano-additives, *Thermochim. Acta*, Vol. 462, pp. 45–55.

Jang, S. P. and Choi, S. U. S. (2004), Role of Brownian motion in the enhanced thermal conductivity of nanofluids, *Appl. Phys. Lett.*, Vol. 84, pp. 4316–4318.

Jang, S. P. and Choi, S. U. S. (2007), Effects of various parameters on nanofluid thermal conductivity, *Trans. ASME, J. Heat Trans.*, Vol. 129, pp. 617–623.

Jha, N. and Ramaprabhu, S. (2008), Synthesis and thermal conductivity of copper nanoparticle decorated multiwalled carbon nanotubes based nanofluids, *J. Phys. Chem. C*, Vol. 112, pp. 9315–9319.

Jha, N. and Ramaprabhu, S. (2009), Thermal conductivity studies of metal dispersed multiwalled carbon nanotubes in water and ethylene glycol based nanofluids, *J. Appl. Phys.*, Vol. 106, p. 084317.

Jiang, W., Ding, G. and Peng, H. (2009), Measurement and model on thermal conductivities of carbon nanotube nanorefrigerants, *Int. J. Therm. Sci.*, Vol. 48, pp. 1108–1115.

Jung, J.-Y. and Yoo, J. Y. (2009), Thermal conductivity enhancement of nanofluids in conjunction with electrical double layer (EDL), *Int. J. Heat Mass Tran.*, Vol. 52, pp. 525–528.

Jwo, C.-S., Teng, T.-P. and Chang, H. (2007), A simple model to estimate thermal conductivity of fluid with acicular nanoparticles, *J. Alloy. Comp.*, Vol. 434–435, pp. 569–571.

Kang, H. U., Kim, S. H. and Oh, J. M. (2006), Estimation of thermal conductivity of nanofluid using experimental effective particle volume, *Exp. Heat Transfer*, Vol. 19, pp. 181–191.

Kao, M. J., Lo, C. H., Tsung, T. T., Wu, Y. Y., Jwo, C. S. and Lin, H. M. (2007), Copper-oxide brake nanofluid manufactured using arc-submerged nanoparticle synthesis system, *J. Alloy. Comp.*, Vol. 434–435, pp. 672–674.

Karthikeyan, N. R., Philip, J. and Raj, B. (2008), Effect of clustering on the thermal conductivity of nanofluids, *Mater. Chem. Phys.*, Vol. 109, pp. 50–55.

Keblinski, P., Phillpot, S. R., Choi, S. U. S. and Eastman, J. A. (2002), Mechanisms of heat flow in suspensions of nano-sized particles (nanofluids), *Int. J. Heat Mass Trans.*, Vol. 45, pp. 855–863.

Kim, H. J., Bang, I. C. and Onoe, J. (2009), Characteristic stability of bare Au-water nanofluids fabricated by pulsed laser ablation in liquids, *Opt. Laser. Eng.*, Vol. 47, pp. 532–538.

Kim, S. H., Choi, S. R. and Kim, D. (2007), Thermal conductivity of metal-oxide nanofluids: particle size dependence and effect of laser irradiation, *Trans. ASME, J. Heat. Trans.*, Vol. 129, pp. 298–307.

Kim, S., Kim, C., Lee, W.-H. and Park, S.-R. (2011), Rheological properties of alumina nanofluids and their implication to the heat transfer enhancement mechanism, *J. Appl. Phys.*, Vol. 110, p. 034316.

Koo, J. and Kleinstreuer, C. (2004), A new thermal conductivity model for nanofluids, *J. Nanopart. Res.*, Vol. 6, pp. 577–588.

Kumar, S. A., Meenakshi, K. S., Narashimhan, B. R. V., Srikanth, S. and Arthanareeswaran, G. (2009), Synthesis and characterization of copper nanofluid by a novel one-step method, *Mater. Chem. Phys.*, Vol. 113, pp. 57–62.

Lee, D., Kim, J.-W. and Kim, B. G. (2006), A new parameter to control heat transport in nanofluids: surface charge state of the particle in suspension, *J. Phys. Chem. B*, Vol. 110, pp. 4323–4328.

Lee, J.-H., Hwang, K. S., Jang, S. P., Lee, B. H., Kim, J. H., Choi, S. U. S. and Choi, C. J. (2008), Effective viscosities and thermal conductivities of aqueous nanofluids containing low volume concentrations of Al_2O_3 nanoparticles, *Int. J. Heat Mass Tran.*, Vol. 51, pp. 2651–2656.

Lee, S., Choi, S. U. S., Li, S. and Eastman, J. A. (1999), Measuring thermal conductivity of fluids containing oxide nanoparticles, *Trans. ASME, J. Heat Trans.*, Vol. 121, pp. 280–289.

Li, C. H. and Peterson, G. P. (2006), Experimental investigation of temperature and volume fraction variations on the effective thermal conductivity of nanoparticle suspensions (nanofluids), *J. Appl. Phys.*, Vol. 99, p. 084314.

Li, X. F., Zhu, D. S., Wang, X. J., Wang, N., Gao, J. W. and Li, H. (2008), Thermal conductivity enhancement dependent pH and chemical surfactant for $Cu-H_2O$ nanofluids, *Thermochim. Acta*, Vol. 469, pp. 98–103.

Liu, M., Lin, M., Tsai, C. Y. and Wang, C. (2006), Enhancement of thermal conductivity with cu for nanofluids using chemical reduction method. *Int. J. Heat Mass Tran.*, Vol. 49, pp. 3028–3033.

Lo, C. H., Tsung, T. T., Chen, L. C., Su, C. H. and Lin, H. M. (2005a), Fabrication of copper oxide nanofluid using submerged arc nanoparticle synthesis system (SANSS), *J. Nanopart. Res.*, Vol. 7, pp. 313–320.

Lo, C.-H., Tsung, T.-T. and Chen, L.-C. (2005b), Shape-controlled synthesis of Cu-based nanofluid using submerged arc nanoparticle synthesis system (SANSS), *J. Cryst. Growth*, Vol. 277, pp. 636–642.

Lo, C.-H., Tsung, T.-T. and Lin, H.-M. (2007), Preparation of silver nanofluid by the submerged arc nanoparticle synthesis system (SANSS), *J. Alloy. Comp.*, Vol. 434–435, pp. 659–662.

Marquis, F. D. S. and Chibante, L. P. F. (2005), Improving the heat transfer of nanofluids and nanolubricants with carbon nanotubes, *J. Miner. Met. Mater. Soc.*, Vol. 57, No. 12, pp. 32–43.

Masuda, H., Ebata, A., Teramae, K. and Fishinuma, N. (1993), Alteration of thermal conductivity and viscosity of liquid by dispersing ultra-fine particles (dispersion of y-Al_2O_3, SO_2, and TiO_2 ultra-fine particles), *Netsu Bussei*, Vol. 7, pp. 227–233.

Maxwell, J.C. (1904), *A Treatise on Electricity and Magnetism*, 2nd edn, Oxford University Press, UK.

Mintsa, H. A., Roy, G., Nguyen, C. T. and Doucet, D. (2009), New temperature dependent thermal conductivity data for water-based nanofluids, *Int. J. Therm. Sci.*, Vol. 48, pp. 363–371.

Mishra, A., Ram, S. and Ghosh, G. (2009), Dynamic light scattering and optical absorption in biological nanofluids of gold nanoparticles in poly(vinyl pyrrolidone) molecules, *J. Phys. Chem. C*, Vol. 113, pp. 6976–6982.

Murshed, S. M. S., Leong, K. C. and Yang, C. (2005), Enhanced thermal conductivity of TiO_2-water based nanofluids, *Int. J. Therm. Sci.*, Vol. 44, pp. 367–373.

Murshed, S. M. S., Leong, K.C. and Yang, C. (2008), Investigations of thermal conductivity and viscosity of nanofluids, *Int. J. Therm. Sci.*, Vol. 47, pp. 560–568.

Murshed, S. M. S., Leong, K.C. and Yang, C. (2009), A combined model for the effective thermal conductivity of nanofluids, *Appl. Therm. Eng.*, Vol. 29, pp. 2477–2483.

Nan, C.-W., Liu, G., Lin, Y. and Li, M. (2004), Interface effect on thermal conductivity of carbon nanotube composites, *Appl. Phys. Lett.*, Vol. 85, pp. 3549–3551.

Oh, D.-W., Jain, A., Eaton, J. K., Goodson, K. E. and Lee, J. S. (2008), Thermal conductivity measurement and sedimentation detection of aluminum oxide nanofluids by using the 3w method, *Int. J. Heat Fluid Flow*, Vol. 29, pp. 1456–1461.

Palabiyik, I., Musina, Z., Witharana, S. and Ding, Y. (2011), Dispersion stability and thermal conductivity of propylene glycol-based nanofluids, *J. Nanopart. Res.*, Vol. 13, pp. 5049–5055.

Patel, H. E., Das, S. K., Sundararajan, T., Nair, A. S., George, B. and Pradeep, T. (2003), Thermal conductivity of naked and monolayer protected metal nanoparticle based nanofluids: manifestation of anomalous enhancement and chemical effects, *Appl. Phys. Lett.*, Vol. 83, pp. 2931–2933.

Patel, K., Kapoor, S., Dave, D. P. and Mukherjee, T. (2005), Synthesis of nanosized silver colloids by microwave dielectric heating, *J. Chem. Sci.*, Vol. 117, No. 1, pp. 53–60.

Paul, G., Pal, T. and Manna, I. (2010), Thermo-physical property measurement of nano-gold dispersed water based nanofluids prepared by chemical precipitation technique, *J. Colloid Interf. Sci.*, Vol. 349, pp. 434–437.

Paul, G., Philip, J., Raj, B., Das, P. K. and Manna, I. (2011), Synthesis, characterization, and thermal property measurement of nano-Al95Zn05 dispersed nanofluid prepared by a two-step process, *Int. J. Heat Mass Tran.*, Vol. 54, pp. 3783–3788.

Peñas, J. R. V., de Zárate, J. M. O. and Khayet, M. (2008), Measurement of the thermal conductivity of nanofluids by the multicurrent hot-wire method, *J. Appl. Phys.*, Vol. 104, p. 044314.

Philip, J., Shima, P. D. and Raj, B. (2007), Enhancement of thermal conductivity in magnetite based nanofluid due to chainlike structures, *Appl. Phys. Lett.*, Vol. 91, p. 203108.

Phuoc, T. X., Soong, Y. and Chyu, M. K. (2007), Synthesis of Ag-deionized water nanofluids using multi-beam laser ablation in liquids, *Opt. Laser. Eng.*, Vol. 45, pp. 1099–1106.

Phuoc, T. X., Massoudi, M. and Chen, R.-H. (2011), Viscosity and thermal conductivity of nanofluids containing multi-walled carbon nanotubes stabilized by chitosan, *Int. J. Therm. Sci.*, Vol. 50, pp. 12–18.

Prasher, R., Phelan, P. E. and Bhattacharya, P. (2006a), Effect of aggregation kinetics on the thermal conductivity of nanoscale colloidal solutions (nanofluid), *Nano Lett.*, Vol. 6, No. 7, pp. 1529–1534.

Prasher, R., Bhattacharya, P. and Phelan, P. E. (2006b), Brownian-motion-based convective-conductive model for the thermal conductivity of nanofluids, *Trans. ASME, J. Heat Trans.*, Vol. 128, pp. 588–595.

Prasher, R., Evans, W., Meakin, P., Fish, J., Phelan, P. and Keblinski, P. (2006c), Effect of aggregation on thermal conduction in colloidal nanofluids, *Appl. Phys. Lett.*, Vol. 89, p. 143119.

Putnam, S. A., Cahill, D. G., Braun, P. V., Ge, Z. and Shimmin, R. G. (2006), Thermal conductivity of nanoparticle suspensions, *J. Appl. Phys.*, Vol. 99, p. 084308.

Sankar, N., Mathew, N. and Sobhan, C.B. (2008), Molecular dynamics modeling of thermal conductivity enhancement in metal nanoparticle suspensions, *Int. Commun. Heat Mass*, Vol. 35, pp. 867–872.

Sarkar, S. and Selvam, R. P. (2007), Molecular dynamics simulation of effective thermal conductivity and study of enhanced thermal transport mechanism in nanofluids, *J. Appl. Phys.*, Vol. 102, p. 074302.

Shaikh, S., Lafdi, K. and Ponnappan, R. (2007), Thermal conductivity improvement in carbon nanoparticle doped PAO oil: An experimental study, *J. Appl. Phys.*, Vol. 101, p. 064302.

Shima, P. D. and Philip, J. (2011), Tuning of thermal conductivity and rheology of nanofluids using an external stimulus, *J. Phys. Chem. C*, Vol. 115, pp. 20097–20104.

Shima, P. D., Philip, J. and Raj, B. (2009), Role of microconvection induced by Brownian motion of nanoparticles in the enhanced thermal conductivity of stable nanofluids, *Appl. Phys. Lett.*, Vol. 94, p. 223101.

Singh, A. K. and Raykar, V. S. (2008), Microwave synthesis of silver nanofluids with polyvinylpyrrolidone (PVP) and their transport properties, *Colloid Polym. Sci.*, Vol. 286, pp. 1667–1673.

Singh, D., Timofeeva, E., Yu, W., Routbort, J., France, D., Smith, D. and Lopez-Cepero, J. M. (2009), An investigation of silicon carbide-water nanofluid for heat transfer applications, *J. Appl. Phys.*, Vol. 105, p. 064306.

Sinha, K., Kavlicoglu, B., Liu, Y., Gordaninejad, F. and Graeve, O. A. (2009), A comparative study of thermal behavior of iron and copper nanofluids, *J. Appl. Phys.*, Vol. 106, p. 064307.

Slistan-Grijalva, A., Herrera-Urbina, R., Rivas-Silva, J. F., Valos-Borja, M. A., Castillón-Barraza, F. F. and Posada-Amarillas, A. (2005), Assessment of growth of silver nanoparticles synthesized from an ethylene glycol–silver nitrate polyvinylpyrrolidone solution, *Physica E*, Vol. 25, pp. 438–448.

Sundar, L. S. and Sharma, K. V. (2008), Thermal conductivity enhancement of nanoparticles in distilled water, *Int. J. Nanoparticles*, Vol. 1, No. 1, pp. 66–77.

Tamjid, E. and Guenther, B. H. (2010), Rheology and colloidal structure of silver nanoparticles dispersed in diethylene glycol, *Powder Technol.*, Vol. 197, pp. 49–53.

Tao, W., Zhong-yang, L., Shim-song G. and Ke-fa, C. (2007), Preparation of controllable nanofluids and research on thermal conductivity, *J. Zhejiang Un. Eng. Sc.*, Vol. 41, No. 3, pp. 514–518.

Tillman, P. and Hill, J. M. (2006), A new model for thermal conductivity in nanofluids, *ICONN, IEEE*, 673–676. Available [Online] http://ieeexplore.ieee.org/iel5/4143299/4140639/04143487.pdf?arnumber=4143487.

Timofeeva, E. V., Routbort, J. L. and Singh, D. (2009), Particle shape effects on thermophysical properties of alumina nanofluids, *J. Appl. Phys.*, Vol. 106, p. 014304.

Tsai, T.-H., Kuo, L.-S., Chen, P.-H. and Yang, C.-T. (2008), Effect of viscosity of base fluid on thermal conductivity of nanofluids, *Appl. Phys. Lett.*, Vol. 93, p. 233121.

Venerus, D. C. and Jiang, Y. (2011), Investigation of thermal transport in colloidal silica dispersions (nanofluids), *J. Nanopart. Res.*, Vol. 13, pp. 3075–3083.

Wagener, M. and Gunther, B. (1999), Sputtering on liquids – a versatile process for the production of magnetic suspensions, *J. Magn. Magn. Mater.*, Vol. 201, pp. 41–44.

Wang, B.-X., Zhou, L.-P. and Peng, X.-F. (2003), A fractal model for predicting the effective thermal conductivity of liquid with suspension of nanoparticles, *Int. J. Heat Mass Tran.*, Vol. 46, pp. 2665–2672.

Wang, X., Xu, X. and Choi, S. U. S. (1999), Thermal conductivity of nanoparticle–fluid mixture. *J. Thermophys. Heat Tran.*, Vol. 13, pp. 474–480.

Wang, X.-J., Zhu, D.-S. and Yang, S. (2009a), Investigation of pH and SDBS on enhancement of thermal conductivity in nanofluids, *Chem. Phys. Lett.*, Vol. 470, pp. 107–111.

Wang, L. and Wei, X. (2009), Nanofluids: Synthesis, heat conduction, and extension, *Trans. ASME, J. Heat Trans.*, Vol. 131, No. 3, pp. 033102, doi:10.1115/1.3056597.

Wang, B.-X., Sheng, W.-Y. and Peng, X.-F. (2009c), A Novel statistical clustering model for predicting thermal conductivity of nanofluid, *Int. J. Thermophys.*, Vol. 30, pp. 1992–1998.

Wang, Z. L., Tang, D.W., Liu, S., Zheng, X. H. and Araki, N. (2007), Thermal-conductivity and thermal-diffusivity measurements of nanofluids by 3ω method and mechanism analysis of heat transport, *Int. J. Thermophys.*, Vol. 28, pp. 1255–1268.

Wasp, E. J., Kenny, J. P. and Gandhi, R. L. (1977), *Solid–Liquid Flow Slurry Pipeline Transportation, Series on Bulk Materials Handling*. Trans. Tech. Publications, Vol. 1, Clausthal, Germany.

Wen, D. and Ding, Y. (2004a), Experimental investigation into convective heat transfer of nanofluids at the entrance region under laminar flow conditions. *Int. J. Heat Mass Tran.*, Vol. 47, pp. 5181–5188.

Wen, D. and Ding, Y. (2004b), Effective thermal conductivity of aqueous suspensions of carbon nanotubes (carbon nanotube nanofluids). *J. Thermophys. Heat Tran.*, Vol. 18, pp. 481–485.

Wen, D. and Ding, Y. (2005), Formulation of nanofluids for natural convective heat transfer applications. *Int. J. Heat Fluid Fl.*, Vol. 26, pp. 855–864.

Wen, D. and Ding, Y. (2006), Natural convective heat transfer of suspensions of titanium dioxide nanoparticles (nanofluids). *IEEE T. Nanotechnol.*, Vol. 5, pp. 220–227.

Xie, H., Wang, J., Xi, T. and Liu, Y. (2002a), Thermal conductivity of suspensions containing nanosized SiC particles, *Int. J. Thermophys.*, Vol. 23, pp. 571–580.

Xie, H., Wang, J., Xi, T. and Ai, F. (2002b), Thermal conductivity enhancement of suspensions containing nano sized alumina particles, *J. Appl. Phys.*, Vol. 91, pp. 4568–4572.

Xie, H., Wang, J., Xi, T., Liu, Y. and Ai, F. (2002c), Dependence of the thermal conductivity of nanoparticle–fluid mixture on the base fluid, *J. Mater. Sci. Lett.*, Vol. 21, pp. 1469–1471.

Xie, H., Lee, H., Youn, W. and Choi, M. (2003), Nanofluids containing multiwalled carbon nanotubes and their enhanced thermal conductivities, *J. Appl. Phys.*, Vol. 94, pp. 4967–4971.

Xie, H., Yu, W. and Chen, W. (2010), MgO nanofluids: higher thermal conductivity and lower viscosity among ethylene glycol-based nanofluids containing oxide nanoparticles, *J. Exp. Nanosci.*, Vol. 55, pp. 463–472.

Xuan, Y. and Li, Q. (2000), Heat transfer enhancement of nanofluids, *Int. J. Heat Fluid Fl.*, Vol. 21, pp. 58–64.

Xuan, Y. and Roetzel W. (2000), Conceptions for heat transfer correlation of nanofluids, *Int. J. Heat Mass Tran.*, Vol. 43, pp. 3701–3707.

Xue, Q. and Xu, W.-M. (2005), A model of thermal conductivity of nanofluids with interfacial shells, *Mater. Chem. Phys.*, Vol. 90, pp. 298–301.

Yang, B. and Han, Z. H. (2006). Temperature-dependent thermal conductivity of nanorod- based nanofluids, *Appl. Phys. Lett.*, Vol. 89, p. 083111.

Yang, Y., Grulke, E. A., Zhang, Z. G. and Wu, G. (2006), Thermal and rheological properties of carbon nanotube-in-oil dispersions, *J. Appl. Phys.*, Vol. 99, p. 114307.

Yatsuya, S., Tsukasaki, Y., Mihama, K. and Uyeda, R. (1978), Preparation of extremely fine particles by vacuum evaporation onto a running oil substrate, *J. Cryst. Growth*, Vol. 45, pp. 490–494.

Yoo, D.-H., Hong, K. S. and Yang, H.-S. (2007), Study of thermal conductivity of nanofluids for the application of heat transfer fluids, *Thermochim. Acta*, Vol. 455, pp. 66–69.

Yu, Q., Kim, Y. J. and Ma, H. (2008a), Nanofluids with plasma treated diamond nanoparticles, *Appl. Phys. Lett.*, Vol. 92, p. 103111.

Yu, W. and Choi, S. U. S. (2003), The role of interfacial layers in the enhanced thermal of nanofluids: a renovated Maxwell model, *J. Nanopart. Res.*, Vol. 5, No. 1–2, pp. 167–171.

Yu, W. and Choi, S. U. S. (2004), The role of interfacial ayers in the enhanced thermal conductivity of nanofluids: A renovated Hamilton–Crosser model. *J. Nanopart. Res.*, Vol. 6, No. 4, pp. 355–361.

Yu, W., France, D. M., Choi, S. U. S. and Rourborr, J. L. (2007), Review and assessment of nanofluid technology for transportation and other applications. Available [Online] http://www.ipd.anl.gov/anlpubs/2007/05/59282.pdf.

Zhang, H., Wu, Q., Lin, J., Chen, J. and Xu, Z. (2010), Thermal conductivity of polyethylene glycol nanofluids containing carbon coated metal nanoparticles, *J. Appl. Phys.*, Vol. 108, p. 124304.

Zhang, X., Gu, H. and Fujii, M. (2006a), Effective thermal conductivity and thermal diffusivity of nanofluids containing spherical and cylindrical nanoparticles, *J. Appl. Phys.*, Vol. 100, p. 044325.

Zhang, X., Gu, H. and Fujii, M. (2006b), Experimental study on the effective thermal conductivity and thermal diffusivity of nanofluids, *Int. J. Thermophys.*, Vol. 27, No. 2, pp. 569–580.

Zhu, H. T., Lin, Y.S. and Yin, Y.S. (2004), A novel one-step chemical method for preparation of copper nanofluids, *J. Colloid Interf. Sci.*, Vol. 277, pp. 100–103.

Zhu, H., Zhang, C., Liu, S., Tang, Y. and Yin, Y. (2006), Effects of nanoparticle clustering and alignment on thermal conductivities of Fe_3O_4 aqueous nanofluids, *Appl. Phys. Lett.*, Vol. 89, p. 023123.

Zhu, H. T., Zhang, C. Y., Tang, Y. M. and Wang, J. X. (2007), Novel synthesis and thermal conductivity of CuO nanofluid, *J. Phys. Chem. C*, Vol. 111, pp. 1646–1650.

11.7 Appendix: thermal conductivity details of nanofluids prepared by two-step process

Reference	Nanofluid	Concentration (vol %)	Enhancement ratio	Synthesis method	Thermal conductivity measurement technique	Effect studied/notes
Masuda et al. (1993)	Al_2O_3–water (31.85°C)	1.30–4.3	1.11–1.32	Two-step method		Temperature effect
	Al_2O_3–water (46.85°C)	1.30–4.3	1.10–1.23			
	Al_2O_3–water 66.85°C)	1.30–4.3	1.1–1.26			
	SiO_2–water (31.85°C)	1.1–2.30	1.01–1.10			
	SiO_2–water (46.85°C)	1.1–2.30	1.009–1.01			
	SiO_2–water (66.85°C)	1.1–2.40	1.005–1.007			
	TiO_2–water (31.85°C)	3.25–4.30	1.080–1.11			
	TiO_2–water (66.85°C)	3.25–4.30	1.08–1.11			
	TiO_2–water (88.85°C)	3.10–4.30	1.075–1.099			
Lee et al. (1999)	Al_2O_3–water	1.00–4.30	1.03–1.10	Two-step method	Transient hot-wire method	—
	CuO–water	1.00–3.41	1.03–1.12			
	Al_2O_3–ethylene glycol	1.00–5.00	1.03–1.18			
	CuO–ethylene glycol	1.00–4.00	1.05–1.23			
Wang et al. (1999)	Al_2O_3–water	3.00–5.50	1.11–1.16	Two-step method	Steady-state parallel-plate technique	—
	CuO–water	4.50–9.70	1.17–1.34			
	Al_2O_3–ethylene glycol	5.00–8.00	1.25–1.41			
	CuO–ethylene glycol	6.20–14.80	1.24–1.54			
	Al_2O_3–engine oil	2.25–7.40	1.05–1.30			
	Al_2O_3–pump oil	5.00–7.10	1.13–1.20			
Xuan and Li (2000)	Cu(+ laurate salt)–water	2.50–7.50	1.22–1.75	Two-step method	Transient hot-wire method	—
	Cu (+ oleic acid)–transformer oil	2.50–7.50	1.12–1.43			

Reference	Nanofluid	Concentration	Enhancement ratio	Synthesis method	Measurement method	Effect
Choi et al. (2001)	MWCNT (+ dispersant)–polyalphaolefin	0.04–1.02	1.02–2.57	Two-step method	Transient hot-wire method	—
Eastman et al. (2001)	Cu (old)–ethylene glycol	0.10–0.56	1.016–1.10	One-step physical method	Transient hot-wire method	—
	Cu (fresh)–ethylene glycol	0.11–0.56	1.031–1.14			
	Cu (+ thioglycolic acid)–ethylene glycol	0.01–0.28	1.002–1.41	One-step physical method		One-step physical method
Xie et al. (2002a)	SiC–water	0.78–4.18	1.03–1.17	Two-step method	Transient hot-wire method	—
	SiC–water	1.00–4.00	1.06–1.24			
	SiC–ethylene glycol	0.89–3.50	1.04–1.13			
	SiC–ethylene glycol	1.00–4.00	1.06–1.23			
Xie et al. (2002b)	Al_2O_3–water	1.80–5.00	1.07–1.21	Two-step method	Transient hot-wire method	Solid crystalline phase effect, morphology effect, pH value effect, base fluid effect
	Al_2O_3–ethylene glycol	1.80–5.00	1.06–1.17			
	Al_2O_3–ethylene glycol	1.80–5.00	1.06–1.18			
	Al_2O_3–ethylene glycol	1.80–5.00	1.10–1.30			
	Al_2O_3–ethylene glycol	1.80–5.00	1.08–1.25			
	Al_2O_3–pump oil	5.00	1.39			
Xie et al. (2002c)	Al_2O_3–water	5.00	1.23	Two-step method	Hot-wire method	Base fluid effect
	Al_2O_3–ethylene glycol	5.00	1.29			
	Al_2O_3–pump oil	5.00	1.38			
	Al_2O_3–glycerol	5.00	1.27			
Das et al. (2003)	Al_2O_3–water (21°C)	1.00–4.00	1.02–1.09	Two-step method	Temperature oscillation technique	Temperature effect
	Al_2O_3–water (36°C)	1.00–4.00	1.07–1.16			
	Al_2O_3–water (51°C)	1.00–4.00	1.10–1.24			
	CuO–water (21°C)	1.00–4.00	1.07–1.14			
	CuO–water (36°C)	1.00–4.00	1.22–1.26			
	CuO–water (51°C)	1.00–4.00	1.29–1.36			

Reference	Nanofluid	Concentration (vol %)	Enhancement ratio	Synthesis method	Thermal conductivity measurement technique	Effect studied/notes
Patel et al. (2003)	Citrate-reduced Ag–water (30°C)	0.001	1.03	Two-step method	Transient hot-wire method	Temperature effect
	Citrate-reduced Ag–water (60°C)	0.001	1.04			
	Citrate-reduced Au–water (30°C)	0.00013	1.03			
	Citrate-reduced Au–water (60°C)	0.00013	1.05			
	Citrate-reduced Au–water (30°C)	0.00026	1.05			
	Citrate-reduced Au–water (60°C)	0.00026	1.08			
	Thiolate-covered Au–toluene (30°C)	0.005	1.03			
	Thiolate-covered Au–toluene (60°C)	0.005	1.05			
	Thiolate-covered Au–toluene (30°C)	0.008	1.06			
	Thiolate-covered Au–toluene (60°C)	0.008	1.07			
	Thiolate-covered Au–toluene (30°C)	0.011	1.06			
	Thiolate-covered Au–toluene (60°C)	0.011	1.09			

Reference	Nanofluid	Concentration	Ratio	Synthesis method	Measurement method	Effects studied
Xie et al (2003)	MWCNT–water	0.40–1.00	1.03–1.07	Two-step method	Transient hot-wire method	Nitric acid treatment
	MWCNT–ethylene glycol	0.23–1.00	1.02–1.13			
	MWCNT (+ 01eylamine)–decene	0.25–1.00	1.04–1.20			
Assael et al. (2004)	MWCNT (+ sodium dodecyl sulfate)–water	0.60	1.07–1.38	Two-step method	Transient hot-wire method	Treatment effect, dispersant concentration effect, sonication time effect
Wen and Ding (2004a)	Al$_2$O$_3$ (+ sodium dodecylbenzene sulfonate)–water	0.19–1.59	1.01–1.10	Two-step method	Transient hot-wire method	—
Wen and Ding (2004b)	MWCNT (+ sodium dodecyl benzene)–water (20°C)	0.04–0.84	1.04–1.24	Two-step method	Not available	Temperature effect
	MWCNT (+ sodium dodecyl benzene)–water (45°C)	0.04–0.84	1.05–1.31			

Reference	Nanofluid	Concentration (vol %)	Enhancement ratio	Synthesis method	Thermal conductivity measurement technique	Effect studied/notes
Assael et al. (2005)	DWCNT (+ hexadecyltrimethyl ammonium bromide)–water	0.75	1.03	Two-step method	Transient hot-wire method	Dispersant effect, sonication time effect
	DWCNT (+ hexadecyltrimethyl ammonium bromide)–water	1.00	1.08			
	MWCNT (+ hexadecyltrimethyl ammonium bromide)–water	0.60	1.34			
	MWCNT (+ nanosperse AQ)–water	0.60	1.28			
Chon et al. (2005)	Al_2O_3–water (21°C)	1.00	1.09	Two-step method	Transient hot-wire method	Temperature effect
	Al_2O_3–water (71°C)	1.00	1.15			
	Al_2O_3–water (21°C)	1.00	1.03			
	Al_2O_3–water (71°C))	1.00	1.10			
	Al_2O_3–water (21°C)	1.00	1.004			
	Al_2O_3–water (71°C))	1.00	1.09			
	Al_2O_3–water (21°C)	1.00	1.08			
	Al_2O_3–water (71°C))	1.00	1.29			
Hong et al. (2005)	Fe–ethylene glycol	0.20–0.55	1.13–1.18	Two-step method	Transient hot-wire method	Sonication time effect

Reference	Nanofluid				Treatment effect
Marquis and Chibante (2005)	SWCNT (+ dispersant)–diesel oil (Shell Rotella 15W-40)	0.25–1.00	1.10–1.46	Two-step method	Thermal constants analyzer technique	Treatment effect
	MWCNT (I) (+sucinimide–poly alpha olefin) (BP Amoco DS-166)	0.25–1.00	1.30–2.17			
	MWCNT (II) (+sucinimide–poly alpha olefin) (BP Amoco DS-166)	1.00	2.83			
Murshed et al. (2005)	TiO₂ (+ cetyl trimethyl ammonium bromide)–water	0.50–5.00	1.05–1.30	Two-step method	Transient hot-wire method	—
	TiO₂ (+ cetyl trimethyl ammonium bromide)–water	0.50–5.00	1.08–1.33			
Wen and Ding (2005)	Al₂O₃–water	0.31–0.72	1.02–1.06	Two-step method	Not specified	—
Ding et al. (2006)	MWCNT (+ gum arabic)–water (20 °C)	0.05–0.49	1.00–1.10	Two-step method	Transient hot-wire method	Temperature effect
	MWCNT (+ gum arabic)–water (25 °C)	0.05–0.49	1.07–1.27			
	MWCNT (+ gum arabic)–water (30 °C)	0.05–0.49	1.18–1.79			
Hong et al. (2006)	Fe–ethylene glycol	0.10–0.55	1.05–1.18	Two-step method	Transient hot-wire method	Cluster size effect

Reference	Nanofluid	Concentration (vol %)	Enhancement ratio	Synthesis method	Thermal conductivity measurement technique	Effect studied/notes
Hwang et al. (2006)	CuO–water	1.00	1.05	Two-step method	Transient hot-wire method	—
	SiO$_2$–water	1.00	1.03			
	MWCNT–water	1.00	1.07			
	CuO–ethylene glycol	1.00	1.09			
	MWCNT–mineral oil	5.00	1.09			
Kang et al. (2006)	Ag–water	0.10–0.39	1.03–1.11	Two-step method	Transient hot-wire method	—
	SiO$_2$–water	1.00–4.00	1.02–1.05			
	Diamond–ethylene glycol	0.13–1.33	1.03–1.75			
Li and Peterson (2006)	Al$_2$O$_3$–water (27.5°C)	2.00–10.00	1.08–1.11	Two-step method	Steady-state cut-bar method	Temperature effect
	Al$_2$O$_3$–water (32.5°C)	2.00–10.00	1.15–1.22			
	Al$_2$O$_3$–water (37.7°C)	2.00–10.00	1.18–1.29			
	CuO–water (28.9°C)	2.00–6.00	1.35–1.36			
	CuO–water (31.3°C)	2.00–6.00	1.35–1.50			
	CuO–water (33.4°C)	2.00–6.00	1.38–1.51			
Liu et al.(2006)	Cu–water	0.05	1.04	One-step chemical method	Transient hot-wire method	Settlement time effect
	Cu–water	0.10	1.24			
	Cu–water	0.10	1.24			
	Cu–water	0.05	1.12			
	Cu–water	0.10	1.11			
	Cu–water	0.05	1.09			
	Cu–water	0.20	1.10			
	Cu–water	0.20	1.04			
	Cu–water	0.20	1.13			

Reference	Nanofluid			Method	Technique	Notes
Putnam et al. (2006)	11-mercapto-1-undecanol functionalized Au (+ alkenethiolate)–ethanol	0.01-0.07	1.003–1.013	Two-step method	Micron-scale beam deflection technique	—
	Dodecanethiol functionalized Au (+ alkenethiolate)–toluene	0.1 1–0.36	1.000–1.015			
	C_{60}–C_{70} fullerenes–toluene	0.15–0.60	1.002–1.009			
Wen and Ding (2006)	TiO_2–water (pH=3)	0.29–0.68	1.02–1.06	Two-step method	Hot-wire method	Dispersant HNO_3 and NaOH
Yang and Han (2006)	Bi_2Te_3–hexadecane oil (20°C)	0.80	1.06	Two-step method	3ω technique	Surfactant used
	Bi_2Te_3–hexadecane oil (50°C)	0.80	1.04			
	Bi_2Te_3–perfluoro-n-hexane (3°C)	0.80	1.08			
	Bi_2Te_3–perfluoro-n-hexane (50°C)	0.80	1.06			
Yang et al. (2006a)	MWCNT (+ poly isobutene succinimide)–poly alpha olefin	0.04–0.34	486	Two-step method	Transient hot-wire method	Dispersing energy effect, aspect ratio effect, dispersant concentration effect

Reference	Nanofluid	Concentration (vol %)	Enhancement ratio	Synthesis method	Thermal conductivity measurement technique	Effect studied/notes
Prasher et al. (2006a)	Al_2O_3–water(radius 7.5 nm)	0.5	1.24 (70°C) 1.20 (85°C)	Two-step process	Temperature oscillation technique	—
	Al_2O_3–water(radius 10 nm)	0.5	1.39 (70°C) 2.01 (85°C)			
	Al_2O_3–water(radius 13.5 nm)	0.5	1.29 (70°C) 1.6 (85°C) 1.26 (70°C) 1.34 (85°C)			
	Al_2O_3–water(radius 20 nm)	0.5	1.26 (70°C) 1.34 (85°C)			
Chopkar et al. (2006)	$Al_{70}Cu_{30}$–EG	0.2–2.5	1.06–2.26	Two-step process	Thermal comparator technique	—
	$Al_{70}Ag_{30}$–EG	0.2–2.5	1.04–2.47			
	$Al_{70}Cu_{30}$–EG(crystallite size 83–90 nm)	0.5	1.03–1.38			
Zhang et al. (2006a)	Au–toluene	0.003	1.08	One- and two-step methods	Transient short hot-wire method	Surfactant used for CNFs and temperature effect
	Al_2O_3–water (10°C)	1.3–15.0	1.04–1.20			
	Al_2O_3–water (30°C)	1.3–15.0	1.05–1.23			
	Al_2O_3–water (50°C)	1.3–15.0	1.04–1.24			
	CNT–water	0.1–0.9	1.03–1.41			
Zhu et al. (2006)	Fe_3O_4–water	0.5–5.0	1.17–1.41	One-step method	Transient hot-wire method	—

Reference	Nanofluid	Concentration	Ratio	Synthesis	Measurement	Focus
Assael et al. (2006)	Cu–EG	Up to 0.48	1.03	Two-step process	Transient hot-wire method	Dispersant used for preparation
	MWCNT–EG	Up to 0.25	1.09			
	MWCNT (+ sodium dodecyl sulfate)–EG	0.6	1.14–1.21			
	MWCNT (+ sodium dodecyl sulfate)–water	0.6	1.07–1.39			
	MWCNT (+ sodium dodecyl sulfate)–vacuum oil	0.6	1.09			
	MWCNT (+ cetyl trimethyl ammonium bromide)–water	0.6-1.0	1.02–1.34			
	MWCNT (+ Triton X-100)–water	0.6	1.11–1.13			
	MWCNT (+ nanosphere)–water	0.6	1.28			
Zhang et al. (2006b)	Al_2O_3–water (10°C)	1.3–15.0	1.04–1.20	Two-step process	Transient short hot-wire method	Temperature effect
	Al_2O_3–water (30°C)	1.3–15.0	1.05–1.23			
	Al_2O_3–water (50°C)	1.3–15.0	1.04–1.24			
	CuO–water (10°C)	2.55–2.65	1.06–1.09			
	CuO–water (23°C)	2.6–5.2	1.06–1.16			
	CuO–water (30°C)	5.2	1.17			
Lee et al. (2006)	CuO–water (pH=6)	0.03–0.30	1.02–1.07	Two-step process	Transient hot-wire method	pH effect
	CuO–water (pH=3)	0.03–0.30	1.04–1.12			
Hwang et al. (2007)	Fullerene–oil	1.5–5.0	1.02–1.06	Two-step method	Transient hot-wire method	Surfactant used for water nanofluids
	MWCNT–oil	0.2–0.5	1.05–1.09			
	MWCNT–water	0.25–1.00	1.01–1.06			
	Mixed fullerene–water	0.5–1.5	0.99–0.97			

Reference	Nanofluid	Concentration (vol %)	Enhancement ratio	Synthesis method	Thermal conductivity measurement technique	Effect studied/notes
Yoo et al. (2007)	Fe–ethylene glycol	0.20–0.55	1.15–1.18	Two-step process	Transient hot-wire method	—
	WO$_3$–ethylene glycol	0.05–0.30	1.05–1.13			
	TiO$_2$–deionized water	0.1–1.0	1.10–1.15			
	Al$_2$O$_3$–deionized water	0.3–1.0	1.01–1.05			
Beck et al. (2007)	Al$_2$O$_3$–ethylene glycol	0.01–0.04	1.03–1.14	Two-step process	Liquid metal transient hot-wire method	—
Chen et al. (2007)	TiO$_2$–ethylene glycol (20°C)	0.1–1.8	1.01–1.13	Two-step process	Transient hot-wire method	Temperature effect
	TiO$_2$–ethylene glycol (40°C)	0.1–1.8	1.02–1.14			
Jwo et al (2007)	CuO–DI water	0.1–0.4	1.02–1.10	SANSS	Transient hot-wire method	—
Jana et al. (2007)	CNT–water	0.3–0.8	1.24–1.34	Two-step process	Modified hot-wire technique	—
	Cu–water	0.05–0.30	1.17–1.74			
Wang et al. (2007)	TiO$_2$–water (18°C)	1.0–4.0	1.03–1.11	Two-step process	3ω technique	Temperature effect
	TiO$_2$–water (36°C)	1.0–4.0	1.05–1.14			
	TiO$_2$–water (43°C)	1.0–4.0	1.06–1.15			
	TiO$_2$–water (52°C)	1.0–4.0	1.08–1.17			
	TiO$_2$–water (65°C)	1.0–4.0	1.10–1.20			
	SiO$_2$–water (20°C)	1.0	1.03			
	SiO$_2$–EG (20°C)	1.0	1.04			
	SiO$_2$–ethanol (20°C)	1.0	1.05			

Reference	Material (size)	Concentration	Ratio	Process	Method	Surfactant used for nanofluid preparation
Kim et al. (2007)	Al_2O_3 water (38 nm)	0.3–3.0	1.01–1.08	Two-step process	Transient hot-wire method	Surfactant used for nanofluid preparation
	Al_2O_3–EG (38 nm)	1.0–3.0	1.03–1.11			
	ZnO–water (10 nm)	1.0–3.0	1.05–1.15			
	ZnO–water (30 nm)	1.0–3.0	1.03–1.11			
	ZnO–water (60 nm)	1.0–3.0	1.02–1.07			
	ZnO–EG (30 nm)	1.0–3.0	1.06–1.21			
	ZnO–EG (60 nm)	1.0–3.0	1.03–1.11			
	TiO_2–water (10 nm)	1.0–3.0	1.03–1.11			
	TiO_2–water (34 nm)	1.0–3.0	1.028–1.09			
	TiO_2–water (70 nm)	1.0–3.0	1.02–1.06			
	TiO_2–EG (10 nm)	1.0–3.0	1.05–1.15			
	TiO_2–water (34 nm)	1.0–3.0	1.04–1.12			
	TiO_2–water (70 nm)	1.0–3.0	1.02–1.08			
Shaikh et al. (2007)	CNT–PAO oil	0.1–1.0	1.34–2.61	Two-step process	Indirect method	Surfactant used
	EXG–PAO oil	0.1–1.0	1.18–2.31			
	HTT–PAO oil	0.1–1.0	1.11–2.03			
Kao et al. (2007)	CuO–brake fluid	2.0	1.6	One-step process	Transient hot-wire method	—
Philip et al. (2007)	Fe_3O_4–kerosene	0.02–7.80	1.01–1.23	Two-step process	Transient hot-wire method	Surfactant used
He et al. (2007)	TiO_2–water (95–210 nm)	0.6	1.03–1.01	Two-step process	Transient hot-wire method	Particle size effect
	TiO_2–water	0.18–1.92	1.01–1.06			

Reference	Nanofluid	Concentration (vol %)	Enhancement ratio	Synthesis method	Thermal conductivity measurement technique	Effect studied/notes
Chopkar et al. (2007)	$Al_{70}Cu_{30}$–EG	0.2–2.5	1.06–2.26	Two-step process	Thermal comparator method	Particle size effect
	$Al_{70}Cu_{30}$–water	0.2–2.5	1.16–2.62			
	$Al_{70}Ag_{30}$–EG	0.2–2.5	1.04–2.47			
	$Al_{70}Ag_{30}$–water	0.2–2.5	1.19–2.77			
	$Al_{70}Cu_{30}$–EG(crystallite size 83–9 nm)	0.5	1.03–1.38			
	$Al_{70}Cu_{30}$–water (crystallite size 83–9 nm)	0.5	1.10–1.45			
Zhu et al. (2007)	CuO–water	1.0–5.0	1.18–1.31	One-step process	Transient hot-wire method	—
Phuoc et al. (2007)	Ag–water	0.01	1.03–1.05	One-step process	Transient hot-wire method	—
Karthikeyan et al. (2008)	CuO–ethylene glycol	0.005–1.005	1.13–1.54	One-step process	Transient hot-wire method	—
	CuO–water	0.03–1.00	1.01–1.32			
Chen et al. (2008a)	Titanate nanotubes–water(20°C)	0.1–0.6	1.001–1.040	Two-step process	Transient hot-wire method	Temperature effect
	Titanate nanotubes–water(40°C)	0.1–0.6	1.02–1.05			
Murshed et al. (2008)	Al–ethylene glycol	0.01–0.05	1.11–1.45	Two-step process	Transient hot-wire method	—
	TiO_2–ethylene glycol	0.01–0.05	1.04–1.18			
	CNT–engine oil	0.001–0.01	1.02–2.57			

Reference	Nanofluid	Concentration	Ratio	Synthesis	Measurement technique	Remarks
Chen et al. (2008b)	CNT–DI water	0.1–1.0	1.001–1.12	Two-step process	Transient short hot-wire method	—
	CNT–ethylene glycol	0.2–1.0	1.04–1.18			
Choi et al. (2008)	Al$_2$O$_3$–transformer oil (sphere)	0.5–4.0	1.05–1.21	Two-step process	Transient hot-wire method	—
	Al$_2$O$_3$–transformer oil (fibre)	0.5	1.05			
	AlN–transformer oil (sphere)	0.5	1.08			
Li et al. (2008)	Cu–water	0.02–0.80	1.08–1.18	Two-step method	Thermal constants analyzer technique	—
Glory et al. (2008)	MWCNT (+ gum Arabic 1 wt%)–water(15–75 C)	0.5	1.14–1.09	Two-step process	Cylindrical cell method	Gum arabic stabilizer used, temperature effect
		1.0	1.26–1.15			
		1.5	1.34–1.27			
	MWCNT (+ gum Arabic 2 wt%)–water(15–75 C)	0.01	1.019–1.002			
		0.10	1.04–1.02			
		0.20	1.08–1.06			
		1.00	1.24–1.17			
		2.00	1.36–1.38			
		3.00	1.47–1.63			
Sundar and Sharma (2008)	Al$_2$O$_3$–water	0.1–0.8	1.02–1.07	Two-step process	Transient hot-wire method	Temperature effect
	CuO–water	0.1–0.8	1.07–1.24			
	Al$_2$O$_3$–water (30–60 C)	0.2	1.02–1.09			
		0.4	1.04–1.13			
		0.8	1.05–1.23			
	CuO–water (30–60 C)	0.2	1.19–1.36			
		0.4	1.24–1.42			
		0.8	1.33–1.50			

Reference	Nanofluid	Concentration (vol %)	Enhancement ratio	Synthesis method	Thermal conductivity measurement technique	Effect studied/notes
Oh et al. (2008)	Al_2O_3–water	1.0–5.5	1.04–1.13	Two-step process	3ω technique	—
	Al_2O_3–EG	1.0–4.0	1.02–1.10			
Han et al. (2008)	Indium–PAO oil (30–90°C)	8.0	1.11–1.13	One-step process	3ω-wire technique	Temperature effect
Jha and Ramaprabhu (2008)	MWCNT–water	0.04	1.15	Two-step procedure	Transient hot-wire method	—
	MWCNT–EG	0.04	1.07			
	Cu/MWCNT–water	0.03	1.35			
	Cu/MWCNT–EG	0.03	1.10			
Chen et al. (2008a)	TiO_2–water	0.22–1.18	1.02–1.06	Two-step procedure	Transient hot-wire method	—
	TiO_2–EG	0.08–1.80	1.02–1.15			
	TNT–water	0.12–0.60	1.003–1.036			
	TNT–EG	0.1–1.8	1.01–1.14			
Yu et al. (2008a)	Diamond (plasma treated)-water (20–50°C)	0.15	1.17–1.25	Two-step process	Transient hot-wire method	Stabilizer used, temperature effect
	Diamond (untreated)-water (20–50°C)	0.15	1.06–1.10			
Peñas et al. (2008)	SiO_2–water (20–60°C)	2.2	1.03–1.01	Two-step process	Transient multi-current hot-wire method	Temperature effect
	SiO_2–EG (20–60°C)	1.2	1.007–1.015			
		2.5	1.036–1.044			
	CuO–water (20–60°C)	0.4	1.004–1.030			
		0.8	1.016–1.047			
	CuO–EG (20–60°C)	0.4	1.025–1.032			
		0.8	1.059–1.062			

Reference	Nanofluid	Concentration	Enhancement	Process	Method	Remarks
Singh and Raykar (2008)	Ag–water (10–50°C)	1.12	1.10–1.47	One-step process	Transient hot-wire method	Temperature effect
	Ag–water (10–50°C)	1.14	1.16–1.60			
Garg et al. (2008)	Cu–EG	0.4–2.0	1.02–1.13	Two-step process	Transient hot-wire method	—
Tsai et al. (2008)	Fe_3O_4–diesel oil	1.1–4.5	1.04–1.19	Two-step process	Transient hot-wire method	Surfactant used to stabilize
	Fe_3O_4–diesel oil/PDMS (75/25)	1.0–2.0	1.04–1.08			
	Fe_3O_4–diesel oil/PDMS (50/50)	1.0–2.3	1.03–1.06			
	Fe_3O_4–diesel oil/PDMS (25/75)	1.0–2.3	1.03–1.07			
Lee et al. (2008)	Al_2O_3–water	0.01–0.30	1.002–1.014	Two-step process	Transient hot-wire method	—
Mintsa et al. (2009)	Al_2O_3–water (particle size = 36 nm)	0.08–17.50	1.001–1.29	Two step process	Transient hot-wire method	Particle size effect
	Al_2O_3–water (particle size = 47 nm)	2.0–18.0	1.04–1.28			
	CuO–water (particle size = 29 nm)	1.0–13.5	1.01–1.22			
Murshed et al. (2009)	Al_2O_3–DI water	0.01–0.04	1.03–1.42	Two step process	Transient hot-wire method	—
	TiO_2–DI water	0.01–0.04	1.18–1.27			

Reference	Nanofluid	Concentration (vol %)	Enhancement ratio	Synthesis method	Thermal conductivity measurement technique	Effect studied/notes
Jha and Ramaprabhu (2009)	MWNT–DI water	0.04	1.15	Two-step process	Transient hot-wire method	—
	Pd/MWNT–DI water	0.03	1.16			
	Au/MWNT–DI water	0.03	1.28			
	Ag/MWNT–DI water	0.03	1.37			
	MWNT–EG	0.04	1.07			
	Pd/MWNT–EG	0.03	1.09			
	Au/MWNT–EG	0.03	1.10			
	Ag/MWNT–EG	0.03	1.11			
Jung and Yoo (2009)	Al_2O_3(150 nm)–water (20–70°C)	1.0	1.07–1.10	Two-step process	Not specified	Temperature effect
	Al_2O_3(47 nm)–water(20–70°C)	1.0	1.08–1.12			
		4.0	1.14–1.24			
Shima et al. (2009)	Magnetite–water(2.8–9.5 nm)	1.0	1.007	One-step procedure	Thermal property analyzer	Surfactant used, particle size effect
		5.5	1.05–1.25			
Timofeeva et al. (2009)	AlOOH–water/EG (50/50)(shape platelets)	1.0–7.0	1.02–1.18	Two-step process	Transient hot-wire method	Particle shape effect
	AlOOH–water/EG (50/50)(shape blades)	1.0–7.0	1.03–1.18			
	AlOOH–water/EG (50/50)(shape bricks)	1.0–7.0	1.04–1.24			
	AlOOH–water/EG (50/50)(shape cylinders)	1.0–7.0	1.05–1.34			

Reference	Nanofluid					Remarks
Wang et al. (2009a)	Cu–water	0.02–0.80	1.08–1.18	Two-step process	Thermal constants analyzer technique	Surfactant used
	Al$_2$O$_3$–water	0.02–0.08	1.07–1.17			
Jiang et al. (2009)	CNT–R113(d=80 nm AR=18.8)	1.0	1.43	Two-step process	Thermal constants analyzer technique	Size and shape effect
	CNT–R113(d=80 nm AR=125.0)	1.0	1.50			
	CNT–R113(d=15 nm AR=100.0)	1.0	1.82			
	CNT–R113(d=15 nm AR=666.7)	1.0	2.04			
Singh et al. (2009)	SiC–water	0.01–0.08	1.04–1.30	Two-step process	Transient hot-wire method	—
	SiC–water (23–70°C)	1.0	1.04			
		2.0	1.09–1.12			
		4.0	1.22–1.23			
Hwang et al. (2009)	Al$_2$O$_3$–water	0.30	1.0144	Two-step procedure	Transient hot-wire method	—
Kim et al. (2009)	Au–water	0.018	1.10	One-step process	Transient hot-wire method	—
Sinha et al. (2009)	Cu–EG	0.55–1.00	1.30–1.62	Two-step process	Guarded hot parallel-plate method	—
Ali et al. (2010)	Al–water	0.42	1.18	One-step process	Hot-wire laser probe beam deflection technique	Surfactant used, particle clustering observed
	Al–EG		1.21			
	Al–ethanol		1.24			

Reference	Nanofluid	Concentration (vol %)	Enhancement ratio	Synthesis method	Thermal conductivity measurement technique	Effect studied/notes
Zhang et al. (2010)	Carbon-coated Al–polyethylene glycol	0.1–1.5 (without dispersant)	1.15–1.37	Two-step process	Transient plane source method	Nanofluid dispersed by ball milling exhibits the best stability, followed by nanofluids dispersed by ultrasonic dispersion way and magnetic stirring way
		0.1–1.5 (with glycerin dispersant)	1.16–1.40			
	Carbon-coated Cu–polyethylene glycol	0.1–1.5 (without dispersant)	1.15–1.45			
		0.1–1.5 (with glycerin dispersant)	1.16–1.49			
	Carbon coated Fe–polyethylene glycol	0.1–1.5 (without dispersant)	1.15–1.27			
		0.1–1.5 (with glycerin dispersant)	1.16–1.30			
Xie et al. (2010)	MgO–EG	0.45–5.00	1.06–1.40	Two-step process	Transient short hot-wire method	—
Paul et al. (2010)	Au–water	0.00006–0.00025	1.12–1.48	One-step process	Transient hot-wire method	Nanoparticle size varied
Abareshi et al. (2010)	Fe_3O_4–DI water	0.25–3.00	1.05–1.11	One-step process	Transient hot-wire method	Surfactant used

Reference	Material					
Chang et al. (2011)	CuO–water	0.01 0.4	1.07–1.11	Two-step process	Transient hot-wire method	Surfactant used, NP prepared by the HiGee system
Phuoc et al. (2011)	MWCNT–water (dispersing agent chitosan 0.1wt%)	0.24–1.40	1.02–1.13	Two-step process	Transient hot-wire method	Surfactant stabilized
	MWCNT–water (dispersing agent chitosan 0.2wt%)	0.24–1.40	1.02–1.13			
	MWCNT–water (dispersing agent chitosan 0.5wt%)	0.24–1.40	1.03–1.04			
Palabiyik et al. (2011)	TiO_2–PG	0.24–2.4	1.01–1.09	Two-step process	Transient hot-wire method	Clustering theory
	Al_2O_3–PG	0.26–2.52	1.03–1.11			
Venerus and Jiang (2011)	SiO_2–water	0.025–0.158	1.04–1.15	Two-step process	Optical technique	Data consistent with EMT
Kim et al. (2011)	Al_2O_3–water	0.01–0.05	1.02–1.11	Two-step process	Transient hot-wire method	Rod type nanoparticles used, NP bead milled
		0.01–0.05 (pH 4)	1.01–1.11			
		0.01–0.05(pH 7)	1.0–1.04			
		0.01–0.05(pH 11)	1.01–1.05			
Paul et al. (2011)	$Al_{95}Zn_{05}$–EG	0.01–0.10	1.10–1.16	Two-step process	Transient hot-wire method	Nanoparticle size varied, variation of time studied

Reference	Nanofluid	Concentration (vol %)	Enhancement ratio	Synthesis method	Thermal conductivity measurement technique	Effect studied/notes
Gupta et al. (2011)	Graphene–water (30°C)	0.01–0.20	1.02–1.11	One-step process	Transient hot-wire method	Enhancement due to Brownian motion and micro-convection effects
	Graphene–water (35°C)	0.01–0.20	1.02–1.14			
	Graphene–water (40°C)	0.01–0.20	1.02–1.17			
	Graphene–water (45°C)	0.01–0.20	1.02–1.21			
	Graphene–water (50°C)	0.01–0.20	1.05–1.27			
	Al_2O_3–water	0.5–1.0	1.02–1.04			
	CuO–water	0.5–1.0	1.03–1.06			
	Cu–water	0.1–1.0	1.04–1.12			
	CNT–water	0.1–1.0	1.06–1.40			
	Graphene–water	0.02–0.20	1.04–1.18			

Part III

Processing

12

Mechanochemical synthesis of metallic–ceramic composite powders

K. W I E C Z O R E K - C I U R O W A,
Cracow University of Technology, Poland

DOI: 10.1533/9780857093493.3.399

Abstract: The chapter discusses several variants of mechanosyntheses for composite powder formation of metal alloy matrices with ceramic particles. The necessity of following progress in mechanochemical processes using different analytical methods is shown. Based on the results of experimental studies on Cu–Al/Al$_2$O$_3$ and Ni–Al/Al$_2$O$_3$ nanocomposite powder formation through mechanochemical synthesis, the method of transforming combustive processes to progressive ones by applying hydroxosalts instead of metal oxides as precursors of composites is described.

Key words: metallic–ceramic composites, high-energy ball milling, metallothermic reactions, combustive self-propagating reactions, progressive reactions, soft mechanochemistry, matrices of Cu–Al and Ni–Al alloys.

Note: This chapter is adapted from Chapter 9 'Mechanochemical synthesis of metallic–ceramic composite powders' by K. Wieczorek-Ciurowa, originally published in *High energy ball milling: mechanochemical processing of nanopowders*, ed. M. Sopicka-Lizer, Woodhead Publishing Limited, 2010, ISBN: 978-1-84569-531-6.

12.1 Introduction

Technical progress is dependent on the development of new materials. Recently a lot of attention has been given to technologies which create and test functional materials, both constructive and tools. Materials properties, especially on the nanometric scale, are specific, comparing well with poly- or

399

monocrystalline ones for materials with the same chemical composition (e.g. Fernández-Bertran, 1999; Ivanov and Suryanarayana, 2000; Suryanarayana, 2001a; Šepelak, 2002; Kelsall et al., 2005; Baláž, 2008). From a practical point of view, the desired features of nanocrystalline powders can be intensified by creating multiphase composites from them. Therefore, bulk composites are engineering materials made from two or more components with significantly different physical and/or chemical properties that remain separate and distinct on the macroscopic level within the finished structures. One of the components is a matrix and the other provides structural support. Three types of composite categories can be defined based on the characteristics of the matrix:

- metallic matrix composites (MMCs)
- ceramic matrix composites (CMCs)
- polymer matrix composites (PMCs).

Application of the relevant composite powders to produce the final bulk product or any other composite product is one of the most attractive features in their fabrication. These composite powders need to be consolidated at elevated temperatures by thermomechanical processing and then their microstructure stability and microhardness should be determined in order to estimate the optimal parameters for the consolidation process. Actual metal–ceramic composite powders usually incorporate light alloys such as Al, Mg and Ti. Other metals commonly used are Cu, Ni, Fe, Co, Zn and Mo. Metals, ceramics or polymers can be used to support MMC powders.

Metal matrix reinforced phases can exist as dispersed particles of (Kaczmar, 2000; Koch, 2001; Agrawal and Sun, 2004):

- metal oxides (ZrO_2, Al_2O_3, ThO_2, Cr_2O_3)
- carbides (SiC, TaC, WC, B_4C)
- nitrides (Si_3N_4, TaN, ZrN, TiN)
- borides (TaB_2, ZrB_2, TiB_2, WB).

Cermetals are composed of metallic and ceramic components, which combine the physical properties of metals and ceramics. The properties of metals include:

- ductility
- tensile strength
- thermal and electrical conductivity.

Ceramics possess physical properties such as:

- high melting point
- chemical stability

• resistance to oxidation.

Milling of soft metals/alloys and hard ceramic materials leads to the formation of valuable composite powders with important new features. Changing the ratio of the cermetal components produces final composite products with widely different properties. These materials have much higher mechanical, temperature and aggressive media resistance properties compared to ceramics properties. The hardness of cermetals increases with decreasing particle size (Chang *et al.*, 1999; Zhang *et al.*, 2000; Portnoy *et al.*, 2002).

12.2   Composite powder formation: bottom-up and top-down techniques

Composite powders can be produced by two techniques: bottom-up and top-down.

The first method is based on building nanostructures atom-by-atom, layer-by-layer. These chemical processes could take place in liquid, solid as well as in gas phases (Tjong and Chen, 2004). Examples of these are the sol–gel method, melt spinning–melt quenching (MQ), chemical vapour deposition (CVD), physical vapour deposition (PVD), plasma ablation, laser pyrolysis, and so on. Unfortunately, there are often some disadvantages, for example multi-stage processes, the high cost of the alkoxide precursors, solvent evaporation and necessity for thermal treatment at high temperatures to provide coarse-grained products.

The top-down technique begins with macrostructured materials and uses mechanical, chemical or other forms of energy to 'break' them into smaller pieces. However, it is very important to develop methods which minimize damage to the environment. One promising candidate is mechanochemistry, often referred to as 'green' processes. Mechanical energy can be easily explored for chemical syntheses of new functional materials like composite powders. No doubt, this method is fast, economical and gives high-purity products, which can often take nanostructure forms (Heegn *et al.*, 2003). In mechanochemical syntheses, the course of solid-state reactions can be effectively controlled and regulated by choosing the precursors of syntheses and/or milling variables.

12.2.1 Mechanical treatment

Mechanical treatment of solids uses mechanical energy through high-energy ball milling, abrasion, fracture or welding to produce new materials or generate products with desired features, for example powder mixtures consisting of metal and ceramics (Courtney, 1994; Boldyrev *et al.*, 1996).

Thus, mechanical treatment of reactants can form metallic–ceramic composite powder particles. The structure quality, that is whether such composite particles are micro- or nanosized, depends on the type of reagents and milling conditions.

12.2.2 Mechanochemical synthesis: mechanical activation, mechanical alloying, reactive milling

Mechanochemical synthesis includes mechanical alloying together with reactive milling which follows from mechanical activation of reagents. This is also done by high-energy ball milling (Murty and Ranganathan, 1998; Avvakumov et al., 2001; Suryanarayana, 2001; Suryanarayana et al., 2001; Boldyrev, 2006; Smolyakov et al., 2007; Baláž, 2008) which is a more complex but very effective process of metallic–ceramic composite formation. There are several variants of mechanochemical syntheses, which depend on the nature of the initial reagents nature (e.g. two different metals with ceramics, metal oxide with active metal or metal salt with active metal, and others).

Mechanical activation is responsible for enhancing the reactivity of solids by enlarging the surface area or accelerating the reaction by correct mixing of the reagents. In the processes of plastic deformation and fracture and friction during ball collisions, the impact energy is converted into other forms of energy. These induce structural defects, broken bonds and other forms of excess energy. These accumulate and a new, active state of the substances is produced. Consequently, the chemical reactivity of the solids increases significantly (e.g. Chen et al., 1997).

Mechanical alloying is the process in which mixtures of powders are milled to achieve alloying at the atomic level. During this process, when high-energy impulses are used, metallic powders can form alloys like different solid solutions, intermetallic phases, mixtures of metals or amorphic materials with properties that are very often different from those found when using traditional methods (McCormick et al., 1989; Maurice and Courtney, 1990; Schwarz, 1996; Urakaev and Boldyrev, 2000).

It is important to note that any form of metal alloy that comprises composite components has better physical and mechanical properties than pure metals. For example, consolidated materials based on intermetallics of nickel, iron, titanium with aluminium are treated as functional materials that have specific physical properties and as constructive materials that have unique mechanical features such as structural stability at high temperature. Especially important are their high melting temperature, high mechanical resistance and low density. These properties allow them to be used in the automobile industry (turbo compressor rotors, valves, combustion chamber and exhauster details), in the aircraft industry, in nuclear power and in metallurgy. However, the low yield point at room temperature (what causes difficulties in mechanical treatment) is one of the main hindrances to using them more widely. Modification of features of the inter-metallic phase can be made, for example, by adding ceramic particles into their structure. Inserting ceramic particles (type: Al_2O_3, ZrO_2 or TiC) into the intermetallic matrix (Ni, Al) leads to improvements in some mechanical properties like hardness and abrasion resistance at high temperatures and a significant rise of utility features, for example an increase in the corrosion temperature and in oxidation and erosion resistance.

The mechanisms of metal alloy formation by mechanical treatment of metal powders are categorized into three different systems depending on the mechanical features: ductile–ductile, ductile–brittle and brittle–brittle. In the first case in the early stages of milling, the components become flatted to platelets by cold welding. In the next stage, they are cold-welded together and form a composite lamellar geometry of the constituent metals. Further true alloying occurs at the atomic level until the homogeneous structure of the powders is attained.

The mechanisms of metallic–ceramic composite formation fall into the ductile–brittle components category because the brittle oxide particles are dispersed in a ductile metallic matrix. The ceramic reagent is closely spaced along the interlamellar spacing. With further milling, the ductile powder of alloy particles becomes work hardened and the lamellae become convoluted and refined. With continued milling the lamellae are further refined and the brittle particles are uniformly dispersed in the ductile metallic matrix. The ductile components are flattened by a micro-forging process while the brittle ones are fragmented. One can assume that the brittle particles of the materials are dispersed in the ductile matrix. However, if both milled materials are brittle, this phenomenon is not observed.

A typical example of ductile–ductile microstructures is $Cu/Al–Al_2O_3$ composite powder formed during milling of the Cu–hydroxocarbonate and Al mixture, described in Section 12.5 (Wieczorek-Ciurowa et al., 2000, 2003a, 2003b). The microstructure of the products after the first stage of the mechanochemical synthesis is shown in Fig. 12.1, which presents the SEM micro-photograph. The material reveals a lamellar microstructure. Based on quantitative energy dispersive X-ray elemental analysis (EDX) (Table 12.1) it is possible to estimate that mechanically alloyed composite particles consist of Cu–Al intermetallic phases as a matrix, and aluminium oxide.

The darker network corresponds to the higher amount of Al_2O_3 and the brighter phase to the higher amount of metallic phases. The final form of $Cu–Al/Al_2O_3$ composite particles shown in Fig. 12.2 reveals its homogeneous microstructure.

Reactive milling of solids relates to the process in which chemical

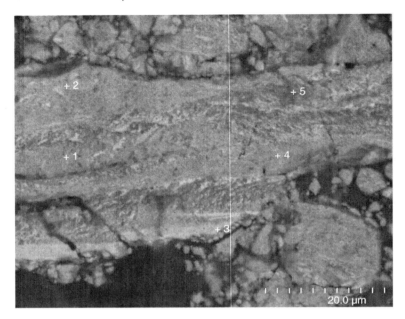

12.1 SEM microphotograph (BSE image) illustrating the lamellar structure of Cu–Al/Al$_2$O$_3$ composite powder following aluminothermic reaction with Cu$_2$(OH)$_2$CO$_3$ at the first stage of reactive milling. The darker network corresponds to a higher amount of Al$_2$O$_3$, the brighter one to a higher content of Cu–Al intermetallics (SEM Hitachi S-4700 with EDX Noran Vantage) (Wieczorek-Ciurowa *et al.*, 2005).

Table 12.1 Results of EDX elemental microanalysis of the composite powder synthesized by milling a Cu$_2$(OH)$_2$CO$_3$–Al$_2$O$_3$ mixture, from the corresponding SEM microphotograph (BSE image) in Fig. 12.1

Point	Concentration/mass (%)			BSE image
	Cu	O	Al	
1	32.7	30.1	37.2	Dark grey
2	39.0	25.5	35.5	Grey
3	47.2	22.2	30.6	Light grey
4	38.4	25.3	35.5	Grey
5	23.3	32.6	44.1	Dark grey

Source: Wieczorek-Ciurowa *et al.*, 2005.

reactions occur. This is a stage of mechanochemical synthesis that can occur in two different kinetic ways:

- a reaction that develops slowly with each collision, which results in a gradual transformation of the substrates

12.2 SEM microphotograph (BSE image) illustrating a high level of structure homogeneity in a Cu–Al/Al_2O_3 composite powder following aluminothermic reaction with $Cu_2(OH)_2CO_3$ at the final stage of reactive milling. The darker spots are finely dispersed Al_2O_3 particles whose average size is no larger then 1 μm. The brighter spots correspond to the higher content of Cu–Al intermetallics (see Fig. 12.1) (SEM Hitachi S-4700 with EDX Noran Vantage) (Wieczorek-Ciurowa and Gamrat, 2007b).

- a self-propagating reaction initiated when the reaction enthalpy is sufficiently high.

These kinetics models depend on the nature, type of reactants and their properties (e.g. hardness), thermodynamics of the reactions and technical parameters of milling (Butyagin, 2000, 2003; Takacs *et al.*, 2001; Takacs and Mandal, 2001; Takacs, 2002; Lyakhov *et al.*, 2008).

The first type of mechanical synthesis concerns reactions, which proceed more slowly up to the point at which processes become a function of the milling time. If ignition does not occur, the collisions between milled material and the grinding medium contribute to comminution, mixing and defect formation. The formation of the final product occurs gradually. Figure 12.3 shows the first and final step of mechanochemical synthesis of Cu–Al/Al_2O_3 composite powder as a function of temperature and pressure inside the mill (Wieczorek-Ciurowa *et al.*, 2007b).

The second type requires a critical time for ignition of the reaction. When the temperature of the vial was recorded during milling, it was observed that initially the temperature increased slowly with time. After a certain period of

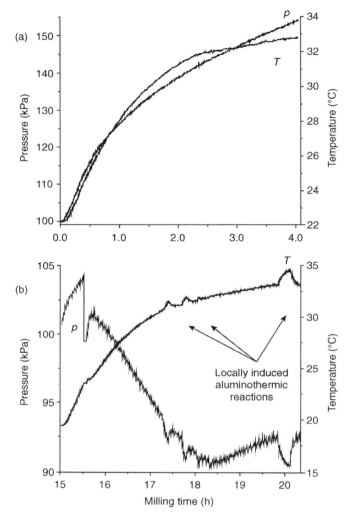

12.3 Variations in temperature and pressure (GTM results) during mechanosynthesis of Cu–Al/Al₂O₃ composite powder from Cu₂(OH)₂CO₃–Al mixture: (a) beginning of mechanical treatment and (b) final steps of mechanosynthesis (planetary ball mill Fritsch GmbH *Pulverisette 6*) (Wieczorek-Ciurowa *et al.*, 2007b).

milling the temperature increased abruptly, which confirmed the fact that ignition had occurred. The time at which a sudden increase in temperature occurs is referred to as the ignition time. Beyond this time, the reaction takes place within seconds (see Fig. 12.4). The example in the figure is an illustration of Cu–Al/Al₂O₃ composite powder formation in the CuO–Al system (Wieczorek-Ciurowa and Gamrat, 2007a; Wieczorek-Ciurowa *et al.*,

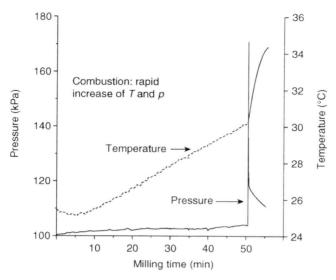

12.4 Variations in temperature and pressure (GTM results) during self-propagating high-temperature synthesis of Cu–Al/Al$_2$O$_3$ composite powder from a CuO–Al mixture. The milling conditions are the same as used for a Cu$_2$(OH)$_2$CO$_3$–Al mixture treatment (see Fig. 12.3) (planetary ball mill Fritsch GmbH *Pulverisette 6*) (Wieczorek-Ciurowa *et al.*, 2007a).

2007a). The ignition temperature is a function of the enthalpy change and the microstructure parameters of the particles, for example the inter-facial area between the reactants. Continuing milling is necessary to form the correct desired structure of synthesis products. Generally, a self-propagating high-temperature synthesis (SHS) reaction consists of three stages: component activation, chemical reaction initiation and a synthesis process in bulk with product formation. Schaffer and McCormick (1992) stated that SHS reactions occur when the temperature created by the colliding balls in a ball mill is higher than initiation temperature, which depends on enthalpy changes in the process. The decreasing of this temperature can be the effect of mechanical activation by diminution of particle size and increasing the surface between interfacial contacts.

The explanation of this phenomenon is as follows. It was assumed that after a period of comminution, mixing and activation, agglomerates begin to form and increase in size. The reaction starts in a single agglomerate or in the powder layer coating a milling ball or the wall of the vial. One reaction front propagates into other parts of the powder layer. The powder layer can be attached to the surface of a milling ball or the inner wall of the container. When a ball hits this layer, part of the kinetic energy is transferred to the powder as heat and a local increase of temperature can occur. The stresses

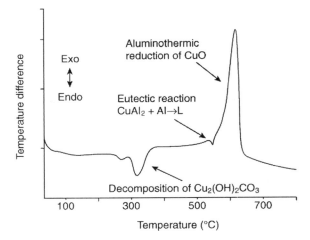

12.5 DTA curve (non-oxidizing atmosphere) of a $Cu_2(OH)_2CO_3$–Al mixture after the first step in mechanical treatment. The product contains an undecomposed part of $Cu_2(OH)_2CO_3$ and unreacted Al. This also represents a simulation of the reactions in the system that may occur under mechanical action (SDT 2960 TA Instrument) (Wieczorek-Ciurowa *et al.*, 2005).

inside the powder are not uniform but are concentrated at a few points. This results in the formation of 'hot spots', where the reaction can start even if the average temperature of the powder is not sufficient to initiate a reaction front. Intimate contact between the reactant phases is an essential requirement for self-propagating synthesis. This condition is easy to achieve when mechanical activation is conducted in a system of ductile–brittle substances.

The temperature of initiation of a metallothermic reaction can be recognized by differential thermal analysis (DTA) (strong exoeffects). This is illustrated by the DTA curve (Fig. 12.5) for the product of the CuO–Al powder after the first step of mechanochemical synthesis. The CuO formed by the mechanical decomposition of $Cu_2(OH)_2CO_3$ reacts with Al, forming copper in a combustible manner at about 600°C. This procedure can be considered a simulation of a mechanochemical reaction (Wieczorek-Ciurowa and Gamrat, 2007a).

An example of self-propagation reactions are processes occurring during milling aluminium and magnetite powders in argon, given by Botta *et al.* (2000) as follows:

$$3Fe_3O_4 + 8Al \rightarrow 9Fe + 4Al_2O_3 \qquad [12.1]$$

There are two different effects, which are due to the time period of mechanical activation. After 30 min milling, a decrease in the starting

Table 12.2 Examples of induction time intervals for selected SHS mechanochemical reactions occurring under different milling conditions

Reaction	Type of mill	BPR	Induction time (min)	Reference
$ZnO + Mg \rightarrow Zn + MgO$	Vibratory	7:1	45	Yang and McCormick, 1993
$2FeO + Ti \rightarrow 2Fe + TiO_2$	Vibratory	8:1	20	Takacs, 2002
$8CuO + 3Fe \rightarrow 4Cu_2O + Fe_3O_4$	Planetary	30:1	153	Shen *et al.*, 1992
$2CuO + Zn \rightarrow Cu_2O + ZnO$	Vibratory	–	44	Takacs, 2002
$Cu_2S + Fe \rightarrow 2Cu + FeS$	Vibratory	8:1	–	McCormick *et al.*, 1989

Source: literature data.

reactant crystallinity could be observed as well as a small amount of product, whereas after 37 min milling time the aluminothermic reaction was completed. Traditionally, this reaction would be thermally initiated during high-temperature treatment. However, under mechanical activation conditions a 'flash point' is the direct consequence of previous energy accumulation in the crystalline structure of powder particles.

In the case of milling time less than 30 min a certain amount of active product precursor forms from the energy accumulated in the solid phase. The precursor releases this energy by reacting at lower temperatures during thermal treatment. This effect is close to the one observed in the TiO_2–Al system, but in this case, the aluminothermic reaction does not occur during milling. This phenomenon of solid activation, because of milling, can be seen in syntheses other than in metallic–ceramic composites ones.

The kinetics of mechanochemical reactions depends not only on the reactant properties but also on the synthesis conditions. The synthesis can be explosive or can occur in a controlled steady way. Changing the conditions of the activation process can prevent the reaction taking place at all. As well as the ball-to-powder ratio (BPR), the type of ball material is also important. Moreover, too low heat conductivity of the milling medium can hinder formation of hot spots in which the reaction is initiated. Table 12.2 gives induction times (in minutes) for SHS reactions carried out under the given conditions of mechanochemical syntheses.

On the other hand, in order to control explosive effects during self-propagating reactions, that is to suppress their kinetics, it is possible for example to use a process control agent (PCA) such as toluene, which does not stop the reduction reaction but allows it to run in a controlled manner (Murty and Ranganathan, 1998; Takacs *et al.*, 2001; Takacs and Mandal, 2001; Takacs, 2002).

The mechanochemical syntheses of metallic–ceramic nanocomposite powders carried out in a non-explosive way based on copper–aluminium and nickel–aluminium alloys reinforced by aluminium oxide using their salts instead of oxides are described in Sections 12.5 and 12.6. For comparison, the explosive reactions in metal oxide with active metal systems are also presented.

12.3   Monitoring mechanochemical processes

A comprehensive study of the physical and chemical processes that occur during mechanical treatment by high-energy ball milling appears only to be possible if a reliable identification of solids and the quantitative phase analysis of activated products is made. Because of the complexity of mechanochemical reactions, the nature of the solids obtained closely relates to the milling conditions and so they should be well defined. Moreover, it is very important to determine the factors that influence the activation effects. Another difficulty arises from the fact that the reactions are composed of many successive stages which vary in different cases. The experimental methods required to identify and characterize materials synthesized mechanochemically involve not only techniques that are applicable to solids, but also others, more particularly adapted to the nanostructured character of the milling products. Thus, different types of analytical methods must be applied (e.g. Wieczorek-Ciurowa et al., 2000, 2001; Baláž, 2008).

The best way to study the kinetics of mechanochemical transformations would be to analyse continuously the milled products in situ. However, until now this has been a difficult task. The only method that has been used in some laboratories is the gas pressure–temperature measuring system (GTM) to acquire in situ data during planetary ball milling; for example, Fritsch GmbH mills, which enable indirect control and provision of synthesis of chemical compounds by selection of appropriate parameters for milling (http://www.fritsch.com). Recording from the beginning of the milling process is especially useful when gaseous products evolve from the system caused by mechanical decomposition of the compounds. It is also useful when the self-propagating high-temperature syntheses occur in an explosive manner owing to local overheating, for example by two impacting balls (e.g. Murty and Ranganathan, 1998; Kwon et al., 2002). In such cases applying less intensive milling could be considered, for example by using lower values of rotation per minute (rpm) and/or ball-to-powder mass ratio and/or the size and material of grinding balls, as well as controlling the milling atmosphere and the type of control agent. Optimization of the milling parameters chosen experimentally must be taken into account when selecting mechanochemical syntheses conditions because not all these variables are completely independent.

Solid phase characterization is realized *ex situ* when a sample of milled powder is picked up from the vial after a defined milling period. Thus, in the case of a material in crystalline form, the X-ray diffraction (XRD) method is advantageous; it is fast and allows the progress of the syntheses to be followed. However, this is not convenient after some stages of milling where the mechanical treatment caused comminution and prolonged activation destroyed the crystallinity of the solid powder particles leading to their complete or partial amorphization and simultaneously to consumption of the initial mixture components, making them undetectable by XRD because the amounts were too small.

Thermal analysis (TA) with registration of thermogravimetric (TG/DTG) and differential thermal analysis (DTA) curves facilitates the determination of an amount of undecomposed and/or unreacted reagents, oxidation/reduction processes and mechanically induced physical transformations in/or between the reagents. The example shown in Fig. 12.5 allows us readily to see that the analysed sample consists of mechanically undecomposed salt, that is Cu–hydroxocarbonate with the remaining amount being aluminium. The endothermic effect at 350°C reveals that the $Cu_2(OH)_2CO_3$–Al mixture decomposed into CuO, which was then reduced by Al to metallic Cu, generating a large amount of heat at about 600°C. This heat may accelerate the alloying of Cu with Al, for example into intermetallic phases. The effect at 520°C can be related to the eutectoid transition of $CuAl_2$ with Al → L at 528°C, according to the phase diagram of the binary Cu–Al system (Massalski, 1992), because the sample used in this thermo-analytical experiment was a product of the early stage of salt–metal mixture milling.

In addition to the above-mentioned analytical techniques for solids, versatile scanning and transmission electron microscopy (SEM and TEM) with backscattered electron imaging and quantitative energy dispersive X-ray elemental microanalysis (EDS) are very helpful in estimating the microstructure of synthesized composite particles for a relatively coarse powder and for fine powders. By combining the grey tone levels with the results of EDS some compounds can be identified and localized (Yakowitz, 1975; Wieczorek-Ciurowa *et al.*, 2003c). To illustrate the utility of transmission electron microscopy in estimating the microstructures of mechanochemically synthesized composite powders, Fig. 12.6 shows a set of TEM microphotographs of $Cu_2(OH)_2CO_3$–Al powder after mechanochemical synthesis showing nanocrystalline copper. Electron diffraction patterns unquestionably confirm that the detected Cu, $CuAl_2$ phases are in the nanocrystalline form.

Moreover, it is worthwhile adding methods such as neutron diffraction, magic angle spinning solid-state nuclear magnetic resonance (MAS NMR), Fourier transform infrared (FT-IR), ultraviolet (UV) or X-ray photo-

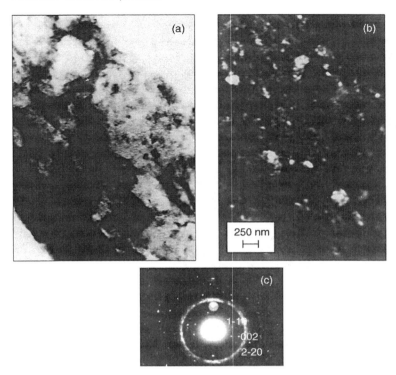

12.6 Set of TEM microphotographs of the $Cu_2(OH)_2CO_3$–Al system after mechanochemical synthesis showing nanocrystalline copper: (a) bright field, (b) dark field, (c) electron diffraction patterns (TEM Philips CM 20 with EDX) (Wieczorek-Ciurowa and Gamrat, 2007a).

electron spectroscopy (XPS). All these techniques have been described in standard textbooks on solid-state chemistry and they are not described in detail here. To determine the real phase constitution with characterization of milled powders, a combination of some methods is necessary.

12.4   Examples of applied high-energy milling in the synthesis of selected metallic–ceramic composite powders

Mechanochemical synthesis is almost an ideal method for preparing nano-sized metal matrix composite powders because of its simplicity and the possibility of forming composite powder particles with a uniform distribution of grain sizes. Moreover, such *in situ* routes of synthesis (Sections 12.5 and 12.6) result in the production of powdered materials that have microstructures that are more homogeneous than those synthesized using conventional *ex situ* techniques (Section 12.7).

In metallic–ceramic composite powder formation, the preparation of metals and alloys by reducing their salts or oxides with more reactive metals is commonly known as metallothermic reduction. This reaction is expressed in general by the equation:

$$M_A X + M_B > M_A + M_B X \qquad [12.2]$$

where a metal M_A is reduced by a more reactive metal M_B (reductant) to the pure metal MA. $M_A X$ and $M_B X$ are oxides, chlorides, sulphides and other salts.

Metallothermic reactions are characterized by a large negative free-energy change and therefore they are thermodynamically feasible at room temperature (see for example reactions 12.4 and 12.5. Moreover, the mechanical activation of reagents significantly increases the synthesis kinetics. A salt such as M_A hydroxocarbonate can be used as a 'source' of $M_A O$, as described in Sections 12.5 and 12.6.

12.5 Copper-based composite powders with Al₂O₃

The formation of a copper-based composite powder with aluminium is a consequence of many complex, simultaneous and subsequent chemical reactions that occur in the milling of a $Cu_2(OH)_2CO_3$ and Al mixture.

They are schematically reported below (equations 12.3–12.5) together with the enthalpy values (ΔH_{298}/kJ mol^{-1}; Barin *et al.*, 1977):

• mechanochemical decomposition of salt: $+82$

$$Cu_2(OH)_2CO_{3(s)} \rightarrow 2CuO_{(s)} + CO_{2(G)} + H_2O_{(G)} \qquad [12.3]$$

• aluminothermic reduction of CuO to Cu with Al₂O₃ formation: -1179

$$3CuO_{(s)} + 2Al^0_{(s)} \rightarrow 3Cu^0_{(s)} + Al_2O_{3(s)} \qquad [12.4]$$

• mechanical alloying of Cu and Al (in the presence of Al₂O₃): -8 (Cu_9Al_4) and -4 ($CuAl_2$)

$$Cu^0_{(s)} + Al_2O_{3(s)} + Al^0_{(s)} \rightarrow Cu^0 - Al^0 + Al_2O_3 \qquad [12.5]$$

subscript (S) indicates solid and (G) indicates gas. The product is a metallic–ceramic composite powder. $Cu^0–Al^0$ alloy can exist as intermetallics and/or a solid solution.

The formation enthalpy of oxide $M_B X$ should be higher than that for $M_A X$ (see equation 12.2). The values for Cu, Ni and Al oxides are given in

Table 12.3 Formation enthalpy values of selected metal oxides

Oxide	$-\Delta H$ (kJ mol^{-1})
Al_2O_3	1675.7
NiO	239.7
Cu_2O	173.2
CuO	161.9

Source: Kubashewski *et al.*, 1993

Table 12.3. From a thermodynamic point of view, the aluminothermic reactions in the systems studied are significantly favoured.

In Sections 12.5.1 and 12.5.2, the in situ formation of a nanostructural metallic–ceramics nanocomposite $Cu–Al/Al_2O_3$ powder is described starting from two different mixtures, $Cu_2(OH)_2CO_3$ and CuO with Al, respectively (Wieczorek-Ciurowa *et al.*, 2002a; Wieczorek-Ciurowa and Gamrat, 2007a, b; Wieczorek-Ciurowa *et al.*, 2007a,b; 2008). Composite powders were synthesized by high-energy ball milling in a laboratory planetary mill with vial and balls made of hardened steel. The component systems were prepared as physical mixtures in which the amount of Al was calculated assuming the formation Cu_9Al_4 with Al_2O_3 and $CuAl_2$ with Al_2O_3 as components of the synthesis products. Among the intermetallics in the binary Al–Cu phase diagram (Hansen, 1958) are $CuAl_2$, Cu_9Al_4, Cu(Al) and Al(Cu) solid solutions. It was found that if Cu_9Al_4 was the expected phase, mechanosynthesis in both reagents used brings about the formation of a composite material consisting of Cu(Al) solid solution and Al_2O_3. The amount of Al dissolved into the Cu matrix is considerably higher for the CuO–Al system than for $Cu_2(OH)_2CO_3$–Al. This is probably caused by the different kinetics of mechanochemical reactions in these two systems.

According to Avvakumov *et al.* (2001), compounds containing groups of atoms with oxygen and/or hydrogen (like acidic or basic salts, hydrates) take part in reactions evolving water and/or volatile compounds. These compounds are characterized by 3–4 times lower hardness in comparison to their anhydrous oxides, probably providing the mechanochemical process with lower mechanical loading. Processes taking place with these types of substances are known as soft mechanochemical reactions.

Mechanochemical syntheses where the mixtures are provided with stoichiometric proportions of reagents to form $CuAl_2$ result in the formation of two intermetallic phases, $CuAl_2$ and Cu_9Al_4, as well as Al_2O_3. It is evident that in the systems with CuO as a reagent for aluminothermic reaction $CuAl_2$ appears after only 1 hour of milling. This indicates the rapid alloying process of both metals: copper and aluminium. The experimental results are described in detail in Sections 12.5.1 and 12.5.2.

12.5.1 Effects of mechanosynthesis in the $Cu_2(OH)_2CO_3$–Al reagent system

Figure 12.7 shows the XRD patterns (Philips X'Pert diffractometer Cu Kα) of the $Cu_2(OH)_2CO_3$–Al system mechanically treated at different time intervals (*Pulverissette 6* Fritsch GmbH). The results indicate that the initial components of the mixture studied still exist in the system after up to 10 hours of mechanical treatment. However, their intensities reduced significantly in the fourth hour of milling. This indicates that $Cu_2(OH)_2CO_3$ started to decompose and metallothermic reduction of CuO with Al proceeded gradually with formation of metallic copper. Furthermore, a broad diffraction peak appears in the range of $2\theta = 42°$ to $45°$ after 10 hours of milling. Its intensity increases significantly with a lengthening of the milling time. The shape of this X-ray indicates its complexity and the deconvolution confirms the presence of copper in addition to part of the Al_2O_3 phase. After 15 hours and 20 hours of milling, the position of the Cu(1 1 1) peak shifted from $2\theta = 43.32°$ to $43.25°$, respectively, while Al peaks completely disappeared. This may suggest that a solid solution of Cu(Al) has been formed as a result of mechanical alloying of two metals, Cu and Al. This is con-firmed by an increase of metallic copper lattice parameter from 3.6151 Å to 3.6203 Å (Pearson, 1958). According to Vegard's law, which states that the lattice parameter of the solid solution is a function of the amount of solute and using parameters for pure Cu and Al, the contents of the Al solute in the Cu matrix for the mechanically treated $Cu_2(OH)_2CO_3$–Al system are estimated on a level of about 5%.

From a thermodynamic point of view, the reactions occurring in the tested system are highly exothermic, for example the enthalpy of aluminothermic reduction of CuO with Al equals -1179 kJ mol^{-1}. However, in the hydroxosalt–active metal system, processes in the mill occur in a controlled manner. Coming back to Fig. 12.3 illustrating the temperature and pressure in a vial during milling process, it can be seen that in the first hours of milling the temperature and pressure increase simultaneously. The rise in pressure is due to mechanical decomposition of $Cu_2(OH)_2CO_3$ to CuO with H_2O and CO_2 evolving, while the temperature increase is caused by the friction and impact of the balls. When Cu–hydroxocarbonate is completely decomposed, any rise in pressure is observed, whereas only a transitory rise in temperature occurs (Fig. 12.3b). On the T versus milling time curve, one can observe several small rises in the gas temperature. However, all of them take place below 40°C. This indicates that the aluminothermic reduction occurs locally, in a repeatable fashion. Based on these data we can conclude that in this case mechanochemical synthesis takes place in a controlled way, gradually, so

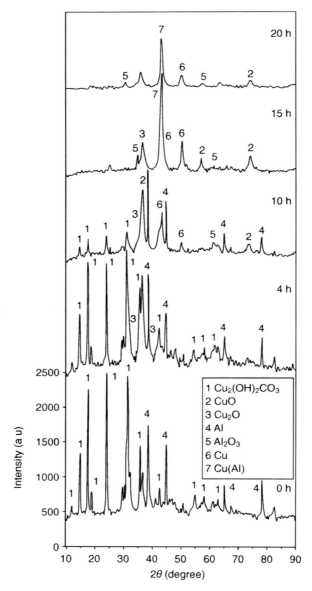

12.7 X-ray diffraction patterns of the $Cu_2(OH)_2CO_3$–Al system after different mechanochemical treatment times in a *Pulverisette 6* Fritsch GmbH mill (intensities are on the same scale) (Philips X'Pert diffractometer, CuKα) (Wieczorek-Ciurowa *et al.*, 2007b).

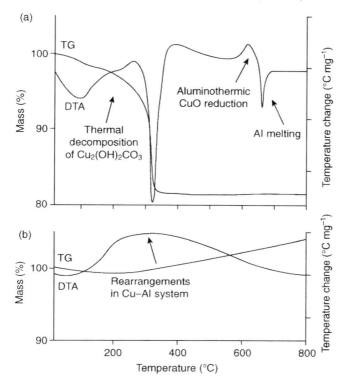

12.8 TG–DTA curves of the $Cu_2(OH)_2CO_3$–Al system after mechanochemical synthesis for: (a) 0 hours, (b) 20 hours (SDT 2960 TA Instrument) (Wieczorek-Ciurowa and Gamrat, 2006).

CuO is reduced by Al in many steps proceeding locally, in isolated parts of treated powders. This process of aluminothermic reduction may be explained by the fact that the occurrence of a highly exothermic reaction via combustion depends not only on thermodynamics but also on other factors such as the nature of reactants, their mechanical properties, crystalline structures and their stability during milling.

The thermoanalytical curves, TG–DTA, presented in Fig. 12.8 (Wieczorek-Ciurowa and Gamrat, 2006) for an untreated mixture of $Cu_2(OH)_2CO_3$ with Al, and after milling for 20 hours, respectively, indicate that Cu–hydroxocarbonate completely decomposes during mechanosynthesis and aluminium is consumed because of the lack of an endo effect at 640°C. Moreover, any sharp exothermic effect caused by the reaction of CuO with Al is detected on the DTA curve. This means that the aluminothermic reaction is completed during that time. It is interesting that there is no endothermic peak before 450°C, which normally corresponds to the transformation of Cu_9Al_4 into a $CuAl_2$ phase. This is

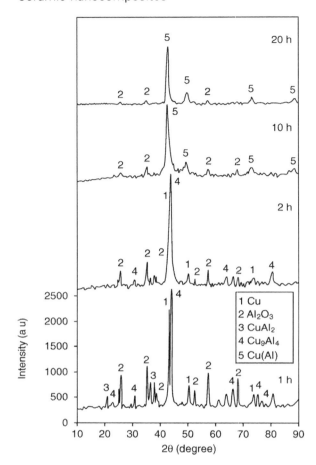

12.9 X-ray diffraction patterns for a CuO–Al system after different mechanochemical treatment times in a *Pulverisette 6* Fritsch GmbH mill (Philips X'Pert diffractometer, CuKα, intensities are on the same scale) (Wieczorek-Ciurowa *et al.*, 2008).

also confirmed by the results of XRD patterns, indicating that this phase did not form.

12.5.2 Effects of mechanosynthesis in the CuO–Al reagent system

Mechanosynthesis in a copper oxide–aluminium system under the same conditions as in the case where $Cu_2(OH)_2CO_3$ was used as a source of CuO caused, within the first hour of milling, the formation of two intermetallic phases, $CuAl_2$ and Cu_9Al_4 (Fig. 12.9) (Wieczorek-Ciurowa *et al.*, 2007a,

2008). Lengthening the milling time results in the disappearance of both intermetallics, although with different kinetics. The first, $CuAl_2$, decays after 1 hour of milling, while Cu_9Al_4 remains in the system up to 2 hours of milling. Further milling caused the consumption of aluminium from these phases to form a Cu(Al) solid solution. This is confirmed by the shift of the Cu(1 1 1) peak from $2\theta = 42.57°$ to $43.33°$ observed in the system treated for 20 hours. The calculated lattice parameter of this phase is equal to 3.6564 Å. Comparison of this parameter with its value for pure Cu suggests expansion of the copper lattice by aluminium substitution, forming a solid solution of Al in the Cu matrix. In fact, the content of the Al solute was estimated to be at a level close to 10%. Conventionally Al dissolves in Cu at the temperature of eutectoid transformation (565°C) at 9.4%.

An increase in the temperature in the milling vial registered by the GTM system indicates that the combustion process occurs during the first hour of milling (see Fig. 12.4). In this case, the reaction between CuO and Al is completed, confirmed by the absence of peaks corresponding to the initial components. Considering phase evolution in the CuO–Al mixed powders, one can suggest that intermetallic phases are formed only as intermediate products under the applied conditions. During further milling they transform to the Cu(Al) solid solution, that is the expected Cu_9Al_4 phase was not formed in final products.

The characteristics of the milling product microstructure were provided using a scanning electron microscopy. Figure 12.10 shows the typical morphology of composite powder after milling for 20 hours. EDS analysis showed that the dark grey phase is Al_2O_3 while the bright grey one is a Cu (Al) matrix. The microphotograph confirmed that alumina particles are evenly embedded in the Cu–Al matrix, in size range from 100 nm to 500 nm. This indicates that milling the CuO–Al system with a stoichiometry close to Cu_9Al_4 brings about formation of metal matrix composite powder particles in which Cu(Al) forms a matrix while alumina grains act as its reinforcement.

12.6   Nickel-based composite powders with Al_2O_3

Formation *in situ* of composite powders of nickel–aluminium metals strengthened by alumina is very important from both a theoretical and a practical point of view. Analogous to the mechanochemical synthesis of Cu–Al/Al_2O_3 composite described in Section 12.5, which is based on the type of reactants, three modifications of the process are possible. Nickel in the form of its salt, oxide or simply elemental metal with aluminium can be used.

Because nickel and its alloys have good plastic fatigue features but are not sufficiently resistant to high temperature oxidation and abrasive wear, intermetallics such as NiAl and Ni_3Al are a better choice for a composite

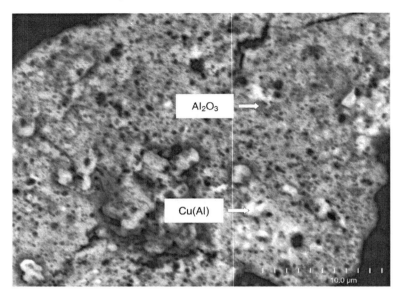

12.10 SEM microphotograph (BSE image) illustrating that Al_2O_3 particles (ranging from 100 nm to 500 nm) are embedded in the Cu–Al matrix formed during reactive milling of the CuO–Al system. The dark spots correspond to Al_2O_3, the brighter grey ones to Cu(Al) (SEM Hitachi S-4700 with EDX Noran Vantage) (Wieczorek-Ciurowa *et al.*, 2007a).

matrix. They are resistant to very high temperature, corrosion, mechanical and abrasive actions and they have a relatively low density. However, their low intergranular cohesion is responsible for their great brittleness. For this reason, a ceramic component is often present in the composite.

In Sections 12.6.1 and 12.6.2, *in situ* mechanochemical syntheses of Ni–Al/Al_2O_3 are described, starting from two different mixtures, $Ni_2(OH)_2CO_3 \cdot xH_2O$ or NiO with Al, respectively (Wieczorek-Ciurowa and Oleszak, 2008). The milling procedure was similar to the one described in Section 12.5. The component systems were prepared as physical mixtures in which the amount of Al was calculated assuming that NiAl with Al_2O_3 or Ni_3Al with Al_2O_3 formed as final products of the synthesis. Among the intermetallics in the binary Ni–Al phase diagram (Hansen, 1958), the main ones are Ni_3Al, NiAl, Ni_2Al_3, $NiAl_3$ and solid solutions. It was found that if Ni_3Al was the expected phase, mechanosynthesis brings about the formation of composite powder particles consisting of Ni_3Al matrix and Al_2O_3.

In mechanochemical syntheses in the NiO–Al system where reagent proportions were calculated for the NiAl phase, the expected NiAl and additionally the Ni_3Al intermetallic phase were formed as well as Al_2O_3. However during the mechanical treatment of $Ni_2(OH)_2CO_3 \cdot xH_2O$ with aluminium, the alloying process does not form NiAl but mainly proceeds to Ni_3Al formation. Presumably, the reason for this is the different nature and structure of the nickel salt compared with copper salt. Ni-hydroxocarbonate belongs to hydrated salts with low crystallinity (Wieczorek-Ciurowa et al., 2002b; Wieczorek-Ciurowa and Gamrat, 2005). Moreover, crystalline water stabilizes the structure of the nickel salt, making its mechanical decomposition more difficult, in effect slowing down the delivery of NiO in the aluminothermic reaction [12.7]. The examples of two kinds of salts presented here, $Ni_2(OH)_2CO_3 \cdot xH_2O$, almost amorphous, and $Cu_2(OH)_2CO_3$ with high crystallinity showed that crystalline structures are more vulnerable to mechanical treatment because their activity becomes higher, for example creating structural defects and new surfaces, and undergoing mass transfer (mixing).

Special attention should be paid to the important part being played by water vapour in the milling system of $Ni_2(OH)_2CO_3.xH_2O$–Al during mechanical decomposition of salt. The compound contains both crystalline water (x = about three molecules) and OH groups. The evolved water reacting rapidly (ΔH_{298} = −949 kJ mol^{-1}) with mechanically activated aluminium powder, see equation 12.6 decreases the amount of Al needed to synthesise the intermetallics (see Section 12.6.1).

$$2Al + 6H_2O \rightarrow 2Al_2O_3 + 3H_2 \qquad [12.6]$$

Too low a concentration of Al in the desired NiAl phase of the $Ni_2(OH)_2CO_3.xH_2O$–Al system gave Ni_3Al intermetallics in the composite powder and a phase with a lower amount of aluminium.

Similar to the formation of copper–aluminium matrix/Al_2O_3 composite particles, which are the result of strong exothermic reactions, formation of composite particles of nickel–aluminium/Al_2O_3 is also typical of a self-propagating high-temperature synthesis (with a combustible nature). The enthalpy changes for reactions 12.7 and 12.8 are negative (Barin et al., 1977; Kubaschewski et al., 1993; Takacs, 2002):

$$3NiO + 2Al \rightarrow Ni + Al_2O_3, \Delta H_{298} = -955 \text{ kJ mol}^{-1} \qquad [12.7]$$

$$xNi + Al \rightarrow Ni_xAl$$

$$\text{where } x = 1 \text{ or } 3, \Delta H_{298} = -177 \text{ KJ mol}^{-1} \text{ (NiAl) and}$$
$$- 153 \text{ KJ mol}^{-1} \text{ (Ni}_3\text{Al)} \qquad [12.8]$$

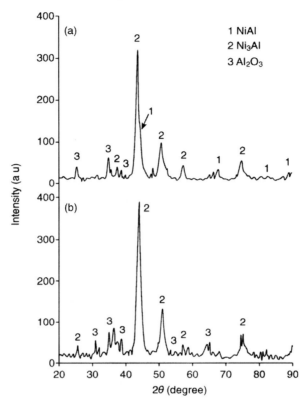

12.11 X-ray diffraction patterns (Philips X'Pert diffractometer, CuKα) for a $Ni_2(OH)_2CO_3.xH_2O$–Al system after 20 hours of mechanochemical treatment (*Pulverisette 6* Fritsch GmbH mill) with reagent proportions calculated for intermetallic phase formation of (a) NiAl and (b) Ni_3Al (Wieczorek-Ciurowa and Oleszak, 2008).

12.6.1 Effects of mechanosynthesis in the $Ni_2(OH)_2CO_3.xH_2O$–Al reagent system

The XRD patterns of the $Ni_2(OH)_2CO_3.xH_2O$–Al systems with reagent proportions calculated for NiAl or Ni_3Al intermetallics mechanically treated for 20 hours are shown in Fig. 12.11. It is seen that, instead of the desired NiAl, the Ni_3Al phase is formed with traces of NiAl. Furthermore, up to the fourth hour of milling, reagents such as NiO, Ni and Al still remain and after 10 hours of milling the reaction was still incomplete, about 15% salt was undecomposed and mechanical alloying of Ni_3Al was in its first stage. The results of thermal analysis measurements also confirmed these effects (Wieczorek-Ciurowa and Oleszak, 2008). After 20 hours of milling the composite powder is fully formed.

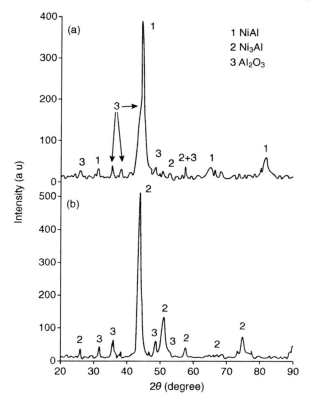

12.12 X-ray diffraction patterns (Philips X'Pert diffractometer, CuKα) of the NiO–Al system after 20 hours mechanochemical treatment (*Pulverisette 6* Fritsch GmbH mill) with reagent proportions calculated for intermetallic phase formation of (a) NiAl and (b) Ni_3Al (Wieczorek-Ciurowa and Oleszak, 2008).

Figure 12.11b shows the synthesis of Ni_3Al/Al_2O_3 composite powder formed after 20 hours of milling in the system with reactant proportions for the Ni_3Al phase. However, after up to 4 hours of mechanical treatment, Ni and Al were present and after 10 hours milling, NiO was still detectable. It is assumed that the slower decomposition kinetics of the hydrate of nickel hydroxocarbonate in comparison to anhydrous $Cu_2(OH)_2CO_3$ is responsible for this behaviour.

12.6.2 Effects of mechanosynthesis in the NiO–Al reagent system

The effect of 20 hours mechanochemical synthesis in the NiO–Al reagent system was to produce a NiAl phase with Al_2O_3, when the ratio of the

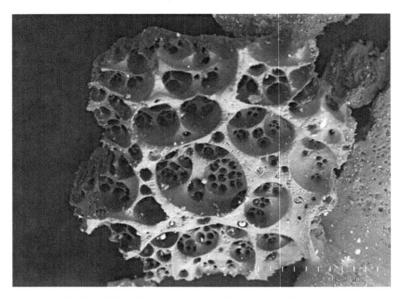

12.13 SEM microphotograph illustrating the product structure of a combustion aluminothermic reaction during mechanical treatment of a NiO–Al mixture (SEM Hitachi S-4700) (Wieczorek-Ciurowa and Oleszak, 2008).

masses of the reagents was used to provide NiAl stoichiometry (Fig. 12.12a). A small amount of Ni_3Al was still present.

The composite powder Ni_3Al/Al_2O_3 is formed explosively in the NiO–Al reagent system (Fig. 12.12b). The SEM microphotograph of the product is shown in Fig. 12.13.

12.7   Other possible variants of the synthesis of metal matrix–ceramic composites in Cu–Al–O and Ni–Al–O elemental systems using mechanical treatment *ex situ* and *in situ*

It has already been reported that copper and nickel as well as other transition metals and intermetallics with these metals are important matrices in composite materials with fine dispersion of Al_2O_3 particles because of their applications. For instance, cermetals such as $Ni–Al/Al_2O_3$ are candidates for military and civil aero engines as exhaust nozzle materials. This type of composite is used in dentistry as a substance for fillings, in prosthetics, as vacuum tube coatings for solar hot water systems, transistors, capacitors, and so on (see, e.g. Grahle and Arzt, 1997; Ying and Zhang, 2000a,b; 2003; Morsi, 2001; Portnoy *et al.*, 2002).

To produce these materials, combustive reactions need to be suppressed during mechanochemical synthesis. Ying and Zhang (2000a,b; 2003) show that this can be achieved utilizing a technique in which the Al powder is first diluted by Cu and then the resultant Cu–Al alloy powder is reacted with CuO. The dilution was done through mechanical alloying of Al and Cu into a Cu(Al) solid solution or Cu–Al intermetallics compounds (depending on the amount of Al needed in the mixture). In the next step, powders of Cu–Al alloy and CuO were milled to form a composite structured powder. Finally, this composite powder was heated ($<450°$C) causing the reaction between CuO and Cu(Al) or Cu_9Al_4. Subsequent heating at a temperature below $800°$C led to formation of Cu and Al_2O_3. Mechanical treatment enhances the kinetics of both reactions by refining the composite particles structure.

Another technique, also a combination of oxidation and mechanical alloying, has been used to produce Cu–Al_2O_3, metal matrix–ceramic composite powders. With this method, Cu powder is first partially oxidized and then the powder obtained is mechanically alloyed with Al powder to facilitate the aluminothermic reaction of Al_2O_3 particles formation (Bobrova and Besterci, 1994).

Particulate reinforced Cu/Al_2O_3 metal matrix–ceramic composites can be mechanochemically synthesized by using several methods which include mixing of Cu melt and Al_2O_3 powder followed by calcination and internal oxidation of Cu–Al alloy powders (Shi and Wang, 1998). The calcination process is limited (Liang et al., 2004) because the Al_2O_3 particle size has to be large enough to allow effective milling, while internal oxidation can only produce composites with a low volume fraction of Al_2O_3 particles, such as oxide dispersion strengthened alloys. Oxidation was used in conjunction with mechanical alloying (Ogbuji, 2004). In this process, Cu–Al alloy powder was milled under an oxidizing atmosphere to produce the composite powder. The advantage of this process is that very small oxide particles can be achieved.

Another approach that avoids too high activities of the reactants has been made by Venugopal et al. (2005) who introduced toluene as the milling medium (PCA) in copper–alumina composite synthesis by reactive milling of CuO/Cu_2O and Al. In fact, the transformation of a combustible reaction to a progressive one yields nanocomposite particles of Cu and Al_2O_3 with both components with a crystallite size in the range of about 20 nm.

Hwang and Lee (2005) demonstrated that Cu/Al_2O_3 nanocomposite powders with various vol% of alumina as a reinforcement phase have been successfully produced by mechanochemical synthesis in a high-energy attritor mill using mixtures of Cu, Cu_2O and Al powder components. It is important to note that in this case the heat generated during milling was removed by a cooling tube coil attached to the outside of the milling chamber. Moreover, in every stoichiometric reaction, excess Cu was added to the system. The role of excess Cu powder was not only as a diluent to control the concentration of Al_2O_3 but also as a thermal conductor to remove heat from the system during milling.

12.8   Conclusions

- Metallic–ceramic nanostructured composite powders can be produced *in situ* simply and successfully at room temperature through a mechanochemical route using high-energy ball milling. The subsequent consolidation produces nanocomposites with the desired microstructure and mechanical characteristics.

- Despite a number of studies having been carried out in different laboratories to synthesize nanocomposite powders, direct comparison and prediction of the course of the process and/or product properties is still very difficult (even when the precursors have a similar character). Using different kinds of mills and procedures for milling influences the input of mechanical energy to the treated material. Moreover, until now, the mechanisms and kinetics of mechanochemical syntheses in solid–solid systems are not fully understood. Therefore it seems to be reasonable to carry out experimental studies to determine the progress of reactions and quality of milling products at various milling stages in relation to the milling conditions and to collect the data in order to formulate a theory of mechanochemical processes.

- In the case of composite mechanosyntheses with a metal matrix it is especially important to inhibit self-propagating explosive reactions and to transform to the progressive ones by wider use of soft mechanochemistry (acidic, basic, hydrated salts) or, for example, different PCA.

Based on the results of two practical examples of syntheses, the transformation of combustion reactions to progressive reactions is shown by applying copper and nickel hydroxocarbonates as composite precursors (even if different in nature) instead of metal oxides, CuO and NiO.

12.9   Acknowledgements

The author would like to thank Professor Yu. G. Shirokov of the Ivanovo State University of Chemistry and Technology, Russian Federation, who in 1995 inspired the study of mechanochemical processes.

The Polish Ministry of Science and Higher Education (Project Nos PB 4 T09B 07024 and PB 1 T09B 023 30) and the State Committee for Scientific Research (Grants CUT/C-1/DS) provided financial support (2002–2008).

12.10 References

Agrawal P and Sun C T (2004), 'Fracture in metal-ceramic composites', *Comp Sci Technol*, **64**, 1167–78.

Avvakumov E G, Senna M and Kosova N (2001), *Soft Mechanochemical Synthesis, A Basis for New Chemical Technologies*, Kluwer Academic Publishers, Boston.

Baláž P (2008), *Mechanochemistry in Nanoscience and Minerals Engineering*, Springer, Berlin.

Barin I, Knacke O and Kubaschewski O (1977), *Thermochemical Properties of Inorganic Substances, Supplement*, Springer, Berlin.

Bobrova E and Besterci M (1994), 'Processing of Cu–Al_2O_3 metal matrix nanocomposite materials by high-energy ball milling', *Powder Metall Sci Tech*, **6**, 7.

Boldyrev V V (2006), 'Mechanochemistry and mechanical activation of solids', *Russian Chem Rev*, **75** (3), 177–89.

Boldyrev V V, Pavlov S V and Goldberg E L (1996), 'Interrelation between fine grinding and mechanical activation', *Int J Mineral Process*, **44–5**, 181–5.

Botta P M, Aglietti E F and Porto Lopez J M (2000), 'Thermal and phase evolution of mechanochemical reactions in the Al–Fe_3O_4 system', *Thermochim Acta*, **363**, 143–7.

Butyagin P YU (2000), 'Mechanochemical synthesis: mechanical and chemical factors', *J Mater Synth Process*, **8** (3/4), 205–11.

Butyagin P YU (2003), 'Diffusion and deformation models of mechanochemical synthesis', *Colloid J*, **65** (5), 648–51.

Chang S T, Tuan W H, You H C and Lin I C (1999), 'Effect of surface grinding on the strength of NiAl and Al_2O_3/NiAl composites', *Mater Chem Phys*, **59**, 220–4.

Chen Y, Hwang T, Marsh M and Williams J S (1997), 'Study on mechanism of mechanical activation', *Mater Sci Eng*, **A226–8**, 95–8.

Courtney T H (1994), 'Modelling of mechanical milling and mechanical alloying', *Rev in Particulate Mater*, **2**, 63–116.

Fernández-Bertran J F (1999), 'Mechanochemistry: an overview', *Pure Appl Chem*, **71** (4), 581–6.

Grahle P and Arzt E (1997), 'Microstructural development in dispersion strengthened NiAl produced by mechanical alloying and secondary recrystallization', *Acta Mater*, **45**, 201–11.

Hansen H (1958), *Constitution of Binary Alloys*, McGraw-Hill, New York.

Heegn H, Birkeneder F and Kmptner A (2003), 'Mechanical activation of precursors for nanocrystalline materials', *Cryst Res Technol*, **38** (1), 7–20.

http://www.fritsch.com Laboratory planetary mill Pulverisette 6 (Fritsch GmbH, Germany).

Hwang S J and Lee J-H (2005), 'Mechanochemical synthesis of Cu–Al_2O_3 nanocomposites', *Mater Sci Eng*, **A405**, 140–6.

Ivanov E and Suryanarayana C (2000), 'Materials and process design through mechanochemical routes', *J Mater Synth Process*, **8** (3/4), 235–44.

Kaczmar J W (2000), 'Production and application of metal matrix composite materials', *J Mater Process Technol*, **106**, 58–67.

Kelsall R W, Hamley J W and Geoghegan M (eds) (2005), *Nanoscale Science and Technology*, J Wiley & Sons, New York.

Koch C C (2001), *Nanostructured Materials Processing, Properties and Potential Applications*, Wiliam Andrew, Norwich, New York.

Kubashewski O, Alcock C B and Spencer P J (1993), *Materials Thermochemistry*, 6th edition, Pergamon Press, Boston.

Kwon Y-S, Gerasimov K B and Yoon S-K (2002), 'Ball temperatures during mechanical alloying in planetary mills', *J Alloys Comp*, **346**, 276–81.

Liang S, Fan Z, Xu L and Fang L (2004), 'Kinetic analysis on Al_2O_3/Cu composite prepared by mechanical activation and internal oxidation', *Composites*, **A35**, 1441–6.

Lyakhov N Z, Vityaz P A, Grigoryeva T F, Talako T L, Barinova A P, Vorsina I A, Letzko A I and Cherepanova S V (2008), 'Nanocomposites intermetallics/oxides, produced by MA SHS', *Rev Adv Mater Sci*, **18**, 326–8.

Massalski T B (ed.) (1992), *Binary Alloy Phase Diagrams, 1*, 2nd edition, ASM Int., Ohio, USA.

Maurice D R and Courtney T H (1990), 'The physics of mechanical alloying: A first report', *Metallurgical Trans A*, **21A**, 289.

McCormick P G, Wharton V N and Schaffer G B (1989), *Physical Chemistry of Powder Metals Production and Processing*, TMS, Warrendale, PA.

Morsi K (2001), 'Review: reaction synthesis processing of Ni-Al intermetallic materials', *Mater Sci Eng*, **A299**, 1–15.

Murty B S and Ranganathan S (1998), 'Novel materials synthesis by mechanical alloying/milling', *Int Mater Rev*, **43**, 101–41.

Ogbuji L U (2004), 'The oxidation behavior of an ODS copper alloy $Cu-Al_2O_3$'. *Oxidation of metals*, **62** (3/4), 141–51.

Pearson W B (1958), *A Handbook of Lattice Spacings and Structures of Metals and Alloys*, Pergamon Press, London.

Portnoy V K, Blinov A M, Tomilin I A, Kuznetsov V N and Kulik T (2002), 'Formation of nickel aluminides by mechanical alloying and thermodynamics of interaction', *J Alloys Comp*, **336**, 196–201.

Schaffer G B and McCormick P G (1992), 'The direct synthesis of metals and alloys by mechanical alloying', *Mater Sci Forum*, **88–90**, 779–86.

Schwarz R B (1996), 'Introduction to the viewpoint set on: mechanical alloying', *Scripta Mater*, **34** (1), 1–4.

Šepelak V (2002), 'Nanocrystalline materials prepared by homogeneous and heterogeneous mechanochemical reactions', *Ann Chim Sci Mater*, **27** (6), 61–76.

Shen T D, Wang K Y and Quan M X (1992), 'Solid state displacement reaction of Fe and CuO induced by mechanical alloying', *Mater Sci Eng*, **A151**, 189–95.

Shi Z and Wang D (1998), 'Processing of $Cu-Al_2O_3$ metal matrix nanocomposite materials by using high energy ball milling', *J Mater Sci Lett*, **17**, 477–81.

Smolyakov V K, Lapshin O V and Boldyrev V V (2007), 'Macroscopic theory of mechanochemical synthesis in heterogeneous systems'. *Int J Self-Propagating High-Temp Synth*, **16** (1), 1–11.

Suryanarayana C (2001), 'Mechanical alloying and milling', *Progress Mat Sci*, **46**, 1–184.

Suryanarayana C, Ivanov E and Boldyrev V V (2001), 'The science and technology of mechanical alloying', *Mater Sci Eng*, **A304–6**, 151–8.

Takacs L (2002), 'Self-sustaining reactions induced by ball milling', *Progr Mater Sci*, **47**, 355–414 (and literature cited herein).

Takacs L and Mandal S K (2001), 'Preparation of some metal phosphides by ball milling', *Mater Sci Eng*, **A304–6**, 429–33.

Takacs L, Sojka V and Baláž P (2001), 'The effect of mechanical activation on highly exothermic powder mixtures', *Solid State Ionics*, **141–2**, 641–7.

Tjong S C and Chen H (2004), 'Nanocrystalline materials and coating', *Mater Sci Eng*, **R45**, 1–88.

Urakaev F KH and Boldyrev V V (2000), 'Mechanism and kinetics of mechanochemical processes in comminuting devices: 2. Applications of the theory. Experiment', *Powder Technol*, **107**, 197–206.

Venugopal T, Prasad Rao K and Murty B S (2005), 'Synthesis of copper-alumina nanocomposite by reactive milling', *Mater Sci Eng*, **A393**, 382–6.

Wieczorek-Ciurowa K and Gamrat K (2005), 'NiAl/Ni$_3$Al-Al$_2$O$_3$ composite formation by reactive ball milling', *J Therm Anal Calorim*, **82** (3), 719–24.

Wieczorek-Ciurowa K and Gamrat K (2006), 'Mechanochemical synthesis of nanocrystalline cermets', *Pol J Chem Technol*, **8** (3), 51–3.

Wieczorek-Ciurowa K and Gamrat K (2007a), 'Some aspects of mechanochemical reactions', *Mater Sci Pol*, **25** (1), 219–32.

Wieczorek-Ciurowa K and Gamrat K (2007b), 'Mechanochemical syntheses as an example of green processes', *J Therm Anal Calorim*, **88** (1), 213–17.

Wieczorek-Ciurowa K and Oleszak D (2008), 'Mechanochemical synthesis of Ni–Al/Al$_2$O$_3$ and Cu–Al/Al$_2$O$_3$ nanocomposite powder', *15th International Symposium on Metastable and NanoMaterials*, ISMANAM2008, Buenos Aires, 26–30 June 2008, Argentina, Book of Abstracts.

Wieczorek-Ciurowa K, Shirokov JU G and Paryło M (2000), 'Use of thermogravimetry to assess the effect of mechanical activation of selected inorganic salts', *J Therm Anal Calorim*, **60**, 59–65.

Wieczorek-Ciurowa K, Paryło M, Shirokov JU G and Gamrat K (2001), 'The characteristics of mechanically activated mixtures of copper hydroxocarbonate with aluminium', *J Therm Anal Calorim*, **65**, 359–66.

Wieczorek-Ciurowa K, Gamrat K, Paryło M and Shirokov JU G (2002a), 'Alloy formation in mechanically activated mixtures hydroxocarbonate with Al^{0}', *J Therm Anal Calorim*, **70**, 165–72.

Wieczorek-Ciurowa K, Gamrat K, Paryło M and Shirokov JU G (2002b), 'The influence of aluminium and aluminium oxide on the effects of mechanical activation of nickel hydroxocarbonate', *J Therm Anal Calorim*, **69**, 237–43.

Wieczorek-Ciurowa K, Gamrat K and Shirokov JU G (2003a), 'Chemical reactions during high energy ball milling of the Cu$_2$(OH)$_2$CO$_3$–Al system', *Solid State Ionics*, **164**, 193–8.

Wieczorek-Ciurowa K, Gamrat K and Shirokov JU G (2003b), 'Mechanism of mechanochemical reactions in malachite–active metal systems', *Thermochim Acta*, **400**, 221–5.

Wieczorek-Ciurowa K, Gamrat K, Fela K and Sawłowicz Z (2003c), 'Structural and morphological changes induced by mechanical activation in copper hydroxocarbonate–aluminium system', *Pol J Chem Technol*, **5**, 72–4.

Wieczorek-Ciurowa K, Gamrat K and Sawłowicz Z (2005), 'Characteristics of CuAl$_2$-Cu$_9$Al$_4$/Al$_2$O$_3$ nanocomposites synthesized by mechanical treatment', *J Therm Anal Calorim*, **80** (3), 619–23.

Wieczorek-Ciurowa K, Oleszak D and Gamrat K (2007a), 'Mechanochemical synthesis of Cu–Al/Al$_2$O$_3$ composite in CuO–Al system under different conditions', *Chem for Sustainable Development*, **15**, 255–8.

Wieczorek-Ciurowa K, Oleszak D and Gamrat K (2007b), 'Mechanosynthesis and process characterization of some nanostructured intermetallics–ceramics composites', *J Alloys Comp*, **434–5**, 501–4.

Wieczorek-Ciurowa K, Oleszak D and Gamrat K (2008), 'Cu-Al/Al$_2$O$_3$ cermet synthesized by reactive ball milling of CuOAl system', *Rev Adv Mater Sci*, **18**, 248–52.

Yakowitz H (1975), 'Methods of quantitative X-ray analysis used in electron probe microanalysis and scanning electron microscopy', In: *Practical Scanning Electron Microscopy*, Goldstein J I and Yakowitz H (eds), Plenum Press, New York.

Yang H and McCormick P G (1993), 'Combustion reaction of zinc oxide with magnesium during milling', *J Solid State Chem*, **107**, 258–63.

Ying D Y and Zhang D L (2000a), 'Processing of Cu-Al$_2$O$_3$ metal matrix nanocomposite materials by using high energy ball milling', *Mater Sci Eng*, **A286**, 152–6.

Ying D Y and Zhang D L (2000b), 'Solid-state reactions between Cu and Al during mechanical alloying and heat treatment'. *J Alloys Comp*, **311**, 275–82.

Ying D Y and Zhang D L (2003), 'Solid state reactions between CuO and Cu(Al) or Cu$_9$Al$_4$ in mechanically milled composite powders', *Mater Sci Eng*, **A361**, 321–30.

Zhang F, Kaczmarek W A, Lu L and Lai M O (2000), 'Formation of Al–TiN metal matrix composite via mechanochemical route', *Scripta Mater*, **43**, 1097–102.

13
Sintering of ultrafine and nanosized ceramic and metallic particles

Z. Z. FANG and H. WANG, University of Utah, USA

DOI: 10.1533/9780857093493.3.431

Abstract: The sintering of nanosized particles exhibits a number of distinctively unique phenomena compared to the sintering of coarse powders, e.g. low sintering temperatures and rapid grain growth. This chapter aims to bring into focus the understanding of the fundamental issues of nanosinteirng, including the thermodynamic driving force of nanosintering, non-linear diffusion and the kinetics of nanosintering, and the relationships between agglomeration, densification and grain growth. This chapter will also examine the effects of microstructure and processing variables.

Key words: nano-particle sintering, sintering, densification, grain growth, coarsening, size effect.

Note: This chapter is adapted from Chapter 17 'Sintering of ultrafine and nanosized particles' by Z. Z. Fang and H. Wang, originally published in *Sintering of advanced materials: fundamentals and processes,* ed. Z. Z. Fang, Woodhead Publishing Limited, 2010, ISBN: 978-1-84569-562-0.

13.1 Introduction

Since the emergence of nanoscaled science and technology, the sintering of nanosized particles has been a topic of both scientific and technological importance. Sintering is a phenomenon that occurs in a broad range of nano materials processes, including the synthesis of nano particles and fabrication of bulk nanocrystalline materials. Moreover, sintering is an important factor in determining the stability of nano materials, nanoscaled coatings, and nano devices. Although sintering of nano particles shares the same basic principles as that of sintering of coarser particles, a number of issues and

431

challenges are unique to sintering of nano particles. For example, the thermodynamic driving force for sintering of nano particles is extremely large, calling into question the use of conventional sintering doctrines based on linear diffusion theories.

In the context of engineering processes, sintering implies the bonding of one solid particle to another. Sintering consists of two intertwined processes: densification and grain growth. A unique issue of sintering of nano particles is that nano particles almost always experience extremely rapid grain growth, rendering the loss of nanocrystalline characteristics in fully sintered states. With respect to manufacturing bulk nanocrystalline materials from nanoscaled particles, the objective of sintering of nano particles is to achieve maximum densification while retaining nanoscaled grain sizes. This goal, however, has been very difficult to reach. Fundamentally, there are two main reasons for this technological impasse. One is that the same factors that result in densification also cause grain growth. In other words, both densification and grain growth processes share the same driving force and mass transport mechanisms. Furthermore, in many cases, grain growth is required in order to break the local interfacial energy balance necessary to sustain continuous elimination of pores.[1]

Innovative sintering technologies include the two-step sintering technique and various pressure-assisted sintering techniques. The two-step sintering technique was designed to decouple the densification and grain growth processes that occur during conventional pressureless sintering. Conventional pressure-assisted processes including hot pressing and hot isostatic pressing and unique processes such as microwave sintering and spark plasma sintering (SPS) are all applied in the research of sintering of nano particles.

Since the early 1990s, there have been a few comprehensive reviews that deal with the topic of sintering of nano particles or processing of nanosized particles.[2,3] The issues of sintering of nano particles and the difficulties of manufacturing bulk nanocrystalline materials from nanoscaled powders are also discussed in other reviews that cover broader topics of nano materials.[4,5] These reviews collectively provide a strong foundation for understanding the science and technology involved in sintering of nanosized particles. This chapter, based on a comprehensive review by the authors,[6] focuses on the size-dependent properties and their effects on sintering. The relationship between grain growth and densification is highlighted.

13.2 Thermodynamic driving force for the sintering of nanosized particles

In general, the sintering of nanosized or nanocrystalline powders follows the same path as larger grain powders. However, compared to conventional micron-sized or submicron-sized particles, the densification behavior of nano particles during sintering is notably different. From the perspectives of thermodynamics, the driving force for sintering particles of any size is the reduction of surface energy. Based on conventional sintering theories, the driving force of sintering can be given by[7]

$$\sigma = \gamma \kappa = \gamma \left(\frac{1}{R_1} + \frac{1}{R_2} \right) \qquad [13.1]$$

where γ is the surface energy of the material, κ is the curvature of a surface, which is defined by $\kappa = \frac{1}{R_1} + \frac{1}{R_2}$ (for a convex surface, it is taken to be positive; for a concave surface, it is taken to be negative), and where R_1 and R_2 are the principal radii of the curvature. The driving force for the sintering of nanosized particles is, therefore, inversely proportional to the size of the particles. This relationship would lead to a much higher driving force for the sintering of nanosized particles compared to micron-sized particles. For example, based on equation 13.1, the driving force for a ten nanometer particle is two magnitudes higher than that for a one micron particle.

The large driving force of sintering of nano particles can be even higher than the result of equation 13.1 if the non-linear dependency of vacancy concentrations on the particle size is considered. During sintering, mass transport, usually mediated by vacancies, is driven by the difference in vacancy concentration $\Delta C_v = C_v - C_{v0}$, where C_v is the vacancy concentration for a surface with curvature of κ, and C_{v0} is vacancy concentration for a flat surface. Based on the Gibbs–Thomson equation,[8]

$$C_v = C_{v0} \exp\left(-\frac{\gamma \kappa \Omega}{kT} \right) \qquad [13.2]$$

where Ω is the atomic volume, k is Boltzmann's constant, and T is the absolute temperature. For micron-sized particles, the term $\frac{\gamma \kappa \Omega}{kT} \ll 1$ in equation 13.2 becomes linear, $C_v \approx C_{v0}\left(1 - \frac{\gamma \kappa \Omega}{kT}\right)$, therefore

$$\Delta C_v \approx -C_{v0} \frac{\gamma \kappa \Omega}{kT} \qquad [13.3]$$

However, when particle size approaches nanoscale, the linear approximation is no longer valid. The correct expression for the driving force of mass

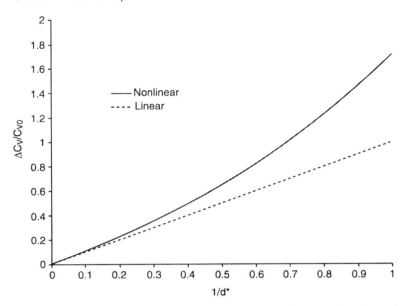

13.1 Schematic comparison of vacancy concentration as a function of particle size between linear approximation and non-linear equation.

transport should be given as

$$\Delta C_v = C_{v0} \exp\left(-\frac{\gamma\kappa\Omega}{kT}\right) - C_{v0} = C_{v0}\left(\exp\left(-\frac{\gamma\kappa\Omega}{kT}\right) - 1\right) \quad [13.4]$$

Equation 13.4 shows that the driving force for mass transport during sintering is a nonlinear function of the surface curvature, and it increases exponentially when particle size decreases to nanoscale. Figure 13.1 schematically illustrates the relationship of $\frac{\Delta C_v}{C_{v0}}$ vs. $-\frac{\gamma\kappa\Omega}{kT} = \frac{1}{d^*}$, where d^* is related to the particle size. This nonlinear relationship of the driving force for sintering of nano particles is expected to have a dramatic effect on the kinetics of sintering.

The driving force of sintering of nano particles is also affected by specific surface energy – γ. The value of γ is also a function of the particle size. Campbell et al.[9] studied the effect of size-dependent nano particle energetics on catalyst sintering. By using microcalorimetric measuring the heat of adsorption of Pb onto MgO (100), they showed that the surface energy increases substantially as the radius decreases below ~3 nm, as shown in Fig. 13.2.

Independently, Nanda et al.[10] showed that the surface energy of nano particles is significantly higher than that of the bulk, as demonstrated by studying size-dependent evaporation of Ag nano particles.

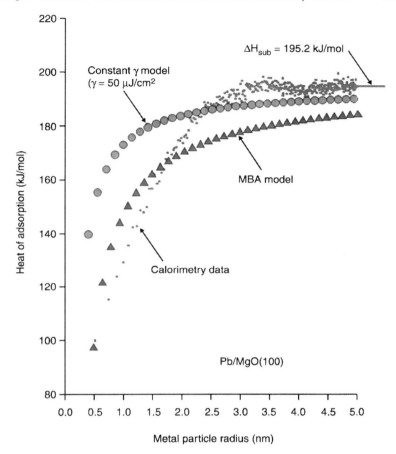

13.2 Illustration of the discrepancy between the measured differential heat of adsorption of Pb onto MgO(100) and the prediction based on constant γ model. Surface energy increases substantially as the radius decrease below ~3 nm.

From another aspect, the surface energy is also a function of the crystal orientations, which could affect the driving force of sintering of nano particles. Groza[2] pointed out that because nanocrystals have significant surface area, the problem of anisotropy becomes even more critical. Although there are numerous studies reporting that nano particles have anisotropic faceted morphology, direct correlation of their effect on surface energy and sintering is not available in the literature and, understandably, is very difficult to perform. One phenomenon that has been reported regarding nanocrystals with faceted morphology is that they form an oriented-attached assembly as shown in Fig. 13.3.[11] The oriented attachment of nano

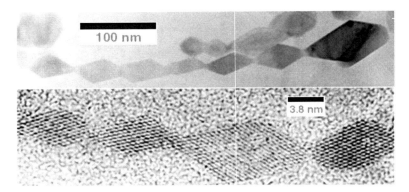

13.3 Oriented attachment of nanosized titania under hydrothermal conditions.

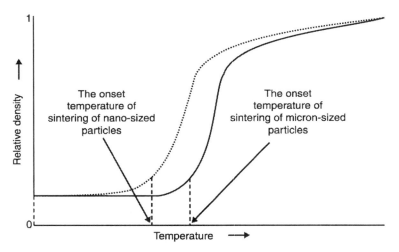

13.4 Schematic diagram illustrating different onset temperatures of sintering of nano and micron-sized particles.

particles could affect the coalescence of these particles and the densification and grain growth during sintering.

13.3   Kinetics of the sintering of nanosized particles

13.3.1 Sintering temperature

Notably, the sintering of nanosized particles occurs at lower temperatures than the sintering of conventional micron-sized or submicron-sized powders. Sintering temperature is, in general, a loose concept, referring to the entire temperature range of densification. In order to be specific and

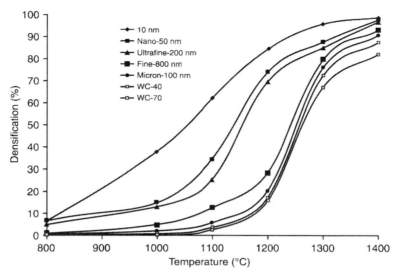

13.5 The percentage of densification of WC-Co as a function of the continuous heating temperature for various initial particle sizes.

quantitative, the starting temperature is often used for comparison. However, because sintering is a continuous kinetic processes, rigorously speaking, a single point of demarcation for the starting temperature of sintering does not exist. Based on typical experimental behavior, the starting temperature can be defined as the temperature at which the rapid densification stage initiates, as marked on Fig. 13.4. In general, the densification versus temperature plot shifts to the left (lower temperature) when nanosized powders are used rather than micron-sized powders.

For example, several studies on the sintering of nano yttrium stabilized zirconia (YSZ) have shown that the sintering temperature of nanocrystalline ZrO_2 initiates at a temperature 200°C lower than that of the micro-crystalline powders.[12–14] An even greater temperature difference of sintering − 400°C − was reported by Mayo[15] for nanosized titania compared to commercial TiO_2 powders. Similar results were observed when sintering nano ceria[16] and nano titanium nitride powders.[17]

A comprehensive study on the sintering of nano tungsten carbide and cobalt (WC-Co) powders was conducted by Maheshwari et al.[18] Figure 13.5 shows the percentage of densification as a function of the continuous heating temperature for various initial particle sizes. Clearly, the entire sintering temperature decreases steadily as the initial average particle size decreases from 30 microns to 10 nanometers. It seems, however, that there is little difference between the onset temperatures of the sintering of particles greater than one micron.

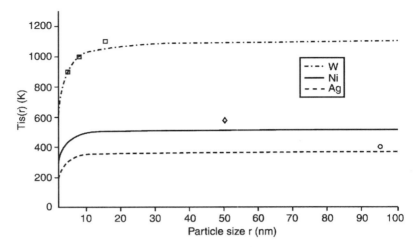

13.6 Theoretical prediction of initial sintering temperature –
$T_{is}(r)$ – for selected metals in terms of equation 13.6. The experimental
initial sintering temperature of W, Ni and Ag are also plotted in
for comparison.[37,38]

General sintering theories hold that a material's sintering temperature is
often correlated with the material's melting point. It has long been known
that the melting temperature of very fine particles decreases with the size of
particles.[19–34] In the case of sintering of nano particles, the decreasing onset
temperature of sintering can be understood, therefore, on the basis of the
lower melting temperature of nano particles. Troitskii et al.[35] studied the
initial sintering temperature of different-sized TiN powders and found the
relationship between initial sintering temperature T_{is} and particle size r is:

$$T_{is}(r) = T(\infty) \exp[-c \cdot (r' - r)/r] \qquad [13.5]$$

where $T(\infty)$ is the initial sintering temperature of coarse particles, c is a
constant determined by the properties of the material and the energetic state
of the surface layer, and r' is arbitrary size.

Jiang and Shi[36] ascribed the size-dependent initial sintering temperature
of nanosized particles to the decreased melting temperature of nanosized
particles based on the relationship:

$$T_{is}(r) = 0.3 T_m(r) = 0.3 \cdot T_m(\infty) \exp\left[-\frac{2S_m(\infty)}{3k} \frac{1}{(r/r_0) - 1}\right] \qquad [13.6]$$

where $T_m(\infty)$ is the melting temperature of the bulk material, $S_{vib}(\infty)$ is the
bulk melting entropy, r is particle radius, $r_0 = 3h$ (h is atomic diameter) for
nano particles and k is Boltzman constant. Figure 13.6 shows the predicted
size-dependent initial sintering temperature of some metallic powders by

using equation 13.6.[36] Note that the significant changes of the initial sintering temperature do not occur until the particle size is less than approximately 20 nm.

Scaling law – dependence of sintering on particle size

In conventional sintering theories, the dependence of densification behavior on the size of particles is described by the scaling law. In 1950, Herring[39] first introduced the scaling law as follows:

$$\Delta t_2 = \lambda^n \Delta t_1 \qquad\qquad [13.7]$$

where $\lambda = R_2/R_1$, R_1 and R_2 are particle radius, n depends on specific diffusion mechanisms of the densification. Specifically, $n = 1$ for viscous flow, 2 for evaporation and condensation, 3 for volume diffusion, and 4 for surface diffusion or grain boundary diffusion. The scaling law states that the time required to sinter powders with particle radii of R_1 and R_2 is proportional to the ratio of the particle radius. Although the densification behavior of nano powders can be qualitatively understood on the basis of the scaling law, few direct analyses of experimental data exist in the literature. The few studies that did apply the scaling law used the following expression to analyze the activation energies of the sintering of nano powders:[40,41]

$$n \ln\left(\frac{d_1}{d_2}\right) = \frac{Q}{R}\left[\frac{1}{T_2} - \frac{1}{T_1}\right] \qquad\qquad [13.8]$$

where d_1, d_2 are particle sizes, T_1 and T_2 are corresponding sintering temperatures, R is the gas constant, and Q is the activation energy. By using the above equation, some studies obtained activation energy values that are closer to grain boundary diffusion, while others obtained values closer to volume diffusion, which is believed to be unlikely at low temperatures. These discrepancies can in part be attributed to the assumption inherent in the scaling law that the particle size of two different powders does not change during sintering and microstructural changes remain geometrically similar for the two systems.[42] However, if the values of n and Q/R can be evaluated by other methods, e.g. curve-fitting experimental data to densification equations, equation 13.8 can be used to estimate the sintering temperature of different particle-sized powders.

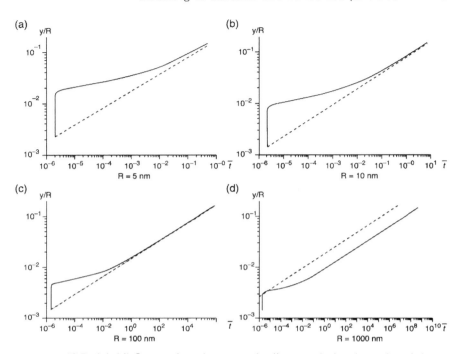

13.7 (a)–(d) Comparison between the linear solution (- - - -) and the nonlinear solutions (——) for the shrinkage between two spherical particles as functions of time.

13.3.2 Kinetic theories, modeling, and simulations of sintering of nano particles

Given the unique physics that presents when sintering nano particles, Pan recognized that the rapid kinetic rate of sintering is a direct result of the large driving force for sintering of nanosized particles, and revised the two-sphere sintering model by using non-linear diffusion law.[43] Because the diffusion is the result of jumping atoms, the flux of diffusion as a function of the frequency of jumping (f), volume atomic concentration (C_{solid}) and the atomic spacing (a) can be given by

$$J = \frac{2D}{a\Omega} \sinh\left(\frac{aF}{2kT}\right)$$
[13.9]

where D is the diffusion coefficient, Ω is the atomic volume, a is the atomic spacing, F is the driving force for diffusion, and k and T are the Boltzmann constant and absolute temperature respectively. Pan pointed out that this equation reduces to linear diffusion law, when $aF \leq kT$, then $\sinh(aF/2kT) \approx$

$aF/2kT$. However, when particle sizes are in the range of nanometers, the linear approximation is no longer reasonable. Then, the diffusion equation becomes a nonlinear equation that can only be solved via numerical methods. Applying this approach to sintering two particles, the ratio of the neck-to-particle radius as a function of the length of time at a given temperature was calculated and shown by Fig. 13.7. The differences between prediction by linear solutions and nonlinear solutions are significant during the initial stage of sintering and diminish as sintering time increases. The distinction between linear and nonlinear solutions also diminishes as particle size increases.

The rapid rate of sintering due to the rapid rate of diffusion is also supported by recent studies which indicate that the coefficient of diffusion, D, is size dependent as shown in equation 13.10,[44]

$$D(r, T) = D_0 \exp\left[\frac{-E(\infty)}{RT} \exp\left[\frac{-2S_{vib}(\infty)}{3R} \frac{1}{r/r_0 - 1}\right]\right] \qquad [13.10]$$

where D_0 is pre-exponential constant, $E(\infty)$ is bulk activation energy, $S_{vib}(\infty)$ is bulk melting entropy, r is particle radius, $r_0 = 3h$ (h is atomic diameter) for nano particles, R is ideal gas constant, and T is absolute temperature.

The dependence of the coefficient of diffusion on particle size is attributed to the theory that, as the size of the nanocrystals decreases, the activation energy of diffusion decreases and the corresponding coefficient of diffusion increases based on the Arrhenius relationship between them. Together, these theories, based on nonlinear diffusion law and the increase of the coefficient of diffusion with decreasing particle size, convincingly explain the rapid formation of necks bonding with neighboring particles.

The rapid kinetics of the sintering of nano particles were also demonstrated by using molecular dynamic simulations (MD).[45–52] The basic approach for simulating sintering using the MD method involves tracking the motion of atoms under stress caused by surface or interfacial energy. The kinetics of sintering is given as the rate of decreasing distance between two atoms in the middle of two particles in contact. It was shown that sintering of nano particles at the atomic level can be accomplished by dislocation motion and grain boundary rotation, as well as other mechanisms. It was further predicted that the sintering time of nano particles would be in the range of a few hundred picoseconds. Although the predicted sintering time is far from engineering reality, the results of the simulation can be used as a basis for understanding the initial bonding or formation of the necks between nano particles.

13.3.3 Densification mechanisms during sintering of nano particles

Using the various methods for figuring activation energies, activation energies for sintering various nanosized powders were reported. For example, a very low activation energy for densification is observed in initial sintering – about $234\,kJmol^{-1}$ for nanocrystalline Al_2O_3 and $96.2\,kJmol^{-1}$ for nanocrystalline TiO_2,[53] $268\,kJmol^{-1}$ for nanocrystalline ZnO,[54] $66.2\,kJmol^{-1}$ for nanocrystalline nickel,[55] $82\,kJmol^{-1}$ for nanocrystalline α titanium and $49\,kJmol^{-1}$ for nanocrystalline β titanium.[56]

It can be seen from the data that the majority of studies point toward lower activation energies for early stages of sintering. This is reasonable for the obvious reason of the huge surface areas and the expected high activity of nano particles. Surface diffusion is one of the most cited mechanisms that contribute to the sintering of nanosized particles. However, in conventional sintering theories, surface diffusion is believed to induce initial neck formation between particles, but not densification. This seemingly conflicting theory of the effects of surface diffusion on sintering of nano particles can be understood from a perspective of the indirect role of surface diffusion to densification.

First of all, the rapid and active surface diffusion may lead to grain boundary slip and rotation of particles that may result in the rearrangement of particles, hence the increased density of the compact. The possibility of grain boundary slip and rotation was mentioned in numerous sintering studies.[2,3,57] Evidence of nano particle coalescence via surface diffusion was presented in Shi's study on barium titanate[58] and Bonevich and Marks's study of Al_2O_3[59] using TEM. Figure 13.8 shows the formation of the neck between two Al_2O_3 particles after sintering at 1000°C for just a fraction of a second. This study shows that surface diffusion is the predominant mechanism for sintering, as evidenced by the fact that the faceted interfaces are similar to ledge growth, and the sintered particles retain their initial adhesion structure with no reorientation occurring during sintering. The driving force for sintering can be considered a chemical potential difference between facet surfaces and the neck region.

The indirect role of surface diffusion on densification can also be understood based on theories of the relationship between coarsening and sintering of particles.[1,58,60,61] As discussed earlier, effects of pores on sintering of nano particles, according to theories first proposed by Kingery and Francois and further elaborated by Lange et al.,[62,63] a pore will shrink during sintering only if the coordination number of the pore is smaller than a critical value $n < n_c$ because only then the surface of the pore is concave. Thermodynamic driving force dictates that mass will diffuse from convex surfaces to concave surfaces. Initial sintering of a compact will develop an

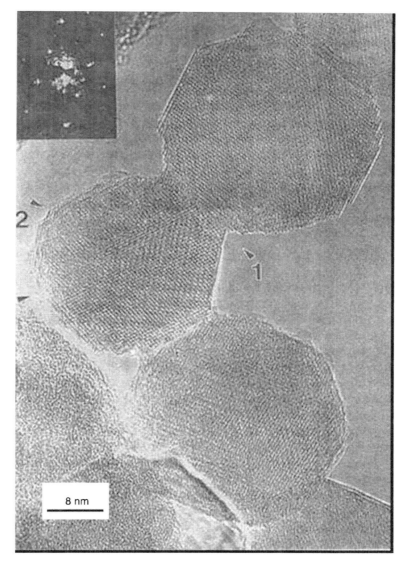

13.8 Particle chain sintered with no reorientation. Gap between the particles (▸1) was filled by surface diffusion that has also roughened the middle particle's surfaces (▸2).[59]

equilibrium configuration at which the driving force for local sintering is zero. Grain growth, by coarsening, will perturb the equilibrium configuration to reinitiate densification. In other words, when the coordination number reaches a critical value, the pore is at equilibrium. The shrinkage of the pore cannot progress until the equilibrium condition can be tipped in favor of sintering by grain growth. With respect to the densification mechanisms of nano particles, surface diffusion can cause coarsening of nano particles which, in turn, contributes to the process of densification. Therefore, it can be stated that by inducing coarsening, surface diffusion will contribute indirectly to densification.

Surface pre-melting is another mechanism that could lead to rapid densification at low temperatures during sintering of nano particles. Due to a large surface to volume ratio in nano particles, surface pre-melting can happen at low temperature, and as a result, particle rearrangement is facilitated by sliding, rotation or viscous flow. Alymov et al.[38] calculated the dependence of the melting point of a particle as a function of its size using the following equation:

$$T_m/T_0 = 1 - 2Q^{-1}\rho^{-1}{}_s[\sigma_{sl}/(r - \delta) + \sigma_{lg}r^{-1}(1 - \rho_s/\rho_l)] \qquad [13.11]$$

where T_0 is the bulk melting point of the solid, Q is its latent heat of fusion, σ_{sl} and σ_{lg} are the interfacial surface tensions between the solid and the liquid and between the liquid and its vapor respectively, ρ_s and ρ_l are the densities of the solid and liquid respectively, r is the radius of particle, and δ is the thickness of melted layer on a particle surface.

Given that the sintering temperature is proportional to the melting point, it is generally understood that as the melting point decreases, the sintering temperature decreases. It has been demonstrated that the melting of a particle with diameter d will result in coagulation with its neighbors and will become the center of a new, larger particle. In an independent study of the sintering of nanometric Fe and Cu, Dominguez et al.[64] attributed the initial densification to surface melting mechanisms because the activation energies that were obtained from either constant heating or isothermal experiments were too small to ascertain lattice diffusion mechanisms. Therefore, it was reasoned that the presence of a liquid-like layer on the surface of the nanometric particles during sintering could simultaneously explain such phenomena as high diffusivity and enhanced grain growth at a narrow temperature range.

A more generally applicable theory that explains the rapid densification of nano particles is based on the hypothesis of non-equilibrium high concentration of vacancies at the inter-particle grain boundaries. In 1974, Vergnon et al. studied the 'initial stage for the sintering of ultrafine particles TiO_2 and Al_2O_3'.[53] Using flash sintering and isothermal experimental techniques, he showed that during the first 20 seconds, a fraction of the total observed shrinkage, up to 95%, was registered.[53] There was an initial loss of surface area, before the shrinkage starts during the heating of the compact to the desired temperature, a process which requires only a few seconds. It was reasoned that this almost instantaneous loss of the surface area corresponds to the formation of junction zones between particles of the compact. The fast formation of the junctions between particles, before the shrinkage onset, involves the creation of a high concentration of vacancies inside these junctions. The shrinkage of the compact results then from a decrease of the distance between the centers of particles due to annihilation of the trapped vacancies in the junction zone. Because the concentration of trapped vacancies inside the junction zone largely exceeds the thermo-dynamic equilibrium concentration, the diffusion can be considered as independent of time and controlled only by the probability of jumping of ions, as long as the concentration of vacancies exceeds the equilibrium content. Any further sintering, after the initial non-equilibrated concentra-tion of vacancies is exhausted, corresponds with the diffusion of equilibrated vacancies.

Furthermore, based on the theory that excessive concentration of vacancies exist ($c > 10^{-4}$), Trusov et al.[65] stipulated that there is a possibility of liquid-like merging (coalescence) of particles into large ones. Liquid-like coalescence, as well as slippage, causes the compact shrinkage of ultrafine particles.

In another study focusing on size-dependent grain growth kinetics observed in nanocrystalline Fe, Krill et al.[66] also established their model on the basis of existence of excess volume at the grain boundaries. The 'excess' volume is in the form of vacancies, which leads to a non-equilibrium vacancy concentration. The issues of grain growth of nano particles during sintering will be further discussed in later sections of this chapter.

Finally, the rapid densification mechanisms of nano particles are also related to the preferential crystalline orientations. It has been observed that in loose nanocrystalline powders, the first neck formation occurs not randomly between particles, but by the orderly mating of parallel, crystallographically aligned facets on the particle surfaces.[59,67] Some nanocrystalline powder compacts also appear to reflect a kind of ordered structure resulting from less than random type matings of particles during the initial stage of sintering.[68]

13.3.4 Effect of green density, agglomeration and pore

In the practice of sintering of nano particles, the densification behavior of nano particles is affected not only by the intrinsic nature of the nanoscale size of the particles, but also by the processing conditions and related difficulties, such as green density and agglomeration.

First, similar to powder compacts of micron-sized powders, the densification of a powder compact depends significantly on the green density of the compact. Green density must be sufficiently high in order to achieve adequate densification under similar sintering conditions. On the

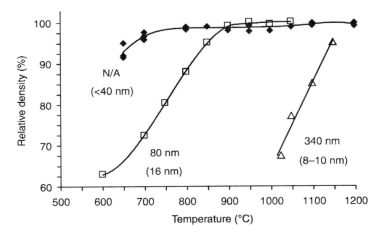

13.9 Densification behavior of nanocrystalline TiO₂ with three different agglomerate sizes: note that the larger the agglomerate size, the higher the sintering temperature (agglomerate size in bold, crystallite size in light). For the non-agglomerated (N/A) powder, sintering time is 120 min;[69] for the 80 and 340 nm agglomerate powders, sintering time is 30 min.[70,71]

other hand, the finer the particle sizes, the lower the green density of powder compacts, assuming the compaction pressure is the same. It should be noted, however, that nanosized particles can be sintered from a green density that is much lower than is possible for sintering coarse (micron or submicron) particles.

It has been widely recognized that agglomeration of nano particles has a critical impact on the sintering of nano particles. Due to the extremely fine size and the strong interactive force between particles, nano particles tend to form agglomerates. The size and strength of the agglomerated particles affect the densification rate. The most direct investigation of the agglomeration of densification was summarized by Mayo,[3] whose data were based on numerous published experimental results as shown in Fig. 13.9.

In essence, a powder compact can be viewed as consisting of a bi-level hierarchical structure: the compact is made of agglomerates which consist of nanosized particles. There is, therefore, a bi-model pore size distribution. The pores existing within agglomerates are finer than the pores between agglomerates. The densification of an individual agglomerate is relatively easy, while the elimination of the inter-agglomerate pores is more difficult. By tracking the evolution of pore size distributions, Petersson and Ågren[72] studied the sintering of fine grain cemented tungsten carbide and cobalt system (WC-Co). They showed that during the intermediate stage of sintering, the considerable densification obtained is primarily connected to removal of small pores rather than shrinkage of larger ones.

From the perspective of achieving full densification and elimination of pores, it is logically desirable to de-agglomerate powders or to avoid the formation of agglomerates in the first place. Fundamentally, the formation of agglomerates is attributed to balance of the inter-particle forces, specifically the van der Waals force, which acts to bind particles, and the electrostatic repulsion which opposes agglomeration. To de-agglomerate, opposite measures must be taken in order to stabilize a colloidal solution. In other words, the repulsive forces must be boosted to achieve a balance between the attractive and repulsive forces such that the dispersion of particles can be stabilized. Methods for nano particle dispersion include electrostatic charge stabilization, steric stabilization, or a combination of the two. Details of the principles of stabilizing colloidal solutions can be found elsewhere.[73] These techniques can be implemented in the processes, including mixing and milling of the powders that are necessary prior to sintering of these materials.

To sinter nano particles for the fabrication of bulk engineering components, a colloidal solution must be dried; agglomerates will inevitably form. Ideally, the agglomerates are soft and the inter-agglomerate pores are small. Lange provided a more extensive discussion of ceramic powder processing techniques for avoiding agglomeration and achieving uniform pore distributions within a powder compact.[74]

Effect of pores on sintering of nano particles

A common thread for the effect of green density and agglomeration on sintering is the effect of pores on densification.[63] A compact consists of particles and pores, and each pore has a volume, shape and coordination number. The pore coordination number is defined as the number of touching particles surrounding and defining each void space. A pore's surface morphology is determined by the dihedral angle and the pore's coordination number. In general, for a given dihedral angle, a critical coordination number, n_c, exists that defines the transition of the pore surface morphology from convex $(n > n_c)$ to concave $(n < n_c)$. Kingery and Francois[62] first recognized that only those pores with $n < n_c$ are able to shrink during sintering because the concave surface morphology with negative chemical potential is thermodynamically unstable. As a result, atoms will diffuse to the pore surface and fill the void space. Figure 13.10 illustrates the stability of pores and their dependence on both the dihedral angles and the coordination number. For any given dihedral angle, which is dictated by the material, a critical coordination number exists below which the pores will shrink and above which the pores will grow.

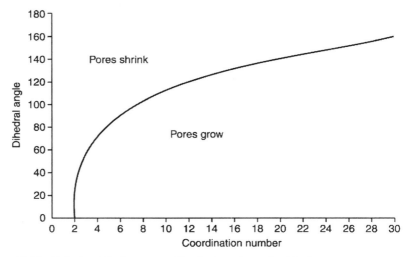

13.10 Relationship between dihedral angle and critical pore coordination number.[62]

The effects of green density and agglomeration on densification can be explained by the pore coordination number theory. Higher green density and less agglomeration result in fine and uniform pores that shift the pore coordination number distribution from high values to low values, i.e. more pores fall into the category below the critical pore coordination number. These pores are easily removed during sintering and thus lead to denser products. As for the large pores that are thermodynamically stable and have coordination numbers higher than the critical value, a process by which the coordination number can be reduced during sintering is essential, since the pores will again become unstable and the densification will then continue. Particle rearrangement and grain growth are the two processes that can play this role, creating a dilemma, of course, and difficulty for any attempt to achieve maximum densification without grain growth.

13.4   Grain growth during sintering of nano particles

13.4.1 The unique issue of grain growth during sintering of nano particles – significance of non-isothermal grain growth

A primary motivation for studying sintering of nanosized particles is rooted in the issue of rapid grain growth during sintering. In many cases, particularly when the goal is to produce nanocrystalline bulk materials, the objective of sintering of nano particles is to achieve full densification as well as the retention of nanoscaled grain structure in the sintered material.

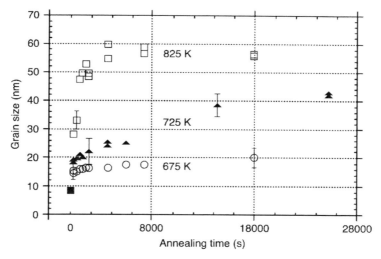

13.11 Evolution of the grain size as a function of the annealing time at three annealing temperatures for nanocrystalline iron. The grain size was determined by the Scherrer equation.

Research has generally shown that after sintering, nanosized particles lose nanoscale characteristics because grain size grows to greater than 100 nm. Therefore, understanding and controlling grain growth is a critical scientific and technical issue of sintering of nano particles.

In a systematic study of the stability of nanosized metal powders, Malow and Koch[75–77] reported that the rate of grain growth of nanocrystalline iron (Fe) powders made by ball milling is initially very rapid (<5 min) when annealed at various temperatures. Grain growth then stabilizes during extended isothermal holding (up to 142 hours). During isothermal holding, grain growth follows a generalized parabolic grain growth law and is similar to that found in bulk materials. It is noted, based on Fig. 13.11, that at the first data point of the isothermal annealing curves at higher annealing temperatures (825 and 875 K), the grain sizes are already several times (3–6×) greater than the original as-milled grain size (~8 nm) (Fig. 13.13). In other words, grains grow rapidly during heat-up, prior to reaching the pre-selected isothermal holding temperature.

In another study of the grain growth of nanocrystalline Fe using in-situ synchrotron X-ray diffraction techniques, Krill et al.[66] further demonstrated that grain growth of nano Fe particles comprises three steps: the 'initial growth spurt,' a linear growth stage, and the normal parabolic stage, as shown in Fig. 13.12. Once again, the normal parabolic stage can be modeled using the classic grain growth parabolic law; however, the 'initial growth spurt' of nanocrystalline Fe during annealing was not captured by isothermal studies.

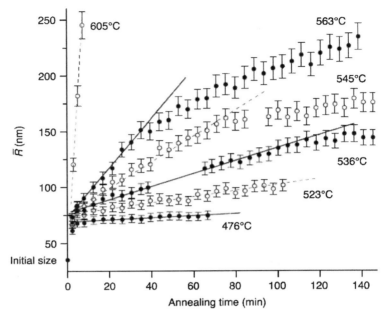

13.12 Size-dependent grain growth kinetics observed in nanocrystalline Fe.

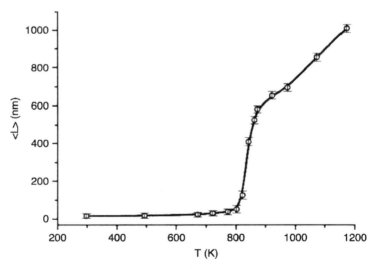

13.13 Change of the mean grain size (the linear intercept) with annealing temperature, measured in pure nanocrystalline Co.

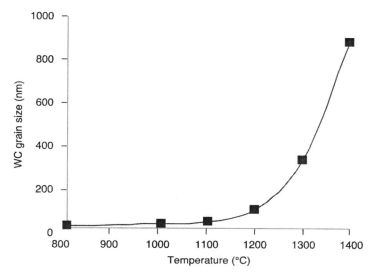

13.14 Grain size vs. temperature during heating up of nanocrystalline WC-Co powder at a heating rate of 10°C/min.

Grain growth during sintering of nano particles is also a strong function of temperature. Figure 13.13[78] shows the relationship between grain size and temperature during the heat treatment of nanocrystalline cobalt powder. It is obvious that the grain growth is initially slow at very low temperatures and that it accelerates dramatically when the temperature is above an apparent critical temperature range. Figure 13.14 provides another example of the relationship between grain size and temperature during heating up of nanocrystalline WC-Co powder at a heating rate of 10°C/min.[79] It shows that the original 20 nm grain size has increased almost 45-fold to 900 nm. This 'explosive' grain growth occurs almost instantly during heat-up, with no significant holding time. Similar behavior has also been reported for sintering of other nanocrystalline ceramics, as well as for metallic powders.[80–86] It appears that a critical temperature exists, above which the grain growth accelerates dramatically as a function of temperature.

The issue of grain growth during sintering can be studied by examining the grain size versus relative density relationship. This approach has been applied to the study of sintering of nano particles. A typical relationship between grain size and density during sintering of nano particles is schematically shown in Fig. 13.15. In one of the earliest studies of the sintering and grain growth of nanosized ceramic powders in the 1990s, Owen and Chokshi[87] and Averback et al.[88] showed that oxides densify without significant grain growth until the density reaches approximately 90% of the bulk density. Then the grain growth becomes very rapid. This

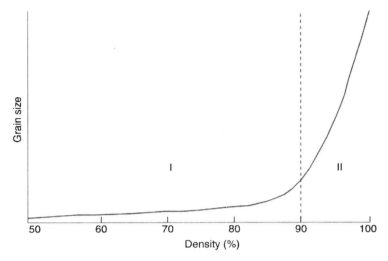

13.15 Relationship between grain growth and densification during sintering of nano particles. I: early stage of grain growth; II: late stage of grain growth.

phenomenon is observed in many different materials.[2,3,89,90] This relationship implies that the grain growth during sintering consists of two stages: the early stages of sintering, before the powder compact reaches 90% relative density; and the late stages of sintering, when relative density is greater than 90%. It is believed that the late stages of grain growth can be viewed as *'normal'* grain growth, similar to that found in bulk materials by boundary migration, but incorporating the effect of pinning by closed pores. In contrast, the early stage of grain growth during sintering is often referred to as *'coarsening.'*

13.4.2 Initial grain growth – coarsening – of nano particles during early stages of sintering (rel. density < 90%)

The above discussion provides evidence of an initial stage of grain growth. This part of grain growth occurs in the beginning of the sintering, often during heating up when the relative density is less than 90%. Thus, the initial grain growth during sintering of nanosized particles can be treated as non-isothermal grain growth. In conventional sintering of micron-sized powders, the contribution of the initial grain growth to the final grain size is relatively minor, compared to that of the normal grain growth during late stages of sintering. For sintering of nano particles, however, the amount of the initial grain growth is significant and sufficient in many cases to cause the material to lose its nanocrystalline characteristics.

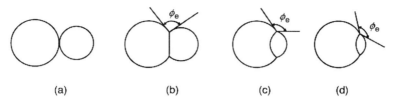

13.16 A linear array of two spheres of initial radii of r_1 and r_2 $(r_1 > r_2)$: (a) just in touch without the formation of interface, (b) when r_1/r_2 R_c, (c) $r_1/r_2 = R_c$, and (d) $r_1/r_2 > R_c$.[60]

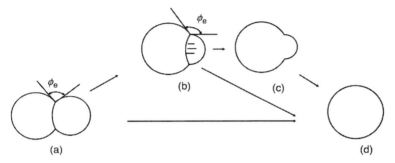

13.17 Particle configuration change after the formation of a dihedral angle shown in Fig. 13.20: (a) the configuration when r_1/r_2 R_c (boundary cannot move); (b) the configuration resulted from the mass transport between particles before boundary motion; (c) the transient configuration after boundary motion where r_1/r_2 becomes $> R_c$; (d) final configuration either directly by mass transport or by combined mass transport and boundary motion.

Neck formation and coarsening of contacting nano particles

To understand initial grain growth, the key issue is the interaction between ultrafine particles at the start of sintering. According to classical sintering theories by Kuczynski,[91] Kingery,[92] Coble[93] and Johnson,[94] necks will form and grow between adjacent particles, which are assumed to have equal diameter. Densification is modeled as the approach of the centers of the two particles. In this situation, no grain growth occurs at the beginning of sintering. In practice, however, there are always wide particle size distributions. The densification and grain growth behavior will be markedly different from two-sphere models. Figure 13.16 illustrates that when very fine particles are in contact, where the particle sizes are not uniform, inter-particle diffusion will lead to coarsening of particles, in addition to formation of the neck. Large particles will grow at the expense of small particles. The coarsening of particles can be understood using the criteria shown by equation 13.12, which was first expressed by Lange[1] based on

Kingery's initial concept of pore stability.[62]

$$R_c = -\frac{1}{\cos \phi_e} \qquad [13.12]$$

R_c is called critical particle size ratio for boundary migration, ϕ_e is the dihedral angle relating surface energy and grain boundary energy. Lange explained that when the size ratio between two particles is larger than the critical size ratio R_c, grain boundary migration will occur, resulting in grain growth. When actual size ratio is less than R_c, boundary migration will yield an increase in the grain boundary area and is energetically unfavorable. In this situation, inter-particle mass transport will happen first, in order to increase the size ratio between adjacent particles. This coarsening process will not stop until the size ratio $R = r_1/r_2$ reaches R_c. Then grain boundary migration will take over because the condition for grain boundary migration is now energetically satisfied.

The studies by Lange and Kingery aimed to explain the stability of pores in the intermediate stage of sintering. Shi further applied the critical size ratio criteria to the initial sintering of ultrafine particles.[60,61] It was shown that the driving force for neck growth and inter-particle diffusion are given respectively as follows:

$$\Delta\mu_n = \gamma_s \Omega \left(\frac{1}{X} - \frac{1}{r} \right) \qquad [13.13]$$

$$\Delta\mu_c = 2\gamma_s \Omega \left(\frac{1}{r_1} - \frac{1}{r_2} \right) \qquad [13.14]$$

$\Delta\mu_n$ and $\Delta\mu_c$ are chemical potential for neck formation and mass transport between two particles; γ_s is surface energy; Ω is atomic volume; X is radius of the neck; r is radius of particles (r_1 and r_2 are radii of two particles with different sizes).

Equation 13.14 indicates that if a difference in the radius of curvature exists, mass transport would take place from the area of larger curvature to the area of smaller curvature. This process is related to the particle coarsening.

Considering equations 13.13 and 13.14 together, both the neck growth and coarsening, driven by the surface tension between the particles, can take place concurrently. However, the magnitude of the driving force for the two processes is different. Assuming the interface energy is not considered, then $|\Delta\mu_n| > |\Delta\mu_c|$, which implies that neck formation takes place before coarsening.

On the other hand, if the interface energy between particles is considered in the analysis of the driving forces as an energy barrier to neck growth,

Shi[50] showed that equation 13.13 becomes

$$\Delta\mu_n' = \gamma_s\Omega\left(\frac{1}{X} - \frac{1}{r}\right) - \gamma_b f(r, \phi, \phi_e)$$ [13.15]

where γ_b is boundary energy, ϕ is the contact angle and ϕ_e is the equilibrium dihedral angle. From a thermodynamic point of view, when $\phi = \phi_e$, the driving force for the neck growth is zero. Intuitively, it is possible under certain conditions when $\phi < \phi_e$, driving force for coarsening may equal that for neck growth. Hence, coarsening by inter-particle mass transport may take place significantly prior to the achievement of the equilibrium dihedral angle and the beginning of grain boundary migrations.

Considering the coarsening mechanisms described above, the initial grain growth can, therefore, be described by a two-step qualitative growth model.[95] When particles of different sizes are in contact, the first step in grain growth is coarsening due to inter-particle mass transport via the growth of larger particles into smaller particles, which results in the increase of the material's average grain size regardless of whether the size ratio r_1/r_2 is larger or smaller than R_c. During the coarsening and sintering progress, the size ratio between particles can increase. When the condition of size ratio $r_1/r_2 > R_c$ is reached, grain boundary migration will occur, leading to the second step of grain growth by the grain boundary migration. Figure 13.16 and 13.17 schematically illustrate the two-step process.

13.4.3 Initial grain growth mechanisms

From the very porous structure at the start of sintering, several possible mechanisms for grain growth during sintering of nano particles exist, including: 1) coarsening as the result of inter-particle diffusion; 2) grain boundary migration; 3) solution and reprecipitation (two-phase system); and 4) coalescence.

Generally, the initial grain growth during sintering is attributed to the coarsening of nano particles due to inter-particle diffusions. Surface diffusion especially plays a major role for inter-particle mass transport. In a study of the sintering of $BaTiO_3$, Shi et al. observed that the contacting particles become one particle via surface diffusion, as shown in Fig 13.18.[58] Surface diffusion transported the atoms from the dissolving small particle to be re-deposited on the surface of the larger particle. This is a direct evidence of the role of surface diffusion in the coarsening of nano particles at the beginning of sintering. It is noted that surface diffusion causes coarsening of larger particles by consuming small particles, i.e. grain growth without requiring either grain boundary migration and rotation, or grain boundary diffusion.

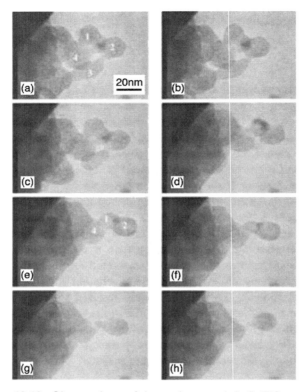

13.18 Observations of the grain growth in BaTiO$_3$ powder at different temperatures from 940°C (a), 950°C (b, c) to 960°C (d to o). Grains grow through reduction of smaller grains and enlargement of larger ones. The distance between the particle centers decreases simultaneously.

Considering that nano particles are usually not at equilibrium states and are likely to contain excess amounts of various defects that are created during the production of nano particles, there will be a relaxation period for migration, redistribution and annihilation of the defects.[96,97] Owing to the non-equilibrium structure of nano particles, diffusivity is dramatically enhanced during the relaxation process,[98-101] which may contribute to dynamic grain growth at the beginning of sintering. Dynamic grain growth usually dominates during the heat-up stage and for the first few minutes after reaching a preset isothermal holding temperature. Therefore, rapid dynamic grain growth accounts for the experimental observation that the first data point during isothermal holding is several times that of the initial grain size. The relaxation time depends on materials, nano particle production methods and temperature.

The role of grain boundary migration should also be considered in

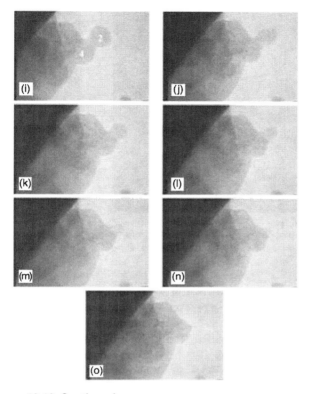

13.18 Continued

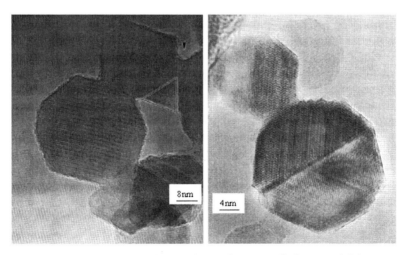

13.19 Alumina particles cluster sintered at 1200°C. One particle's 'grain' has grown, outlined, and has distinct grain boundary (left). The grain boundary migrated into small particles (right).

13.20 Coalescence of two platelet-shaped grains of a nanocrystalline WC-Co compact heated up to 1200 C at a heating rate of 10 C/min. and held for 1 min.

discussing the initial grain growth during sintering of nano particles. As discussed earlier, for single-phase materials at late stages of sintering when relative density is greater than 90%, grain boundary migration is the most logical mechanism of grain growth found in bulk single-phase materials. Grain boundary migration has also been observed during early stages of sintering nanosized Al_2O_3.[59] Figure 13.19 shows that when the nanosized Al_2O_3 particles were subjected to high temperatures in a flash sintering set-up, instant grain growth was observed and grain boundary migration was believed to be part of the process. This confirms the analysis of the coarsening of nano particles that when r_1/r_2 is greater than R_c, grain boundary migration will take place.

Coalescence is another grain growth mechanism that is often cited to qualitatively explain rapid grain growth. Coalescence is a term that is often loosely used to describe various phenomena. For example, coalescence is sometimes used interchangeably with the term 'sintering' to describe the growth of particles during particle synthesis and growth process.[102–105] For clarity in this chapter, coalescence is used strictly to describe the increase of grain size due to the merging of two grains by eliminating the common grain boundaries between them. Differing from other grain growth processes, which may also be described as the merging of two grains, the two original grains should not demonstrate significant change from their morphology prior to coalescence.

The term coalescence, as defined above, describes a unique method of grain growth, which can be accomplished only through various diffusion mechanisms. Possible mechanisms for coalescence include grain boundary diffusion, dislocation climb along grain boundaries, or even grain rotations.

13.21 Densification and grain growth within individual aggregated particles prior to bulk densification.

In liquid-phase sintering systems, it is believed that the solution–reprecipitation mechanism may also help facilitate the coalescence of grains. Direct evidence of coalescence is, however, very difficult to identify. Fang et al.[79] studied the grain growth of nano WC during sintering and found the growth of nanosized tungsten carbide grains within aggregates via coalescence, as shown in Fig. 13.20. An analysis of the sintering of WC-Co composites suggests that the lattice shift along low-energy CSL grain boundaries is a viable mechanism for materials with a high degree of crystallographic anisotropy.[106]

Effects of agglomerates on initial grain growth

Another important factor in grain growth mechanisms during sintering of nano particles is the role of agglomerates in grain growth. Agglomerates are defined as loosely-packed particles forming fractals, while aggregates are particles packed together in a more defined equiaxial shape. Mayo[3] pointed out that grain size is often related to the size of agglomerates at the beginning of sintering. As Mayo summarized, the larger the agglomerate size, the higher the sintering temperature required to eliminate large inter-agglomerate pores. By contrast, the crystallite size has little effect on the temperature required to reach full density. The same temperatures, however, promote grain growth to such an extent that the grain size can easily balloon to the agglomerate size.

Fang[107] observed a similar phenomenon. Figure 13.21 shows an

13.22 Microstructure of same sample as Fig. 13.21 at 1200°C.
Agglomerates were transformed into individual grains.

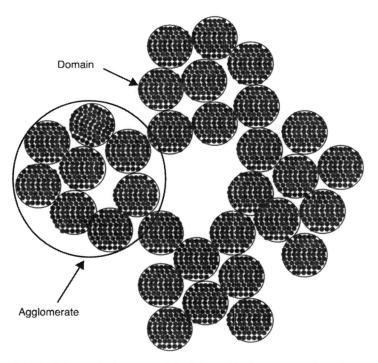

13.23 Schematic diagram of the hierarchical structure of agglomerates
(large circle), domains (small circle) and primary particles (dots within
small circles).

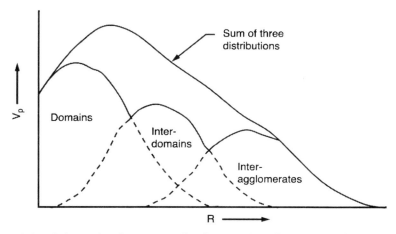

13.24 Schematic of pore coordination number distribution of agglomerated powder indicating three classes of pores, i.e., those within domains, those between domains, and those between agglomerates (*R* stands for coordination number).

agglomerate of WC-10%Co when heated to 800°C within a powder compact, while Fig. 13.22 shows the structure when the same compact is heated to 1200°C. It can be seen that the original agglomerates, within which the WC grains are visible at 800°C, no longer exist at 1200°C. Instead, the individual grains with sizes similar to those of the agglomerates at lower temperatures constitute the microstructure. It is thus deduced that the densification and grain growth processes during sintering of nano particles progressed via consolidation and grain growth within individual agglomerates, and then proceeded to the consolidation and elimination of porosities between agglomerates. This mechanistic process of sintering was also observed and discussed by Petersson and Ågren.[72] The process that first takes place within individual agglomerates was characterized as 'nucleation' sites.

To explain the effect of agglomerates, Lange[63] classified the structure of a powder compact as a hierarchical structure of agglomerates, domains and primary particles, as shown by Fig. 13.23. Defining the coordination number as the number of particles surrounding the pore, Lange explained that pores within domains have the lowest coordination number, pores between domains have higher, and pores between agglomerates have the highest coordination number. Figure 13.24 shows schematically the volume distribution of the three classes of pores as a function of coordination number. When $N < N_c$, a pore is unstable. Otherwise, grain growth, or coarsening of particles within agglomerates, will be necessary for elimination of the pore and continuation of the sintering. This explains the correlation between grain growth and the size of agglomerates.

13.5   Techniques for controlling grain growth while achieving full densification

With regard to processes for sintering nanosized particles, in principle, all conventional and advanced sintering techniques can be applied. However, if maximum densification is to be achieved while retaining the nanoscaled grain sizes, then special processes and techniques are necessary. This section will highlight techniques that are particularly useful for sintering nanosized particles.

13.5.1 Two-step sintering: decoupling of grain growth from densification

As an example of understanding and controlling normal grain growth during sintering of nano particles, Chen and Wang developed a clever approach to decouple grain growth from densificaton of nanosized particles,[108] using a pressureless sintering process to fully densify nanocrystalline Y_2O_3. In a simple two-step process, the compact is briefly heated to 1310°C; the temperature is then lowered to 1150°C and held at that temperature for an extended period of time. As a result, the material can be sintered to full density with minimum grain growth. If the lower temperature is applied at the onset, complete densification would not be possible. It is reasoned, then, that suppression of the final-stage grain growth is achieved by exploiting the difference in kinetics between the grain-boundary diffusion and the grain boundary migration. Grain growth requires grain boundary migration, which requires higher activation energy than grain boundary diffusion. At a temperature that is high enough to overcome the energy hurdles for grain boundary diffusion, but low enough to deactivate grain boundary migration, the densification will proceed via grain boundary diffusion without triggering significant grain growth. This phenomenon was further studied in multiple publications of Kim et al.[109,110]

It is noted, once again, that in this successful work to decouple grain growth from densification by exploiting differences in grain boundary mechanisms, the authors explicitly showed that at the beginning of the second sintering step, the grain size increases to four to six times larger than the original size of the powder, which is attributed to coarsening during the first sintering step (Fig. 13.25). It is necessary for the first step to be carried out at a higher temperature in order to quickly achieve a high relative density. The second isothermal step at the lower temperature is selected so

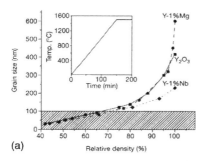 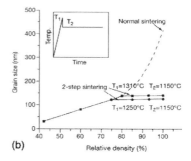

13.25 (a) Increasing grain size of Y_2O_3 with density in normal sintering (heating schedule shown in inset). Even with fine starting powders (30 nm), the final grain size of dense ceramics is well over 200 nm regardless of whether dopant was used. The shaded area indicates the grain size regime commonly defined as nanostructured materials. At lower densities, the mean grain (particle) size was estimated on the fracture surface. At higher densities, the grain size was obtained by multiplying by 1.56 the average linear intercept length of at least 500 grains. (b) Grain size of Y_2O_3 in two-step sintering (heating schedule shown in inset). Note that the grain size remains constant in the second sintering step, despite density improvement to 100%.

that the densification can continue, whereas grain growth is limited because grain boundary migration is suppressed. This significant discovery proves that it is possible to decouple grain growth from densification and, hence, to achieve full densification while retaining nanoscaled grain sizes.

13.5.2 Use of grain growth inhibitors

The use of grain growth inhibitors is a common method for controlling grain growth during sintering. For example, with the addition of SiC to Al_2O_3,[111] ZrO_2 to $\beta''-Al_2O_3$,[112] and Al_2O_3 to cubic ZrO_2,[113] large grain growth can be effectively prevented. The use of grain growth inhibitor can also be found during sintering of other materials.[87,114–116]

Grain growth inhibitors are widely used in manufacturing fine and ultrafine grained cemented tungsten carbide (WC-Co) materials.[117–119] When VC, or VC combined with Cr_2C_3, is used in liquid-phase sintering of nanosized WC-Co powder, grain size after sintering is dramatically finer than if grain growth inhibitors had not been used. Researchers[18,120] found that the effect of VC on grain growth during sintering of WC-Co significantly inhibits the rapid grain growth during the solid state as well as the liquid phase sintering stage. However, the finest grain size that has been reportedly achieved using grain growth inhibitors in pressureless liquid-phase sintering processes is approximately 100 to 200 nm or larger, which is significantly larger than the original grain size (10 nm) of the nanosized powder. One explanation for the limited effect of grain growth inhibitors has to do with the size of agglomerates. If the mixing and distribution scale of grain growth inhibitors are larger than the original nanoscaled grain sizes and closer to the size of the agglomerate, then the grain growth inhibitors are effective only in that same dimensional scale.

13.5.3 Deagglomeration

Deagglomerating agglomerated particles prior to sintering is another critical strategy for minimizing the initial dynamic grain growth. Based on the understanding that a pore's stability is dependent upon its coordination numbers, a powder compact that has uniformly distributed fine pores would have the most efficient densification without relying on coarsening. An ideal scenario for minimizing grain growth while, at the same time, achieving full densification is the utilization of green powder compacts composed of monosized, spherical nano particles without agglomerates. The pores within such a compact would be evenly distributed and would have uniform size. To achieve this type of green compact structure, Lange proposed a methodology in which the powders would be treated prior to sintering according to colloidal processing principles.[74] First, in order to deagglome-rate the particles, dry powders are dispersed in fluid containing a surfactant that produces inter-particle repulsive forces. After removing large particles that cannot be deagglomerated, the powder slurry is flocked by changing the inter-particle forces from repulsion to attraction. As a result, the powder becomes a weak, continuous network of touching primary particles. Colloidal treated slurries could be used directly for consolidation. The effects of deagglomeration on the density of green compacts and sintering of nanosized particles are experimentally demonstrated in many reported studies.[74,121–126] In particular, Ahn et al.[127] showed that the grain size of sintered nano SnO_2 particles that have a dense green structure is dramatically smaller than that of a compact with loose structure.

13.5.4 Pressure-assisted sintering

The use of pressure-assisted processes is another straightforward approach for minimizing grain growth while achieving maximum densification. A variety of pressure-assisted sintering processes has been used in sintering nanosized powders, including hot pressing (HP), hot isostatic pressing (HIP), spark plasma sintering (SPS) and sinter forging. It is generally believed that the applied pressure is able not only to enhance densification by increasing sintering driving force, assisting particle rearrangement and promoting diffusion creep,[2,128–130] but also to inhibit grain growth by decreasing the diffusivity and thus the grain boundary mobility.[128] The total sintering driving force when an external pressure is applied includes both the intrinsic curvature-driven sintering stress and the applied external stress. The significance of the applied pressure on sintering depends on the relative magnitudes of the two components. The applied pressure is independent of the particle size, while the intrinsic sintering pressure increases when the particle size is reduced, and it can reach a very high value as particle size is in nanoscale. In order to make the effect of applied pressure on densification become dominant, the applied pressure has to be larger than the intrinsic curvature-driven pressure.[2,130] Therefore, a threshold pressure, which is dependent on the particle size, must be exceeded so that the applied pressure can have a significant influence on sintering. The existence of the threshold pressure had been confirmed experimentally by Skandan et al. on sintering of nanosized zirconia powder.[14,131] The logic for this approach is based on the belief that densification can be achieved at lower-than-normal sintering temperatures with the aid of pressure. The lower temperature would, of course, slow the kinetic rate of grain growth. For example, Hayashi and Etoh[132] studied sintering behavior of nanosized Fe, Co, Ni and Cu metal powders under pressures ranging from 400 to 500 MPa. The nanosized powders were fabricated by evaporation and condensation method in an inert gas. The starting average particle size of Fe, Co and Ni were about 20 nm; that of Cu was about 50 nm. The results showed that sintering temperatures were effectively decreased by increasing pressure. Under 400 MPa, the sintering temperatures for Fe, Co, Ni and Cu powders were about 590, 640, 450 and 450 K respectively, which were 380–620 K lower than those without pressure. The minimum average grain sizes of Fe, Co, Ni and Cu in the fully densified compacts after sintering were about 80 nm, 210 nm, 120 nm and 400 nm, which were much smaller compared to those obtained by pressureless sintering.

Pressure-induced phase transformation is a phenomenon that was observed in several ceramic materials during the consolidation of nanosized powders under high pressure. It was also characterized as 'transformation-assisted consolidation' (TAC).[133] TAC has also been used to densify several materials, including Al_2O_3 and Si_3N_4.[134] The key criterion for the suitability of this technique is that the starting material must be a metastable phase, which transforms into the desired stable phase in a controlled way during pressing and sintering. The combination of increased nucleation and controlled grain growth produces fully consolidated material with nanosized grain structure. As an example, nanosized TiO_2 undergoes phase transformation from anatase to rutile when pressure is higher than 1 GPa during a pressure-assisted consolidation process.[128,135] The grain size of the product phase (rutile) is smaller than that of the parent phase (anatase) after consolidation. Under such high pressure, sintering temperature is as low as $\frac{1}{3}$ of T_m. The theory regarding the effect of high pressure on the sintering behavior of these materials has two key points. One is that the pressure reduces the nucleation barrier for a phase transformation that is accompanied by volume reduction, as in the case of anatase to rutile of TiO_2, and γ to α phase for Al_2O_3. The second point is that the high hydrostatic pressure reduces the diffusion rate and, thus, grain growth rate.

Sinter-forging is another unique technique that has been used to produce fully dense materials with nanosized grains from nanosized ceramic or metal powders. First of all, sinter-forging, also termed powder-forging, is a routine process that has been used to manufacture ferrous powder metallurgy automobile parts. In this case, green compacts are first sintered via a standard atmospheric sintering process, and then the sintered compacts are placed in a forging die and forged at an elevated temperature. The forging process results in fully dense metal parts.

In recent years, the term 'sinter-forging' has been used to describe the processing of nanoscaled powder materials in a different way from conventional sinter-forging. In this process, powders are placed in a die without lateral constraint. The die and powder are heated via techniques similar to the hot pressing process and a uniaxial load is applied. During sinter-forging, the powder compact is allowed to 'upset,' i.e. bulge laterally, and is densified under load at specific temperature. In most of the published research regarding sintering of nanosized powders using the sinter-forging process, a typical hot press is adapted for experiments. Load capacity and loading rate are thus comparable to a conventional hot pressing process. In a study on the consolidation of nano 3Y-TZP (3mol% yttria-stabilized tetragonal zirconia polycrystals) using sinter-forging process, the loading rate varied from 0.005 to $0.1\,kN^{-1}$, and sintering temperature was around 1000–1100°C. Maximum load was not allowed to exceed 30 kN.[136] In another example, Ma et al.[137] applied a 'constrained sinter-forging' or 'upset hot forging' process scheme to consolidate nanocrystalline Fe and Fe–Cu alloyed powders. Consolidated nanocrystalline Fe with grain sizes within the typical range of nanocrystalline (<100 nm) was achieved.

Spark plasma sintering (SPS) is a new pressure-assisted sintering process that quickly gained popularity with researchers looking for ways to consolidate materials with nanoscale or simply very fine grain sizes, or with other non-equilibrium microstructures. The history of development, the principles and applications of SPS are discussed in the literature; readers can also refer to for reviews.[130,138]

13.6   Conclusion

Sintering of nanosized particles is a uniquely important topic that is both scientifically and technologically challenging. From a scientific perspective, the markedly different sintering behavior of nanosized powders, compared to micron-sized powders, raises fundamental issues that deserve detailed studies. It is notable that the driving force for sintering nano particles is significantly higher than for micron-sized particles. The linear approximation used in conventional theories of sintering, with regard to modeling the driving force and the diffusion equations, is no longer valid. The rate of sintering predicted by the nonlinear diffusion model is much higher than that predicted by the conventional linear diffusion model.

With regard to the complex mechanisms of sintering of nano particles, several possible mechanisms discussed in this chapter appear to contribute to the initial densification, including the rapid diffusion rate due to non-equilibrium defect concentrations in nanosized powders, the indirect role of mass transport by surface diffusion, and the possible surface melting of nano particles, all of which contribute to the densification as well as coarsening of the nano particles.

From a technology perspective, proof that the sintering temperature drastically decreases as particle size decreases to nanoscale represents an actionable knowledge that can be exploited in the production of engineering materials from nanosized powders.

The greatest challenge for sintering nanosized powders is the ability to retain nanoscale grain sizes while achieving full densification. The grain growth of nanosized powders is characterized by two parts of grain growth: the initial dynamic growth and the normal grain growth, which is reminiscent of that in bulk materials. The initial grain growth is the result of the coarsening of particles via the inter-particle mass transport. For nanosized powders, the initial grain growth causes the material to lose nanocrystalline characteristics. Therefore, if at least part of the goal of sintering is the retention of nanoscaled grain sizes, the initial grain growth must be controlled and minimized. On the other hand, in the absence of the need to retain nanoscaled grain size, the initial grain growth or coarsening is one of the mechanisms that can be exploited to aid densification.

The methods for retaining grain growth include the use of grain growth inhibitors, various high-pressure hot consolidation processes, and decoupling grain growth from densification by manipulating different diffusion mechanisms at different temperatures. The popular use of SPS for consolidation of nanosized powders combines the advantages of rapid heating rate and pressure. However, regardless of the sintering technique used, powder processing and green compact fabrication techniques are crucial for controlling grain growth and densification. In general, it is desirable to have minimum agglomeration of particles, minimum pore size and uniformly distributed pores.

13.7   References

1. F. F. Lange and B. J. Kellett: *Journal of the American Ceramic Society*, 1989, **72**, 735–41.
2. J. R. Groza: 'Nanocrystalline Powder Consolidation Methods', in *Nanostructured Materials – Processing, Properties and Potential Applications* (ed. C. C. Koch), 1st edn, 115–78; 2002, William Andrew Publishing/Noyes
3. M. J. Mayo: *International Materials Reviews*, 1996, **41**, 85–115.
4. V. Viswanathan, T. Laha, K. Balani, A. Agarwal and S. Seal: *Materials Science & Engineering R-Reports*, 2006, **54**, 121–285.
5. C. C. Koch: *Journal of Materials Science*, 2007, **42**, 1403–14.
6. Z. Z. Fang and H. Wang: *International Materials Reviews*, 2008, **53**, 326–52.
7. R. M. German: *Sintering Theory and Practice*, 1996, Wiley–Interscience.
8. D. A. Porter and K. E. Easterling: *Phase Transformation in Metal and Alloys*, 1992, CRC.
9. C. T. Campbell, S. C. Parker and D. E. Starr: *Science*, 2002, **298**, 811–14.
10. K. K. Nanda, A. Maisels, F. E. Kruis, H. Fissan and S. Stappert: *Physical Review Letters*, 2003, **91**, 106102.
11. R. L. Penn and J. F. Banfield: *Geochimica et Cosmochimica Acta*, 1999, **63**, 1549–57.
12. H.-Y. Lee, W. Riehemann and B. L. Mordike: *Journal of the European Ceramic Society*, 1992, **10**, 245–53.
13. H. Hahn: *Nanostructured Materials*, 1993, **2**, 251–65.
14. G. Skandan: *Nanostructured Materials*, 1995, **5**, 111–26.
15. M. J. Mayo: *Materials & Design*, 1993, **14**, 323–29.
16. Y. C. Zhou and M. N. Rahaman: *Journal of Materials Research*, 1993, **8**, 1680–96.
17. T. Rabe and R. Waesche: *Nanostructured Materials*, 1995, **6**, 357–60.
18. P. Maheshwari, Z. Z. Fang and H. Y. Sohn: *International Journal of Powder Metallurgy (Princeton, New Jersey)*, 2007, **43**, 41–7.
19. M. Attarian Shandiz, A. Safaei, S. Sanjabi and Z. H. Barber: *Journal of Physics and Chemistry of Solids*, 2007, **68**, 1396–9.
20. P. Buffat and J. P. Borel: *Physical Review A (General Physics)*, 1976, **13**, 2287–98.
21. T. Castro, R. Reifenberger, E. Choi and R. P. Andres: *Physical Review B (Condensed Matter)*, 1990, **42**, 8548–56.
22. P. R. Couchman: *Philosophical Magazine A (Physics of Condensed Matter, Defects and Mechanical Properties)*, 1979, **40**, 637–43.
23. P. R. Couchman and W. A. Jesser: *Nature*, 1977, **269**, 481–3.
24. P. R. Couchman and C. L. Ryan: *Philosophical Magazine A (Physics of Condensed Matter, Defects and Mechanical Properties)*, 1978, **37**, 369–73.
25. Q. Jiang, S. Zhang, and M. Zhao: *Materials Chemistry and Physics*, 2003, **82**, 225–27.
26. E. A. Olson, M. Y. Efremov, M. Zhang, Z. Zhang and L. H. Allen: *Journal of Applied Physics*, 2005, **97**, 034304–034301.
27. W. H. Qi and M. P. Wang: *Materials Chemistry and Physics*, 2004, **88**, 280–84.
28. J. Ross and R. P. Andres: *Surface Science*, 1981, **106**, 11–17.
29. F. G. Shi: *Journal of Materials Research*, 1994, **9**, 1307–13.

30. C. Solliard: *Solid State Communications*, 1984, **51**, 947–9.
31. M. Wautelet: *Solid State Communications*, 1990, **74**, 1237–9.
32. M. Wautelet: *European Journal of Physics*, 1995, **16**, 283–4.
33. M. Wautelet: *Physics Letters A*, 1998, **246**, 341–2.
34. P. Yang, Q. Jiang and X. Liu: *Materials Chemistry and Physics*, 2007, **103**, 1–4.
35. V. N. Troitskii, A. Z. Rakhmatullina, V. I. Berestenko and S. V. Gurov: *Soviet Powder Metallurgy and Metal Ceramics (English translation of Poroshkovaya Metallurgiya)*, 1983, **22**, 12–14.
36. Q. Jiang and F. G. Shi: *Journal of Materials Science and Technology*, 1998, **14**, 171–172.
37. H. Ezaki, T. Nambu, M. Morinaga, M. Udaka and K. Kawasaki: *International Journal of Hydrogen Energy*, 1996, **21**, 877–81.
38. M. I. Alymov, E. I. Maltina and Y. N. Stepanov: *Nanostructured Materials*, 1994, **4**, 737–42.
39. C. Herring: *Journal of Applied Physics*, 1950, **21**, 301–3.
40. G. L. Messing and M. Kumagai: *American Ceramic Society Bulletin*, 1994, **73**, 88–91.
41. M. F. Yan and W. W. Rhodes: *Material Science and Engineering*, 1983, **61**, 59–66.
42. M. N. Rahaman: *Ceramic Processing and Sintering*, 2nd Edition, 2003, New York, CRC.
43. J. Pan: *Philosophical Magazine Letters*, 2004, **84**, 303–10.
44. Q. Jiang, S. H. Zhang and J. C. Li: *Solid State Communications*, 2004, **130**, 581–84.
45. H. Zhu and R. S. Averback: *Journal of Engineering and Applied Science*, 1995, **A204**, 96–100.
46. H. Zhu and R. S. Averback: *Materials and Manufacturing Processes*, 1996, **11**, 905–23.
47. Z. Xing, W. Shaoqing and Z. Caibei: *Journal of Materials Science & Technology*, 2006, **22**, 123–26.
48. V. N. Koparde and P. T. Cummings: *Journal of Physical Chemistry B*, 2005, **109**, 24280–24287.
49. K. Tsuruta, A. Omeltchenko, R. K. Kalia and P. Vashishta: *Europhysics Letters*, 1996, **33**, 441–6.
50. J. S. Raut, R. B. Bhagat and K. A. Fichthorn: *Nanostructured Materials*, 1998, **10**, 837–51.
51. C. Win-Jin, F. TeHua and C. Jun-Wei: *Microelectronics Journal*, 2006, **37**, 722–27.
52. D. Hai, M. KyoungSik and C. P. Wong: *Journal of Electronic Materials*, 2004, **33**, 1326–30.
53. P. Vergnon, M. Astier and S. J. Teichner: *Polymer Preprints, Division of Polymer Chemistry, American Chemical Society*, 1974, 299–307.
54. K. G. Ewsuk, D. T. Ellerby and C. B. DiAntonio: *Journal of the American Ceramic Society*, 2006, **89**, 2003–9.
55. B. B. Panigrahi: *Materials Science and Engineering A*, 2007, **460–461**, 7–13.
56. V. V. Dabhade, T. R. Rama Mohan and P. Ramakrishnan: *Materials Science and Engineering: A*, 2007, **452–453**, 386–94.

57. G. S. A. M. Theunissen, A. J. A. Winnubst and A. J. Burggraaf: *Journal of the European Ceramic Society*, 1993, **11**, 315–24.
58. J.-L. Shi, Y. Deguchi and Y. Sakabe: *Journal of Materials Science*, 2005, **40**, 5711–19.
59. J. E. Bonevich and L. D. Marks: *Materials Research Society Symposium Proceedings*, 1993, **286**, 3–8.
60. J. L. Shi: *Journal of Materials Research*, 1999, **14**, 1378–88.
61. J. L. Shi: *Journal of Materials Research*, 1999, **14**, 1389–97.
62. W. D. Kingery and B. Francois: 'Sintering of Crystalline Oxide, I. Interactions Between Grain Boundaries and Pores', in *Sintering and Related Phenomena* (ed. G. C. Kuczynski, N. A. Hooton, and G. F. Gibbon), 471–498; 1967, New York, Gordon and Breach.
63. F. F. Lange: *Journal of the American Ceramic Society*, 1984, **67**, 83–89.
64. O. Dominguez, Y. Champion and J. Bigot: *Metallurgical and Materials Transactions A: Physical Metallurgy and Materials Science*, 1998, **29A**, 2941–9.
65. L. I. Trusov, V. N. Lapovok and V. I. Novikov: 'Problems of Sintering Metallic Ultrafine Powders', in *Science of Sintering* (ed. D. P. Uskokovic, H. Plamour III, and R. M. Spriggs), 185–192; 1989, New York, Plenum Press.
66. C. E. Krill, III, L. Helfen, D. Michels, H. Natter, A. Fitch, O. Masson and R. Birringer: *Physical Review Letters*, 2001, **86**, 842–45.
67. A. H. Carim: 'Microstructure in Nanocrystalline Zirconia Powders and Sintered Compacts', in *Nanophase materials: synthesis, properties, applications* (ed. G. C. Hadjipanayis and R. W. Siegel), 283–286; 1994, Dordrecht, The Netherlands, Kluwer Academic Publishers.
68. J. Y. Ying, L. F. Chi, H. Fuchs and H. Gleiter: *Nanostructured Materials*, 1993, **3**, 273.
69. M. F. Yan and W. W. Rhodes: *Materials Science and Engineering*, 1983, **61**, 59–66.
70. E. A. Barringer, R. Brook and H. K. Bowen: 'The Sintering of Monodisperse TiO$_2$', in *Sintering and Heterogeneous Catalysis* (ed. G. C. Kuczynske, A. E. Miller, and G. A. Sargent), 1–21; 1984, New York, Plenum Press.
71. D. C. Hague: 'Chemical Precipitation, Densification, and Grain Growth in Nanocrystalline Titania Systems', MS thesis, The Pennsylvania State University, Pittsburgh, 1992.
72. A. Petersson and J. Ågren: *Acta Materialia*, 2005, **53**, 1673–83.
73. W. B. Russel D. A. Saville and W. R. Schowalter: *Colloidal Dispersions*, 1992, Cambridge University Press.
74. F. F. Lange: *Journal of the American Ceramic Society*, 1989, **72**, 3–15.
75. T. R. Malow and C. C. Koch: 'Grain growth of nanocrystalline materials – a review', TMS Annual Meeting – Synthesis and Processing of Nanocrystalline Powder, Anaheim, CA, USA, 1996, 33–44.
76. T. R. Malow and C. C. Koch: *Materials Science Forum*, 1996, **225–7**, 595–604.
77. T. R. Malow and C. C. Koch: *Acta Materialia*, 1997, **45**, 2177–86.
78. X. Song, J. Zhang, L. Li, K. Yang and G. Liu: *Acta Materialia*, 2006, **54**, 5541–50.
79. Z. Fang, P. Maheshwari, X. Wang, H. Y. Sohn, A. Griffo and R. Riley: *International Journal of Refractory Metals and Hard Materials*, 2005, **23**, 249–57.

80. Z. J. Shen, H. Peng, J. Liu and M. Nygren: *Journal of the European Ceramic Society*, 2004, **24**, 3447–52.
81. S. Okuda, M. Kobiyama, T. Inami and S. Takamura: *Scripta Materialia*, 2001, **44**, 2009–12.
82. D. J. Chen and M. J. Mayo: *Nanostructured Materials*, 1993, **2**, 469.
83. W. Dickenscheid, R. Birringer, H. Gleiter, O. Kanert, B. Michel and B. Guenther: *Solid State Communications*, 1991, **79**, 683–86.
84. G. Hibbard, K. T. Aust, G. Palumbo and U. Erb: *Scripta Materialia*, 2001, **44**, 513–18.
85. R. Klemm, E. Thiele, C. Holste, J. Eckert and N. Schell: *Scripta Materialia*, 2002, **46**, 685–90.
86. F. Zhou, J. Lee and E. J. Lavernia: *Scripta Materialia*, 2001, **44**, 2013–17.
87. D. M. Owen and A. H. Chokshi: *Nanostructured Materials*, 1993, **2**, 181–7.
88. R. S. Averback: *Zeitschrift für Physik D Atoms, Molecules and Clusters*, 1993, **26**, 84–8.
89. J. G. Li and Y. P. Ye: *Journal of the American Ceramic Society*, 2006, **89**, 139–43.
90. R. Vassen: *CFI – Ceramic Forum International – Berichte der Deutschen Keramischen Gesellschaft*, 1999, **76**, 19–22.
91. G. C. Kuczynski: *Journal of Metals*, 1949, **1**, 169–78.
92. W. D. Kingery and M. Berg: *Journal of Applied Physics*, 1955, **26**, 1205–12.
93. R. L. Coble: *Journal of the American Ceramic Society*, 1958, **41**, 55–62.
94. D. L. Johnson and I. B. Cutler: *Journal of the American Ceramic Society*, 1963, **46**, 541–5.
95. C. Greskovich and K. W. Lay: *Journal of the American Ceramic Society*, 1972, **55**, 142–6.
96. R. A. Andrievski: *Journal of Materials Science*, 2003, **38**, 1367–75.
97. Roland Wuschum, Simone Herth and U. Brossmann: 'Diffusion in Nanocrystalline Metals and Alloys – A Status Report', in *Nanomaterials by Severe Plastic Deformation* (ed. Michael Zehetbauer and R. Z. Valiev), 755–66; 2002, Wiley–VCH.
98. L. G. Kornelyuk, A. Y. Lozovoi and I. M. Razumovskii: *Diffusion and Defect Data. Pt A Defect and Diffusion Forum*, 1997, **143/1**, 1481.
99. R. Wuerschum, K. Reimann and P. Farber: *Diffusion and Defect Data. Pt A Defect and Diffusion Forum*, 1997, **143/1**, 1463.
100. V. N. Perevezentsev: 'Theoretical Investigation of Nonequilibrium Grain Boundary Diffusion Properties', in *Nanomaterials by Severe Plastic Deformation* (ed. Michael Zehetbauer and R. Z. Valiev), 773–779; 2002, Wiley–VCH.
101. A. Nazarov: *Physics of the Solid State*, 2003, **45**, 1166–9.
102. K. Nakaso, M. Shimada, K. Okuyama and K. Deppert: *Journal of Aerosol Science*, 2002, **33**, 1061–74.
103. K. E. J. Lehtinen and M. R. Zachariah: *Journal of Aerosol Science*, 2002, **33**, 357–68.
104. W. Koch and S. K. Friedlander: *Journal of Aerosol Science*, 1989, **20**, 891–4.
105. D. Mukherjee, C. G. Sonwane and M. R. Zachariah: *Journal of Chemical Physics*, 2003, **119**, 3391–404.

106. V. Kumar, Z. Fang, S. Wright and M. Nowell: *Metallurgical and Materials Transactions A*, 2006, **37**, 599–607.
107. Z. Fang, unpublished work.
108. I. W. Chen and X. H. Wang: *Nature*, 2000, **404**, 168–71.
109. H.-D. Kim, Y.-J. Park, B.-D. Han, M.-W. Park, W.-T. Bae, Y.-W. Kim, H.-T. Lin and P. F. Becher: *Scripta Materialia*, 2006, **54**, 615–19.
110. Y.-I. Lee, Y.-W. Kim, M. Mitomo and D.-Y. Kim: *Journal of the American Ceramic Society*, 2003, **86**, 1803–5.
111. F. F. Lange and M. Claussen: 'Some Processing Requirements for Transformation-Toughened Ceramics', in *Ultrastructure Processing of Ceramics, Glasses, and Composites* (ed. L. L. Hench and D. R. Ulrich), 493; 1984, New York, Wiley.
112. D. J. Green: *Journal of Materials Science*, 1985, **20**, 2639–46.
113. F. J. Esper, K. H. Friese and H. Geier: 'Mechanical, Thermal, and Electrical Properties in the System of Stabilized $ZrO_2(Y_2O_3)/$ alpha-Al_2O_3', Science and Technology of Zirconia II, Stuttgart, Austria, 1984, 528–36.
114. T. K. Gupta: *Journal of the American Ceramic Society*, 1972, **55**, 276–7.
115. T.-G. Nieh and J. Wadsworth: *Journal of the American Ceramic Society*, 1989, **72**, 1469–72.
116. R. S. Averback, H. Hahn, H. J. Hofler, J. L. Logas and T. C. Shen: 'Kinetic and thermodynamic properties of nanocrystalline materials', Interfaces Between Polymers, Metals and Ceramics Symposium, San Diego, CA, USA, 1989, 3–12.
117. C. W. Morton, D. J. Wills and K. Stjernberg: *International Journal of Refractory Metals and Hard Materials*, 2005, **23**, 287–93.
118. R. K. Sadangi, L. E. McCandlish, B. H. Kear and P. Seegopaul: *The International Journal of Powder Metallurgy*, 1999, **35**, 27–33.
119. O. Seo, S. Kang and E. J. Lavernia: *Materials Transactions*, 2003, **44**, 2339–45.
120. X. Wang and Y. Sakka: *Scripta Materialia*, 2001, **44**, 2219–23.
121. O. Vasylkiv Z. Zak Fang and H.Y. Sohn: *International Journal of Refractory Metals and Hard Materials*, 2008, **26**, 232–41.
122. O. Vasylkiv and Y. Sakka: *Journal of the American Ceramic Society*, 2001, **84**, 2489–94.
123. F. F. Lange: *Current Opinion in Solid State & Materials Science*, 1998, **3**, 496–500.
124. L. Bergstrom, K. Shinozaki, H. Tomiyama and N. Mizutani: *Journal of the American Ceramic Society*, 1997, **80**, 291–300.
125. C. Duran, Y. Jia, Y. Hotta, K. Sato and K. Watari: *Journal of Materials Research*, 2005, **20**, 1348–55.
126. P. Bowen, C. Carry, D. Luxembourg and H. Hofmann: *Powder Technology*, 2005, **157**, 100–7.
127. J.-P. Ahn, M.-Y. Huh and J.-K. Park: *Nanostructured Materials*, 1997, **8**, 637–43.
128. S. C. Liao, W. E. Mayo and K. D. Pae: *Acta Materialia*, 1997, **45**, 4027–40.
129. B. J. Kellett and F. F. Lange: *Journal of the American Ceramic Society*, 1988, **71**, 7–12.
130. Z. A. Munir, U. AnselmiTamburini and M. Ohyanagi: *Journal of Materials Science*, 2006, **41**, 763–77.

131. G. Skandan, H. Hahn, B. H. Kear, M. Roddy and W. R. Cannon: *Materials Letters*, 1994, **20**, 305–9.
132. K. Hayashi and H. Etoh: *Materials Transactions, JIM*, 1989, **30**, 925–31.
133. B. H. Kear, J. Colaizzi, W. E. Mayo and S. C. Liao: *Scripta Materialia*, 2001, **44**, 2065–8.
134. J. Colaizzi, W. E. Mayo, B. H. Kear and S.-C. Liao: *The International Journal of Powder Metallurgy*, 2001, **37**, 45–54.
135. L. Shih-Chieh, K. D. Pae and W. E. Mayo: *Nanostructured Materials*, 1997, **8**, 645–56.
136. D. C. Hague and M. J. Mayo: *Materials Science and Engineering A*, 1995, **204**, 83–9.
137. E. Ma, D. Jia and K. T. Ramesh: 'Nanophase and ultrafine-grained powders prepared by mechanical milling: Full density processing and unusual mechanical behavior', *Powder Materials: Current Research and Industrial Practices*, Indianapolis, IN, USA, 2001, 257–66.
138. J. R. Groza and A. Zavaliangos: *Reviews on Advanced Materials Science*, 2003, **5**, 24–33.

14

Surface treatment of carbon nanotubes using plasma technology

B. RUELLE, C. BITTENCOURT and P. DUBOIS,
University of Mons and Materia Nova Research Centre, Belgium

DOI: 10.1533/9780857093493.3.474

Abstract: The chapter begins by discussing carbon nanotube (CNT) surface chemistry and solution-based functionalization. It then reviews promising alternative ways to modify the CNT surface through plasma processes. Plasma treatments have the advantage of being non-polluting and provide a wide range of grafted functional groups, depending on the plasma parameters such as power, gases used, duration of treatment and gas pressure. However, the interaction of high energy particles with the CNT surface can induce damage in the CNT structure. A plasma post-discharge treatment, which can prevent the degradation of CNTs, is presented.

Key words: carbon nanotubes, functionalization, plasma, post-discharge treatment.

Note: This chapter is adapted from Chapter 2 'Surface treatment of carbon nanotubes via plasma technology' by B. Ruelle, C. Bittencourt and P. Dubois, originally published in *Polymer–carbon nanotube composites: preparation, properties and applications,* ed. T. McNally and P. Pötschke, Woodhead Publishing Limited, 2011, ISBN: 978-1-84569-761-7.

14.1 Introduction

Since their observation by Iijima (1991), carbon nanotubes (CNTs) have attracted the attention of many researchers due to their outstanding properties (Thostenson *et al.*, 2001). Their unique physical properties result from their structure, which at atomic scale can be thought of as a hexagonal

474

sheet of carbon atoms rolled into a seamless one-dimensional cylindrical shape. Besides their extremely small (nanometric) size, their excellent electrical and thermal conducting performances, combined with their strong toughness and transverse flexibility make them promising for use in a wide variety of applications such as electronic components, chemical and biological sensors, chemical and genetic probes, field emission tips or mechanical memories (Collins and Avouris, 2000). Carbon nanotubes are thus expected to expand their unique properties in multidisciplinary fields and should play a key role in nanotechnology.

Characterized by both high aspect ratio (length-to-diameter) and low density, this allotropic variety of carbon is an ideal candidate to be used as filler materials in composites. In fact, CNTs can provide a three-dimensional conductive network through the polymer matrix with exceedingly low percolation thresholds (Kilbride *et al.*, 2002). Furthermore, it has been suggested that their high thermal conductivity can be exploited to make thermally conductive composites (Bagchi and Nomura, 2006). In addition, the mechanical enhancement of polymer materials using CNTs as reinforcing nanofillers will be exploited in the very near future (Wang *et al.*, 2008).

A key parameter for producing high quality nanocomposite materials with improved physical properties is the homogeneous dispersion of the individual CNTs. Unfortunately, the CNTs tend to form large bundles thermodynamically stabilized by van der Waals forces and physical entanglements between the tubes, which occur during their synthesis. The presence of these aggregates in addition to the CNT's low solubility in water and organic solvents represent a drawback for the engineering of CNT in polymer nanocomposites. It has been reported that the homogeneous dispersion of CNTs is relatively difficult to achieve in the large majority of polymers (Andrews and Weisenberger, 2004).

The chemical functionalization of the CNT sidewalls, i.e. the CNT surface-anchoring of functional (reactive) groups, represents a solution for the tuning of interactions between CNTs and the host polymer matrix, improving their dispersion ability (Hirsch and Vostrowsky, 2005). This surface modification can be divided into two main approaches.

First is the *non-covalent functionalization*, used in different techniques such as the addition of surfactants (Hirsch, 2002), polymer wrapping (Liu, 2005) or polymerization-filling technique (PFT) (Bonduel *et al.*, 2005), which allows the unaltered CNTs to preserve their physical properties. However, the interaction between the wrapping molecules and the nanotubes remains generally weak, limiting the efficiency of the property transfer between the nanotubes and the host polymer, in particular, there may be a low load transfer for mechanical properties.

The other approach relies upon the *covalent grafting* of functional chemical groups along the CNT sidewalls. These chemical functions can be used as anchoring sites for polymer chains (Hirsch and Vostrowsky, 2005) and to improve the interaction with the polymer matrix (Thostenson *et al.*, 2005). The covalent bonding between CNTs and the polymer chains ensures an optimal interfacial strength and thus a better load transfer. Nevertheless, the covalent functionalization can influence the CNTs' properties, depending on the nature and density of the functional sites. Indeed, the functionalization can take place on already existing CNT structural defects, or active sites can be created by grafting chemical functions. In the latter case, functionalization can degrade the CNTs' physical properties if the density of the active site is significant, leading to damaged CNT structure (Fu *et al.*, 2001). The majority of developed covalent sidewall functionalizations are carried out in organic solvent. Nevertheless, as the CNT surface is largely inert, rather harsh conditions are needed for wet chemical functionalization of CNTs, which rarely results in a controlled covalent functionalization process (Hirsch and Vostrowsky, 2005).

To overcome these drawbacks, 'dry' processes are being developed. For example, the ball-milling of CNTs in a reactive atmosphere was shown to functionalize nanotubes with different chemical groups, such as amides, amines, carbonyls and thiols, for instance, depending on the reactant gas (Konya *et al.*, 2002). Using this method, CNTs are broken, and dangling bonds are formed on their newly created extremities that can then react with a selected gas. A relatively long period of time, a minimum of 24 hours, is usually necessary to graft a significant amount of functional groups onto CNTs. However, the average length of the CNTs decreases as the treatment time increases and some amorphous carbon was also observed after the breaking process (Ma *et al.*, 2008). Another drawback of this technique is the presence of functional groups exclusively at the ends of cup-stacked carbon nanotubes.

Alternative 'dry' approaches to chemical modification of CNTs using plasma discharges have been also proposed and developed. Considered the fourth state of matter, plasma is an ionized gas that constitutes a highly unusual and highly reactive chemical environment. Plasma treatments have evolved into a valuable technique to modify material surface properties without altering the bulk properties, allowing their use in various applications such as plasma cleaning, plasma sterilization or biomedical applications (D'Agostino *et al.*, 2005). Recently, plasmas were also applied to modify the surface properties of CNTs. Characterized by a relatively low reaction time, surface plasma treatment is an environmentally friendly process without any use of organic solvents. The plasma technique allows the grafting of a wide range of different functional groups on the CNT surface depending on plasma parameters such as power applied to the discharge, nature of the gas used, duration of treatment and gas pressure.

14.2   Carbon nanotube surface chemistry and solution-based functionalization

As mentioned above, even though chemical functionalization displays some drawbacks, it still represents a very important method investigated by several research groups. Hereafter are discussed some important results recently reported in the literature.

Actually, the chemical functionalization of CNTs via covalent grafting reaction can be divided into: (a) end-caps and 'defect-groups' chemistry; and (b) sidewall functionalization.

14.2.1 End-caps and defect-sites chemistry

Functionalization via 'end-caps and defect-site' chemistry consists of grafting functional groups directly on the already existing defects in the structure of CNTs using wet chemistry. Carbon nanotubes are in general described as perfect graphite sheets rolled into nanocylinders. In reality, CNTs present structural defects: typically, ca. 1–3% of the carbon atoms of a nanotube are located at a defect site (Hu *et al.*, 2001).

The end-caps of nanotubes are composed of highly curved fullerene-like hemispheres, which are much more reactive when compared to the sidewalls (Sinnott, 2002). The sidewalls themselves contain defect-sites such as pentagon–heptagon pairs called Stone-Wales defects, sp^3-hybridized defects or holes in the carbon sheet (Charlier, 2002). The most frequently encountered type of defect is the Stone-Wales (or 7–5–5–7) defect, which leads to a local deformation of the nanotube curvature. Grafting reactions are most favored at the carbon–carbon double bonds in these positions. The different defects commonly observed in CNTs are shown in Fig. 14.1 (Hirsch, 2002).

Frequently, the techniques applied for the purification of the raw material, such as acid oxidation, induce the opening of the tube caps as well as the formation of holes in the sidewalls. These vacancies and tube extremities are therefore functionalized with oxygenated functional groups such as carboxylic acid, ketone, alcohol and ester groups (Chen *et al.*, 1998). The introduced carboxyl groups can present useful sites for further modifications in organic solvents such as the coupling of molecules through the creation of amide or ester bonds. Figure 14.2 represents common functionalization routes on single-wall CNT ends and defect-sites through solution-based chemistry (Banerjee *et al.*, 2005).

Using solution-based chemistry, nanotubes can be grafted with a wide range of functional moieties via amidation or esterification reactions, for which bifunctional molecules are often utilized as linkers/coupling agents. This method was used to graft amine moieties onto carbon nanotubes via

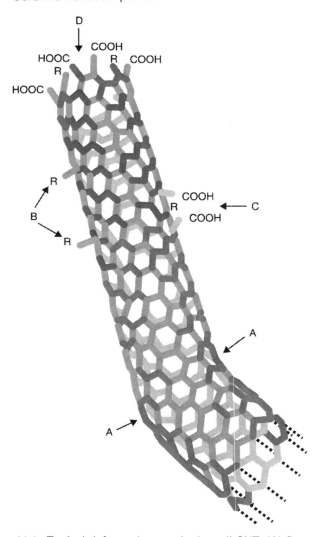

14.1 Typical defects along a single-wall CNT: (A) five- and seven-membered rings forming a bend in the tube; (B) dangling bonds; (C) hole in the carbon framework; and (D) open end of CNT terminated with –COOH groups (Hirsch, 2002).

the reaction with diamines such as triethylenetetramine (Gojny *et al.*, 2003), ethylenediamine (Meng *et al.*, 2008) or 1,6-hexamethylenediamine (Li *et al.*, 2007). Another approach involves the reduction of the carboxyl groups to hydroxyls, followed by transformation into amino groups via phthalimide coupling and hydrolysis (Ramanathan *et al.*, 2005). Gromov *et al.* (2005) have developed two other approaches, using amino-decarboxylation

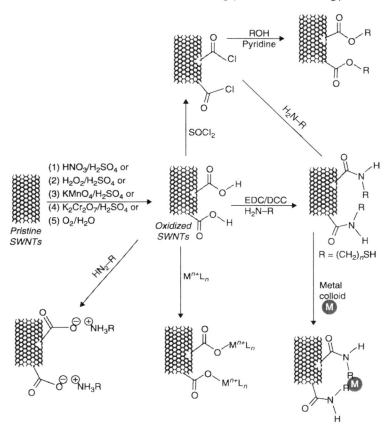

14.2 Some representative functionalization reactions at CNT ends and defect-sites (Banerjee *et al.*, 2005). SWNTs, single-walled nanotubes.

substitution, to make primary amino groups directly attach onto the carbon nanotube surface. The carboxylic groups, the majority grafted at the open ends of CNTs, are replaced with amino groups via Hofmann rearrangement of carboxylic amide or via Curtius reaction of carboxylic acid chloride with sodium azide.

One advantage of the defect-site chemistry is that functionalized carbon nanotubes mostly retain their pristine electronic and mechanical properties (Zhang *et al.*, 2003). However, as mentioned earlier, the spatial distribution of grafted functions is not homogeneous. In fact, the majority of carboxylic groups, used as anchoring sites, are localized at the extremities of the carbon nanotubes.

14.2.2 Sidewall functionalization

The other approach for covalent grafting reactions is the 'sidewall functionalization', which consists of grafting chemical groups through covalent reactions onto the π-conjugated skeleton of CNTs (Hirsch, 2002). Unlike defect-site chemistry, which takes advantage of defects already present in the CNT structure, the direct covalent sidewall functionalization is associated with a change in hybridization from sp^2 to sp^3. This simultaneous loss of conjugation of CNTs influences their physical properties, and more particularly their electrical conductivity, depending on the density of functionalization.

First, covalent sidewall functionalization was carried out on the basis of well-developed grafting chemistry on fullerenes whose reactivity depends strongly on the curvature of the carbon framework. However, the sidewall reaction chemistry of CNTs differs from that of fullerenes as the chemical reactivity in carbon systems arises from two factors that induce local strain: the pyramidalization at the carbon atoms and the π-orbital misalignment between adjacent pairs of conjugated carbon atoms (Chen et al., 2003). The fullerene structures and the CNT end-caps present a pronounced pyramidalization of the carbon atoms further improving chemical reactions. In the CNT sidewalls, the pyramidalization strain is not as acute and, thus, π-orbital misalignment has a greater influence on sidewall chemical reactivity (Hamon et al., 2001). In Fig. 14.3, the reactivity of the C–C bond in CNT structure is presented in function of its angle to the tube

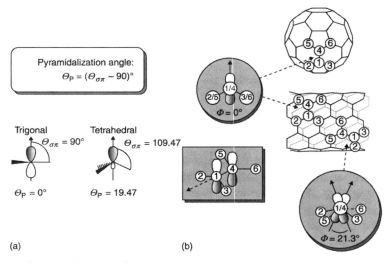

(a)   (b)

14.3 (a, b) Pyramidalization angles (Θ_P) and the π-orbital misalignment angles (Φ) along C1–C4 in the CNT framework and its capping fullerene C_{60} (Hirsch and Vostrowsky, 2005).

circumference. Because the deformation energy of the sp^2 bond is inversely proportional to the diameter of the CNT, nanotubes with a smaller diameter have higher reactivity (Hirsch, 2002).

However, whereas the nanotube end-caps are quite reactive due to their fullerene-like structure, even taking into account the folding of the graphene sheet, the reactivity of CNT sidewalls remains low and sidewall-functionalization is only successful if a highly reactive reagent is used. As mentioned, an additional constraint on sidewall functionalization is the tendency of CNTs to form bundles, which limits the available nanotube surface for the grafting of chemical reagents.

Typically, the covalent sidewall functionalization is carried out in organic solvent, which allows the utilization of the sonication process to improve the dispersion of CNTs and, thus, to increase the available surface of carbon nanotubes. However, precipitation immediately occurs when this process is interrupted (Tasis *et al.*, 2006). The required reactive species such as carbenes, nitrenes or radicals are in general made available through thermally activated reactions (Balasubramanian and Burghard, 2005). Normally, the grafting reaction can be initiated exclusively on the intact sidewall or in parallel at defect-sites. The most common sidewall functionalizations using organic solvents, such as carbene (Lee *et al.*, 2001) or nitrene (Holzinger *et al.*, 2004) [2 + 1] in cycloaddition reactions or radical additions via diazonium salts (Bahr *et al.*, 2001), are shown in Fig. 14.4.

The first sidewall functionalization studied was the fluorination of CNTs. Carbon nanotubes were fluorinated using fluorine in the range between room temperature and 600°C (Mickelson *et al.*, 1998). This reaction is very useful because further nucleophilic substitutions can easily be accomplished and make possible a flexible approach to provide the CNT sidewalls with various types of functional groups (Khabashesku *et al.*, 2002).

Among these reactions, several diamines are reported to react with 'fluoronanotubes' via nucleophilic substitution reactions (Stevens *et al.*, 2003), leading to the formation of amino-functionalized CNTs. However, if bifunctional reagents such as diamines with sufficiently long carbon chains are used, the nanotubes can be covalently cross-linked with each other. Modification with amino-containing substituents was developed using photolysis of acetonitrile (Nakamura *et al.*, 2008). The 1,3-dipolar cycloaddition reaction can be also used to graft linkers, with amino groups at their ends, uniformly distributed along the sidewalls (Pantarotto *et al.*, 2003).

As discussed before, the carbon nanotube covalent functionalizations carried out in organic solvent present some drawbacks. The need to agitate carbon nanotubes by sonication to improve their dispersion in organic solvents can trigger severe damage to the tube walls and the average length

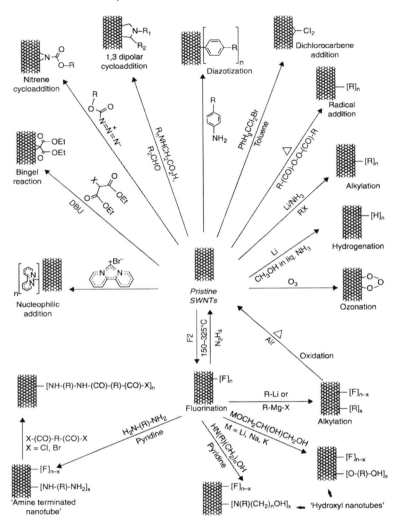

14.4 Representative covalent sidewall functionalization reactions of CNTs as carried out in organic solvents (Banerjee *et al.*, 2005).

of CNTs can be reduced (Monthioux *et al.*, 2001). Moreover, certain highly reactive chemical products also involve harsh conditions causing detrimental damage to the tips and sidewalls of CNTs, which decreases their stability and can cut them into short pieces (Liu *et al.*, 1998). In addition, the solvent functionalization process generally requires time-consuming multistep reactions. Finally, the use of organic solvent pollutes and is difficult to upgrade for industrial processes.

In the next section, the use of plasmas as an optimal alternative for grafting functional groups at the CNT surface will be presented and discussed.

14.3   Plasma treatment of carbon nanotubes

14.3.1 Introduction to plasma: the fourth state of matter

Plasma is physically defined as an ionized gas with an equal number of positively and negatively charged particles. It consists of free electrons, ions, radicals, UV radiation and various highly excited neutral and charged species (Chapman, 1980). The entire plasma is electrically neutral; however, the displacement of the species is controlled by electric and magnetic fields. According to the gas temperature, the plasma can be classified into thermal plasma, which is characterized by being fully ionized (gas temperature T_g = electron temperature T_e), or non-thermal (or cold) plasma where the gas is only partially ionized (T_g is much lower than T_e).

Thermal plasma implies that the temperature of all active species (electrons, ions and neutral species) is the same. This is true for stars, for example, as well as for fusion plasma. High temperatures are required to form these equilibrium plasmas, typically ranging from 4000K to 20 000K (Bogaerts et al., 2002).

In non-thermal or cold plasmas, the temperature of neutral and positively charged species is low, with the electrons having a much higher temperature than the other particles because they are light and easily accelerated by the applied electromagnetic field. As a result, the plasma is in the non-equilibrium state and the reactions may proceed at low temperature. Indeed, the electrons can reach temperatures of 10^4–10^5 K (1–9 eV) while the temperature of the gas can be as low as room temperature (Grill, 1994). Figure 14.5 shows an image of a cold plasma discharge.

14.3.2 Characteristics and principal applications of plasma

To reach the plasma state of atoms and molecules, energy for the ionization must be input into the atoms and the molecules from an external energy source. Ionization occurs when an atom or a molecule gains enough energy from the excitation source or via collisions with other particles. There are many kinds of plasma sources differing greatly from one another. The electrically induced discharge in gas is the most common method for plasma ignition because of handling convenience. Direct current (DC) discharges, pulsed DC discharges, radio frequency (RF) discharges (13.56 MHz) and microwave (μ-wave) discharges (2.45 GHz) represent the plasma categorization based on electric apparatuses (Denes and Manolache, 2004). These are the common standard frequencies used for RF, and μ-wave plasmas are

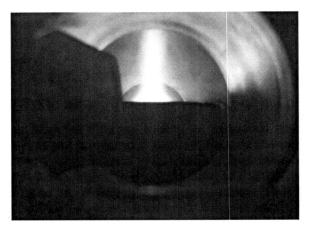

14.5 Radio frequency cathodic magnetron sputtering of titanium target in an argon plasma.

authorized frequencies for industrial, scientific and medical applications worldwide in order to avoid interfering with telecommunication. The basic feature of a variety of electrical discharges is that they produce plasmas in which the majority of the electrical energy primarily goes into the production of energetic electrons, instead of heating the entire gas stream (Fridman *et al.*, 2005). These energetic electrons induce ionization, excitation and molecular fragmentation of the background gas molecules to produce excited species that create a 'chemically-rich' environment. Due to their essential role, the electrons are therefore considered to be the primary agents in the plasma.

Plasma processes provide a cost-effective and environmentally friendly alternative to many important industrial processes because this reliable method produces no waste products and in most cases exposes operators to no significant hazards. Plasmas thus find well-established use in industrial applications, however, they are also gaining more interest in the field of life sciences, related to environmental issues (Hammer, 1999) and biomedical applications (Frauchiger *et al.*, 2004). Recently, the production of CNTs via plasma-enhanced chemical vapor deposition has been reported as well (Boskovic *et al.*, 2005).

Gas discharge plasmas are used in a large variety of applications requiring surface modification. Plasma processing is generally used for film deposition and may also be used for resistant materials development (McOmber and Nair, 1991). Surface modification by plasmas also plays a crucial role in the microelectronics industry, in the microfabrication of integrated circuits and in materials technology (Hino and Akiba, 2000). Actually, the versatility of plasma technologies stems from its many advantages: different sizes, shapes, geometries and type of materials can be treated, surface topography and bulk properties are usually not affected and they exhibit high reproducibility. For surface modification, plasma approaches can be classified into different categories: plasma etching (Inoue and Kajikawa, 2003), plasma polymerization and plasma functionalization (Chu et al., 2002) based on the outcomes of plasma interactions. Etching, deposition or grafting of chemical groups can dominate in modifications of the material's surface, depending on various factors such as the gas used, the nature of the surface and the plasma parameters.

Plasma etching occurs when the surface material removal effect, due to ions and active neutral species as well as vacuum ultraviolet radiation, is prominent in the modification effect. During this process, the plasma generates volatile etched products from the chemical reactions between the elements present on the material surface and the plasma species.

In the plasma polymerization process, the plasma interacts with organic molecules (monomers) and involves their fragmentation and subsequent deposition. The plasma polymer formed thus is deposited in the form of a thin film. Notwithstanding the use of the word polymer, 'plasma polymer' refers to a new class of material that has little in common with the conventional polymer. In the case of the plasma polymer, the chains are short and randomly branched and terminate with a high degree of cross-linking, and are not constituted by regularly repeating units.

Plasma treatment using gases such as O_2, N_2, NH_3 or CF_4 allows reactive chemical functionalities to be inserted onto the material's surface. Compared with other chemical modification methods, plasma-induced functionalization presents interesting characteristics such as being a solvent-free and time-efficient technique. Moreover, this treatment allows the grafting of a wide range of different functional groups depending on plasma parameters such as power, nature of the gas used and its pressure, duration of treatment, etc. This method also provides the possibility of scaling up to produce large quantities necessary for commercial use. Plasma treatment is widely used for surface activation of various materials, ranging from organic polymers to ceramics and metals. In this field, we can cite the polymer surface functionalization in pulsed and continuous nitrogen microwave plasma, under admixture of hydrogen (Meyer-Plath et al., 2003). This process makes it possible to graft nitrogen functional groups onto polystyrene with a high selectivity in primary amine groups.

14.3.3 Plasma-based treatment of carbon nanotubes

The surface modification of carbon nanotubes can be carried out through a wide range of plasma processes. In addition to the already cited advantages of plasma treatments, the number of functional groups grafted on the CNT

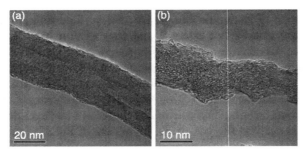

14.6 HRTEM images of MWNTs (a) before and (b) after Ar microwave plasma treatment for 3 minutes (Qin and Hu, 2008).

surface can also be tailored. This is an important characteristic of this type of surface treatment since having saturation of these groups on the surface can alter the electronic conductivity of nanotube materials.

To facilitate the discussion, we will classify the plasma treatment of CNTs into: (1) CNT surface etching, which is generally carried out via an argon (Ar) plasma treatment; (2) coating of CNTs through plasma polymerization; and (3) plasma functionalization, which allows the grafting of reactive functional groups on to CNT surfaces and opens the way to a wide variety of subsequent chemical reactions.

Carbon nanotube surface etching

The plasma surface modification of CNTs is a non-reactive treatment when Ar gas is used. The effects of argon RF plasma treatment on the surface of vertically grown CNTs have been investigated. The inert Ar plasma produces an efficient etching and cleaning process of CNT films, causing structural changes in the CNTs and leading to an increase in their field emission ability (Ahn *et al.*, 2003). An observed improvement in gas ionization in comparison with that of untreated aligned CNTs was also observed after Ar RF plasma treatment (Yan *et al.*, 2005). The Ar plasma was also used to activate the CNT surface, allowing the subsequent grafting of maleic anhydride (Tseng *et al.*, 2007) or 1-vinylimidazole (Yan *et al.*, 2007).

Nevertheless, the destruction of carbon nanotube sidewalls was observed after treatment with an Ar non-reactive μ-wave plasma (Qin and Hu, 2008). Figure 14.6 shows high resolution transmission electronic microscopy (HRTEM) images of multi-wall CNTs recorded before and after Ar μ-wave plasma treatment. We can observe that ion bombardment and irradiation in Ar plasma deform the graphite walls and destroy the CNT layered structure; a reduction in the power supplied to the plasma or in the treatment duration can reduce the destruction of sidewalls.

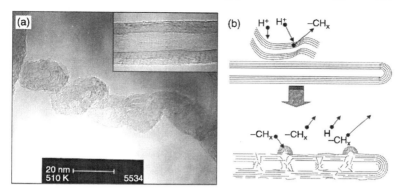

14.7 HRTEM images of a formed nodular CNT and an untreated CNT in the insert (a) and schematic diagram of the formation of nodular CNTs (b) (Zhang *et al.*, 2004).

Hydrogenation of CNT films through H_2 RF plasma processes improved the CNT field emission properties via creation of defects along the CNT structure (Yu *et al.*, 2004; Feng *et al.*, 2007). The enhancement of emission properties was attributed to apparition of nodular CNTs due to bending of graphene sheets along the CNT wall (Fig. 14.7 (a)) (Zhang *et al.*, 2004). During the H_2 RF plasma treatment, it was suggested that hydrogen ions bombarding the tube sidewall remove carbon atoms from the CNT surface under $-CH_x$ radical forms. Meanwhile, a small fraction of $-CH_x$ redeposited on the remaining CNT surface induces the formation of nanoscale particles with an onion-like structure distributed along the tubes (Fig. 14.7 (b)). The improvement of the emission characteristics was attributed to the change in the electronic (formation of sp^3 defects in sp^2 graphite network) and/or geometrical CNT structure (Zhang *et al.*, 2004).

Vertically aligned CNTs were hydrogenated in a H_2 + Ar pulsed DC plasma, and correlations between the plasma characteristics and the CNT surface chemistry were discussed (Jones *et al.*, 2008). Actually, scanning electron microscope (SEM) morphological analysis of CNTs treated in this way showed the etching of the tangled nanotubes and the 'welding' of nanoparticles. The extent of the etching effect was correlated with the quantity of plasma-excited hydrogen H* interacting with CNTs.

Plasma polymerization

The plasma polymerization process has been used to coat CNTs with plasma polymer films. Several examples are reported in the literature. A plasma polymerization method spraying a mixture of aniline and CNTs was developed to deposit plasma polyaniline onto CNTs leading to the formation of composite films with improved electrical properties in

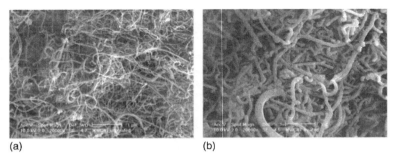

(a)   (b)

14.8 SEM images of multi-walled CNTs before treatment (a) and after PMMA plasma polymerization treatment (b) (Gorga *et al.*, 2006).

comparison to neat plasma (Nastase *et al.*, 2006). Chen *et al.* (2001) developed an approach based on RF plasma activation, followed by chemical reactions between derivatized dextran- and plasma-generated functional groups such as aldehyde groups from acetaldehyde and amino groups from ethylenediamine. The resulting polysaccharide-grafted CNTs proved highly hydrophilic.

Multi-walled CNTs were also modified using plasma polymerization with ethylene glycol (EG) as the monomer (Avila-Orta *et al.*, 2009). The plasma-polymerized EG-coated CNTs showed very stable dispersion with water, methanol and ethylene glycol, confirming the hydrophilic behavior of the treated CNTs. This plasma polymerization technique was successfully used to produce a plasma poly(methyl methacrylate) (PMMA) coating on CNTs (Gorga *et al.*, 2006). Excellent suspensions of CNTs were achieved in toluene after plasma treatment, and plasma PMMA coating was observed using SEM (Fig. 14.8). The coated CNTs were incorporated into the PMMA matrix via melt mixing and the mechanical properties of the nanocomposites formed in this way were determined with tensile measurements. The CNT coating slightly improved the load transfer from the PMMA matrix to the nanotubes, over the uncoated CNTs, but it did not significantly influence the dispersion of CNTs. The mechanical properties were not dramatically improved.

Styrene plasma was also reported to deposit polystyrene (PS) plasma polymer coating onto multi-walled CNTs (Felten *et al.*, 2007). It was shown that polystyrene nanocomposite filled with plasma PS-coated CNTs presented better mechanical properties than the ones filled with unmodified CNTs (Shi and He, 2004).

Plasma chemical functionalization

CNT sidewalls can be fluorinated via CF_4 plasma treatment (Plank *et al.*, 2003). Fluorination is one of the most effective modifications of the CNT

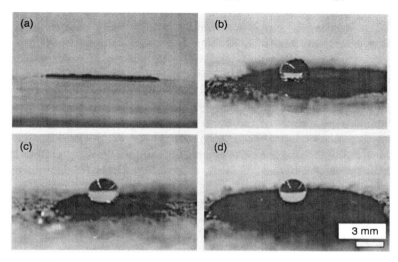

14.9 Photographic images of liquid droplets. Untreated CNT powder (a) completely absorbed liquid droplets and (b)–(d) the droplets of PEG, glycerol and distilled water on 10-min plasma-treated CNTs, respectively (Hong *et al.*, 2006).

surface and makes possible a wide range of subsequent derivatization reactions (Lee *et al.*, 2003). It has been observed that bundles of pristine CNTs can be transformed into unroped fluorinated CNTs without structural deformation (Valentini *et al.*, 2005). Moreover, unlike other fluorination procedures, the CF_4 plasma treatment was demonstrated to enhance the reactivity of fluoronanotubes with aliphatic amines at room temperature. Two minutes of CF_4 plasma treatment was reported to increase the field emission current (Zhu *et al.*, 2005). Fluorine atoms were grafted on the CNT sidewalls when these were exposed to a CF_4 plasma that had been generated in a surface wave microwave plasma reactor (Kalita *et al.*, 2008). The amount of grafted fluorine (up to 24 atomic %) was determined by X-ray photoelectron spectroscopy (XPS) and Raman studies revealed a structure change in the treated CNTs. HRTEM showed modification of the outer wall with fluorine incorporation while the inside walls remained unaffected. Super-hydrophobic CNTs were obtained by NF_3 glow-discharge plasma supplied by alternating current (AC) power (Hong *et al.*, 2006). The treatment time to obtain hydrophobic CNT powders was shorter than 1 minute. Figure 14.9 illustrates liquid droplets on CNT cushions of about 1 mm thickness. Figure 14.9 (a) was taken after a water droplet fell onto untreated CNTs. The other photographs (Figs 14.9 (b)–(d)) show respectively the droplets of polyethylene glycol (PEG), glycerol and water placed on CNTs that had been exposed for 10 minutes to NF_3 plasma treatment. These images reveal a significant enhancement of the hydrophobicity.

The grafting of oxygen-containing groups using plasma processes has been extensively reported. Hydroxyl, carbonyl and carboxyl groups were grafted onto CNTs via O_2 and CO_2 RF plasma in order to form polar groups that could improve the overall adhesion of CNTs to a polyamid matrix (Bubert et al., 2003). Oxygen concentration of up to 14% was determined by XPS, which was considered by the authors as equivalent to a complete saturation of the outer CNT surface. The efficiency of oxygen grafting in function of plasma frequency (RF or μ-wave) was determined through contact angle measurement, showing that μ-wave plasma treatment leads to better wettability (Chirila et al., 2005). Other plasma processes were proposed to readily functionalize CNTs with oxygenated groups such as air-atmospheric pressure dielectric barrier discharge (Okpalugo et al., 2005) or air μ-wave oven-generated plasma (HojatiTalemi et al., 2009).

Zschoerper et al. (2009) have characterized Ar/O_2 and Ar/H_2O RF plasma-treated CNTs by XPS. To overcome the limitations of XPS analysis in distinguishing between functional groups with similar binding energies, alcohol, keto/aldehyde and carboxyl groups were tagged using derivatization techniques with fluorine-containing reagents. Trifluoroacetic anhydride (TFAA) was used for the derivatization of alcohol groups, (trifluoromethyl) phenylhydrazine (TFMPH) for keto/aldehyde groups and trifluoroethanol for carboxyl groups through reactions carried out in the saturated vapor phase. Despite the fact that the total oxygen content is almost identical, variations in different functional group concentration are observed in the function of treatment parameters such as pressure or treatment time. Based on this work, CNT bucky papers were modified using Ar/O_2 plasma and thereafter melt-mixed into polycarbonate (Pötschke et al., 2009). Carboxylic acid and ester groups were formed on the CNT surface, allowing better macrodispersion and better phase adhesion to the matrix, as shown by morphological investigations.

A flexible amperometric biosensor based on O_2 RF plasma-functionalized CNTs films on polydimethylsiloxane substrates has been developed (Lee et al., 2009). The plasma-treated samples presented better glucose response than non-treated ones, showing their potential application as biosensors. The O_2 RF plasma treatment was also found to improve the sensing potential of CNTs to detect NO_2 (Ionescu et al., 2006) or to clean CNT surfaces by removing amorphous carbon (Rawat et al., 2006). In fact, it has been observed that purification of carbon nanotube powder can be attained after CNT treatment in glow discharges (RF or μ-wave). Amorphous carbon domains are eliminated and the impurities are removed by ion bombardment and irradiation in an O_2 RF plasma (Xu et al., 2007). However, it was also reported that the average diameter of multi-walled CNTs decreases with treatment duration. Figure 14.10 shows a three-step model proposed to explain this decrease. First, ion bombardment causes the

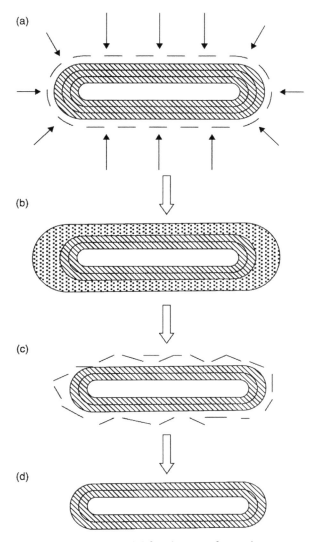

14.10 Proposed model for the transformation process of multi-walled CNTs under O_2 plasma treatment. (a) Oxygen functional groups are grafted by O_2 plasma; (b) defects and carbon nanowires are created by ion bombardment, losing the superficial structure; (c) ions continue to react with amorphous carbon which are peeled totally from the carbon nanotubes; (d) amorphous carbon is oxidized under O_2 plasma which forms carbon dioxide, the diameter of the multi-walled carbon nanotubes decreases due to the disappearance of the outer wall (Xu *et al.*, 2007).

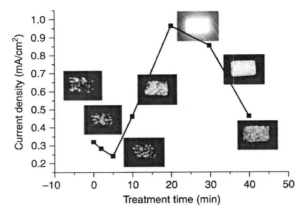

14.11 The emission current density and fluorescent photos of the CNT films at 6.4 V/μm after various N_2 plasma treatment times (Feng *et al.*, 2007).

creation of vacancies and interstitials in CNTs; the superficial structure, including the amorphous carbon that was produced, is thus lost. In a second step, the ion beams continue to react with amorphous carbon until they are peeled totally away from the CNTs. Finally, the oxidation of amorphous carbon in CO_2 occurs, inducing the decrease of the CNT's average diameter.

Plasma-oxidized CNTs were reported to present improved dispersion in epoxy resin in comparison with non-treated CNTs (Kim *et al.*, 2006). The epoxy nanocomposites filled with plasma-modified CNTs exhibited higher storage and loss moduli than CNT/epoxy nanocomposites as well as improved tensile strength and elongation at break due to better dispersion and stronger interaction between the CNTs and the polymer matrix.

Felten *et al.* showed that cluster dispersion on thermal evaporation coating of CNTs with various metals can be tuned by increasing the cluster nucleation density through the formation of interaction sites when CNTs are treated in an O_2 RF plasma (Felten *et al.*, 2007). The same effect was also observed when the CNTs were exposed to an NH_3 RF plasma before the metal coating.

The nitrogen-containing groups can be grafted through plasma treatment if gases, such as N_2, N_2/H_2 or NH_3, are used. The treatment of multi-walled CNTs in an N_2/Ar RF plasma was shown to be an effective procedure to enhance the field emission characteristics of CNTs due to the doping of the CNT structure with N atoms (Gohel *et al.*, 2005). The N1s core level peak was fitted with components at 398.2, 398.6, 399.7 and 400.8eV, which corresponds to C–N, C≡N (sp^3 bonding), C=N (sp^2 bonding) and NO. The field emission properties of screen-printed CNT films were also improved after NH_3 or N_2 RF plasma treatment (Feng *et al.*, 2007). For the CNT films treated by N_2 plasma, the emission current density and fluorescent photos at 6.4 V/μm clearly demonstrated the influence of the treatment time on field emission properties (Fig. 14.11).

The improvement of the field emission properties was attributed to the change in CNT morphology after plasma treatment with the formation of nano-protuberances, i.e. 'multi-tip' CNTs. The dependence of the emission current with treatment duration showed an optimum after 20 minutes of treatment. Prolonged treatment was shown to reduce the field emission properties that were associated with the destruction of CNTs and nano-protuberances.

The N_2 RF plasma treatment was also used to improve the Pt catalyst deposition at the CNT surface, resulting in an enhancement of the electrochemical activity (Kim et al., 2008).

Bystrzejewski et al. reported the treatment of single-walled CNTs in a DC hollow cathode glow discharge (HCGD) (Bystrzejewski et al., 2009). Various gases (N_2, H_2O, $N_2 + H_2O$ and $NH_3 + H_2O$) were used to functionalize CNTs. The N_2 and/or H_2O plasma treatment resulted in the presence of amorphous carbon species at the CNT surface while the use of $NH_3 + H_2O$ yielded a very clean product consisting of functionalized CNTs. In-situ optical emission studies showed that the functionalization occurs via radical addition channels with the initial assistance of N_2^+ radical ions. The N_2^+ bombardment breaks the C–C bonds on the CNT surface, after which other chemical radicals are subsequently added and quenched (Fig. 14.12).

An NH_3 μ-wave plasma treatment has been also used to enhance the solubility of CNTs (Wu et al., 2007). The introduction of polar functional groups increases the hydrophilicity of CNTs, making the application of CNTs in the immobilization of biomolecules and construction of biosensors more convenient.

The studies cited above demonstrate the wide range of plasma processes that can be used to modify the CNT surface. The excited species such as electrons, ions and radicals within the plasma interact with the surface of CNTs and break the C=C bonds, which promotes the creation of active sites to bind functional groups as well as some physical modification of the CNT surface. Moreover, it was also reported that UV photons interact with CNTs and create active sites on their sidewalls, however, they can at the same time promote the defunctionalization of moieties grafted onto CNTs (Khare et al., 2002). Due to the interaction of the different plasma reactive species with CNTs, the grafting of functional groups can therefore occur through different simultaneous reactions. This wide range of interactions involves a lack of control on the grafting of wanted chemical function and makes it difficult to determine the weight of each functional mechanism. Moreover, prolonged plasma treatments promote damage of CNTs,

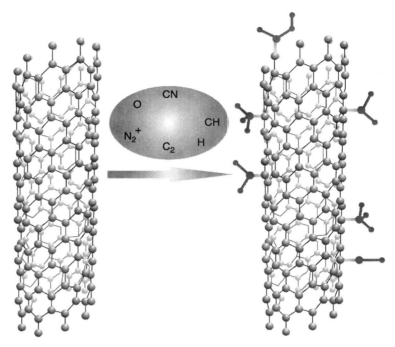

14.12 Radical addition yielding functional groups (–CH₃, –CN, –OH, –NH₂, . . .) covalently attached to CNT sidewall (Bystrzejewski *et al.*, 2009).

resulting in diminution of their diameter with destruction of the outer wall, the forming of onion-like structures, or even their complete destruction.

Plasma post-discharge treatment of carbon nanotubes

To prevent the unwanted effects of plasma functionalization, a solution was proposed by Ruelle *et al.* (2007). These authors proposed avoiding the functionalization of CNTs directly inside the plasma where the density of high energy ions is very high, and instead, placing the CNTs outside the discharge production zone to reduce the detrimental effects associated with ion bombardment and irradiation. In this approach, radicals are the most important reactive species to graft functional groups onto the CNT surface. In this case, the application of microwave discharge sources for the production of intense beams of atomic, radical and metastable species is well established. For example, carbon nanotubes have been hydrogenated by atomic hydrogen generated in a H₂ microwave plasma (Wu *et al.*, 2007). In this type of functionalization, the exact location of the sample and its distance from the plasma discharge zone become a key factor in determining

14.13 An Ar + N_2 microwave discharge sustained by a surface wave launched on the quartz tube. A high quantity of atomic nitrogen is produced in this plasma discharge (Ruelle, 2009).

the density of functionalization. Khare *et al.* studied the exposition of SWNTs to microwave-generated N_2 plasma, by placing CNTs at different distances from discharge (Khare *et al.*, 2005). At the shortest distance (1 cm), they observed the highest concentration in nitrogen groups but also a loss of integrity of SWNTs due to highly reactive species, like N_2^+. At intermediate distances, the incorporation of nitrogen, but also a high quantity of oxygen, onto the CNTs is obtained while functionalization was not observed for the maximal distance of 7 cm due to total recombination of atomic nitrogen before arriving at the CNT surface. It is thus important to place CNTs outside the glow discharge zone at an optimal distance to avoid interaction with highly reactive ions and UV photons. In this treatment configuration, radicals become the most important reactive species for the grafting of functional groups onto carbon nanotubes. The efficiency of this approach depends on the density of the radicals produced.

In this context, a surface wave discharge set-up was developed (Godfroid *et al.*, 2003). The microwave-induced plasmas present a higher density of electrons in comparison with other type of plasmas, such as RF plasmas, for instance, because electrons are more easily created with microwaves (at the same power).

The high electron density of the μ-wave-induced plasmas is the key parameter in the creation of atomic nitrogen in Ar + N_2 microwave plasma (Fig. 14.13). Godfroid *et al.* have demonstrated that production of atomic nitrogen in Ar + N_2 μ-wave discharge is achieved through two electronic mechanisms (Godfroid *et al.*, 2005). The first is dissociation by direct electronic impact, which is achieved by collision between a nitrogen molecule and an electron:

$$e + N_2\left(\sum_g^+ n = 0\right) \rightarrow e + N\left(^4S\right) + N\left(^4S\right) \qquad [14.1]$$

This reaction leads to the production of atomic nitrogen in the fundamental state $N(^4S)$.

The second is called dissociative recombination and is achieved by recombination between an electron and an ionized nitrogen molecule:

$$e + N_2^+ \rightarrow N\left(^4S\right) + N\left(^4S\right) \qquad [14.2]$$

Finally, the set-up and parameters of the microwave discharge were studied to find the higher nitrogen dissociation rate (Godfroid *et al.*, 2003). The most influential parameters proved to be the power supply pulse, the dilution of N_2 in Ar and the total gas flow, which makes it possible to achieve a high homolytic dissociation up to 40% of the molecular N_2.

The CNTs were thus treated in the post-discharge of an Ar + N_2 microwave plasma sustained by a surface wave launched in a quartz tube via a surfaguide supplied by a 2.45 GHz microwave generator, which can be pulsed (Ruelle *et al.*, 2007). In this plasma-induced functionalization, the plasma was not used to directly interact with the surface of CNTs but as a source of atomic nitrogen N· reactive flow. It is important to note that the discharge tube and the post-discharge chamber are separated by a diaphragm that consists of an aluminium ring with a circular aperture whose diameter measures a few millimetres. This diaphragm plays two roles in the post-discharge treatment. First, the separation between the plasma set-up and the post-discharge protects the samples placed in the post-discharge from irradiation of high-energy particles from μ-wave plasma. Moreover, the average distance of 40 cm between the end of the discharge tube and the sample holder promotes the interaction of CNTs with atomic nitrogen species that have enough mean lifetime to reach the sample holder, in contrast to other reactive species.

High resolution photoelectron spectroscopy analysis showed that by exposing multi-walled CNTs to atomic nitrogen N· generated in the μ-wave plasma, nitrogen chemical groups were grafted onto the CNT surface, altering the density of electronic states (Ruelle *et al.*, 2008). High-resolution TEM and SEM images revealed that the atomic nitrogen exposition did not damage the surface of the CNTs (Fig. 14.14). The absence of damage in the structure of carbon nanotubes after the plasma treatment also indicated that nitrogen group grafting resulted only from chemical reaction between CNT

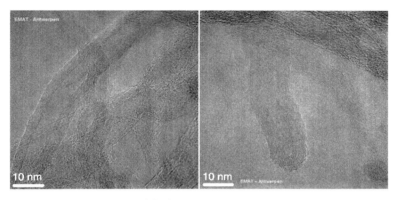

14.14 HRTEM images of CNTs treated in the post-discharge chamber of Ar + N_2 μ-wave plasma (Ruelle *et al.*, 2008).

14.15 TEM image of one CNT with PCL islets resembling a pearl-necklace structure (Ruelle, 2009).

surfaces and atomic nitrogen species, without the side-effects of irradiation or high energy particle interaction at the CNT surface.

With the addition of H_2 in the post-discharge, an efficient and selective grafting of primary amine groups was developed. Actually, two-thirds of grafted nitrogenated functional groups were primary amine groups, as determined by XPS after (trifluoromethyl)benzaldehyde derivative reaction (Ruelle, 2009). These grafted amine groups have been used as initiation sites for promoting the ring opening polymerization (ROP) of ε-caprolactone (ε-CL) yielding polyester-grafted CNT nanohybrids (Ruelle *et al.*, 2007). The morphology of the recovered nanohybrids has been characterized by TEM, showing that PCL islets, and so the initiator primary amines were homogeneously dispersed along the CNT sidewalls (Fig. 14.15). Moreover, the nanohybrids displayed the best dispersion ability in chloroform, a good PCL solvent, in comparison to pristine and amino-CNTs.

The CNT-g-PCL nanohybrids, mixed with free PCL chains, presented a high degree of CNT pre-disaggregation and were thus used as nanofillers in PCL; pristine and amino-CNT-filled nanocomposites were also prepared as comparative materials. TEM images showed that slightly disrupted CNT bundles are present in the CNT-filled PCL nanocomposites while CNTs are homogeneously dispersed when CNT-g-PCL nanohybrids are used (Fig. 14.16). Electrical conductivity measurements of the PCL nanocomposites showed a similar percolation threshold (~0.35 wt%) for each PCL

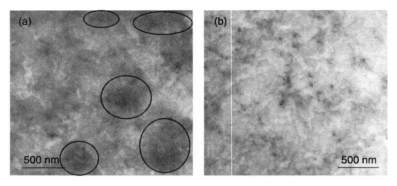

14.16 TEM images of PCL nanocomposites filled with (a) 3 wt% CNTs and (b) 3 wt% CNT-g-PCL nanohybrids. CNTs remain agglomerated (black circles) in the CNTs-filled PCL nanocomposites while very small bundles are observed (circled) when PCL is filled with CNT-g-PCL nanohybrids (Ruelle, 2009).

nanocomposite, corresponding to a theoretical value for optimal dispersion of CNTs, which is equal to the inverse of CNT aspect ratio.

The different nanofillers (CNTs, amino-CNTs and CNT-g-PCL nanohybrids) were also dispersed in a high-density polyethylene (HDPE) matrix. TEM analysis indicated that CNTs and amino-CNTs were poorly dispersed in HDPE while CNT-g-PCL were more homogeneously dispersed, even though PCL and HDPE are known to be non-miscible polymers. Electrical measurements confirmed these observations, showing an electrical percolation threshold of 2.5 wt% for CNTs and amino-CNT-filled HDPE nanocomposites and 1.1 wt% for samples filled with a CNT-g-PCL nanohybrid.

The grafting of primary amine groups on the CNT surface did not influence the dispersion ability of CNTs in PCL and HDPE matrices. However, the electrical measurements showed that the electrical conductivity of amino-CNTs was higher than that of pristine CNTs; functionalization can be used to improve the electrical conductivity. This last result confirmed that the CNT structure is preserved after CNT functionalization in the microwave plasma post-discharge.

14.4 Summary

Offering both a high aspect ratio (length-to-diameter) and a low density, CNTs show strong application potential as advanced filler materials to replace or complement the conventional nanofillers, such as nanoclays, in the fabrication of multifunctional polymer nanocomposites. In this field, the degree of dispersion of the individual CNTs is an essential parameter to produce nanocomposite materials with enhanced properties. In order to homogeneously disperse CNTs throughout polymer matrices, the CNT entanglements and bundles, stabilized by intermolecular van der Waals forces, must be disrupted. One of the most reported techniques to overcome these drawbacks is the surface modification of CNTs. In this context, solvent-based covalent functionalizations are well reported in the literature. Nevertheless, this type of functionalization presents different drawbacks, such as CNT insolubility, which can limit control over the grafting of the functional group onto the CNT surface.

Plasma processes constitute a promising alternative for surface modification. Plasma treatments have the advantage of being non-polluting and provide a wide range of different functional groups depending on plasma parameters such as power, type of gas used, treatment duration and gas pressure.

Generally, CNTs are placed inside the plasma, where their surface interacts with plasma-excited species such as radicals, electrons, ions and UV radiation which induces the breaking of C–C bonds and the creation of active sites for bonding of functional groups on the CNT surface.

Argon and hydrogen plasmas are reported to etch CNT surface and end-caps, improving their field emission properties. When organic monomers are introduced in the plasma, CNTs are coated with plasma polymer films. This results in the pre-disaggregation of CNT bundles, which can be well dispersed in polymer matrices. Fluorinated, oxygenated and nitrogenated gases were also used to generate functional groups at the CNT surface. This surface modification of CNTs opens the way for compatibilization with polymer matrices or for further attachment of other molecules in order to form polymer nanohybrids. Nevertheless, total control over the grafting of functional groups on the CNT surface is difficult to achieve when CNTs are treated directly in the plasma discharge chamber.

To prevent the degradation of the CNTs during plasma treatment, it was proposed to use plasma discharges as a source of reactive species, such as radicals, which can interact with the CNT surface outside the plasma discharge, avoiding structural damage and destruction due to high energy species bombardment. This process can be considered as a 'dry' chemical reaction at the CNT surface. The grafted functional groups can be used as initiator sites for polymerization, yielding CNT nanohybrids that can be introduced into polymer matrices, forming high-performance polymer nanocomposites and nanohybrids.

14.5   References

Ahn K. S., Kim J. S., Kim C. O. and Hong J. P. (2003) 'Non-reactive RF treatment of multiwall carbon nanotube with inert argon plasma for enhanced field emission', *Carbon*, 41, 2481–2485.

Andrews R. and Weisenberger M. C. (2004) 'Carbon nanotube polymer composites', *Curr. Opin. Solid State Mater. Sci.*, 8, 31–37.

Avila-Orta C. A., Cruz-Delgado V. J., Neira-Velazquez M. G., Hernandez-Hernandez E., Mendez-Padilla M. G. and Medellin-Rodriguez F. J. (2009) 'Surface modification of carbon nanotubes with ethylene glycol plasma', *Carbon*, 47, 1916–1921.

Bagchi A. and Nomura S. (2006) 'On the effective thermal conductivity of carbon nanotube reinforced polymer composites', *Compos. Sci. Technol.*, 66, 1703–1712.

Bahr J. L., Yang J., Kosynkin D. V., Bronikowski M. J., Smalley R. E. and Tour J. M. (2001) 'Functionalization of carbon nanotubes by electrochemical reduction of aryl diazonium salts: a bucky paper electrode', *J. Am. Chem. Soc.*, 123, 6536–6542.

Balasubramanian K. and Burghard M. (2005) 'Chemically functionalized carbon nanotubes', *Small*, 1, 180–192.

Banerjee S., Hemraj-Benny T. and Wong S. S. (2005) 'Covalent surface chemistry of single-walled carbon nanotubes', *Adv. Mater.*, 17, 17–29.

Bogaerts A., Neyts E., Gijbels R. and van der Mullen J. (2002) 'Gas discharge plasmas and their applications', *Spectrochimica Acta B*, 57, 609–658.

Bonduel D., Mainil M., Alexandre M., Monteverde F. and Dubois Ph. (2005) 'Supported coordination polymerization: a unique way to potent polyolefin carbon nanotube nanocomposites', *Chem. Commun.*, 781–783.

Boskovic B. O., Golovko V. B., Cantoro M., Kleinsorge B., Chuang A. T. H., Ducati C., Hofmann S., Robertson J. and Johnson B. F. G. (2005) 'Low temperature synthesis of carbon nanofibres on carbon fibre matrices', *Carbon*, 43, 2643–2648.

Bubert H., Haiber S., Brandl W., Marginean G., Heintze M. and Brüser V. (2003) 'Characterization of the uppermost layer of plasma-treated carbon nanotubes', *Diam. Rel. Mater.*, 12, 811–815.

Bystrzejewski M., Rummeli M. H., Gemming T., Pichler T., Huczko A. and Lange H. (2009) 'Functionalizing single-wall carbon nanotubes in hollow cathode glow discharges', *Plasma Chem. Plasma Process*, 29, 79–90.

Chapman B. (1980) *Glow Discharge Processes*, New York: Wiley Interscience.

Charlier J.-C. (2002) 'Defects in carbon nanotubes', *Acc. Chem. Res.*, 35, 1063–1069.

Chen J., Hamon M. A., Hu H., Chen Y., Rao A. M., Eklund P. C. and Haddon R. C. (1998) 'Solution properties of single-walled carbon nanotubes', *Science*, 282, 95–98.

Chen Q. D., Dai L. M., Gao M., Huang S. M. and Mau A. (2001) 'Plasma activation of carbon nanotubes for chemical modification', *J. Phys. Chem. B*, 105, 618–622.

Chen Z., Thiel W. and Hirsch A. (2003) 'Reactivity of the convex and concave surfaces of single-walled carbon nanotubes (SWCNTs) towards addition reactions: dependence on the carbon-atom pyramidalization', *Chem. Phys. Chem.*, 4, 93–97.

Chirila V., Marginean G. and Brandl W. (2005) 'Effect of the oxygen plasma treatment parameters on the carbon nanotubes' surface properties', *Surf. Coat. Technol.*, 200, 548–551.

Chu P. K., Chen J. Y., Wang L. P. and Huang N. (2002) 'Plasma-surface modification of biomaterials', *Mater. Sci. and Eng. R*, 36, 143–206.

Collins P. G. and Avouris P. (2000) 'Nanotubes for electronics', *Sci. Am.*, 283, 62–69.

D'Agostino R., Favia P., Oehr C. and Wertheimer M. R. (2005) 'Low-temperature plasma processing of materials: past, present and future', *Plasma Process Polym.*, 2, 7–15.

Denes F. S. and Manolache S. (2004) 'Macromolecular plasma-chemistry: an emerging field of polymer science', *Prog. Polym. Sci.*, 29, 815–885.

Felten A., Bittencourt C., Colomer J.-F., Van Tendeloo G. and Pireaux J.-J. (2007) 'Nucleation of metal clusters on plasma treated multi wall carbon nanotubes', *Carbon*, 45, 110–116.

Feng T., Zhang J., Li Q., Wang X., Yu K. and Zou S. (2007) 'Effects of plasma treatment on microstructure and electron field emission properties of screen-printed carbon nanotube films', *Physica E*, 36, 28–33.

Frauchiger V. M., Schlottig F., Gasser B. and Textor M. (2004) 'Anodic plasma-chemical treatment of CP titanium surfaces for biomedical applications', *Biomaterials*, 25, 593–606.

Fridman A., Chirokov A. and Gutsol A. (2005) 'Non-thermal atmospheric pressure discharges', *J. Phys. D: Appl. Phys.*, 38, R1–R24.

Fu K., Huang W., Lin Y., Riddle L. A., Carroll D. L. and Sun Y. P. (2001) 'Defunctionalization of functionalized carbon nanotubes', *Nano. Lett.*, 1(8), 439–441.

Godfroid T., Dauchot J.-P. and Hecq M. (2003) 'Atomic nitrogen source for reactive magnetron sputtering', *Surf. Coat. Technol.*, 174–175, 1276–1281.

Godfroid T., Dauchot J.-P. and Hecq M. (2005) 'Effect of plasma temperature and plasma pulsation frequency on atomic nitrogen production', *Surf. Coat. Technol.*, 200, 649–654.

Gohel A., Chin K. C., Zhu Y. W., Sow C. H. and Wee A. T. S. (2005) 'Field emission properties of N_2 and Ar plasma-treated multi-wall carbon nanotubes', *Carbon*, 43, 2530–2535.

Gojny F.H., Nastalczyk J., Roslaniec Z. and Schulte K. (2003) 'Surface modified multi-walled carbon nanotubes in CNT/epoxy-composites', *Chem. Phys. Lett.*, 370, 820–824.

Gorga R. E., Lau K. K. S., Gleason K. K. and Cohen R. E. (2006) 'The importance of interfacial design at the carbon nanotube/polymer composite interface', *J. Appl. Polym. Sci.*, 102, 1413–1418.

Grill A. (1994) *Cold Plasma in Materials Fabrication*, Boca Raton, FL: IEEE Press.

Gromov A., Dittmer S., Svensson J., Nerushev O. A., Perez-Garcia S. A., Licea-Jiménez L., Rychwalski R. and Campbell E. E. B. (2005) 'Covalent amino-functionalisation of single-wall carbon nanotubes', *J. Mater. Chem.*, 15, 3334–3339.

Hammer T. (1999) 'Application of plasma technology in environment', *Contrib. Plasma Phys.*, 39, 441–462.

Hamon M. A., Itkis M. E., Niyogi S., Alvaraez T., Kuper C., Menon M. and Haddon R. C. (2001) 'Effect of rehybridization on the electronic structure of single-walled carbon nanotubes', *J. Am. Chem. Soc.*, 123, 11292–11293.

Hino T. and Akiba M. (2000) 'Japanese development of fusion reaction plasma components', *Fus. Eng. Des.*, 49–50, 97–105.

Hirsch A. (2002) 'Functionalization of single-walled carbon nanotubes', *Angew. Chem. Int. Ed.*, 41, 1853–1859.

Hirsch A. and Vostrowsky O. (2005) 'Functionalization of carbon nanotubes', *Top. Curr. Chem.*, 245, 193–237.

Hojati-Talemi P., Cervini R. and Simon G. P. (2009) 'Effect of different microwave-based treatments on multi-walled carbon nanotubes', *J. Nanopart. Res.*, DOI 10.1007/ s11051–009–9609-y.

Holzinger M., Steinmetz J., Samaille D., Glerup M., Paillet M., Bernier P., Ley L. and Graupner R. (2004) '2 + 1 Cycloaddition for cross-linking SWCNT', *Carbon*, 42, 941–947.

Hong Y. C., Shin D. H., Cho S. C. and Uhm H. S. (2006) 'Surface transformation of carbon nanotube powder into super-hydrophobic and measurement of wettability', *Chem. Phys. Lett.*, 427, 390–393.

Hu H., Bhowmik P., Zhao B., Hamon M. A., Itkis M. E. and Haddon R. C. (2001) 'Determination of the acidic sites of purified single-walled carbon nanotubes by acid-base titration', *Chem. Phys. Lett.*, 345, 25–28.

Iijima S. (1991) 'Helical microtubules of graphite carbon', *Nature*, 354, 56–58.

Inoue S. and Kajikawa K. (2003) 'Inductivity coupled plasma etching to fabricate the nonlinear optical polymer photonic crystal waveguides', *Mat. Sci. Eng. B*, 103, 170–178.

Ionescu R., Espinosa E.H., Sotter E., Llobet E., Vilanova X., Correig X., Felten A., Bittencourt C., Van Lier G., Charlier J.-C. and Pireaux J. J. (2006) 'Oxygen functionalisation of MWNT and their use as gas sensitive thick-film layers', *Sensors and Actuators B*, 113, 36–46.

Jones J. G., Waite A. R., Muratore C. and Voevodin A. A. (2008) 'Nitrogen and hydrogen plasma treatments of multiwalled carbon nanotubes', *J. Vac. Sci. Technol. B*, 26(3), 995–1000.

Kalita G., Adhikari S., Aryal H. R., Ghimre D. C., Afre R., Soga T., Sharon M. and Umeno M. (2008) 'Fluorination of multi-walled carbon nanotubes (MWNTs) via surface wave microwave (SW-MW) plasma treatment', *Physica E*, 41, 299–303.

Khabashesku V. N., Billups W. E. and Margrave J. L. (2002) 'Fluorination of single-wall carbon nanotubes and subsequent derivatization reactions', *Acc. Chem. Res.*, 35, 1087–1095.

Khare B. N., Meyyappan M., Cassell A. M., Nguyen C. V. and Han J. (2002) 'Functionalization of carbon nanotubes using atomic hydrogen from a glow discharge', *Nano. Lett.*, 2, 73–77.

Khare B., Wilhite P., Tran B., Teixeira E., Fresquez K., Mvondo D. N., Bauschlicher C. and Meyyappan M. (2005) 'Functionalization of carbon nanotubes via nitrogen glow discharge', *J. Phys. Chem. B*, 109, 23466–23472.

Kilbride B. E., Coleman J. N., Fraysse J., Fournet P., Cadek M. and Drury A. (2002) 'Experimental observation of scaling laws for alternating current and direct current conductivity in polymer-carbon nanotube composite thin films', *J. Appl. Phys.*, 92, 4024–4030.

Kim J. A., Seong D. G., Kang T. J. and Youn J. R. (2006) 'Effects of surface modification on rheological and mechanical properties of CNT/epoxy composites', *Carbon*, 44, 1898–1905.

Kim S., Jung Y. and Park S.-J. (2008) 'Preparation and electrochemical behaviors of platinum nanocluster catalysts deposited on plasma-treated carbon nanotube supports', *Coll. Surf. A: Physicochem. Eng. Aspects*, 313–314, 189–192.

Konya Z., Vesselenyi I., Niesz K., Kukovecz A., Demortier A., Fonseca A., Delhalle J., Mekhalif Z., Nagy J. B., Koos A. A., Osvath Z., Kocsonya A., Biro L. P. and Kiricsi I. (2002) 'Large scale production of short functionalized carbon nanotubes', *Chem. Phys. Lett.*, 360, 429–435.

Lee J.-Y., Park E.-J., Lee C.-J., Kim S.-W., Pak J. J. and Min N. K. (2009) 'Flexible electrochemical biosensors based on O_2 plasma functionalized MWCNT', *Thin Solid Films*, 517, 3883–3887.

Lee W. H., Kim S. J., Lee W. J., Lee J. G., Haddon R. C. and Reucroft P. J. (2001) 'X-ray photoelectron spectroscopic studies of surface modified single-walled carbon nanotube material', *Appl. Surf. Sci.*, 181, 121–127.

Lee Y. S., Cho T. H., Lee B. K., Rho J. S., An K. H. and Lee Y. H. (2003) 'Surface properties of fluorinated single-walled carbon nanotubes', *J. Fluor. Chem.*, 120, 99–104.

Li J., Fang Z. P., Tong L. F., Gu A. J. and Liu F. (2007) 'Improving dispersion of multiwalled carbon nanotubes in polyamide 6 composites through amino-functionalization', *J. Appl. Polym. Sci.*, 106, 2898–2906.

Liu J., Rinzler A. G., Dai H., Hafner J. H., Bradley R. K., Boul P. J., Lu A., Iverson T., Shelimov K., Huffman C. B., Rodriguez-Marcias F., Shon Y.-S., Lee T. R., Colbert D. T. and Smalley R. E. (1998) 'Fullerene pipes', *Science*, 280, 1253.

Liu P. (2005) 'Modifications of carbon nanotubes with polymers', *Eur. Polym. J.*, 41, 2693–2703.

Ma P. C., Tang B. Z. and Kim J.-K. (2008) 'Conversion of semiconducting behavior of carbon nanotubes using ball milling', *Chem. Phys. Lett.*, 458, 166–169.

McOmber J. I. and Nair R.S. (1991) 'Development of a process to achieve residue-free photoresist removal after high-dose ion implantation', *Nucl. Instrum. Meth. Phys. Resear. B*, 55, 281–286.

Meng H., Sui G. X., Fang P. F. and Yang R. (2008) 'Effects of acid- and diaminemodified MWNTs on the mechanical properties and crystallization behavior of polyamide 6', *Polym.*, 49, 610–620.

Meyer-Plath A. A., Finke B., Schröder K. and Ohl A. (2003) 'Pulsed and CW microwave plasma excitation for surface functionalization in nitrogen-containing gases', *Surf. Coat. Technol.*, 174–175, 877–881.

Mickelson E. T., Huffman C. B., Rinzler A. G., Smalley R. E., Hauge R. H. and Margrave J. L. (1998) 'Fluorination of single-wall carbon nanotubes', *Chem. Phys. Lett.*, 296, 188–194.

Monthioux M., Smith B. W., Burteaux B., Claye A., Fischer J. E. and Luzzi D. E. (2001) 'Sensitivity of single-wall carbon nanotubes to chemical processing: an electron microscopy investigation', *Carbon*, 39, 1251–1272.

Nakamura T., Ohana T., Ishihara M., Hasegawa M. and Koga Y. (2008) 'Photochemical modification of single-walled carbon nanotubes with amino functionalities and their metal nanoparticles attachment', *Diam. Rel. Mater.*, 17, 559–562.

Nastase C., Nastase F., Vaseashta A. and Stamatin I. (2006) 'Nanocomposites based on functionalized nanotubes in polyaniline matrix by plasma polymerization', *Prog. Solid State Chem.*, 34, 181–189.

Niyogi S., Hamon M. A., Hu H., Zhao B., Bhowmik P., Sen R., Itkis M. E. and Haddon R. C. (2002) 'Chemistry of single-walled carbon nanotubes', *Acc. Chem. Res.*, 35, 1105–1113.

Okpalugo T. I. T., Papakonstantinou P., Murphy H., McLaughlin J. and Brown N. M. D. (2005) 'Oxidative functionalization of carbon nanotubes in atmospheric pressure filamentary dielectric barrier discharge (APDBD)', *Carbon*, 43, 2951–2959.

Pantarotto D., Partidos C. D., Graff R., Hoebeke J., Briand J.-P., Prato M. and Bianco A. (2003) 'Synthesis, structural characterization and immunological properties of carbon nanotubes functionalized with peptides', *J. Am. Chem. Soc.*, 125, 6160–6164.

Plank N., Jiang L. and Cheung R. (2003) 'Fluorination of carbon nanotubes in CF plasma', *Appl. Phys. Lett.*, 83, 2426.

Pötschke P., Zschoerper N. P., Moller B. P. and Vohrer U. (2009) 'Plasma functionalization of multiwalled carbon nanotube bucky papers and the effect on properties of melt-mixed composites with polycarbonate', *Macromol. Rapid Comm.*, 30, 1828–1833.

Qin Y. and Hu M. (2008) 'Effects of microwave plasma treatment on the field emission properties of printed carbon nanotubes/Ag nano-particles films', *Appl. Surf. Sc.*, 254, 1757–1762.

Ramanathan T., Fischer F. T., Ruoff R. S. and Brinson L. C. (2005) 'Amino-functionalized carbon nanotubes for binding to polymers and biological systems', *Chem. Mater.*, 17, 1290–1295.

Rawat D. S., Taylor N., Talapatra S., Dhali S. K., Ajayan P. M. and Migone A. D. (2006) 'Effect of surface cleaning and functionalization of nanotubes on gas adsorption', *Phys. Rev. B*, 74, 113403.

Ruelle B. (2009) 'Functionalization of carbon nanotubes via plasma post-discharge surface treatment: Implication as nanofiller in polymeric matrices', PhD thesis, University of Mons.

Ruelle B., Felten A., Ghijsen J., Drube W., Johnson R. L., Liang D., Erni R., Van Tendeloo G., Dubois Ph., Hecq M. and Bittencourt C. (2008) 'Functionalization of MWCNTs with atomic nitrogen: electronic structure', *J. Phys. D Appl. Phys.*, 41, 045202.

Ruelle B., Peeterbroeck S., Gouttebaron R., Godfroid T., Monteverde F., Dauchot J.-P., Alexandre M., Hecq M. and Dubois Ph. (2007) 'Functionalization of carbon nanotubes by atomic nitrogen formed in a microwave plasma Ar + N_2 and subsequent poly(ε-caprolactone) grafting', *J. Mater. Chem.*, 17, 157–159.

Shi D. and He P. (2004) 'Surface modifications of nanoparticles and nanotubes by plasma polymerization', *Rev. Adv. Mater. Sci.*, 7, 97–107.

Sinnott S. B. (2002) 'Chemical functionalization of carbon nanotubes', *J. Nanosci. Nanotechnol.*, 2, 113–123.

Stevens J. L., Huang A. Y., Peng H., Chiang I. W., Khabashesku V. N. and Margrave J. L. (2003) 'Sidewall amino-functionalization of single-walled carbon nanotubes through fluorination and subsequent reactions with terminal diamines', *Nano. Lett.*, 3, 331–336.

Tasis D., Tagmatarchis N., Bianco A. and Prato M. (2006) 'Chemistry of carbon nanotubes', *Chem. Rev.*, 106, 1105–1136.

Thostenson E., Li C. and Chou T. W. (2005) 'Nanocomposites in context', *Compos. Sci. Technol.*, 65, 491–516.

Thostenson E., Ren Z. and Chou T.W. (2001) 'Advances in the science and technology of carbon nanotubes and their composites: a review', *Compos. Sci. Technol.*, 61, 1899–1912.

Tseng C. H., Wang C. C. and Chen C. Y. (2007) 'Functionalizing carbon nanotubes by plasma modification for the preparation of covalent-integrated epoxy composites', *Chem. Mater.*, 19 (2), 308–315.

Valentini L., Puglia D., Armentano I. and Kenny J. M. (2005) 'Sidewall functionalization of single-walled carbon nanotubes through CF_4 plasma treatment and subsequent reaction with aliphatic amines', *Chem. Phys. Lett.*, 403, 385–389.

Wang W., Ciselli P., Kuznetsov E., Peijs T. and Barber A. H. (2008) 'Effective reinforcement in carbon nanotube–polymer composites', *Philos. Trans. Roy. Soc. A*, 366, 1613–1626.

Wu Z., Xu Y., Zhang X., Shen G. and Yu R. (2007) 'Microwave plasma treated carbon nanotubes and their electrochemical biosensing application', *Talanta*, 72, 1336–1341.

Xu T., Yang J., Liu J. and Fu Q. (2007) 'Surface modification of multi-walled carbon nanotubes by O_2 plasma', *Appl. Surf. Sc.*, 253, 8945–8951.

Yan B., Qian K., Zhang Y. and Xu D. (2005) 'Effects of argon plasma treating on surface morphology and gas ionization property of carbon nanotubes', *Physica E*, 28, 88–92.

Yan Y. H., Cui J., Chan-Park M. B., Wang X. and Wu Q. Y. (2007) 'Systematic studies of covalent functionalization of carbon nanotubes via argon plasma-assisted UV grafting', *Nanotechnol.*, 18, 1–7.

Yu K., Zhu Z., Zhang Y., Li Q., Wang W., Luo L., Yu X., Ma H., Li Z. and Feng T. (2004) 'Change of surface morphology and field emission property of carbon nanotube films treated using a hydrogen plasma', *Appl. Surf. Sci.*, 225, 380–388.

Zhang J., Feng T., Yu W., Liu X., Wang X. and Li Q. (2004) 'Enhancement of field emission from hydrogen plasma processed carbon nanotubes', *Diam. Rel. Mater.*, 13, 54–59.

Zhang J., Zou H., Qing Q., Yang Y., Li Q., Liu Z., Guo X. and Du Z. (2003) 'Effect of chemical oxidation on the structure of single-walled carbon nanotubes', *J. Phys. Chem. B*, 107, 3712–3718.

Zhu Y. W., Cheong F. C., Yu T., Xu X. J., Lim C. T., Thong J. T. L., Shen Z. X., Ong C. K., Liu Y. J., Wee A. T. S. and Sow C. H. (2005) 'Effects of CF_4 plasma on the field emission properties of aligned multi-wall carbon nanotube films', *Carbon*, 43, 395–400.

Zschoerper N. P., Katzenmaier V., Vohrer U., Haupt M., Oehr C. and Hirth T. (2009) 'Analytical investigation of the composition of plasma-induced functional groups on carbon nanotube sheets', *Carbon*, 47, 2174–2185.

Part IV

Applications

15

Ceramic nanocomposites for energy storage and power generation

B. KUMAR, University of Dayton Research Institute, USA

DOI: 10.1533/9780857093493.4.509

Abstract: This chapter covers the electrical conductivity of polymer–ceramic and ceramic–ceramic nanocomposites with a major emphasis on the ionic conductivity. These nanocomposites are potentially important materials as electrolytes for power generation and energy storage devices. The nanocomposites offer major enhancements in ionic conductivity because of an underlying basic physical mechanism emanating from their structure. The basic physical mechanism for conductivity enhancement is the space charge effect. The chapter explains applicability and quantification of the effect for lithium ion conducting nanocomposites. The applicability of the effect can be extended to other lithium ion and oxygen ion conductors. Uses of nanocomposites in energy storage and power generation devices are also covered.

Key words: electrical conductivity, nanocomposites, polymer–ceramic, ceramic–ceramic, energy storage.

15.1 Introduction

The evolutionary processes of nature have created three states of matter – solid, liquid, and gas. Each of these states may be composed of a single element or a combination of elements, such as in naturally occurring minerals, which are the most commonly found solid state matter in the earth's crust. The three states of matter can also be synthesized by man-made processes. Nanocomposites are man-made solids tailored with distinctive chemical, mechanical, and electrical properties that are generally designed for a specific application. Nanocomposites can be synthesized using basic principles of materials engineering and developed for a given commercial application. The focus of this chapter is on the electrical

509

properties of nanocomposites for energy storage and power generation applications.

Matter in its solid state form can be further classified as metal, semiconductor, or dielectric. This classification is based on electrical properties. A metal can have very high electrical conductivity – as high as $10^8\,S\,cm^{-1}$, whereas the conductivities of semiconductors and dielectrics are orders of magnitude lower. Typically, a semiconductor has a conductivity around $10\,S\,cm^{-1}$, whereas dielectrics have conductivity values around $10^{-12}\,S\,cm^{-1}$. Man-made nanocomposites are basically a mixture of metals, semiconductors, and dielectrics. When these types of solids (phases) are mixed in a nanocomposite and subsequently processed (annealed, sintered, drawn, extruded, etc.), they tend to exhibit interesting electrical properties at interfaces and also in the bulk. Two examples of different types of nanocomposites are now considered.

When a metallic powder is introduced in a dielectric or ionic conducting matrix, for example, Ni in yttria stabilized zirconia (YSZ), there is a little change in the electrical conductivity of the nanocomposite at lower concentrations of Ni. However, when the concentration of Ni approaches a threshold (of electron percolation), there is a rapid increase (orders of magnitude) in the conductivity. Physically, the ionic conducting phase (YSZ) blocks electron transport at lower Ni concentrations. At higher concentrations, the interparticle spacing between the Ni particles decreases to the extent that electrons can percolate from one Ni particle to another, leading to a conductivity enhancement by orders of magnitude. Further increase in the concentration of the nickel phase increases conductivity, but at a much lower rate. This type of nanocomposite is used as an anode in solid oxide fuel cells (SOFCs).

The addition of a dielectric phase Al_2O_3 in an ionically conducting matrix (LiI) leads to conductivity enhancements by orders of magnitude. In this case, the percolation of ions is stimulated by the introduction of the dielectric phase, Al_2O_3. These types of nanocomposites are potentially important for lithium batteries and electrochemical sensors.

The aforementioned examples of nanocomposites make use of the transport of electrons and ions, which are the basic constituents of solids. These electrons and ions of nanocomposites can also interact with molecular species to catalyze certain chemical reactions, just as some metals (generally precious metals) do. The electrical properties of nanocomposites are an intriguing and important field of research. Regrettably, our basic understanding of these materials remains tentative. It is hoped that enough interest will develop in the future to advance the science and technology of electrically active nanocomposites.

The electrical properties of composites were analyzed by Maxwell (1881) and Rayleigh (1892) – the first attempts to quantitatively explain electrical conduction in heterogeneous solids. Many developments have occurred since these pioneering works. What is perhaps a major development in the last 40 years is the recognition and accumulation of experimental evidence of the existence of electrically active interfaces. These interfaces also have a major influence on the bulk electrical properties of solids.

The application of nanocomposites in energy storage and power generation devices is in its infancy and is expected to grow as we gain a more thorough understanding of their properties and usefulness. The development of many alloys followed the need to manufacture high power density internal combustion engines. Similarly, growth, interest, and development in nanocomposites will follow the need for energy production, transmission, and use in devices for energy harvesting and sensors.

15.2   Electrical properties

15.2.1 Fundamental relationships

The electrical properties of nanocomposites may be approached from two different points of view. Engineers may consider them primarily as components in energy storage batteries, power generating fuel cells, and electrical circuits. Engineers are interested in specific properties and physical characteristics so that the batteries, fuel cells, and circuit will perform at high efficiency, last longer, and be cost efficient. Scientists view electrical properties in terms of quantitative understanding of their electronic and ionic character. A practical researcher must take a position somewhere in the middle to reconcile the two points of view. A researcher should thus be able to correlate and explain the effects of composition, structure, and external environment on the bulk properties.

The application of an electrical field to a nanocomposite leads to a generation of current that reaches an equilibrium direct current value either rapidly or slowly. One can express the equilibrium in terms of number of charged particles present, n, and their drift velocity, v. The current density, j, is defined as the charge transported through a unit area in a unit time. The equilibrium current density is expressed as:

$$j = nzev \qquad [15.1]$$

where z is the valency and e is the electronic charge. The conductivity is defined by:

$$\sigma = j/E \qquad [15.2]$$

where E is the electric field strength. Combining equations 15.1 and 15.2

establishes a relationship between conductivity and field strength:

$$\sigma = nzev/E \qquad\qquad [15.3]$$

The drift velocity, v, is directly proportional to the local electric field strength and the constant relating of the two quantities is defined as mobility. Equation 15.4 is expressed in subscripted form to specify the contribution of the ith species to electrical mobility:

$$\mu_i = v_i/E_i \qquad\qquad [15.4]$$

Equation 15.3 can now be expressed in terms of mobility

$$\sigma_i = (n_i z_i e)\mu_i \qquad\qquad [15.5]$$

At times, it is desirable to make use of absolute mobility, B_i, which is defined as drift velocity per unit of applied force:

$$B_i = v_i/F_i = v_i/Z_i eE \qquad\qquad [15.6]$$

In terms of absolute mobility, the conductivity is expressed by:

$$\sigma_i = n_i z_i^2 e^2 B_i \qquad\qquad [15.7]$$

Equation 15.7 relates conductivity with the most important variable, n_i, and the absolute mobility, B_i; whereas z_i and e remain constant for a given material.

Generally, in solids, more than one charge carrier can contribute to the electrical conduction. In this case, the total conductivity is the sum of partial conductivities attributed to different charge carriers. That is, for the ith charge carrier, equation 15.5 is applicable and then the total conductivity, σ, is expressed by:

$$\sigma = \sigma_1 + \sigma_2 + \cdots + \sigma_i + \cdots \qquad\qquad [15.8]$$

The fraction of total conductivity contributed by each charge carrier is

$$t_i = \sigma_i/\sigma \qquad\qquad [15.9]$$

where t_i is termed the transport number of the ith species. The sum of all the transport numbers must be unity:

$$t_1 + t_2 + \cdots + t_i + \cdots = 1 \qquad\qquad [15.10]$$

15.2.2 Space charge

Space charge refers to an accumulation or depletion of local, uncompensated charges in solids. It may result from the absorption of charged species on a dielectric surface, thermal ionization, and migration (sometimes field assisted) of point defects. Interfaces such as grain boundaries are active sites for the creation and retention of space charges. Depending on the nature of the interface, the space charge could be transient, permanent, temperature dependent, or field sensitive. The complexity of the interface–space charge relationship often makes it difficult to quantify its effect on the bulk electrical property. The space charge can significantly influence the electrical properties (ionic and electronic conductivity) of nanocomposites.

15.2.3 Electronic conductivity

For an electronic conductor such as a metal, conduction occurs by the motion of free electrons. The mobility, μ, of free electrons is limited by scattering:

$$\mu = e\tau/m \qquad [15.11]$$

where τ is the mean free time between scattering events, m is the mass of the electron ($m = 9.11 \times 10^{-31}$ kg), and e is the elementary electron charge.

$$\sigma_e = (n_i z_i e)e\tau/m \qquad [15.12]$$

The electronic conductivity of metals ranges from 10^{16} to $10^8\,\mathrm{S\,cm^{-1}}$ and is temperature dependent. Generally, increasing temperature decreases the electronic conductivity.

15.2.4 Ionic conductivity

In solids, the ionic conduction process is very different from the electronic conduction mechanism in metals. Ion mobility in solids generally takes place by a hopping and diffusion mechanism. In semi-solids such as polymers, a polymer chain motion contributes to the transport of ions. The effectiveness of the hopping mechanism is related to the diffusivity, D, of solids:

$$D = D_0 e^{\Delta G_{act}/RT} \qquad [15.13]$$

where D_0 is a constant representing the attempt frequency of the hopping mechanism, ΔG_{act} is the energy barrier, R is the gas constant, and T is the temperature (K). The mobility of ions in solids is expressed by:

$$\mu_i = (|z_i|FD)/RT \qquad [15.14]$$

where $|z_i|$ is the valence of ions, and F is the Faraday constant.

Substitution of the expression for ion mobility into equation 15.5 for conductivity yields:

$$\sigma = C(z_i F)^2 D / RT \qquad\qquad [15.15]$$

where C is a carrier concentration in solids, which is related to the density of mobile ionic species. Typical ion concentration and diffusivity in a solid are 10^2–10^3 mol m^{-3} and 10^{-8} m^2 s^{-1}, respectively. Inserting these values in equation 15.15 provides ionic conductivity values in the range of 10^{-6}–10^0 S cm^{-1}. It should be noted that the ionic conductivity values are orders of magnitude lower than the electronic conductivity of a typical metal. The ionic charge transport is far more complex than the electronic charge transport. For this reason, much more effort in recent years has been devoted towards the development of ionic conductors.

15.2.5 Mixed ionic–electronic conductors

Beyond the traditional boundaries of electronic and ionic conductors, there are solids known as 'mixed conductors', which can conduct both ions and electrons. Both ionic and electronic species can move through mixed conductors, leading to simultaneous transport of ions and electrons. This attribute of mixed nanocomposites makes them very useful as electrodes – materials critical for energy storage and power generation devices. These mixed conductors can also catalyze electrochemical reactions.

15.3   Ionic nanocomposites

Ionic nanocomposites are of profound interest to electrochemical engineers and scientists because of their potential application in devices such as batteries, fuel cells, sensors, and displays. These state-of-the-art devices employ liquid electrolytes, which are corrosive and hazardous and degrade after brief exposure to high temperatures, thus limiting the life and application of the devices. Ionic nanocomposites provide many advantages over liquid electrolytes such as ease of containment, thermal and chemical stability, and non-flammability. Nanocomposites are potentially useful for electrochemical devices that are expected to be functional over a wide range of temperatures.

A considerable number of papers on ionic conductivity and related properties of nanocomposites have been published in last three decades. A number of review papers document their developmental history and general characteristics. Two major types of nanocomposites, polymer–ceramic and ceramic–ceramic, will be discussed in this section.

15.3.1 Polymer–ceramic nanocomposites

A polymer–ceramic nanocomposite is defined as an ionically conducting solid material derived from a polymer and a ceramic phase. This type of material is a subset of solid electrolytes. The discussion here will be primarily confined to lithium ion conductors, although other types of nanocomposite conductors (silver and sodium) have also been reported.

A broader range of polymer and ceramic chemistries could be combined to synthesize polymer–ceramic nanocomposites. Analyses of the published literature suggest that the incorporation of a ceramic phase into an ionically conducting matrix leads to enhanced conductivity, cationic transport number, and electrode–electrolyte interfacial stability. The ionic conductivity of the nanocomposite is dependent upon several variables such as the characteristics (chemistry, size, and volume fraction) of the ceramic phase, annealing parameters, physical and chemical properties of the polymer matrix, degree of reactivity between polymer and ceramic phases, and temperature.

A number of processing techniques (e.g. solvent casting, melt casting, and hot pressing) may be employed to obtain film and bulk nanocomposite specimens. The most convenient is the blend and hot press technique in which the polymer, salt (of lithium), and ceramic components are mixed in a predetermined proportion, ground in a mortar and pestle or energy milled, hot pressed into pellet form, and then rolled into films of desired thickness. Nanocomposite materials covering a wide range of concentration of ceramic phases in a polymer matrix can be obtained by this technique.

The temperature dependence of conductivity of the PEO:LiBF$_4$ complex and PEO:LiBF$_4$ (8:1)–MgO(10 wt%) nanomaterials containing micro- and nanosize MgO is shown in Fig. 15.1 (Kumar et al., 2001). The lowest conductivity values are associated with the PEO:LiBF$_4$ (8:1) complex. Near the melting point of PEO, 68°C, a precipitous drop in conductivity begins, and at room temperature the conductivity drops to $10^{-9}\,S\,cm^{-1}$. The specimen containing microsize (~5 micron) MgO exhibits much improved conductivity as compared to the PEO:LiBF$_4$ complex. At around ambient temperature, the conductivity is improved by approximately three orders of magnitude by the incorporation of microsize MgO in the polymer complex. The highest conductivity values are associated with the specimen containing nanosize MgO. The conductivity of this specimen is about four orders of magnitude higher than the PEO:LiBF$_4$ complex around the ambient temperature. Furthermore, the temperature dependence of conductivity diminishes as the MgO particles are reduced from micro- to nanosize. At 100°C, all three specimens possess similar conductivity values, whereas the curves diverge as the temperature is lowered to 40°C.

The temperature dependence of conductivity of the PEO:LiBF$_4$ complex

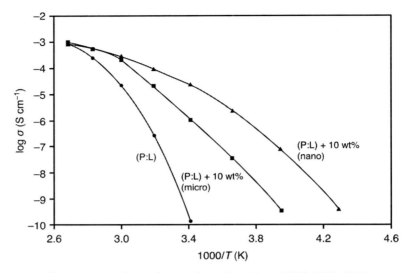

15.1 Temperature dependence of conductivity of PEO:LiBF$_4$ (P:L) complex (8:1), P:L + 10 wt% MgO(micro) and P:L + 10 wt% MgO(nano).

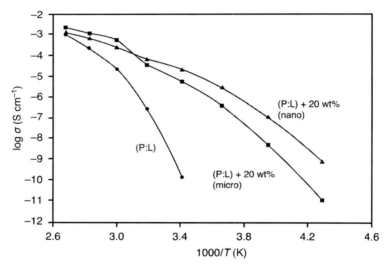

15.2 Temperature dependence of conductivity of P:L + 20 wt% BaTiO$_3$(micro) and P:L + 20 wt% BaTiO$_3$(nano).

and specimens containing micro- and nanosize BaTiO$_3$ is shown in Fig. 15.2. These specimens also exhibit behavior similar to the MgO-containing electrolytes shown in Fig. 15.1. The low-temperature conductivities are greatly improved by the incorporation of BaTiO$_3$. The nano- and microsize BaTiO$_3$ specimens, however, exhibit a crossover at around 50°C. The

Table 15.1 Physical properties of MgO and BaTiO$_3$

Ceramic oxide	Melting temperature (°C)	Density (g cm^{-3})	Dielectric constant (κ)	Mean particle size	Mass of particle (g)
MgO	2852	3.58	9.65	5 µm	2.34×10^{-10}
				20 nm	1.50×10^{-17}
BaTiO$_3$	—	6.02 (tetragonal)	$\kappa_{11} = 3600$ $\kappa_{33} = 150$	1 µm	9.45×10^{-12}
				70 nm	8.65×10^{-15}

nanosize specimen possesses lower conductivity above 50°C but higher conductivity below 50°C as compared to the microsize BaTiO$_3$ specimen. This trend was also noted for MgO in Fig. 15.1, but to a much lower degree. The ambient temperature conductivities of BaTiO$_3$-containing specimens are in the range 10^{-4}–10^{-5} S cm^{-1}.

Table 15.1 presents the physical properties of MgO and BaTiO$_3$ materials. BaTiO$_3$ is an important ferroelectric material and its dielectric constant in the tetragonal phase (ferroelectric state) is much higher and anisotropic. The dielectric constant of BaTiO$_3$ is also known to be particle-size dependent. The two ceramic materials, MgO and BaTiO$_3$, in effect represent the upper and lower limits of dielectric constants of available ceramic materials. The last column of Table 15.1 shows masses corresponding to the two particle sizes used for each material. For MgO and BaTiO$_3$, the masses corresponding to each particle size differ by approximately seven and three orders of magnitude, respectively.

In spite of widely different physical characteristics, including density and dielectric constant, both MgO and BaTiO$_3$ additives exhibit similar effects with regard to particle size, and thus this is believed to be a dominant parameter. Reducing the particle size from the micro- to nano-range increases low-temperature conductivities and decreases their temperature dependence, as shown in Fig. 15.2. The enhancements are pronounced and associated with lower activation energies for the ionic transport.

The average length and mass of a PEO chain are 20 µm and 3×10^{-18} g, respectively. These length and mass values must now be compared with the particle size and mass of the ceramic dopants to develop insight into possible interactions. The proximity and interaction of a polymer chain and a nanosize MgO particle (0.02 µm) are shown schematically in Fig. 15.3. A constant segmental chain motion near the glass transition temperature is expected to cause displacement and distribution of MgO particles. Furthermore, the process can be facilitated by thermal treatments and cycling. The displacement and distribution of MgO particles continues until the polymer chain dipole and MgO dipole interaction takes place. The interaction leads to latching of the polymer chain and MgO particle, and thus stabilization of the structure. When experimentally monitored, this

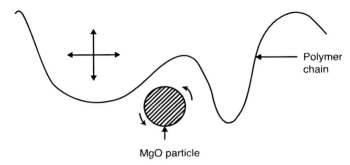

MgO particle

15.3 Schematic representation of a polymer chain segment and MgO particle interaction.

phenomenon is reflected by an increase in conductivity during isothermal stabilization at low temperatures.

The degree of polymer chain and ceramic particle interaction can be reduced by a heavier, fewer, and lower weight or volume percent of MgO. For example, if the weight percent of MgO is maintained constant and the particle size is reduced from 5 μm to 20 nm, over 15 million additional polymer–ceramic interaction sites are created and conductivity enhancements occur. Table 15.1 shows that reducing particle size from 5 μm to 20 nm reduces the mass of a particle by seven orders of magnitude. These lighter MgO particles become far more receptive to the segmental chain motion of the polymer, leading to improved polymer chain and MgO particle interaction. Such an interaction is reflected by an enhanced and time-dependent conductivity.

It is imperative (from equation 15.5) that conductivity is directly related to the mobility and concentration of ions. Because the concentration is maintained constant in a given specimen, the mobility must increase to account for the conductivity enhancement. It has been suggested that the mobility of charge carriers in nanocomposites is increased by the annealing-induced structural changes that occur because of an interaction between polymer chains and ceramic particles (Kumar *et al.*, 2001).

The fact that motions of polymer chains contribute to the transport of ions in polymer electrolytes also has deleterious effects on the transport number. The chain motion also facilitates transport of larger anionic species and thus measured conductivity includes a contribution from both cationic and anionic species. Lithium in PEO-based electrolytes is coordinated with five oxygen atoms, and the probability of finding a similar site for lithium transport is low. The polymer electrolytes have a lithium ion transport number that may be as small as 0.3. As a result, when such materials are used in a lithium cell, extensive concentration gradients are set up during use and these affect their performance.

The cationic transport number in amorphous inorganic and polymeric electrolytes can be approximated through the use of the decoupling constant, R. The larger the value of R, the greater is the structural relaxation time and less is the transport of ions (both anions and cations) mediated by the structure. In polymer–ceramic composite electrolytes, the glass transition temperature, T_g, is proportional to the volume fraction of the ceramic phase. Angell (1986) has shown that for an R value of ~10^2, the cationic transport number could be over 0.9, as the transport of ions mediated by the structure would diminish. An increase of 50°C in the T_g in most polymer electrolytes resulting from the addition of ceramic phase will bring the T/T_g ratio to 1.2 ($R \simeq 10^2$) and the cationic transport number to around 0.9.

The conductivity of a polymer electrolyte originates from two distinct processes: ion hopping and ion transport assisted by polymer chain motion. The addition of a ceramic phase suppresses the chain motion-mediated contribution and thus must increase the contribution associated with ion hopping if the conductivity remains the same (the worst case scenario). The ion hopping process is more favorable for cationic species than anionic species because of their small size and mass. This scenario suggests an enhanced cationic transport number as the volume fraction of ceramic phase increases in the polymer matrix of the polymer–ceramic nanocomposites.

The conductivity and transport number of nanocomposities comprising LiI, PEO, SiO_2, MgO, and Al_2O_3 have been measured and reported by Nagasubramanian et al. (1993) and Peled et al. (1993). They calculated the conductivity from bulk resistance, R_b, measured at high frequency and transport number, t^+, using:

$$t^+ = R_b/(R_b + Z_d)$$ [15.16]

where Z_d is diffusional impedance as measured from a Nyquist plot. For a composite electrolyte film containing 0.05 μm alumina, the bulk conductivity is around $10^{-4}\,S\,cm^{-1}$ and the lithium ion transport number is close to unity at 104°C. Cho and Lin (1997) report a transport number of 0.98 for a glass–polymer composite electrolyte containing 13 vol% PEO:LiN $(CF_3SO_2)_2$ and 87% $0.56Li_2S.0.19B_2S_3.0.25LiI$. Croce et al. (1998) reported a transport number of 0.6 in a PEO:$LiCO_4$–TiO_2 (10 wt%) composite electrolyte in the 45–90°C temperature range.

15.3.2 Ceramic–ceramic nanocomposite

Lithium ion conductors

This subclass of nanocomposites contains only ceramic components. The matrix is generally an ionic conducting material, whereas the dopant is a

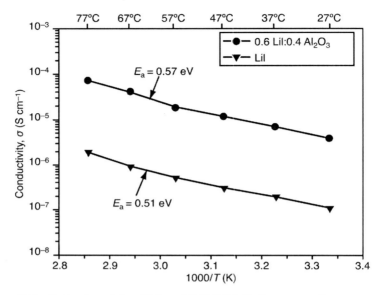

15.4   Arrhenius plots of LiI and 0.6 LiI:0.4 Al$_2$O$_3$ composite.

dielectric phase. In these nanocomposites, the dopant also remains insoluble in the host matrix and exists as a physically distinct phase of the bulk structure even after processing (sintering) at high temperatures. It was Liang who first demonstrated, in 1973, that the intrinsic conductivity of lithium iodide doped with 35–45 mol% Al$_2$O$_3$ was enhanced by orders of magnitude as compared to the conductivity of LiI (Liang, 1973). The enhancement in conductivity has been explained on the basis of space charge formation at the LiI–Al$_2$O$_3$ interface (Maier, 1995). Since the pioneering work of Liang (1973) almost four decades ago, the phenomenon has been demonstrated in several other systems ranging from lithium to oxygen ion conducting heterogeneous solids (Kumar, 2007; Kumar and Thokchom, 2008).

Figure 15.4 shows Arrhenius plots of conductivity of LiI and 0.6 LiI:0.4 Al$_2$O$_3$ nanocomposites in the 27 to 77°C temperature range (Kumar, 2007). The conductivity of the nanocomposite specimen increases by nearly two orders of magnitude by doping LiI with 40 mol% Al$_2$O$_3$. The activation energies associated with the transport of ions in LiI and 0.6 LiI:0.4 Al$_2$O$_3$ specimens are 0.51 and 0.57 eV, respectively. An increase in the activation energy by over 10% reflects a deviation in the transport mechanism of the nanocomposites that may be attributed to the space charge mediation. The conductivity of the nanocomposite is reproducible and stable within the temperature range of 27 to 77°C (Kumar, 2007).

The space charge mediated transport mechanism was elucidated by Kumar and Thokchom (2007) using a glass-ceramic material, lithium

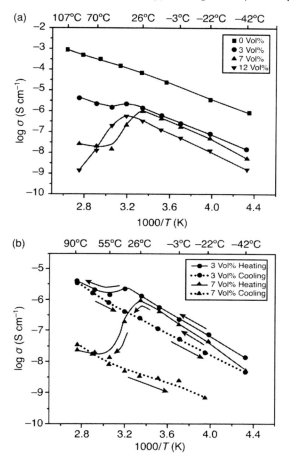

15.5 Arrhenius plots of LATP nanocomposites (a) with 3, 7, and 12 vol% Al_2O_3, (b) with 3 and 7 vol% Al_2O_3 during heating and cooling.

aluminum titanium phosphate (LATP). The LATP glass-ceramic is a heterogeneous polycrystalline solid lithium ion conductor. The bulk structure of LATP is characterized by the presence of a primary $Li_{1+x}Ti_{2-x}Al_x(PO_4)_3$ phase with $x = 0.275$, which is a fast lithium ion conductor, along with impurity phases such as Li_2O and $LiPO_4$. The introduction of a dielectric phase, Al_2O_3, to the LATP provides a characteristic feature – a deviation from linearity in the Arrhenius plots of the nanocomposites. Figure 15.5(a) shows Arrhenius plots of LATP–Al_2O_3 nanocomposites. The introduction of Al_2O_3 in LATP is detrimental to the conductivity across the entire temperature range. Nonetheless, the degree of conductivity reduction varies across the temperature range and it also depends upon the Al_2O_3 concentration. For instance, in the case of LATP–

3 vol% Al_2O_3, an inflection around 27°C is noted. The inflection transformed into a peak for 7 and 12 vol% nanocomposites. The conductivities of the 12 vol% specimen at −40 and 90°C were reduced by approximately two and five orders of magnitude, respectively.

The dielectric phase Al_2O_3 and impurities in LATP such as $AlPO_4$ are the underlying causes for the creation of the space charge. A significant concentration of Al_2O_3 interacts with lithium ions in the LATP. During the interaction, lithium ions are adsorbed onto the Al_2O_3 surface and become a source of space charge. The space charge then affects the transport of the remaining lithium ions. It is evident that the adsorption of ions and resulting space charges occurs below 27°C, implying a very low energy of adsorption (<5 zJ). At temperatures above 27°C, a desorption of lithium ion occurs, and the transport mechanism solely depends upon the movement of ions through the channels of the $Li_{1+x}Ti_{2−x}Al_x(PO)_3$ material. The LATP glass-ceramic is a single lithium ion conductor (lithium transport number ~1) and its primary transport mechanism involves conduction through the channels of an aluminum titanium phosphate network. Thus, it was anticipated that the addition of Al_2O_3 would result in reduced conductivities across the temperature range due to the blocking effect. From the experimental data presented in Fig. 15.5(a), however, it may be inferred that both blocking and space charge effects coexist in the temperature range of −40 to 27°C. Because of a higher thermal energy, the space charge effect is destroyed above 27°C by the desorption of lithium ions.

The nature of the adsorption and desorption processes of lithium onto the Al_2O_3 surface can be investigated by measuring conductivity during heating and cooling cycles. Figure 15.5(b) shows the Arrhenius plots of LATP 3 and 7 vol% Al_2O_3 specimens during heating and cooling cycles. During the heating cycle, once the lithium ions have been desorbed from the Al_2O_3 surface above 27°C, the space charge contribution to conductivity is eliminated. Therefore, the specimens exhibit lower conductivity across the entire temperature range during the cooling cycle. It should also be noted that the lithium ion diffusion coefficient at these sub-ambient temperatures is low but significant; therefore, a re-formation of the space charge is a distinct possibility if the specimen is kept below 27°C, but it may take a long time (days). In the case of polymer specimens, it has been shown that complete recovery of the conductivities at lower temperature takes from tens to hundreds of hours (Kumar and Scanlon, 1999). It should be appreciated that the contribution of space charge to conductivity can be quantitatively determined by analyzing the Arrhenius plot as well as measuring conductivity during heating and cooling cycles.

It is now evident that the presence of a dielectric phase in an ionic conducting matrix may lead to two antagonistic influences: (a) the blocking effect and (b) the space charge effect. The blocking effect is characterized by a decline (sometimes precipitous) in the ionic conductivity of nanocomposites with the gradual addition of the dielectric phase and temperature. Under appropriate conditions, the space charge effect enhances conductivity. Collectively, these two factors determine whether or not a given dopant will have a positive space charge effect, i.e. a conductivity enhancement.

Kumar *et al.* (2006) have shown that blocking and space charge effects can be delineated and their effects quantified in different temperature regions, especially in simple systems such as a single lithium ion conductor doped with dielectric phases, such as the LATP–Al_2O_3 system. A delineation of the effects in more complex systems where several conducting charge species contribute to the conductivity may be cumbersome.

Oxygen ion conducting nanocomposites

Oxygen ion conductors are of significant commercial interest as yttria stabilized zirconia (YSZ) is the electrolyte of choice for solid oxide fuel cells (SOFCs). The oxygen ion transport determines, to a large extent, the operating temperature and usable power of a SOFC. The oxygen ion transport in YSZ has been known and investigated for decades. In fact, the high oxygen ion conductivity of YSZ electrolytes has been a prime motivator for the development and commercialization of SOFCs based on the YSZ electrolyte. Yet, the SOFC community is seeking even higher oxygen ion conductivity, especially at lower temperatures (~650°C) for alleviating chemical degradation and material compatibility issues. A number of different approaches lead to enhanced conductivity of the YSZ electrolyte. For example, a rare earth dopant such as scandium rather than yttrium in ZrO_2 is known to enhance the bulk conductivity of stabilized zirconia (Dixon *et al.*, 1963). An alternative route to enhance the bulk conductivity of YSZ is to employ a heterogeneous dopant. These heterogeneous dopants are insoluble in the host YSZ and remain as a physically distinct phase in the bulk structure.

Arrhenius plots of bulk conductivities (including grain and grain boundaries) of YSZ and YSZ–Al_2O_3 nanocomposites are shown in Fig. 15.6. The YSZ–Al_2O_3 nanocomposite exhibits higher conductivity by factors ranging from 3 to 7. Again, in this case the activation energy (2.14 eV) for oxygen ion transport in the nanocomposite is greater than the activation energy (1.67 eV) for oxygen ion transport in YSZ. The higher activation in the YSZ–Al_2O_3 system is attributed to the tortuous path that oxygen vacancies must travel to participate in the conduction process. Generally, higher ionic conductivity of a material is associated with lower activation energy, but in the case of space charge mediated transport, the data generally show noncompliance with conventional wisdom.

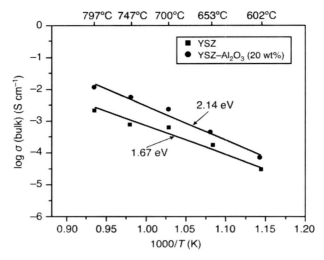

15.6 Arrhenius plots of bulk conductivities for yttria-stabilized zirconia (YSZ) and YSZ–Al$_2$O$_3$ specimens.

15.3.3 Charge carrier and mobility measurements

An ionic conductor responds to an electric field, leading to polarization of charges. This basic property of ionic conductors is used to calculate the number of charge carriers and their mobility (Kumar *et al.*, 1994). The technique basically involves polarizing a specimen using a DC field for a given period and subsequently monitoring the decay of current resulting from polarized charges. The polarized nanocomposite acts like a concentration cell whose potential can be expressed by the Nernst equation:

$$E = \ln[C_2]/[C_1] \qquad\qquad [15.17]$$

where E is the thermodynamic potential, C_1 is the anodic concentration, and C_2 is the cathodic concentration.

If the polarized cell is allowed to reach equilibrium, the potential drops to zero. The area under the current decay curves yields the capacity (µAh) of the cell, which is also related to the number of charge carriers (n_i) displaced during the application of DC potential. The mobility, μ_i, is calculated using the basic equation of conductivity as expressed by equation 15.5.

Table 15.2 lists the bulk conductivity (σ_b), number of charge carriers (n_i), and mobility (cm^2 V^{-1} s^{-1}) of oxygen vacancies in YSZ and YSZ–Al$_2$O$_3$ specimens at 500, 600, and 700 °C. At a given temperature, the YSZ–Al$_2$O$_3$ specimen shows a decreased number of charge carriers as compared with the YSZ specimen. The ratio $n_{(YSZ–Al2O3)}/n_{YSZ}$ decreases from 0.95 at 500 °C to 0.16 at 700 °C. The decrease in the number of charge carriers by a factor of

Table 15.2 Bulk conductivity (σ_b), number of charge carriers (n_i), and mobility (μ_i) in yttria stabilized zirconia (YSZ) and YSZ–Al$_2$O$_3$ specimens at 500, 600, and 700°C

	YSZ			YSZ–Al$_2$O$_3$		
	500°C	600°C	700°C	500°C	600°C	700°C
σ_b	1.0963×10^{-5}	2.18×10^{-5}	2.913×10^{-3}	5.127×10^{-5}	2.02×10^{-4}	4.03×10^{-3}
n_i	4.40×10^{14}	8.71×10^{15}	1.10×10^{17}	4.16×10^{14}	1.79×10^{15}	1.810×10^{16}
μ_i	7.77×10^{-2}	7.82×10^{-2}	8.24×10^{-2}	28.513×10^{-2}	35.31×10^{-2}	69.4210^{-2}
$\frac{n_{YSZ-Al_2O_3}}{n_{YSZ}}$	—	—	—	0.95	0.21	0.16
$\frac{\mu_{YSZ-Al_2O_3}}{\mu_{YSZ}}$	—	—	—	3.67	4.52	8.42

six at 700°C suggests that there is a temperature-dependent interaction between Al$_2$O$_3$ and oxygen vacancies. Higher temperatures lead to more interaction and a reduction in the number of charge carriers. The mobility, conductivity, and number of charge carriers are related by equation 15.5. The calculated mobility shows an increase for the YSZ–Al$_2$O$_3$ specimens. The mobility ratio, $\mu_{(YSZ-Al2O3)}/\mu_{YSZ}$, also shows an increase from 3.67 at 500°C to 8.42 at 700°C.

Analysis of Table 15.2 suggests that a significant fraction of oxygen vacancies in the YSZ–Al$_2$O$_3$ specimens are immobilized at interfaces that serve as a source of local field and influence the transport of the remaining conducting vacancies. The interaction of oxygen vacancies and Al$_2$O$_3$ is illustrated by:

$$Y_2O_3 \rightarrow 2Y^*{}_{Zr} + 3O^x + Vo^{\cdot\cdot} \qquad [15.18]$$

$$Al_2O_3 + Vo^{\cdot\cdot} \rightarrow Al_2O_3 : Vo^{\cdot\cdot} \qquad [15.19]$$

Equation 15.18 expresses a well-recognized structural disorder that shows a generation of oxygen vacancies (Vo$^{\cdot\cdot}$) in ZrO$_2$ after doping with Y$_2$O$_3$. The oxygen vacancies interact with Al$_2$O$_3$ particles and form a thermally stable complex (Al$_2$O$_3$: Vo$^{\cdot\cdot}$) after the interaction as shown in equation 15.19. The equation is expected to be reversible if enough thermal energy becomes available to dissociate the A$_2$O$_3$: Vo$^{\cdot\cdot}$ complex; however, in the temperature range under investigation (500–800°C), such a phenomenon was not observed.

The number of conducting oxygen vacancies (Vo$^{\cdot\cdot}$) is reduced and mobility is increased after interaction of the vacancies with Al$_2$O$_3$. The proposed explanation is schematically illustrated in Fig. 15.7. The conducting vacancies are represented by arrows and Al$_2$O$_3$ particles are depicted by solid circles. The oxygen vacancies that are immobilized by the

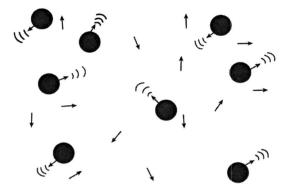

15.7 Schematic representation of interaction of oxygen ions with Al_2O_3 in YSZ–Al_2O_3 specimens.

Al_2O_3 particles act as the source of local electric field (shown by the dark arcs).

15.4   Energy storage and power generation devices

15.4.1 Energy storage

Societal energy storage needs call for concerted measures to deal with issues of human transportation, efficient uses of energy, and global warming. In a broader sense, all of these issues are interlinked, and a satisfactory solution will require technological innovations in a number of different disciplines. Energy-related issues include efficient storage, generation, and transmission of electricity and its conversion to different usable forms. The storage aspect of energy-related issues could be addressed by lithium rechargeable batteries. Lithium is the lightest solid element and possesses the highest oxidation potential. These attributes allow lithium batteries to offer higher energy density compared to the standard, state-of-the-art lead acid and nickel–metal hydride batteries.

The use of lithium in a battery also brings some interesting challenges. This section identifies an energy storage device, its challenges, and presents applications of emerging nanocomposites. Nanocomposites developed at the University of Dayton have allowed the fabrication of potentially very high energy density Li–O_2/air cells. These cells have yet to be introduced in the market; nonetheless, the rechargeable high energy density storage devices illustrate the importance of nanocomposites in developing the next generation of energy storage devices.

A typical rechargeable, solid-state Li–O_2/air cell is shown in Fig. 15.8: a lithium anode and a composite cathode (LAGP–C) are separated by an electrolyte laminate composed of a glass ceramic (GC) and two polymer–

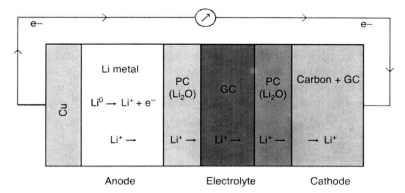

15.8 Schematic illustration of Li–O₂/air cell and its component materials. Lithium is protected by copper foil, which is also used as a current collector. The current collector on the cathode side is a metal casing not shown in the diagram.

ceramic (PC) membranes. A PC membrane of PEO:LiBETI (8.5:1)–(1 wt% Li₂O) electrochemically couples the lithium anode and GC membrane. The PC membrane that couples the cathode with the GC membrane is also a PEO-based lithium ion conductor containing Li₂O. Details on the development of the PC membranes can be found elsewhere (Kumar and Scanlon, 1994; Kumar *et al.*, 2001; Kumar and Kumar, 2009; Kumar *et al.*, 2010). The PC membranes are nanocomposites of the type discussed in Section 15.3. These nanocomposites reduce impedance of the cell, enhance charge transfer reaction ($Li \leftrightarrow Li^+ + e^-$), passivate lithium, and electrochemically couple the cathode to the GC membrane. The central part of the membrane laminate (GC) is also a nanocomposite. The cathode of the cell is a porous nanocomposite made from predominantly LAGP and carbon. The cells of Fig. 15.8 reversibly carry the reaction:

$$2Li + O_2 \leftrightarrow Li_2O_2 \qquad [15.20]$$

The open circuit voltage of this reaction is 2.90 V. Li–O₂ cells that exhibit a lifetime greater than 500 cycles have been developed.

15.4.2 Power generation

SOFCs are believed to be a critical element of future power generation technologies due to their high fuel-to-power conversion efficiency with minimal adverse influence on the environment. However, significant hurdles remain before they can become widely acceptable. The hurdles include operating temperatures, mechanical integrity, thermal instability, and cost. Nanocomposites based on the YSZ–Al₂O₃ system were successfully used to fabricate a direct oxidation SOFC (Thokchom *et al.*, 2008). The anodic structure comprises CeO_2 and Cu, contained in a porous YSZ–Al_2O_3 (20 wt %) composite layer. An interlayer consisting of YSZ–Al_2O_3 (20 wt%) electrolyte and lanthanum strontium manganate (LSM), and a cathode consisting of LSM were employed. The rationale for the use of a nanocomposite electrolyte from the YSZ–Al_2O_3 system was based on its enhanced conductivity, improved mechanical integrity, and cost as compared to YSZ.

15.5   Future trends

The emerging field of renewable energy technologies requires materials for fabricating various components with specific electrical properties. Superionic conductors are needed for efficient energy storage and power generation devices. Inexpensive mixed conductors can facilitate energy harvesting from ambient vibrations. Catalytic materials that can enhance reaction kinetics at lower temperatures are sought for energy conversion and other industrial applications. The basic elements for power generation from renewable energy sources are production, transmission, distribution, and storage. Each of these elements requires materials with a specific combination of electrical, mechanical, and thermal properties. Nanocomposites provide intriguing ionic and electronic conductivities. Tailored electrical conductivity of nanocomposites in combination with desired mechanical and thermal properties can satisfy these needs. Future trends in this field will be guided by the needs of renewable energy technologies.

15.6   References

Angell C A (1986), 'Recent developments in fast ion transport in glassy and amorphous materials,' *Solid State Ionics*, **18–19**, 72–88.

Cho J and Liu M (1997), 'Preparation and electrochemical properties of glass-polymer composite electrolytes in lithium battery,' *Electrochem. Acta*, **42(10)**, 1481–1488.

Croce F, Appetecchi G B, Persik L, and Scrosati B (1998), 'Nanocomposite polymer electrolytes for lithium batteries,' *Nature*, **394**, 456–458.

Dixon J M, LaGrange L D, Merten U, Miller C F, and Porter II J T (1963), 'Electrical resistivity of stabilized zirconia at elevated temperatures,' *J. Electrochem. Soc.*, **110(11)**, 276–280.

Kumar B (2007), 'Ionic transport through heterogeneous solids,' *Trans. Ind. Ceram.*, **66(3)**, 123–130.

Kumar B and Scanlon LG (1994), 'Polymer-ceramic composite electrolytes,' *J. Power Sources*, **52**, 261–268.

Kumar B and Scanlon LG (1999), 'Polymer-ceramic composite electrolytes: conductivity and thermal history effects,' *Solid State Ionics*, **124(3–4)**, 239–254.

Kumar B and Thokchom J S (2007), 'Space charge signature and its effect on ionic transport in heterogeneous solids,' *J. Amer. Ceram. Soc.*, **90(10)**, 3323–3325.

Kumar B and Thokchom J S (2008), 'Space charge mediated ionic transport in heterogeneous solids,' *J. Amer. Ceram. Soc.*, **91(4)**, 1175–1181.

Kumar B, Schaffer J D, Nookala M, and Scanlon L G (1994), 'An electrochemical study of PEO:LiBF$_4$-glass composite electrolytes,' *J. Power Sources*, **47(1–2)**, 63–78.

Kumar B, Rodrigues S J, and Scanlon L G (2001), 'Ionic conductivity of polymer-ceramic components,' *J. Electrochem. Soc.*, **148(10)**, A1191–A1195.

Kumar B, Nellutla S, Thokchom J S, and Chem C (2006), 'Ionic conduction through heterogeneous solids: delineation of blocking and space charge effects,' *J. Power Sources*, **160(2)**, 1329–1335.

Kumar J and Kumar B (2009), 'Development of membranes and a study of their interfaces for rechargeable lithium-air battery,' *J. Power Sources*, **194**, 1113–1119.

Kumar J, Rodrigues S J, and Kumar B (2010), 'Interface mediated electrochemical effects in lithium/polymer-ceramic cells,' *J. Power Sources*, **195**, 327–334.

Liang C C (1973), 'Conduction characteristics of the lithium iodide-aluminum oxide solid electrolytes,' *J. Electrochem. Soc.*, **120**, 1289–1292.

Maier J (1995), 'Ionic conduction in space charge regions,' *Prog. Solid State Chem.*, **23**, 171–263.

Maxwell JC (1881), *A Treatise on Electricity and Magnetism*, Vol. **1**, 2nd edition, pp. 435. Clarendon Press, Oxford.

Nagasubramanian G, Ahia A I, Halpert G, and Peled E (1993), 'Composite solid electrolyte for Li battery applications,' *Solid State Ionics*, **67**, 51–56.

Peled E, Golodnitsky D, Menachem C, Ardel G, and Lavi Y (1993), *Extended Abstracts of the 184th Electrochemical Society Meeting (New Orleans, LA)*, Abstract No. 504.

Rayleigh R S (1892), 'On the influence of obstacles arranged in rectangular order upon properties of a medium,' *Philos. Mag.*, **5(34)**, 502.

Thokchom J S, Xiao H, Rottmayer M, Reitz T L, and Kumar B (2008), 'Heterogeneous electrolytes (YSZ-Al$_2$O$_3$) based direct oxidation solid oxide fuel cell,' *J. Power Sources*, **178**, 26–33.

16

Biomedical applications of ceramic nanocomposites

N. GARMENDIA, FideNa, Spain; B. OLALDE and
I. OBIETA, Tecnalia, Spain

DOI: 10.1533/9780857093493.4.530

Abstract: Bioceramics and bioceramic composites have been widely used for biomedical applications for the last 50 years. This chapter discusses the advantages of using ceramic nanocomposites. The application of both inert and bioactive ceramics for orthopaedic and dental implants, as well as in the novel field of tissue engineering, is discussed and future trends are presented.

Key words: bioceramic nanocomposites, inert ceramics, bioactive ceramics, bone tissue engineering, orthopaedic and dental implants.

16.1 Introduction

Over the past 50 years significant progress has been made in the use of bioceramics and bioceramic composites for biomedical applications. The first attempts at implantation of ceramics in the biomedical sector started in the late 18th century, with the use of porcelain for crowns in dental applications. A hundred years later, in the late 19th century, there were studies of the use of plaster for bone repair according to Chevalier and Gremillard (2009). Soon after, in 1920, TCP (tricalcium phosphate), a bioresorbable ceramic, was proposed for use to fill gaps in bone. Some 10 years later, Rock (1933) was the first to consider the use of alumina ceramics as joint replacements. However, it was not until 30 years later (Sandhaus, 1967) when an alumina material, Degussit AL 23, was first patented for hip joints. According to Rieger (2001), this alumina material can be considered as the ancestor of today's high-tech ceramics. Other

530

ceramics were introduced in the following years, including hydroxyapatite (HAp) and TZP (tetragonal zirconia polycrystals) in the 1970s.

Over the past 50 years, therefore, there has been a significant move forward in the use of bioceramics for biomedical applications. The development of bioceramics can be classified into three generations (Vallet-Regi, 2010). The first generation would correspond to bioinert bioceramics, such as alumina and zirconia, which are mostly used for inert orthopaedic and dental implants. The second generation comprises bioactive and bioabsorbable ceramics, such as calcium phosphates or bioglasses. Finally, scaffolds for tissue engineering are the third generation; these aim to drive the regeneration of living tissues. In the meantime, many studies have demonstrated that better and unusual material properties can be achieved by manipulating ceramic length scales in the nano range. For that reason, during the last two decades, nanostructured materials have been widely studied and significant steps forward have been made in their understanding in recent years.

Nanostructured materials are defined as solid materials with at least one characteristic structural length in the order of a few nanometres ($1 \, nm = 10^{-9} \, m$) (Gleiter, 1995). Although some authors (e.g. Meyers *et al.*, 2006) define an upper limit of 250 nm for considering a material as nanometric, in general the nanometric grain size is meant to be below 100 nm (Nazarov and Mulyukov, 2002; Narayan *et al.*, 2004; Tjong and Chen, 2004; Liu and Webster, 2007; Kim and Estrin, 2008). Over this limit, the terms ultrafine grained materials or submicrometric materials (100–300 nm) are used.

This chapter discusses the advantages of using ceramic nanocomposites for biomedical applications. A material is defined as nanocomposite when at least one of the solid phases is in the nanometric range. First, the improvements achieved using nanocomposites are described. The next section focuses on inert ceramic nanocomposites for orthopaedic and dental implants. The applications of bioactive ceramics, such as calcium phosphates, in bone tissue engineering are then reviewed and future trends are presented.

16.2 Why ceramic nanocomposites are used in biomedical applications

Ceramics are broadly used in a large variety of technological applications requiring both structural and functional properties. They have received significant attention as candidate materials for use as structural materials under conditions of high loading rates, high temperature, wear and chemical etching too severe for metals. In this sense, bone-related biomedical applications are the most demanding of bioceramics. However, their inherent brittleness derived from their low fracture toughness has prevented their use in some applications. Moreover, the presence of flaws or defects in the material can lead to catastrophic failure during mechanical loading. Therefore, new kinds of materials have been studied for increasing the performance of ceramic matrix materials. For ceramics used in biomedical applications, which are extensively called bioceramics, this problem still remains. Nanophased ceramics are being investigated as a way of solving some of the structural and bio-related problems. For example, nanometric features in the surface of a prosthesis seem to reduce the risk of rejection and enhance the proliferation of osteoblasts (bone-forming cells). Nanophased or nanostructured ceramics can be obtained either by nanocrystalline materials or with nanocomposites.

Nanocrystalline materials are solids with a nanometric microstructure, consisting of polycrystals with one or several nanometric phases (Gleiter, 2000). Nanocomposites are materials with at least one of the solid phases in the nanometric range. Both nanomaterials are structurally characterized by a large volume fraction of grain boundaries, which can significantly alter their physical, mechanical and chemical properties.

16.2.1 Nanocrystalline ceramics

In the case of nanocrystalline ceramics, as the grain size is reduced, the grain volume at grain boundaries is increased (Meyers *et al.*, 2006). Thus, due to the high density of interfaces, an important fraction of atoms will be at the interface. This fact allows nanocrystalline materials to offer unusual and improved properties when compared to microscale materials.

There are studies (Webster *et al.*, 1999) that provide evidence that nanophase ceramics could promote osseointegration, which is critical for the clinical success of orthopaedic/dental implants. Webster *et al.* (2000) synthesized dense nanophase alumina (Al_2O_3) materials and showed a significant increase in protein absorption and osteoblast adhesion on the nano-sized ceramic materials compared to traditional micron-sized ceramic materials. Other studies (Du *et al.*, 1999) have suggested that better osteoconductivity would be achieved if synthetic HAp could more resemble bone minerals in composition, size and morphology.

The use of nanocrystalline materials can thus offer advantages for use in biomedical applications, such as:

- increased resistance/hardness
- improved toughness
- lower elastic modulus and lower ductility
- reduced risk of rejection

- enhanced proliferation of osteoblasts
- promotion of ossointegration.

With ceramic nanocomposites, even greater improvements can be achieved and the use of new ceramic matrix nanocomposites has been suggested (e.g. Gleiter, 1995; Narayan *et al.*, 2004; Meyers *et al.*, 2006; Liu and Webster, 2007).

16.2.2 Ceramic nanocomposites

Nanocomposites based on ceramic materials have been studied in order to improve mechanical properties and alter functional properties. The ceramic nanocomposites reported until now are either a ceramic nanophase in a ceramic matrix, a carbonaceous nanophase in a ceramic matrix or a ceramic nanophase in a polymer matrix.

Enhancements in stability, hardness, strength, toughness and creep resistance compared to the unreinforced matrix material have been reported in nanocomposites (Narayan *et al.*, 2004). Moreover, the combination of properties can lead to a new generation of medical devices and implants combining mechanical properties with bioactive properties. Some examples of ceramic-based nanocomposite materials are as follows.

- Alumina-based nanocomposites: with the addition of several nano-reinforcements, alumina matrix materials with improved mechanical properties (higher resistance, hardness, wear resistance and fracture toughness) have been obtained.
- Alumina/silicon carbide nanocomposites: the incorporation of SiC nanoparticles to an alumina matrix increases wear resistance.
- Alumina/zirconia nanocomposites: also known as zirconia-toughened alumina (ZTA) nanocomposites, they consist of a fine-grained alumina matrix reinforced with zirconia particles. The addition of the zirconia nanoparticles is intended to increase the toughness of the alumina matrix.
- Alumina/titania nanocomposites: increased hardness, fracture tough-ness and fracture resistance have been achieved.
- Zirconia/alumina nanocomposites: also known as alumina-toughened zirconia (ATZ), they consist of a zirconia matrix reinforced with alumina nanoparticles. They show exceptional resistance and extra-ordinary toughness.
- Silicon nitride/silicon carbide nanocomposites: the obtained results are controversial.
- Ceramic/carbon nanofibre composites: widely used, there is an improvement in properties (Pace *et al.*, 2002).
- Ceramic/carbon nanotube (CNT) composites: mechanical and electrical

properties are enhanced, but biocompatibility issues are still controversial (Streicher *et al.*, 2007; Garmendia *et al.*, 2008, 2009, 2010, 2011).
- Ceramic in polymer composites: especially relevant for tissue engineering applications.

16.3   Orthopaedic and dental implants

Ceramics such as zirconia (ZrO_2) and alumina (Al_2O_3) appear to be ideally
suited for the fabrication of orthopaedic implants because of their hardness,
low wear rates and excellent biocompatibility. For dental applications,
where aesthetic requirements (colour, translucency) are also essential, the
use of zirconia is preferred.

16.3.1 Bearing materials for orthopaedic implants

Alumina and zirconia ceramic materials have been used as joint substitutes
for over 30 years as an alternative to CoCr–UHMWPE bearing pairs. These
ceramics have crystal structures where atoms are joined by the combination
of strong ionic and covalent bonds. Due to the existence of these bonds they
show exceptional mechanical properties (high compressive strength, elastic
modulus and hardness) and they are chemically inert *in vivo*. Their
biocompatibility is also related to their high chemical stability, which
confers resistance to corrosion and reliability in the *in vivo* behaviour during
the lifetime of the implant (Rahaman and Yao, 2007). Also, the surfaces of
these oxides present polar hydroxyl groups (–OH), which promote
interaction with aqueous body fluids, providing a lubricating layer. In
addition, ceramics can be polished to tight tolerances and, due to the
hardness of the material, they are not affected by the wear particles than can
be generated due to the wear of bone cement, for example (Skinner, 2006).

The drawback of alumina and zirconia ceramics is their intrinsic
brittleness, which can lead to catastrophic failure *in vivo* and limits their
use in orthopaedic applications. However, the incidence of brittle failures
decreases with improvements in the quality of materials, manufacturing
techniques and implant design. Nevertheless, their low fracture toughness,
combined with their susceptibility to failure due to slow crack growth under
stresses below their fracture resistance, remains a problem regarding the
reliability of ceramic bearings (Rahaman and Yao, 2007).

The use of ceramics in joint prostheses began in the early 1970s. It was
observed that the low production of polyethylene wear debris in contact
with the ceramic solved the problem of loss of the prosthesis as a result of
osteolysis or bone loss. Boutin in France and Mittelmeier in Germany began
using aluminum oxide or alumina (Al_2O_3) as a constituent material of the
bearing surfaces of total hip prostheses. Alumina shows a resistance, as

measured by bending tests, of more than 550 MPa and a Vickers hardness of more than 1800–2000 HV. These hardness values are much higher than those offered by metals such as CoCr or titanium alloys used in orthopaedics, with hardness values bellow 500 HV. Currently, the most widely used ceramic in total hip arthroplasty is alumina. It is estimated that by the year 2005 more than 5 million Al_2O_3 femoral heads had already been implanted worldwide. But, together with its many advantages, alumina presents some drawbacks. In addition to its low fracture toughness ($\sim 4\,MPa\,m^{1/2}$), it is very sensitive to the surface finish and the sphericity of the bearing surfaces, to the tolerance between them and to the orientation of the components. It is also sensitive to fracture due to fatigue if the assemblage of the metal components is irregular or if there is not perfect adaptation of the dimensions.

Due to these drawbacks, zirconium oxide or zirconia (ZrO_2) was introduced in the mid 1980s. Clarke et al. (2003) estimate that from 1985 to 1995 around 104,000 zirconia femoral heads were implanted. The 3 mol% yttria stabilized tetragonal zirconia, known as YTZP, offers the best values of flexure resistance and fracture toughness among the bioceramics. It presents resistance values in bending tests and fracture toughness almost double those of alumina (Rahaman and Yao, 2007). It also shows an excellent wear behaviour, better when combined with UHMWPE than that of CoCr–UHMWPE couples, or even that of alumina (Cales, 2000; Fernández-Fairen et al., 2005). However, its use in implants has serious limitations. The most important is its sensitivity to low-temperature degradation or aging. The most dramatic case of aging was reported at the beginning of 2002 for zirconia hip joints heads, when several hundreds of implants failed within a short period (Chevalier, 2006). Aging is produced through the slow surface transformation of the zirconia from its high-temperature structure (tetragonal structure), obtained by the stabilization of the ceramic with yttria, into the stable monoclinic phase in the presence of water or water vapour. The phase transition entails a volume increase of the grains, which causes microcracking and, usually, failure and loss of functionality of the implant.

Due to the modest failure resistance of alumina and the problem of YTZP in terms of long-term reliability, there is a trend today to develop new alternatives, especially for critical and/or new designs for which alumina and YTZP do not satisfy all requirements.

Alumina–zirconia nanocomposites

There is a critical zirconia grain size below which no tetragonal to monoclinic phase transformation occurs. When the grain size is above 1 μm, the material behaves unstably and is susceptible to spontaneous tetragonal to monoclinic transformation. When the grain size is below 0.5 μm, a slow transformation occurs. With grain sizes under 0.2 μm the martensitic transformation is not promoted, therefore reducing the possibility of cracking (Evans and Heuer, 1980; Gutknecht *et al.*, 2007). Therefore, by reducing the zirconia grain size, aging resistance is increased. But, on the other hand, the transformation toughening mechanism that gives zirconia its exceptional mechanical properties will be lost. With the development of zirconia–alumina nanocomposites, the combination of both aging resistance and enhanced mechanical properties is promising. During recent years, several zirconia–alumina composites and nanocomposites have been developed and have shown significant improvement in toughness, strength and aging resistance (Menezes and Kiminami, 2008; Nevarez-Rascon *et al.*, 2009).

Two kinds of composites can be prepared in the zirconia–alumina system (De Aza, 2002): an alumina matrix reinforced with zirconia particles (zirconia-toughened alumina, ZTA) or a phase-stabilized zirconia matrix reinforced with alumina particles, known as alumina-toughened zirconia (ATZ). Composites with high fracture toughness are suitable in the ATZ system while composites with high hardness and relatively low fracture toughness belong to the ZTA system (Nevarez-Rascon *et al.*, 2009). Zirconia–alumina composites are available in the market, such as the ZTA Biolox® Delta (CeramTec GmbH, Germany) and the ATZ BIO HIP® (Metoxit AG, Switzerland).

For biomedical applications, ZTA is the most popular system. The dispersion of zirconia grains as a discrete second phase in the alumina matrix is a widely studied way of improving the mechanical properties of alumina (Schehl *et al.*, 2002). Significant improvements in toughness, strength and aging resistance have been shown, which are due to the dispersion of metastable tetragonal zirconia particles in the alumina matrix, which transform into the stable monoclinic phase under loading (Menezes and Kiminami, 2008). Other reinforcement mechanisms have been identified, such as microcrack toughening, compressive surface stresses and crack deflection (Laurent *et al.*, 1996).

Nawa *et al.* (1998) developed 10 mol% Ce–TZP/Al$_2$O$_3$ nanocomposites (with both phases being of nanometre scale) that exhibit high resistance to aging, complete biocompatibility and a high wear resistance (Uchida *et al.*, 2002; Tanaka *et al.*, 2002, 2003). Moreover, Benzaid *et al.* (2008) showed that the cyclic fatigue threshold of these nanocomposites stands above that measured in conventional biomedical-grade alumina and zirconia. Therefore, they state that Ce–TZP/Al$_2$O$_3$ nanocomposites may be considered as an option for future biomedical applications.

Benzaid *et al.* (2008) also proposed a nano–nano Ce–TZP/Al$_2$O$_3$ composite. The starting nanopowders (Ce–TZP and alumina) were coated with an alumina and zirconia precursor, respectively. The obtained nanocomposites show a fine nanostructure where the matrix is formed by Ce–TZP crystals. Nano-sized alumina crystals (<20 nm) are retained inside Ce–TZP grains and other alumina grains (200–300 nm) are homogeneously distributed in the Ce–TZP matrix. They claimed that these nanocomposites show high flexural strength and fracture toughness values.

De Aza *et al.* (2002) obtained ZTA composites with submicrometre alumina grains and mainly intergranular, nano-sized zirconia particles. They stated that this leads to a high portion of tetragonal phase retained at room temperature (after sintering) with the ability to transform under applied stress. Therefore, the dominant toughening mechanism in these composites is considered to be transformation toughening. They concluded that ZTA composites present a higher reliability than monolithics, due to the combination of the advantages of both alumina and zirconia.

The addition of alumina to zirconia can also drastically reduce the aging kinetics. Chevalier *et al.* (2000) obtained composites with 1.7 vol% (2.5 wt%) zirconia nanoparticles. It has been proved (Chevalier *et al.*, 2011) that alumina-zirconia nanocomposites exhibit a very limited surface damage after 7 million cycles in a hip simulator. Moreover, these nanocomposites show crack resistance under static and cyclic fatigue well beyond that of all existing biomedical-grade ceramics. Chevalier *et al.* stated that, associated with expected full stability *in vivo*, the overall set of results ensures potential future development for these kinds of new nanocomposites in the field of orthopaedics.

YTZP–CNT nanocomposites

As has been already mentioned, there is a critical zirconia grain size below which no tetragonal to monoclinic phase transformation occurs. Therefore, by decreasing the grain size, the sensitivity to aging of the material can be avoided but, at the same time, the transformation toughness mechanism that gives zirconia its exceptional mechanical properties will be lost. Garmendia *et al.* (2009, 2010, 2011) studied the addition of carbon nanotubes (CNTs) to a zirconia matrix as reinforcing agents to maintain, or even increase, toughness in spite of the decrease in grain size. The CNTs dispersed in the ceramic matrix act as fracture energy dissipating sites through mechanisms such as crack deviation in the CNT and ceramic matrix interface, crack bonding by the CNT, and CNT pull-out in fracture surfaces. In any case, to really improve the mechanical properties of a ceramic matrix, a good dispersion of the CNTs in the matrix and chemical bonding between the CNTs and the zirconia are needed. In order to reach these requirements, zirconia-coated CNTs have been introduced. In this way, nanozirconia nanocomposites reinforced with CNTs with improved mechanical properties are obtained. Although the biocompatibility of this material has been assessed (Garmendia *et al.*, 2009), the biocompatibility of CNTs still remains a controversial issue.

16.3.2 Dental implants

The oral environment presents extreme working conditions that include humidity, acidic or basic pH, cyclic loading and peak loads that can reach extremely high levels when hard objects are accidentally encountered during mastication. The introduction of new materials and processing techniques has led the technological evolution of ceramics for dental applications, mostly in the last 30 years. New ceramic materials offer greater performance in both strength and toughness, which has made it possible to expand the range of indications to long-span fixed partial prostheses, implant abutments and implants. The high level of crystallinity is responsible for an improvement in mechanical properties through various mechanisms such as crystalline reinforcement or stress-induced transformation. Unfortunately, higher crystallinity is also associated with higher opacity, which is not always desirable for dental ceramics. Other factors such as crystal size and geometry, modulus of elasticity, phase transformation, and thermal expansion mismatch between crystal and glassy phases play a crucial role in determining the final mechanical response of the ceramic (Denry and Holloway, 2010).

YTZP is a very interesting ceramic for dental implants due to its outstanding mechanical and biocompatibility properties. Currently there is a wide range of zirconia block suppliers in the dental implant market. These blocks are used to manufacture posterior bridges using CAD/CAM. But aging is still a problem, as dental implants are placed in a humid environment. Furthermore, YTZP is also one of the most opaque of the ceramic materials currently available (Spear and Holloway, 2008). There are also other important features to be considered; these include: the effect of sandblasting and heat treatment on the microstructure and strength; bonding to veneering porcelains; bonding to cement; the visible light translucency related to aesthetic restoration; X-ray opacity; and clinical survival rates. Nanoscale zirconia–alumina composites are meant to further improve these properties (Ban, 2008).

Philipp *et al.* (2010) examined the clinical performance of veneered ceria-stabilized tetragonal zirconia–alumina nanocomposite frameworks for dental prostheses. Their nanocomposite was found to be a reliable framework material after 1 year of functioning. Nevarez-Rascon *et al.* (2009) studied the performance of different ATZ and ZTA nanocomposites. The reported sintered density, grain size, Vickers hardness and fracture toughness values are in accordance with international standards applied to dental applications. Consequently, ATZ and ZTA nanocomposites appear to be adequate materials for use in the fabrication of implants and abutments instead of the pure oxides currently in use. Moreover, the relatively lower hardness in some of these composites (compared with alumina) is said to be an advantage, since the final form of the implant can be easily processed by machining. Equally, a composite with high hardness would have a disadvantage due to the prolonged milling times needed and high wear rates of the diamond milling CAM machines. Therefore, it is established that ATZ and ZTA composites present an intermediate hardness and higher fracture toughness, which is the ideal combination of mechanical properties for specific dental applications. The use of alumina whisker reinforced alumina–zirconia nanocomposites in dental applications has also been investigated (Nevarez-Rascon *et al.*, 2010). The reported hardness and fracture values of these nanocomposites can compete with commercially available materials for different dental applications.

In summary, while the mechanical properties of zirconia–alumina nanocomposites are very interesting for dental implants, both oxides are bioinert. However, with the addition of bioactive calcium phosphates such as HAp and TCP, the biocompatibility of zirconia–alumina nanocomposites in load-bearing applications is significantly enhanced and osseointegration and bone regeneration are improved. Kong *et al.* (2005) focused their study on a zirconia–alumina matrix composed of nanocomposite powder. HAp was added to the nanocomposite powder and the potential of these nanocomposites for use in load-bearing applications was verified. Yousefpour *et al.* (2011) also developed hydroxyapatite–zirconia–alumina bionanocomposites by the mechanical blending of separately synthesized nano-scaled powders. They state that $10Ce–TZP/Al_2O_3/HA$ nanocomposites with 30 vol% of HAp show the optimal composition for biological applications.

16.4   Tissue engineering

Tissue engineering has shown tremendous promise in creating biological alternatives for tissue repair and regeneration (Mooney and Mikos, 1999). In a general approach, a porous scaffold serves as a temporary template for cells seeding *in vitro* and for the consequent formation of new tissue (Hua *et al.*, 2002). The ideal scaffold should be a three-dimensional biocompatible highly porous network (Karageorgiou and Kaplan, 2005; Rezwana *et al.*, 2006), with an appropriate surface for cell adhesion, proliferation and differentiation (Cima *et al.*, 1991). It should also allow easy invasion of blood vessels in order to supply nutrients to the cells (Mikos *et al.*, 1993). They also ought to provide the necessary mechanical strength or, in other words, to maintain proper mechanical properties until the new tissue grows

(Uemura *et al.*, 2003). The scaffolds should biodegrade with a controlled degradation rate and eventually disappear when the new tissue is fully regenerated. In that sense, the three-dimensional space occupied by the porous scaffolds would be replaced by newly formed tissue (Langer and Vacanti, 1993; Holy *et al.*, 1999). The development of new materials to meet all these specifications is being addressed worldwide.

In recent years, several bioactive ceramics, such as hydroxyapatite, tricalcium phosphate, biphasic calcium phosphate, and bioactive glasses have been used in bone tissue engineering applications (Vallet-Regi, 2001). Hydroxyapatite (HAp) is one of the most widely used synthetic ceramics due to its chemical similarities to the inorganic component of hard tissue (Akao *et al.*, 1981; Dewith *et al.*, 1981). Bioactive ceramic scaffolds can either induce the formation of bone from the surrounding tissue or can act as a carrier or guide for enhanced bone regeneration by cell migration, proliferation and differentiation. But, as has already been highlighted, ceramics are brittle and are not suitable for load-bearing applications (Paul and Sharma, 2006). Therefore, the use of polymer–ceramic composites has been suggested. These composites mimic the inorganic–organic composition of natural bone, with nanometre size inorganic components (mainly bone-like apatite). To make the mechanical properties more similar to those of natural bone, nanoscale calcium phosphates are added to the polymer matrix (Rho *et al.*, 1998). Furthermore, with nanometric components, the scaffold osteoconductivity and bone bonding ability are enhanced, and osteoblasts and osteoprogenitor cells can adhere, migrate inside, differentiate and synthesize new bone matrix (Ma *et al.*, 2001; Zhang and Ma, 1999).

Poly(α-hydroxyl acids) such as poly(lactic acid) (PLA), poly(glycolic acid) (PGA), poly(lactic acid-co-glycolic acid) (PLGA) and poly(ε-caprolactone) (PCL) satisfy many of the scaffold's material requirements and they have already been used as scaffolding material for a variety of tissue engineering applications, including bone (Lo *et al.*, 1995). However, the highly porous polymeric scaffolds are chemically hydrophobic, biologically inert and relatively weak, which limits their use for bone tissue regeneration, especially in the *in vivo* implant site.

Wei and Ma (2004) produced a nano-hydroxyapatite (nHAp)/PLLA composite scaffold for bone tissue engineering. It was proved that the compressive modulus increased significantly when the nHAp proportion reached 30% of the composite. Moreover, the addition of nHAp increased the protein absorption, thus improving cell adhesion. Kothapalli *et al.* (2005) also proposed a nano-sized HA/PLLA composite scaffold. The starting nanopowders had an average size of approximately 25 nm in width and 150 nm in length. The obtained nanocomposite scaffold showed an increase in compression modulus from 4.72 MPa to 9.87 MPa when the nHAp content increased from 0 to 50 wt%. They claim that these scaffolds should be suitable materials for non-load-sharing tissue engineering applications.

More recently, Wang *et al.* (2010b) examined the *in vitro* response of porous nano-hydroxyapatite/polycaprolactone (nHAp/PCL) scaffolds. After 7 days of culture, the bone marrow stromal cells (BMSCs) coalesced to form a large and flat layer of cells and cover on the nHA/PCL scaffold, while they presented as clusters or agglomerates on the PCL scaffold. The poor cell growth on the PCL scaffold is probably due to its hydrophobicity, which may interfere in the cell attachment. These results suggest that nHA/PCL scaffolds could be promising in bone tissue engineering. Moreover, Nejati *et al.* (2009) compared the cell affinity and cytocompatibility of mesenchymal stem cells (MSCs) in nanoHAp and microHAp PCL composites. They stated that cell proliferation was higher in the nanocomposite, which was attributed to the larger surface area of nHAp, proving superior cell viability and cytocompatibility of the nanocomposite.

Wang *et al.* (2010a) synthesized a new silane-modified nano-HAp (mnHAp) in order to improve the interfacial connection of HAp to PLLA. Characterization of the composite scaffold showed that mnHAp was homogeneously distributed in the scaffold. As a result, the compressive modulus and protein adsorption of PLLA/mnHAp (80:20 w/w) composite scaffold increased 4.2-fold and 2.8-fold compared with those of a pure PLLA scaffold. Incorporating mnHAp into PLLA network also buffered the pH reduction and reduced weight loss in *in vitro* degradation significantly.

These nano-hydroxyapatite composites show higher mechanical properties and better cell behaviour than conventional polymeric scaffolds. Their ceramic counterparts are bioactive, but show very low biodegradation rates. Furthermore, if bioresorbable ceramics such as biphasic calcium phosphate (BCP) and bioglass are used for the repair of bone, reconstruction could be more rapid (Habraken *et al.*, 2007; Migliaresi *et al.*, 2007).

BCP possesses sufficient degradation, excellent biocompatibility, osteoconductivity, and even osteoinductivity (Yuan and Groot, 2005). Ebrahimian-Hosseinabadi *et al.* (2011) recently fabricated PLGA/nano-biphasic calcium phosphate (nBCP) composite scaffolds. The size of the nanoparticles was estimated to be less than 100 nm. The elastic modulus and yield strength of the nanocomposite scaffold were significantly enhanced, showing the highest values with 30% wt of the nanopowder. Moreover, from the biodegradability point of view, the authors state that by increasing the amount of nBCP particles in the PLGA/nBCP scaffolds, hydrophilicity increases and causes a larger weight loss.

Some bioactive glasses can directly bond to living bone without the formation of surrounding fibrous tissue. In such cases, a bone-like apatite layer is deposited *in vivo* between the implant and bone (Kokubo *et al.*, 2003). Hong *et al.* (2008) developed a new bioactive glass ceramic (BGC) nanoparticle and PLLA nanocomposite scaffold. The BGC nanoparticles were synthesized using a combination of the sol–gel and coprecipitation methods and were homogeneous nanospheres, 20–40 nm in diameter. The compressive modulus of the scaffold increased from 5.5 to 8.0 MPa, while the compressive strength increased from 0.28 to 0.35 MPa as the BGC content was increased from 0 to 30 wt%. Moreover, PLLA/(20 wt%)–BGC composites exhibited the best mineralization property in simulated body fluid after 1 day of incubation, with apatite clusters covering almost the entire surface of the scaffold.

16.5   Future trends

Current load-bearing implants with osteconductive surfaces or tissue engineering based on natural or synthetic biodegradable scaffolds offer a significant increase in the quality of bone repair and improved mechanical properties. However, they still present limitations and hence there is potential for major advances to be made in the field. Bone can actually heal itself when it is broken or removed. However, this capability is impaired in situations where substantial loss of bone has occurred due to a trauma or tumour resection, leading to non- or delayed unions. This inability in bone healing is also related to disease or old age (Graus, 2006).

Most of the current limitations in this field are related to the lack of 'smart' biomaterials with the capability to elicit specific responses at molecular and cellular level. These biomaterials act as structural support and delivery vehicles, providing cells and bioactive molecules necessary for the formation of new bone tissue (Ma, 2004; Mistry *et al.*, 2006). Ideal biomaterial must possess mechanical properties adequate to support growing bone tissue, good biocompatibility and high porosity (Sitharaman *et al.*, 2008). In addition, it should be able to avoid rejection (e.g. associated with infections), react to changes in the immediate environment and stimulate specific regenerative events at the molecular level, directing cell proliferation, cell differentiation, and extracellular matrix production and organization.

In the field of developmental biology, in-depth knowledge is increasingly available on the key factors that regulate the highly complex processes of bone growth, repair and regeneration. Unfortunately, the tremendous potential of this source of information has not yet been fully exploited in biomaterials science, since this field has developed from an engineering tradition, yielding a biomechanically inspired rather than a bioinspired approach towards tissue integration and regeneration.

Therefore, future research into bioceramic nanocomposites will be focused on the development of new biomaterials in the following areas (Best *et al.*, 2008):

- the development of smart materials capable of combining sensing with bioactivity
- the development of improved biomimetic composites
- enhanced bioactivity in terms of gene activation.

16.6 References

Akao M, Aoki H and Kato K (1981), 'Mechanical properties of sintered hydroxyapatite for prosthetic applications', *J Mater Sci*, 16, 809–812.

Ban J (2008), 'Reliability and properties of core materials for all-ceramic dental restorations', *Jap Dent Sci Rev*, 44, 3–21.

Benzaid R, Chevalier J, Saâdaoui M, Fantozzi G, Nawa M, Diaz L A and Torrecillas R (2008), 'Fracture toughness, strength and slow crack growth in a ceria stabilized zirconia–alumina nanocomposite for medical applications', *Biomaterials*, 29, 3636–3641.

Best S M, Porter A E, Thian E S and Huang J (2008), 'Bioceramics: Past, present and for the future', *J Eur Ceram Soc*, 28, 1319–1327.

Cales B (2000), 'Fractures des têtes de prothèses de hanche en zircone', *Maitr Orthop*, 96, 26–30.

Chevalier J (2006), 'What future for zirconia as a biomaterial?', *Biomaterials*, 27, 535–543.

Chevalier J and Gremillard L (2009), 'Ceramics for medical applications: A picture for the next 20 years', *J Eur Ceram Soc*, 29, 1245–1255.

Chevalier J, De Aza A, Schehl M, Torrecillas R and Fantozzi G (2000), 'Extending the lifetime of ceramic orthopaedic implants', *Adv Mater*, 12, 1619–1621.

Chevalier J, Taddei P, Gremillard L, Deville S, Fantozzi G, Bartolomé J F, Pecharroman C, Moya J S, Diaz L A, Torrecillas R and Affatato S (2011), 'Reliability assessment in advanced nanocomposite materials for orthopaedic applications', J Mech Behavior Biom Mat, 4, 303–314.

Cima L G, Vacanti J P, Vacanti C, Ingber D, Mooney D and Langer R (1991), 'Tissue engineering by cell transplantation using degradable polymer substrates', *J Biomech Eng*, 113, 143–151.

Clarke I C, Manaka M, Green D D, Williams P, Pezzoti G, Kim Y H, Ries M, Sugano N, Sedel L, Delauney C, Ben Nissan B, Donaldson T and Gustafson G A (2003), 'Current status of zirconia used in total hip implants', *J Bone Joint Surg Am*, 85, 73–84.

De Aza A H, Chevalier J, Fantozzi G, Schehl M and Torrecillas R (2002), 'Crack growth resistance of alumina, zirconia and zirconia toughened alumina ceramics for joint prosthesis', *Biomaterials*, 23, 937–945.

Denry I and Holloway J A (2010), 'Ceramics for dental applications: A review', *Materials*, 3, 351–368.

Dewith G, van Dijk H J A, Hattu N and Prijs K (1981), 'Preparation, microstructure and mechanical properties of dense polycrystalline hydroxyapatite', *J Mater Sci*, 6, 1592–1598.

Du C, Cui F Z, Zhu X D and de Groot K (1999), 'Three-dimensional nano-HAp/

collagen matrix loading with osteogenic cells in organ culture', *J Biomed Mater Res*, 44, 407–415.

Ebrahimian-Hosseinabadi M, Ashrafizadeh F, Etemadifar M and Venkatraman S S (2011), 'Preparation and mechanical behavior of PLGA/nano-BCP composite scaffolds during in-vitro degradation for bone tissue engineering', *Pol Degrad Stab*, 96, 1940–1946.

Evans A G and Heuer A H (1980), 'Transformation toughening in ceramics: martensitic transformation in crack-tip stress fields', *J Am Ceram Soc*, 63, 241–248.

Fernández Fairen M, Gil Mur F J, Martínez S, Sala P, Delgado J A and Blanco A (2005), 'Estudio del desgaste catastrófico de la circona', *Biomecánica*, 12(2), 23–34.

Garmendia N, Bilbao L, Muñoz R, Goikoetxea L, García A, Bustero I, Olalde B, Garagorri N and Obieta I (2008), 'Nanozirconia partially coated MWCNT: nanostructural characterization and cytotoxicity and lixiviation study', *Key Eng Mat*, 361–363, 775–778.

Garmendia N, Santacruz I, Moreno R and Obieta I (2009), 'Slip casting of nanozirconia/MWCNT composites using a heterocoagulation process', *J Eur Ceram Soc*, 29, 1939–1945.

Garmendia N, Santacruz I, Moreno R and Obieta I (2010), 'Zirconia-MWCNT nanocomposites for biomedical applications obtained by colloidal processing', *J Mater Sci Mater Med*, 21, 1445–1451.

Garmendia N, Grandjean S, Chevalier J, Diaz L A, Torrecillas R and Obieta I (2011), 'Zirconia–multiwall carbon nanotubes dense nanocomposites with an unusual balance between crack and aging resistance', *J Eur Ceram Soc*, 31(6), 1009–1014.

Gleiter H (1995), 'Nanostructured materials: State of the art and perspectives', *Nanostructured Mater*, 6, 3–14.

Gleiter H (2000), 'Nanostructured materials: basic concepts and microstructure', *Acta Mat*, 48, 1–29.

Graus K H (2006), 'Mesenchymal stem cells and bone regeneration', *C Vet Surg*, 35, 232–242.

Gutknecht D, Chevalier J, Garnier V and Fantozzi G (2007), 'Key role of processing to avoid low temperature aging in alumina zirconia composites for orthopaedic application', *J Eur Ceram Soc*, 27, 1547–1552.

Habraken W, Wolke J G C and Jansen J A (2007), 'Ceramic composites as matrices and scaffolds for drug delivery in tissue engineering', *Adv Drug Deliv Rev*, 59, 234–248.

Holy C E , Dang S M , Davies J E and Shoichet M S (1999), 'In vitro degradation of a novel poly(lactide-co-glycolide) 75/25 foam', *Biomaterials*, 20, 1177–1185.

Hong Z, Reis R Land Mano J F (2008), 'Preparation and in vitro characterization of scaffolds of poly(L-lactic acid) containing bioactive glass ceramic nanoparticles', *Acta Biomaterials*, 4, 1297–1306.

Hua F J, Kim G E, Lee J D, Son Y K and Lee D S (2002), 'Macroporous poly(L-lactide) scaffold. Preparation of a macroporous scaffold by liquid-liquid phase separation of a PLLA-dioxane-water system', *J Biomed Mater Res*, 63, 161–167.

Karageorgiou V and Kaplan D (2005), 'Porosity of 3D biomaterial scaffolds and osteogenesis', *Biomaterials*, 26, 5474–5491.

Kim H S and Estrin Y (2008), 'Strength and strain hardening of nanocrystalline materials', *Mat Sci Eng A*, 483–484, 127–130.

Kokubo T, Kim H M and Kawashita M (2003), 'Novel bioactive materials with different mechanical properties', *Biomaterials*, 24, 2161–2175.

Kong Y-M, Bae C-J, Lee S-H, Kim H-W and Kim H-E (2005), 'Improvement in biocompatibility of ZrO2–Al2O3 nano-composite by addition of HA', *Biomaterials*, 26, 509–517.

Kothapalli C R, Shaw M T and Wei M (2005), 'Biodegradable HA-PLA 3D porous scaffolds: Effect of nano-sized filler content on scaffold properties', *Acta Biomat*, 1, 653–662.

Langer R and Vacanti J P (1993), 'Tissue Engineering', *Science*, 260, 920–926.

Laurent Ch, Rousset A, Bonnefond P, Oquab D and Lavelle B (1996), 'Mechanical properties of alumina-metal-zirconia nano-micro hybrid composites', *J Eur Cer Soc*, 16, 937–943.

Liu H and Webster T J (2007), 'Nanomedicine for implants: A review of studies and necessary experimental tools', *Biomaterials*, 28, 354–369.

Lo H, Ponticiello M S and Leong K W (1995), 'Fabrication of controlled release biodegradable foams by phase separation', *Tissue Eng*, 1, 15–28.

Ma P X (2004), 'Scaffolds for tissue fabrications' *Mat. Today*, May 30–40.

Ma P X, Zhang R, Xiao G and Franceschi R (2001), 'Engineering new bone tissue in vitro on highly porous poly(α-hydroxyl acids)/hydroxyapatite composite scaffolds', *J Biomed Mater Res*, 54, 284–293.

Menezes R R and Kiminami R H G A (2008), 'Microwave sintering of alumina–zirconia nanocomposites', *J Mat Proccess Techn*, 203, 513–517.

Meyers M A, Mishra A and Benso D J (2006), 'Mechanical properties of nanostructured materials', *Prog Mat Sci*, 51, 427–556.

Migliaresi C, Motta A and Dibenedetto A T (2007), 'Injectable scaffolds for bone and cartilage regeneration', in Bronner F, Farach-Carson M C and Mikos A, *Engineering of Functional Skeletal Tissues*, London, Springer, 95–109.

Mikos A G, Sarakinos G, Leite S M, Vacanti J P and Langer R (1993), 'Laminated three-dimensional biodegradable foams for use in tissue engineering', *Biomaterials*, 5, 323–330.

Mistry A S Shi X and Mikos A G (2006), 'Nanocomposite scaffolds for tissue engineering', Brozino J, *Biomedical Engineering Handbook* Boca Raton, FL, CRC Press 40-41–40-11.

Mooney D J and Mikos A G (1999), 'Growing new organs', *Sci Am*, 280, 60–65.

Narayan R J, Kumta R N, Sfeir C, Lee D H, Olton D and Choi D (2004), 'Nanostructured ceramics in medical devices: applications and prospects', *JOM*, 56(10), 38–43.

Nawa M, Nakamoto S, Sekino T and Niihara K (1998), 'Tough and strong Ce-TZP/alumina nanocomposites doped with titania', *Ceram Int.*, 24, 497–506.

Nazarov A A and Mulyukov R R (2002), 'Nanostructured materials', in Goddard W, Brenner D, Lyshevski S and Iafrate G, *Nanoscience, Engineering and Technology Handbook*, Boca Raton, FL, CRC Press 22-1–22-41.

Nejati E, Firouzdor V, Eslaminejad M B and Bagheri F (2009), 'Needle-like nano hydroxyapatite/poly(L-lactide acid) composite scaffold for bone tissue engineering application', *Mater Sci Eng C*, 29(3), 942–949.

Nevarez-Rascon A, Aguilar-Elguezabal A, Orrantia E and Bocanegra-Bernal M H

(2009), 'On the wide range of mechanical properties of ZTA and ATZ based dental ceramic composites by varying the Al2O3 and ZrO2 content', *Int J Refrac Met Hard Mat*, 27, 962–970.

Nevarez-Rascon A, Aguilar-Elguezabal A, Orrantia E and Bocanegra-Bernal M H (2010), 'Al2O3(w)–Al2O3(n)–ZrO2 (TZ-3Y)n multi-scale nanocomposite: An alternative for different dental applications?', *Acta Biomat*, 6, 563–570.

Pace N, Spurio S, Pavan L, Rizzuto G and Streicher R M (2002), 'Clinical trial of a new CF-PEEK acetabular insert in hip arthroplasty', *Hip Int*, 12(2), 212–214.

Paul W and Sharma C P (2006), 'Nanoceramic matrices: biomedical applications', *Am J Biochem Biotech*, 2(2), 41–48.

Philipp A, Fischer J, Hämmerle C H and Sailer I (2010), 'Novel ceria-stabilized tetragonal zirconia/alumina nanocomposite as framework material for posterior fixed dental prosthesis: Preliminary results of a prospective case series at 1 year of function', *Quintessence Int*, 41(4), 313–9.

Rahaman M and Yao A (2007), 'Ceramics for prosthetic hip and knee joint replacement', *J Am Ceram Soc*, 90(7), 1965–1988.

Rezwana K, Chena Q Z, Blakera J J and Boccaccini A R (2006), 'Biodegradable and bioactive porous polymer/inorganic composite scaffolds for bone tissue engineering', *Biomaterials*, 27, 3413–3431.

Rho J-Y, Kuhn-Spearing L and Zioupos P (1998), 'Mechanical properties and the hierarchical structure of bone', *Med Eng Phys*, 20, 92–112.

Rieger W (2001), 'Ceramics in orthopaedics – 30 years of evolution and experience', in Rieker C, Oberholzer S and Wyss U, *World Tribology Forum in Arthroplasty*, Bern, Hans Huber Verlag.

Rock M (1933), *Kuenstliche Ersatzteile fuer das Innere und AEussere des menschlichen und tierischen Koerpers*. German Patent, DRP no. 583589. 1933-Sept-06.

Sandhaus S (1967), *Bone implants and drills and taps for bone surgery*. British Patent No. 1083769. 1967-Sept-20.

Schehl M, Diaz L A and Torrecillas R (2002), 'Alumina nanocomposites from powder-alkoxide mixtures', *Acta Mater*, 50, 1125–1139.

Sitharaman B Xinfeng S Walboomers F X Liao H Cuijpers V Wilson L J Mikos A G and Jansen J A (2008), 'In vivo biocompatibility of ultra-short single walled carbon nanotube/biodegradable polymer nanocomposites for bone tissue engineering', *Bone*, 43 362–370.

Skinner H B (2006), 'Ceramics in total joint surgery – the pros and cons', *Semin Arthroplasty*, 17(3), 196–201.

Spear F and Holloway J A (2008), 'Which all-ceramic system is optimal for anterior esthetics?', *J Am Dent Assoc*, 139, 19S–24S.

Streicher R M, Schmidt M and Fiorito S (2007), 'Nanosurfaces and nanostructures for artificial orthopaedic implants', *Nanomedicine*, 2(6), 861–874.

Tanaka K, Tamura J, Kawanabe K, Nawa M, Oka M, Uchida M, Kokubo T and Nakamura T (2002), 'Ce-TZP/Al2O3 nanocomposites as a bearing material in total joint replacement', *J Biomed Mater Res*, 63, 262–270.

Tanaka K, Tamura J, Kawanabe K, Nawa M, Uchida M, Kokubo T and Nakamura T (2003), 'Phase stability after aging and its influence on pin-on-disk wear properties of Ce-TZP/Al2O3 nanocomposite and conventional Y-TZP', *J Biomed Mater Res*, 67, 200–207.

Tjong S C and Chen H (2004), 'Nanocrystalline materials and coatings', *Mat Sci Eng R*, 45, 1–88.

Uchida M, Kim H M, Kokubo T, Nawa M, Asano T, Tanaka K and Nakamura T (2002), 'Apatite forming ability of a zirconia/alumina nano-composite induced by chemicals treatment', *J Biomed Mater Res*, 60, 277–282.

Uemura T, Dong J, Wang Y, Kojima H, Saito T, Iejima D, Kikuchi M, Tanaka J and Tateishi T (2003), 'Transplantation of cultured bone cells using combinations of scaffolds and culture techniques', *Biomaterials*, 24, 2277–2286.

Vallet-Regi M (2001), 'Ceramics for medical applications', *J Chem Soc*, 97–108

Vallet-Regi M (2010), 'Evolution of bioceramics within the field of biomaterials', *C R Chimie*, 13, 174–185.

Wang X· Song G and Lou T (2010a), 'Fabrication and characterization of nano-composite scaffold of PLLA/silane modified hydroxyapatite'· *Med Eng Phys*, 32 (4), 391–397.

Wang Y, Liu L and Guo S (2010b), 'Characterization of biodegradable and cytocompatible nano-hydroxyapatite/ polycaprolactone porous scaffolds in degradation in vitro', *Polym Degrad Stab*, 95(2), 207–213.

Webster T J, Siegel R W and Bizios R (1999), 'Osteoblast adhesion on nanophase ceramics', *Biomaterials*, 20, 1221–1227.

Webster T J, Ergun C, Doremus R H, Siegel R W and Bizios R (2000), 'Specific proteins mediate enhanced osteoblast adhesion on nanophase ceramics', *J Biomed Mater Res*, 51, 475–483.

Wel G and Ma P X (2004), 'Structure and properties of nano-hydroxyapatite/ polymer composite scaffolds for bone tissue engineering', *Biomaterials*, 25, 4749–4757.

Yousefpour M, Askari N, Abdollah-Pour H, Amanzadeh A and Riahi N (2011), 'Investigation on biological properties of dental implant by Ce-TZP/Al2O3/HA bio-nano-composites', *Digest J Nanomat Biostruc*, 6(2), 675–681

Yuan H and Groot K (2005), 'Calcium phosphate biomaterials: a overview', in Reis R L and Weiner S, *Learning from Nature How to design New Implantable Biomaterials: From Biomineralization Fundamentals to Biomimetic Materials and Processing Routes*, Amsterdam, IOS Press, 37–57.

Zhang R and Ma P X (1999), 'Poly(α-hydroxyl acids)/hydroxyapatite porous composites for bone-tissue engineering. I. Preparation and morphology', *J Biomed Mater Res*, 44, 446–455.

17

Synthetic biopolymer/layered silicate
nanocomposites for tissue engineering
scaffolds

M . O K A M O T O , Toyota Technological Institute, Japan

DOI: 10.1533/9780857093493.4.548

Abstract: Current research trends on nanocomposite materials for tissue engineering, including strategies for the fabrication of nanocomposite scaffolds with highly porous and interconnected pores, are presented. The results of *in-vitro* cell culture, used to analyzed the cell–scaffold interaction considering the colonization of mesenchymal stem cells and degradation of the scaffolds *in-vitro*, are also discussed.

Key words: biopolymer, tissue engineering scaffolds, mesenchymal stem cell, nanoparticles.

17.1 Introduction

Since the industrial revolution, and particularly after World War II, materials research has increased rapidly, and has resulted in the widespread use of materials such as polymers, metals, semiconductors and agricultural chemicals (e.g. pesticides and fertilizers). The production of these materials in increasing quantities to meet the demands of a growing population has led to the significant consumption of fossil fuels and production of waste. Together, these have resulted in regional and global environmental problems ranging from air, water and soil pollution to climate change. These problems limit the extent to which such materials can be used and call for a re-thinking of their design, synthesis and production.

Natural materials not only provide a point of comparison with the performance of man-made materials but also a clue to the development of new materials that are both more environmentally sustainable and functional. Biopolymers are a well-known example of renewable, envir-

548

onmentally-friendly polymeric materials (Smith, 2005). Biopolymers include polysaccharides such as cellulose, starch, alginate and chitin/chitosan, carbohydorate polymers produced by bacteria and fungi (Chandra and Rustgi, 1998), and animal protein-based biopolymers such as wool, silk, gelatin and collagen. Naturally-derived polymers combine biocompatibility and biodegradability. One of the advantages of naturally-derived polymers is their ability to support cell adhesion and function. However, these materials have poor mechanical properties. Many of them are also limited in supply and can therefore be costly (Johnson et al., 2003).

Polyvinyl alcohol (PVA), poly(ε-caprolactone) (PCL), poly(lactic acid) (PLA), poly(glycolic acid) (PGA), poly(hydroxy butyrate) (PHB) and poly (butylene succinate) (PBS) are examples of polymers of synthetic origin which are biodegradable (Platt, 2006). In today's commercial environment, synthetic biopolymers have proven to be relatively expensive and available only in small quantities, with limited applications that include the textile, medical and packaging industries. However, synthetic biopolymers can be produced on a large scale under controlled conditions and with predictable and reproducible mechanical properties, degradation rate and microstructure (Platt, 2006; Sinha Ray and Okamoto, 2003a).

One of the most promising synthetic biopolymers is PLA because it is made from agricultural products. PLA is not a new polymer, but recent developments in the capability to manufacture the monomer economically from agricultural products have placed this material at the forefront of the emerging biodegradable plastics industries.

PLA, PGA and their copolymers, poly(lactic acid-*co*-glycolic acid) (PLGA) are also extensively used in tissue engineering for treating patients suffering from damaged or lost organs or tissue (Ma, 2004; Langer and Vacanti, 1993). They have been demonstrated to be biocompatible, they degrade into non-toxic components and have a long history of degradable surgical sutures with gained FDA (US Food and Drug Administration) approval for clinical use. PCL and PHB are also used in tissue engineering research.

The task of tissue engineering demands a combination of molecular biology and materials engineering since, in many applications, a scaffold is needed to provide a temporary artificial matrix for cell seeding. In general, scaffolds must meet certain specifications such as high porosity, proper pore size, biocompatibility, biodegradability and proper degradation rate (Quirk et al., 2004). The scaffold must provide sufficient mechanical support to maintain stresses and loadings generated during *in-vitro* or *in-vivo* regeneration.

For some of the aforementioned applications, enhancement of the mechanical properties is often needed (Tsivintzelis et al., 2007). This could be achieved by the incorporation of nanoparticles, such as hydroxyapatite

(HA) carbon nanotubes (CNTs) and layered silicates. Polymer/layered silicate nanocomposites have recently become the focus of academic and industrial attention (Sinha Ray and Okamoto, 2003b). The introduction of small quantities of high aspect ratio nano-sized silicate particles can significantly improve the mechanical and physical properties of the polymer matrix (Lee *et al.*, 2003). However, many factors – such as the pore structure, the porosity, the crystallinity and the degradation rate – may alter the mechanical properties and, thus, the efficiency of a scaffold. As a consequence, the scaffold fabrication method should allow for the control of its pore size and shape and should enhance the maintenance of its mechanical properties and biocompatibility (Ma, 2004; Quirk *et al.*, 2004).

Many techniques have been applied for making porous scaffolds. Among the most popular are particulate leaching (Mikos *et al.*, 1993), temperature-induced phase separation (Nam and Park, 1999), phase inversion in the presence of a liquid non-solvent (Van de Witte *et al.*, 1996), emulsion freeze-drying (Whang *et al.*, 1995), electrospinning (Bognitzki *et al.*, 2001) and rapid prototyping (Ma, 2004). On the other hand, foaming of polymers using supercritical fluids is a versatile method for obtaining a porous structure (Quirk *et al.*, 2004; Goel and Beckman, 1994).

This chapter intends to highlight synthetic biopolymer-based nanocomposites used for producing porous scaffolds in tissue engineering applications. The chapter reviews current research trends on nanocomposite materials for tissue engineering, including strategies for the fabrication of nanocomposite scaffolds with highly porous and interconnected pores. The results of *in-vitro* cell culture to analyze the cell–scaffold interaction using the colonization of mesenchymal stem cells (MSCs) and degradation of the scaffolds *in-vitro* are also discussed.

17.2   Tissue engineering applications

Tissue engineering applies methods from materials engineering and life science to create artificial constructs for the regeneration of new tissue (Ma, 2004; Langer and Vacanti, 1993). Tissue engineering can create biological substitutes to repair or replace failing organs or tissues. One of the most promising approaches is to grow cells on scaffolds; such scaffolds are highly engineered structures that act as temporary support for cells, facilitating the regeneration of the target tissues without loss of the three-dimensional (3D) stable structure.

Polymeric scaffolds play a pivotal role in tissue engineering through cell seeding, proliferation and new tissue formation in three dimensions. These scaffolds have shown great promise in the research of engineering a variety of tissues. Pore size, porosity and surface area are widely recognized as important parameters for a tissue engineering scaffold. Other architectural features such as pore shape, pore wall morphology and interconnectivity between pores of the scaffolding materials are also suggested to be important for cell seeding, migration, growth, mass transport and tissue formation (Ma, 2004).

Natural scaffolds made from collagen are fast being replaced with ultraporous scaffolds made from biodegradable polymers. Biodegradable polymers are attractive candidates for scaffolding materials because they degrade as new tissues are formed, eventually leaving nothing foreign in the body. The major challenges in scaffold manufacture lie in the design and fabrication of customizable biodegradable constructs with desirable properties that promote cell adhesion and cell porosity, along with mechanical properties that match the host tissue with predictable degradation rate and biocompatibility (Ma, 2004; Mohanty et al., 2000; Langer and Vacanti, 1993).

The biocompatibility of the materials is imperative. The substrate materials should not elicit an inflammatory response nor demonstrate immunogenicity of cytotoxicity. The scaffolds must be easily sterilizable in both surface and bulk to prevent infection (Gilding and Reed, 1979). For scaffolds used in bone tissue engineering, a typical porosity of 90% with a pore diameter of ca 100 µm is required for cell penetration and a proper vascularization of the ingrown tissue (Karageorgiou and Kaplan, 2005).

Another major class of biomaterials for bone repair is bioactive ceramics such as HA and calcium phosphates (Kim et al., 2004; Hench, 1998). They show appropriate osteoconductivity and biocompatibility because of their chemical and structural similarity to the mineral phase of native bone, but are inherently brittle and have poor shape ability. For this reason, polymer/ bioactive ceramic composite scaffolds have been developed for applications in bone tissue engineering. They exhibit good bioactivity, manipulation and control microstructure in shaping to fit bone defects (Zhang and Ma, 1999).

17.3   Synthetic biopolymers and their nanocomposites for tissue engineering

The extensive research literature on nanocomposites (e.g. polymer/layered silicate) is covered in several reviews (e.g. Okada and Usuki, 2006; Gao, 2004; Sinha Ray and Okamoto, 2003b; Alexandre and Dubois, 2000). The study of nanocomposites has gained momentum. This new class of materials is now being introduced in structural applications such as gas barrier films, flame retardant products and other load-bearing applications (Okamoto, 2006a). Among these nanocomposites, biopolymer-based nancomposites or green nanocomposites are considered to be a stepping stone towards a greener and more sustainable environment. Green nanocomposites are

Table 17.1 Physical properties of synthetic biopolymers used as scaffold materials

Biopolymer	Thermal properties T_m (°C)[a] T_g (°C) [b]		Tensile modulus (GPa)	Biodegradation time (months)
PLLA	173–178	60–65	1.2–3.0	>24
PDLLA	—	55–60	1.9–2.4	12–16
PGA	225–230	35–40	5–7	3–4
PLGA (50/50)	—	50–55	1.4–2.8	Adjustable: 3–6
PCL	58–63	−60	0.4–0.6	>24
PPF	30–50	−60	2–3	>24

[a] Melting temperature.
[b] Glass transition temperature.
Reproduced with permission from Rezwan *et al.* (2006). Copyright (2006) Elsevier.

described in detailed studies and reviews elsewhere (Bordes *et al.*, 2009; Okamoto, 2006b; Sinha Ray and Bousmina, 2005).

The most often utilized synthetic biopolymers for 3D scaffolds in tissue engineering are saturated poly(α-hydroxy esters) including PLA, racemic mixtures of D,L-PLA (PDLLA) and PGA, as well as PLGA (Seal *et al.*, 2001; Jagur-Grodzinski, 1999; Kohn and Langer, 1996). These polymers degrade through hydrolysis of the ester bonds. Once degraded, the monomeric components of each polymer are removed via natural pathways. The body already contains highly regulated mechanisms for removing monomeric components of lactic and glycolic acids. PGA is converted to metabolites or eliminated by other mechanisms. PLA is cleared instead via the tricarboxylic acid cycle. The abrupt release of their acidic degradation products can cause a strong inflammatory response (Martin *et al.*, 1996). In PLA and PGA, their degradation rates and mechanical properties are controlled by changing molecular weights and copolymer composition. Table 17.1 shows an overview of the discussed biopolymers and their physical properties.

PCL degrades at a significantly slower rate than PLA, PGA or PLGA. The slow degradation makes PCL less attractive for biomedical applications, but more attractive for long-term implants and controlled release application (Ma, 2004; Pitt *et al.*, 1981). Recently, PCL has been used as a candidate polymer for bone tissue engineering, where scaffolds are required to maintain physical and mechanical properties for at least 6 months (Pektok *et al.*, 2008).

Poly(propylene fumarate) (PPF) has been developed for orthopedic and dental applications (Shi and Mikos, 2006). PPF is an unsaturated linear polyester. The degradation products are biocompatible and readily removed from the body. The double bond of the PPF main chains leads to *in situ* cross-linking, which causes a moldable composite. The preservation of the double bond and molecular weight are key for control of the final mechanical properties and degradation time.

However, PPF and other biodegradable polymers lack the mechanical strength required for tissue engineering of load-bearing bones (Mistry and Mikos, 2005). The development of composite and nanocomposite materials combining inorganic particles, e.g. apatite component (i.e. the main constituent of the inorganic phase of bone (Ma, 2004)), bioactive glasses, carbon nanostructures (e.g. nanotubes, nanofibers and graphene), and metal nanoparticles has been investigated.

17.3.1 Hydroxyapatite (HA)-based nanocomposites

HA promotes bone ingrowth and is biocompatible because around 65 wt% of bone is made of HA, $Ca_{10}(PO_4)_6(OH)_2$. Natural or synthetic HA has been intensively investigated as a major component of scaffold materials for bone tissue engineering (Knowles, 2003). The Ca/P ratio of 1.50–1.67 is the key to promoting bone regeneration. Recently, much better osteoconductive properties in HA by changing composition, size and morphology have been reported (Gay et al., 2009). Nano-sized HA (nHA) may have other special properties due to its small size and huge specific surface area. A significant increase in protein adsorption and osteoblast adhesion on nano-sized ceramic materials was reported by Webster et al. (2000).

Figure 17.1 shows the rod-shaped morphology of nHA with particle width ranging from 37 to 65 nm and length from 100 to 400 nm (Nejati et al., 2008). The compressive strength of bioceramics increases when their grain size is reduced to the nanolevel (El-Ghannam et al., 2004).

Nanocomposites based on HA particles and biopolymers have attracted much attention for their good osteoconductivity, osteoinductivity, biodegradability and high mechanical strength. Wei and Ma (2004) mimicked the size scale of HA in natural bone and showed that the incorporation of nHA improved the mechanical properties and protein adsorption of the composite scaffolds whilst maintaining high porosity and suitable microarchitecture.

Nejati et al. (2008) reported on the effect of the synthesis nHA on the scaffold's morphology and mechanical properties in poly(L-lactic acid) (PLLA)-based nanocomposites. The morphology and microstructure of the scaffolds were examined using a scanning electron microscope (SEM) (Fig. 17.2). The nanocomposite scaffold (Fig. 17.2(c) and (d)) maintained a regular internal ladder-like pore structure similar to a neat PLLA scaffold (Fig. 17.2(a) and (b)) with a typical morphology processed by thermally-induced phase separation (Nam and Park, 1999). Rod-like nHA particles are distributed within the pore walls and no aggregation appears in the pores (Fig. 17.2(e) and (f)). However, the nanocomposite exhibits little effect

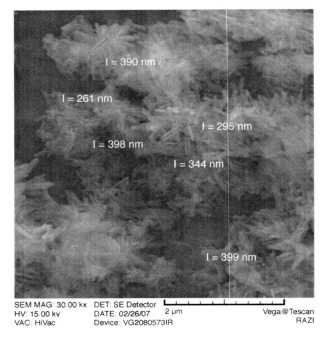

17.1 SEM micrograph of synthetic nHA rods. Reprinted with
permission from Elsevier (Nejati *et al.*, 2008).

of the nHA on the development of the pore morphology as compared with
that of neat PLLA. The pore size of neat PLLA and nanocomposite
scaffolds is in the range of 175 and 97 µm, respectively. In the case of the
microcomposite scaffold (Fig. 17.2(g) and (h)), micro-HA (mHA) particles
are randomly distributed in a PLLA matrix. Some are embedded in the pore
wall and some are piled together between or within the pores. Among the
composite and neat PLLA scaffolds, nanocomposites scaffolds showed the
highest compressive strength (8.67 MPa) with 85.1% porosity, comparable
to the high end of compressive strength of cancellous bone (2–10 MPa)
(Ramay Hassna and Zhang, 2004).

PCL/nHA nanocomposites that combine the osteoconductivity and
biocompatibility shown by HA ceramic with PCL properties have also
been prepared (Bianco *et al.*, 2009; Rezwan *et al.*, 2006; Hong *et al.*, 2005;
Wei and Ma, 2004). The structural characterization of a novel electrospun
nanocomposite and the analysis of cell response by a highly sensitive cell
(embryonic stem cell for PCL/Ca-deficient nHA system) were investigated
by Bianco *et al.* (2009). For higher Ca-deficient nHA contents (~55 wt%),
the mechanical properties significantly decreased, as did the onset
decomposition temperature and crystallinity.

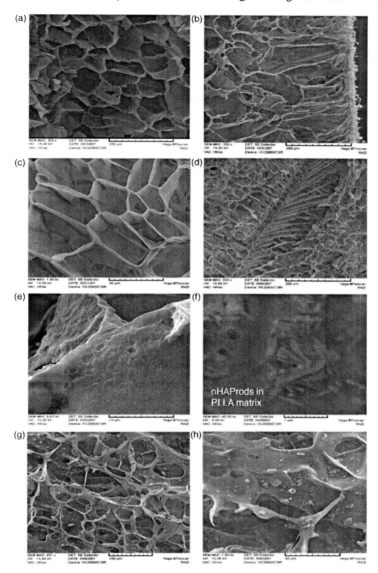

17.2 SEM micrographs of pure PLLA, PLLA/nHA and PLLA/mHAP scaffolds: (a) and (b) neat PLLA and cross-section; (c) and (d) PLLA/nHA: 50/50 and cross-section; (e) and (f) PLLA/nHAP: 50/50; (g) and (h) PLLA/mHAP: 50/50 scaffold. Reprinted with permission from Elsevier (Nejati *et al.*, 2008).

Due to the brittleness of HA and lack of interaction with the polymer matrix, ceramic nanoparticles may present deleterious effects on the mechanical properties when loaded at high amounts. Coupling agents are generally used to overcome the lack of interaction with polymer and nHA aggregation (Li *et al.*, 2008; Hong *et al.*, 2005, 2004). In order to increase the interfacial strength between PLLA and HA, and hence to increase the mechanical properties, nHA particles were surface-grafted (g-HA) with the polymer and further blended with PLLA (Li *et al.*, 2008). The PLLA/g-HA nanocomposites also demonstrated improved cell compatibility due to good biocompatibility of the nHA particles and more uniform distribution of the g-HA nanoparticles on the film surface (Li *et al.*, 2008; Hong *et al.*, 2004). These nanocomposites are of great interest to the biomedical community because the materials have a suitable structure that induces and promotes new bone formation at the required site.

Calcium phosphate biomaterials certainly posses osteoconductive properties and may bind directly to bone under certain conditions (Nejati *et al.*, 2008). Calcium phosphate materials are suitable for calcified tissue generation.

17.3.2 Metal nanoparticle-based nanocomposites

Silver (Ag) is known to have disinfecting properties and has found application in traditional medicine. Ag nanoparticles have been investigated for their antibacterial properties (Rai *et al.*, 2009) and biopolymer-embedded Ag nanoparticles have also been studied (Lee *et al.*, 2006). Nano-sized Ag permits a controlled antibacterial effect due to the high surface contact area. For PLLA-based nanocomposite fibers including Ag nanoparticles, antibacterial effects lasting longer than 20 days have been reported (Xu *et al.*, 2006). PLGA-based nanocomposites have also been reported in the scientific literature (Schneider *et al.*, 2008; Xu *et al.*, 2008). Metal nanoparticles enhance the thermal conductivity of the nanocomposites and enhance the degradation rate (Xu *et al.*, 2006). Furthermore, Ag nanoparticles change the surface wettability and roughness of the nanocomposites. For these reasons it is very difficult to control the bacterial adhesion process.

It is also important to note that Ag nanoparticles have been listed under carcinogenic materials by the World Health Organization (WHO). This requires immediate and thorough action not only from environmental and human health viewpoints, but also from the perspective of socio-economic benefits (WHO, 2007).

17.3.3 Carbon-based nanocomposites

Carbon nanostructures in a polymer matrix have been extensively investigated for biomedical applications (Harrison and Atala, 2007). Carbon nanotubes (CNTs) have the potential for biomedical scaffolds. Honeycomb-like matrices of multi-walled nanotube (MWNTs) were fabricated as potential scaffolds for tissue engineering (Mwenifumbo et al., 2007). Mouse fibroblast cells were cultured on nanotube networks, prepared by treating with an acid solution that generates carboxylic acid groups at the defect of the nanotubes. The carbon networks can be used as a biocompatible mesh for restoring or reinforcing damaged tissues because of the non-cytotoxicity of the networks. The electrical conductivity of nanocomposites, including carbon nanostructures, is a useful tool to direct cell growth because they can conduct an electrical stimulus into the tissue healing process. Osteoblast proliferation on PLLA/MWNT nanocomposites under alternating current stimulation has been investigated (Supronowicz et al., 2002). The results showed an increase in osteoblast proliferation and extra-cellular calcium deposition on the nanocomposites as compared with the control samples. Unfortunately, no comparison was made with currently used orthopedic reference material under electrical stimulation.

Shi et al. (2008) investigated in-vitro cytotoxicity of single-walled CNT (SWNT)/PPF nanocomposites. The results did not reveal any in-vitro cytotoxicity for PPF/SWNT functionalized with 4-tert-butylphenylene nanocomposites. Moreover, nearly 100% cell viability was observed on the nanocomposites and cell attachment on their surfaces was comparable with that on tissue culture polystyrene. The functional group at the CNT surface seems to play an important role in improving the dispersion of CNTs in a polymer matrix and in the mechanism of interaction with the cells. The sidewall carboxylic functionalized SWNTs exhibited a nucleation surface to induce the formation of a biomimetic apatite coating (Armentano et al., 2008).

Nanodiamonds (NDs) synthesized by detonation are one of the most promising materials for multifunctional nanocomposites for various applications, including biomedical (Mochalin et al., 2012; Zhang et al., 2011; Huang et al., 2008). To fully benefit from the advantages of NDs as nanofillers for biopolymeric bone scaffolds, they need to be dispersed into single particles. The quality of the filler dispersion in the matrix is important, because it determines the surface area of the nanoparticles available for interaction with the matrix. When adequately dispersed, NDs increase strength, toughness and thermal stability of the nanocomposites (Karbushev et al., 2008). Purified NDs are composed of particles with 5 nm average diameter. They contain an inert diamond core and include functional groups such as COOH, OH, NH_2, etc. (Mochalin et al., 2012).

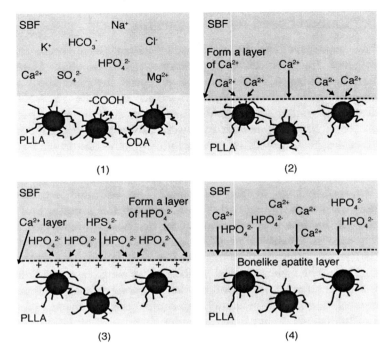

17.3 A schematic representation of a biomineralization process on PLLA/ND-ODA scaffolds in SBF. (1) The initial stage. (2) While in contact with SBF, PLLA is hydrolyzed, resulting in the formation of –COOH groups on the surface of the scaffold. Due to the degradation of PLLA, ND-ODA is exposed to SBF. The exposed –COOH groups of ND-ODA dissociate and form negatively charged –COO– on the surface. In addition, the ND-ODA may speed up the degradation of PLLA to produce more –COOH groups on the PLLA surface. The negatively charged surface attracts Ca^{2+}. (3) The deposited calcium ions, in turn, interact with phosphate ions in the SBF and form bonelike apatite. (4) The bonelike apatite then grows spontaneously, consuming the calcium and phosphate ions to form apatite clusters. Reprinted with permission from Elsevier (Zhang *et al.*, 2012).

In a very recent work, Zhang *et al.* (2012) prepared multifunctional bone scaffold materials composed of PLLA and octadecylamine-functionalized ND (ND-ODA) via solution casting followed by compression molding. Addition of 10 wt% of ND-ODA resulted in a 280% increase in the strain to failure and a 310% increase in fracture energy as compared to neat PLLA. Both of these parameters are crucial for bone tissue engineering and for the manufacture of orthopedic surgical fixation devices. The biomineralization of nanocomposite scaffolds in simulated body fluid (SBF) (Kokubo and Takadama, 2006) was tested. Apatite nucleation and growth occurred more quickly on nanocomposites than on neat PLLA (Fig 17.3) (Zhang *et al.*,

2012). The increased mechanical properties and enhanced biomineralization make PLLA/ND-ODA nanocomposites promising materials for bone surgical fixation devices and regenerative medicine.

Despite research into potential biomedical applications of carbon-based nanocomposites, there have been many published studies on the cytotoxicity of carbon nanostructures (Smart *et al.*, 2006). Some research groups detected high toxicity in both cells (Chen *et al.*, 2006; Margrez *et al.*, 2006; Nimmagadda *et al.*, 2006; Sayes *et al.*, 2006; Cui *et al.*, 2005; Jia *et al.*, 2005; Monterio-Riviere *et al.*, 2005; Shvedora *et al.*, 2003;) and animals (Huczko *et al.*, 2005; Muller *et al.*, 2005; Warheit *et al.*, 2004), and explained mechanisms to cell damage at molecular and gene expression levels (Ding *et al.*, 2005).

Exposure to single-walled carbon nanotubes (SWNTs) resulted in accelerated oxidative stress (increased free radical and peroxide generation and depletion of total antioxidant reserves), loss in cell viability and morphological alterations to cellular structure. It was concluded that these effects were a result of high levels of iron catalyst present in the unrefined SWNT. Possible dermal toxicity in handling unrefined CNT was noted. Similar dermal toxicity warnings were echoed in 2005, in a study that found MWNTs initiated an irritation response in human epidermal keratinocyte (HEK) cells (Monterio-Riviere *et al.*, 2005). Purified MWNT incubated (at doses of 0.1–0.4 mg/ml) with HEK cells for up to 48 h were observed to localize within cells (Fig. 17.4), elicit the production of the pro-inflammatory cytokine release, and decrease cell viability in a time- and dose-dependent manner.

These controversial results reported by different researchers reflect the complex material properties of SWNTs and MWNTs. In addition, different synthesis methods may produce CNTs with different diameters, lengths and impurities. The results urge caution when handling CNTs and the introduction of safety measures in laboratories should be seriously considered. Most importantly, the success of CNT technology is dependent upon the continuation of research into the toxicology of CNT and CNT-based nanocomposites. At the same time, pharmacological development must continue in parallel before providing guidelines for safe use in biomedical applications.

17.4   Three-dimensional porous scaffolds

The development of composite scaffolds is advantageous as the properties of two or more types of materials can be combined to better suit the mechanical and physiological demands of the host tissue. The tissues in the body are organized into 3D structures as a function of organs. Scaffolds with a designed microstructures provide structural support and adequate

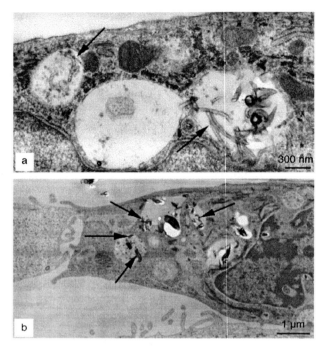

17.4 Transmission electron micrograph of human epidermal keratinocytes (HEKs): (a) intracellular localization of the MWNT – arrows depict the MWNT present within the cytoplasmic vacuoles of a HEK; (b) keratinocyte monolayer grown on a Permanox surface – arrow depicts the intracytoplasmic localization of the MWNT. Reprinted with permission from Elsevier (Smart *et al.*, 2006).

mass transport to guide tissue regeneration. To achieve the goal of tissue regeneration, scaffolds must meet certain specific needs. A high porosity and an adequate pore size are necessary to facilitate cell seeding and diffusion throughout the entire structure of both cells and nutrients (Rezwan *et al.*, 2006).

Nanocomposite 3D scaffolds based on biopolymers have been developed using different processing methods. Popular techniques include solvent casting and porogen (particulate) leaching, gas foaming, emulsion freeze-drying, electrospinning, rapid prototyping and thermally-induced phase separation (Nejati *et al.*, 2008; Kretlow and Mikos, 2007; Rezwan *et al.*, 2006; Ma, 2004; Bognitzki *et al.*, 2001; Nam and Park, 1999; Van de Witte *et al.*, 1996; Whang *et al.*, 1995; Goel and Beckman, 1994; Mikos *et al*, 1993).

17.4.1 Solvent casting and particulate leaching

Organic solvent casting and particulate leaching is an easy method that has been widely used to fabricate biocomposite scaffolds (Mikos *et al.*, 1994). This process involves the dissolution of a polymer in an organic solvent, mixing with nanofillers and porogen particles, and casting the mixture into a predefined 3D mold. The solvent is subsequently allowed to evaporate, and the porogen particles are subsequently removed by leaching (Lu *et al.*, 2000). However, residual solvents in the scaffolds may be harmful to transplanted cells or host tissues. To avoid any toxicity effect of the organic solvent, gas foaming can be used to prepare a highly porous biopolymer foam (Okamoto, 2006b). Kim *et al.* (2007) fabricated PLGA/nHA scaffolds by carbon dioxide (CO_2) foaming and solid porogen (i.e. sodium chloride crystals) leaching (GF/PL) without the use of organic solvents. Selective staining of the nHA indicated that nHA particles exposed to the scaffold surface were observed more abundantly in the GF/PL scaffold than in the conventional solvent casting and particulate leaching scaffold. The GF/PL scaffolds exhibited significant enhanced bone regeneration when compared with a conventional scaffold.

In recent work (Sakai *et al.*, 2012), highly porous cross-linked PLA scaffolds were successfully prepared through particulate leaching and foaming or simple leaching methods. The scaffolds were porous with good interconnectivity and thermal stability. SEM images confirm the pore connectivity and structural stability of the cross-linked PLA scaffold (Fig. 17.5). The scaffolds (Lait-X/b and Lait-X/c) have the same percentage of salt particulates of similar particle size; the former was turned into a scaffold through simple leaching while the latter went through a process of batch foaming followed by leaching. Qualitative evaluation of the SEM images of Lait-X/b and Lait-X/c showed a well-developed porosity and interconnectivity with pore sizes spanning a very wide range, from a few microns to hundreds of microns.

Figure 17.6 shows the relation between pore diameter and the cumulative and differential intrusions of mercury in Lait-X/b and Lait-X/c scaffolds. The maximum intrusion and interconnectivities of Lait-X/b occurred for 0.1 to 1 μm pore diameter and for Lait-X/c it was from 0.1 to 10 μm. The porosity, total intrusion volume, total pore area and median pore diameter (volume) of Lait-X/b calculated by mercury porosimetry were 43%, 0.511 ml/g, 13.4 m^2/g and 0.520 μm, respectively; for Lait-X/c, the values were 49%, 0.688 ml/g, 27.6 m^2/g and 1.26 μm. The shift in values of Lait-X/b and Lait-X/c was due to the effect of batch foaming, which led to the movement of salt particulates, resulting in increased porosity and total intrusion volume. The *in-vitro* cell culture demonstrated the ability of the scaffold to support human mesenchymal stem cells (hMSCs) adhesion,

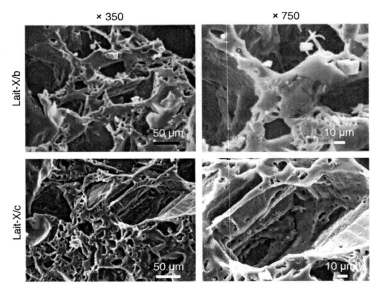

17.5 SEM images of cross-linked PLA porous scaffolds: Lait-X/b and Lait-X/c, for 350 × and 750 × magnifications. Reprinted with permission from Wiley-VCH (Sakai *et al.*, 2012).

confirming biocompatibility through the cell–scaffold interaction. The *in-vitro* degradation of the PLA thermoset scaffolds in a phosphate-buffered solution was faster for samples prepared by foaming and subsequent leaching (Sakai *et al.*, 2012). The agglomeration of the smaller crystal (solid porogen) within the 3D polymer matrix enables the creation of an interconnected pore network with well-defined pore sizes and shapes.

17.4.2 Thermally-induced phase separation (TIPS)

Three-dimensional (3D) resorbable polymer scaffolds with very high porosities (~97%) can be produced using the TIPS technique to give controlled microstructures that form scaffolds for tissues such as nerve, muscle, tendon, intestine, bone and teeth (Boccaccini and Maquet, 2003). The obtained scaffolds are highly porous with an anisotropic tubular morphology and extensive pore interconnectivity. The microporosity of TIPS-produced foams, their pore morphology, mechanical properties, bioactivity and degradation rates can be controlled by varying the polymer concentration in solution, the volume fraction of the secondary phase, quenching temperature and the polymer and solvent used (Boccaccini and Maquet, 2003).

When dioxane alone was used, the porous structure resulted from a solid–liquid phase separation of the polymer solution. During quenching, the

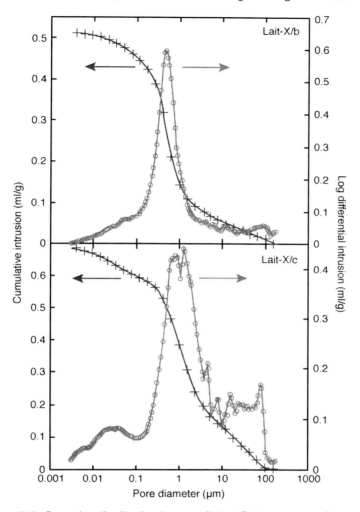

17.6 Pore size distribution in cross-linked PLA porous scaffolds: Lait-X/ b and Lait-X/c. Reprinted with permission from Wiley-VCH (Sakai *et al.*, 2012).

solvent crystallized and the polymer was expelled from the solvent crystallization front. Solvent crystals formed pores after subsequent sublimation. To better mimic the mineral component and the microstructure of natural bone, novel nHA nanocomposite scaffolds with high porosity and well-controlled pore architectures were prepared using the TIPS techniques (Fig. 17.2 (Nejati *et al.*, 2008)). The incorporation of nHA particles into PLLA solution perturbed the solvent crystallization to some extent and thereby made the pore structure more irregular and isotropic. The perturbation due to nHA particles, however, was small even in proportions of up to 50% due to their nanometer scale and uniform distribution. The SEM images showed that the nHA particles were dispersed in the pore walls of the scaffolds and bound to the polymer very well. PLLA/nHA scaffolds prepared using a pure solvent system had a regular anisotropic but open 3D pore structure similar to neat polymer scaffolds, whereas PLLA/mHA scaffolds had an isotropic irregular pore structure.

TiO$_2$ nanoparticles (nTiO$_2$) have recently been proposed as attractive fillers for biodegradable PDLLA matrices (Boccaccini et al., 2006). 3D PDLLA foams containing both nTiO$_2$ and Bioglass$^{\circledR}$ additions have been synthesized by TIPS. The foams demonstrated enhancement of bioactivity and surface nanotopography.

17.4.3 Electrospinning

The electrospinning technique has attracted great interest because it allows the production of fibrous non-woven micro/nano fabrics for tissue engineering, mainly due to structural similarity to the tissue extracellular matrix (ECM) (Bianco et al., 2009; Greiner and Wendorff, 2007). The composition and topology of the ECM was found to affect cell morphology, function and physiological response (Causa et al., 2007). Electrospun nanofibrous scaffolds, which aim to mimic the architecture and biological functions of ECM, are considered very promising substrates for tissue engineering. PCL scaffolds, a bioresorbable aliphatic polyester, have been used to provide a 3D environment for in-vitro embryonic stem cell cultures. Electrospun nanocomposite scaffolds based on bioresorbable polymers and conventional HA allow osteoblast proliferation and differentiation, and are thus considered very promising as bone scaffolding materials (Bianco et al., 2009).

Fibrous PCL/Ca-deficient nHA nanocomposites were obtained by electrospinning. The electrospun mats showed a non-woven architecture, with average fiber size of 1.5 μm, porosity of 80–90% and specific surface area of 16 m^2/g (Fig. 17.7). Murine embryonic stem (ES) cell response to neat PCL and to PCL/Ca-deficient nHA (6.4 wt%) mats was evaluated by analyzing morphological, metabolic and functional markers. Cells growing on either scaffold proliferated and maintained pluripotency markers at essentially the same rate as cells growing on standard tissue culture plates with no detectable signs of cytotoxicity, despite a lower cell adhesion at the beginning of culture. These results indicate that electrospun PCL scaffolds may provide adequate supports for murine ES cell proliferation in a pluripotent state, and that the presence of Ca-deficient nHA within the mat does not interfere with their growth.

Aligned nanocomposite fibers of PLGA/nHA were fabricated using a

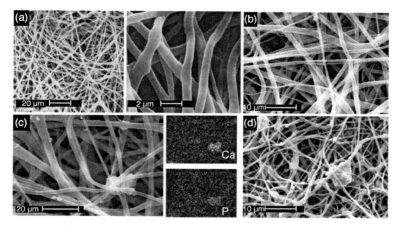

17.7 (a) SEM micrographs of neat PCL mat. (b) SEM micrograph of PCL/Ca-deficient nHA 2.0 wt%. (c) SEM micrograph and EDS mappings of PCL/Ca-deficient nHA 6.4 wt%. (d) SEM micrograph of PCL/Ca-deficient nHA 24.9 wt%. Reprinted with permission from Elsevier (Bianco *et al.*, 2009).

rotating collector by electrospinning. At low concentrations the fibres had no agglomerates and good dispersion was achieved (Jose *et al.*, 2009). The presence of well-dispersed nHA particles reduced the chain mobility and hence helped prevent shrinkage to some degree. The glass transition was affected by the incorporation of nHA into the polymer matrix, which hinders chain mobility.

Interestingly, the electrospinning technique allows the possibility of aligning conductive nanoparticles with a high aspect ratio within the polymeric fibers. Carbo nanofibers (CNFs) can orientate along the axis of electrospun fibers due to sink flow and high extension of the electrospun jet (Ago and Tobita, 2002). The CNF alignment depends on the CNF dispersion in the polymer solution. The idea involves dispersing and aligning carbon nanostructures in a polymer matrix to form highly ordered structures. The mechanical properties of PCL/CNF mats, however, were only slightly affected by CNF introduction (Armentano *et al.*, 2009b).

Ternary nanocomposite scaffolds involving three different materials have been developed (Mei *et al.*, 2007; Misra *et al*, 2007). The addition of MWNTs to the biopolymer makes for a new highly conductive material due to the 3D electrical conducting network. The results showed that combining two different nanostructures (e.g. MWNT/nHA or MWNT/Bioglass®) led to multifunctional biomaterials with tailored bioactivity, structural and mechanical integrity as well as electrical conductivity of the porous scaffolds.

17.5   *In-vitro* degradation

Since tissue engineering aims at the regeneration of new tissues, biomaterials are expected to be degradable and absorbable with a proper rate to match the speed of new tissue formation. The degradation behavior has a crucial impact on the long-term performance of a tissue-engineered cell/polymer construct. The degradation kinetics may affect a range of processes such as cell growth, tissue regeneration and host response. The mechanism of aliphatic polyester biodegradation is the bio-erosion of the material mainly determined by the surface hydrolysis of the polymer. Scaffolds can lead to heterogeneous degradation, due to neutralization of carboxylic end groups located at the surface by the external buffer solution (*in-vitro* or *in-vivo*). These phenomena reduce the acidity at the surface whereas, in the bulk, the degradation rate is enhanced by autocatalysis due to carboxylic end groups of aliphatic polyesters. In general, the amount of absorbed water depends on the diffusion coefficients of chain fragments within the polymer matrix, temperature, buffering capacity, pH, ionic strength, additions in the matrix and the medium and the processing history. Different polyesters can exhibit quite distinct degradation kinetics in aqueous solutions. For example, PGA is a stronger acid and is more hydrophilic than PLA, which is hydrophobic due to its methyl groups.

Of particular significance for application in tissue engineering are debris and crystalline by-products, as well as the acidic degradation products of PLA, PGA, PCL and their copolymers (Niiranen *et al.*, 2004). Several groups have incorporated basic compounds to stabilize the pH of the environment surrounding the polymer and to control its degradation. Bioglass® and calcium phosphates have been introduced (Rich *et al.*, 2002). Naocomposites showed a strongly enhanced polymer degradation rate when compared to the neat polymer (Dunn *et al.*, 2001). As mentioned with regard to PLLA/ND nanocomposites (Fig. 17.3), improvement of osteoconductivity of PLLA nanocomposites (i.e. the deposition of the HA crystal on the surface) was observed. Fast degradation and superior bioactivity make these nanocomposites a promising material for orthopedic applications (Dunn *et al.*, 2001).

In contrast, the degradation rates of biopolyester elastomer/MWNT nanocomposites tended to decrease with the increase of MWNT loadings (above 1 wt%) in SBF solution (Liu *et al.*, 2009). Investigation of the degradation behavior of biopolymer-based nanocomposites when nano-structures are incorporated into the matrices is to be continued.

Allen *et al.* (2008) reported on the biodegradation of SWNTs through natural, enzymatic catalysis. By incubating SWNTs with a natural horse-radish peroxidase (HRP) and a low concentration of H_2O_2 (~40 μM) at 4°C over 12 weeks under static conditions, the degradation of SWNTs proceeded. It is tempting to speculate that other peroxidases in plants and animals may be effective in oxidative degradation of CNTs. More studies are necessary to ascertain the by-products of biodegradation, as well as cellular studies for practical application.

17.6   Stem cell–scaffold interactions

Synthetic biopolymers are widely used for the preparation of porous scaffolds via different techniques. The main limitation in the use of PLA-based systems is represented by their low hydrophilicity, which causes a low affinity for the cells as compared with biological polymers. Therefore the addition of biological components to a synthetic biopolymer represents an interesting way to produce a bioactive scaffold that can be considered as a system showing both adequate mechanical stability and high cell affinity at the same time (Baek et al., 2008; Lazzeri et al., 2007; He et al., 2006). A variety of ECM protein components such as gelatin, collagen, laminin and fibronectin could be immobilized onto a plasma-treated surface of a synthetic biopolymer to enhance cellular adhesion and proliferation. Arg–Gly–Asp (RGD) is the most effective and frequently employed peptide sequence for stimulating cell adhesion on synthetic polymer surfaces. This peptide sequence is present in many ECM proteins and can interact with the integrin receptors at the focal adhesion points. Once the RGD sequence is recognized by and binds to the integrins, it will initiate an integrin-mediated cell adhesion process and active signal transduction between the cell and ECM (Zhang et al., 2009; Hersel et al., 2003). Zhang et al. (2009) reported a simple method to immobilize RGD peptide on PCL 3D scaffold surfaces. They demonstrated that rat bone marrow stromal cell (BMSC) adhesion was significantly improved on the RGD-modified PCL scaffolds in a serum-free culture condition.

Surface treatment techniques such as plasma treatment, ion sputtering, oxidation and corona discharge affect the chemical and physical properties of the surface layer without changing the bulk material properties. The effect of oxygen plasma treatments on the surface of materials has been shown to change wettability and roughness and enable selective interaction between the PLLA surface and the protein, further improving stem cell attachment (Armentano et al., 2009a).

Bone marrow derived hMSCs are an important cell source for cell therapy and tissue engineering applications. The interactions between stem cells and their environment are very complex and not fully clarified. Previous work has shown that cells respond to the mechanical properties of the scaffolds on which they are growing (Discher et al., 2005). Rohman et al. (2007) reported that PLGA and PCL are biocompatible for the growth of normal human urothelial and human bladder smooth muscle cells. Their analysis of the potential mechanism indicated that differences in degradation behaviors between polymers are not significant, but that the elastic modulus is a critical parameter, being relevant to biology at the microscopic (cellular) level and possibly also having an impact at macroscopic (tissue/organ) scales. They concluded that the elastic modulus is a property that should be considered in the development and optimization of synthetic biopolymers for tissue engineering.

Recently, MSCs have provided striking evidence that ECM elasticity influences differentiation. Indeed, multipotent cells are able to start a transdifferentiation process towards very soft tissues, such as nervous tissue, when the elastic modulus (E) of the substrate is about 0.5 kPa. Intermediate stiffness (~10 kPa) addresses cells toward a muscle phenotype and harder E (≥ 30 kPa) to cartilage/bone (Engler et al., 2006). This should address intelligent design of new biopolymers intended for specific applications (Mitragotri and Lahann, 2009). Biopolymers presently used in tissue engineering are extremely stiff. PLA has a bulk elasticity of $E \sim 1$ GPa, ten thousand times stiffer than most soft tissues. Thus, the engineering of soft tissue replacements requires biopolymers softer than those presently available.

Poly(butylene/thiodiethylene succinate) block copolymers (PBSPTDGS) were prepared by reactive blending of the parent homopolymers (PBS and PTDGS) in the presence of Ti(OBu)$_4$ (Soccioa et al., 2008). The random copolymer, characterized by the lowest crystallinity degree, exhibits the lowest elastic modulus and the highest deformation at break. When evaluated for indirect cytotoxicity, films of block PBSPTDGS30 and random PBSPTDGS240 copolymers appeared entirely biocompatible. In addition, the cellular adhesion and proliferation of H9c2 cells (Tantini et al., 2006) (derived from embryonic rat heart) seeded and grown up to 14 days in culture over the same films demonstrated that these new materials might be of interest for tissue engineering applications.

The biocompatibility of neat PLLA, PLLA/nHA and PLLA/mHA composite scaffolds were evaluated in-vitro by observing the behavior of stained MSCs cultured in close contact with scaffolds (Nejati et al., 2008). Cell growth in material-free organ cultures can be separated into four stages. Cells adhered on the surface of the composite in a round shape during the first two days. The round cell then attached, spread and proliferated into the inner pores of the scaffold, exhibiting morphologies ranging from spindle shaped to polygonal. After one week, the cells reached confluence on the material while the material-free group did not reach this status (Fig. 17.8). The representative cell culture micrographs of cell attachment into the scaffolds after seven days were observed. It was seen that the round-shaped cells attached and proliferated to the scaffold's surface, became spindle like and then migrated through the pores. The number of round-shaped cells is

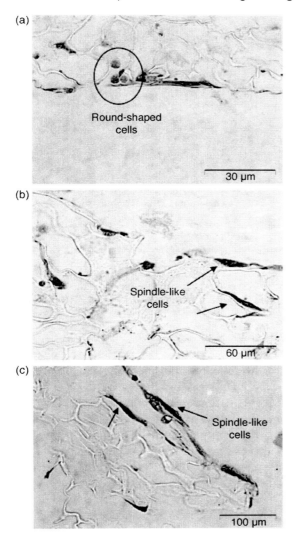

17.8 Optical microscopy photographs of colored MSCs (H&E staining) attached to (a) neat PLLA, (b) PLLA/mHAP and (c) PLLA/nHAP. Reprinted with permission from Elsevier (Nejati *et al.*, 2008).

noticeable on the surface of pure PLLA scaffold (Fig. 17.8(a)) while proliferated cells on the micro- and nanocomposite scaffolds exhibit spindle-shaped morphology (Fig. 17.8(b) and (c)). The PLLA/HA scaffolds appeared to be *in-vitro* biocompatible and non-cytotoxic to cells.

Clinical trials have demonstrated the effectiveness of cell-based therapeutic angiogenesis in patients with severe ischemic diseases, but their

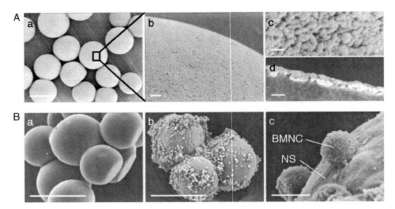

17.9 SEM image of NS (A) and marked cell adhesiveness to NS *in-vitro* (B). (A) NS microspheres approximately 100 μm in diameter (a). The NS surface uniformly coated with nano-scale hydroxyapatite (nHA) crystals was observed at different magnifications (low and high magnification in (b) and (c), respectively). SEM image of NS cross-section indicating a single layer of nHA particles on the NS surface (d). (B) Murine BMNCs incubated with LA (a) or NS (b) and, (c) at 37°C for 8 h. Large numbers of BMNCs adhered to NS ((b), and (c)) but not to LA (a). Scale bars: 100 μm (A(-a), B(-a), B(-b)); 5 μm (B(-c)); 1 μm (A(-b)); 100 nm (A(-c), A(-d)) (Mima *et al.*, 2012). Copyright 2012, Plos ONE.

success remains limited. Maintaining transplanted cells in place is expected to augment cell-based therapeutic angiogenesis.

In 2012, nHA-coated PLLA microspheres, termed a nano-scaffold (NS), were, for the first time, generated as a non-biological, biodegradable and injectable cell scaffold (Mima *et al.*, 2012). Mima *et al.* investigated the effectiveness of the NS on cell-based therapeutic angiogenesis. The NS was formed from microspheres approximately 100 μm in diameter (Fig. 17.9A (a)), the surfaces of which are coated with a monolayer of nHA particles of 50 nm diameter (Fig. 17.9A(b)–(d)). To assess the cell adhesiveness of NS, SEM was performed after incubation of NS, with bare PLLA microspheres (LA) as controls, with murine bone marrow mononuclear cells (BMNCs) at 37°C for 8 h *in-vitro*. The number of cells adhering to NS was much greater than that to LA (Fig. 17.9B(a) and (b)). High-magnification SEM images showed active cell adhesion to the NS (Fig. 17.9B(c)).

BMNCs from enhanced green fluorescent protein (EGFP)-transgenic mice and rhodamine B-containing PLLA microspheres (orange) as the scaffold core or control microspheres were implanted into the ischemic hind limbs of eight-week-old male (C57BL/6NCrSlc) mice to determine the co-localization of implanted cells with injected microspheres (Fig. 17.10A). A few implanted BMNCs were observed around the LA (Fig. 17.10A(a), while markedly larger numbers of cells were seen with the NS (Fig. 17.10A(b)) in

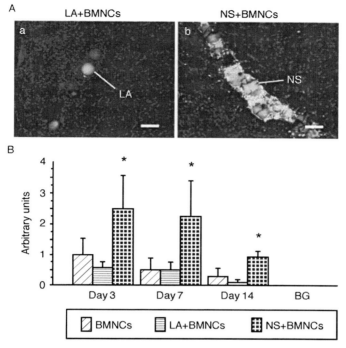

17.10 Prolonged localization of implanted BMNCs in ischemic tissues by NS. (A) Co-localization of BMNCs with NS and LA *in-vivo*. Murine BMNCs derived from EGFP-transgenic mice were transplanted together with LA or NS into the thighs in the hind limb ischemic model. Cores of NS and LA containing rhodamine B (orange) were used to indicate localization of the injected microspheres in ischemic tissues. Tissue sections 7 days after transplantation of LA+BMNCs (a) or NS+BMNCs (b) were counterstained with DAPI (blue), and merged images of DAPI, GFP and rhodamine B are shown. BMNCs (green) were observed as densely clustered around NS (b) but not LA (a). Scale bars: 100 mm. (B) Quantitative evaluation of implanted cells existing in ischemic tissues. Quantitative analysis of intramuscular GFP was performed 3, 7 and 14 days after transplantation. BMNCs were derived from EGFP-transgenic mice. BMNCs were transplanted alone or together with LA or NS into ischemic thigh muscles. Intramuscular GFP values of whole thigh muscles were corrected for total protein and expressed in arbitrary units ($n = 6$ in each group). *$P < 0.05$ for the NS+BMNCs group compared to the BMNCs alone and LA+BMNCs groups. GFP concentration in normal murine muscle was measured as background (BG) (Mima *et al.*, 2012). Copyright 2012, Plos ONE.

ischemic thigh tissue 7 days after transplantation. Intramuscular levels of GFP derived from transplanted BMNCs were consistently and significantly higher in the group injected with NS than that injected with LA or BMNCs alone at 3, 7 and 14 days after implantation, while GFP levels were not significantly different between BMNCs alone and LA + BMNC groups (Fig. 17.10B).

Kaplan–Meier analysis demonstrated that NS + BMNC markedly prevented hind limb necrosis. NS + BMNC revealed much higher induction of angiogenesis in ischemic tissues and collateral blood flow, confirmed by 3D computed tomography angiography, than those of BMNC or LA + BMNC groups (Mima et al., 2012). NS-enhanced therapeutic angiogenesis and arteriogenesis showed good correlations with increased intramuscular levels of vascular endothelial growth factor and fibroblast growth factor-2. NS co-implantation also prevented apoptotic cell death of transplanted cells, resulting in prolonged cell retention.

This nano-scaffold provides a promising local environment for implanted cells with regard to the effects on angiogenesis and arteriogenesis through cell clustering, augmented expression of proangiogenic factors, and supporting cell survival without gene manipulation or artificial ECM.

17.7   Conclusions

The synthetic biopolymer-based nanocomposites reviewed in this chapter are particularly attractive as tissue engineering scaffolds due to their biocompatibility and adjustable biodegradation kinetics. Conventional materials processing methods have been adapted through incorporation of inorganic nanoparticles into porous and interconnected 3D porous scaffolds. The incorporation of nanoparticles and the immobilization of biological components on the surface to enhance cellular adhesion and proliferation show promise and the methodology is currently under research. Current research is focused on the interaction between stromal cells and biopolymer interfaces. Synthetic biopolymer-based nanocomposite scaffolds with bioactive inorganic phases will be highly important, together with stem cell seeding.

The new approach of biopolymer-based nanocomposites enables the scaffold surface to mimic complex local biological functions and may lead in the future to in-vitro and in-vivo growth of tissues and organs.

Bioceramic entities have been used for bone tissue engineering scaffolds and drug delivery (Dvir et al., 2011; Rahaman et al., 2011; Wu et al., 2008). Osteomyelitis is a common medical condition related to bones. Caused by inflammation, it leads to bone destruction caused by infective microorganisms and bone tissue regeneration is required (Gristina et al., 1985). Although bioceramic scaffolds serve the purpose of tissue regeneration and drug release, they present formidable limitations such as a lack of information relating to long-term effects in the body. Bioceramics, especially HA, when resorbed into biological systems over the long term result in secondary fixation (70% remains in dogs after 4 months and 90% remains in humans even after 4 years) (Heisel, 1987). HA crystals released from bone scaffolds will accumulate in the joints and can stimulate inflammatory response in the prosthesis area (White *et al.*, 1972). In order to target clinical and medical applications, *in-vitro* and *in-vivo* studies are required.

17.8 References

Ago H and Tobita M (2002), 'Polymer composites of carbon nanotubes aligned by a magnetic field', *Adv Mater*, 14, 1380–1383.

Alexandre M and Dubois P (2000), 'Polymer-layered silicate nanocomposites: preparation, properties and uses of a new class of materials', *Mater Sci Eng*, 28, 1–63.

Allen BL, Kichambare PD, Gou P, Vlasova II, Kapralov AA, Konduru N, Kagan VE and Star A (2008), 'Biodegradation of single-walled carbon nanotubes through enzymatic catalysis', *Nano Lett*, 8, 3899–3903.

Armentano I, Alvarez-Pérez MA, Carmona-Rodríguez B, Gutiérrez-Ospina I, Kenny JM and Arzate H (2008), 'Analysis of the biomineralization process on SWNT-COOH and F-SWNT films', *Mater Sci Eng C*, 28, 1522–1529.

Armentano I, Ciapetti G, Pennacchi M, Dottori M, Devescovi V, Granchi D, Baldini N, Olalde B, Jurado MJ, Alava JIM and Kenny JM (2009a), 'Role of PLLA plasma surface modification in the interaction with human marrow stromal cells', *J App Polym Sci*, 114, 3602–3611.

Armentano I, Del Gaudio C, Bianco A, Dottori M, Nanni F, Fortunati E and Kenny JM (2009b), 'Processing and properties of poly(e-caprolactone)/carbon nanofibre composite mats and films obtained by electrospinning and solvent casting', *J Mater Sci*, 44, 4789–4795.

Baek HS, Park YH, Ki CS, Park JC and Rah DK (2008), 'Enhanced chondrogenic responses of articular chondrocytes onto porous silk fibroin scaffolds treated with microwave-induced argon plasma', *Surf Coat Technol*, 202, 5794–5797.

Bianco A, Di Federico E, Moscatelli I, Camaioni A, Armentano I, Campagnolo L, Dottori M, Kenny JM, Siracusa G and Gusmano G (2009), 'Electrospun poly(e-caprolactone)/Ca-deficient hydroxyapatite nanohybrids: microstructure, mechanical properties and cell response by murine embryonic stem cells', *Mater Sci Eng C*, 29, 2063–2071.

Boccaccini AR and Maquet V (2003), 'Bioresorbable and bioactive polymer/Bioglass (R) composites with tailored pore structure for tissue engineering applications', *Compos Sci Technol*, 63, 2417–2429.

Boccaccini AR, Blaker JJ, Maquet M, Chung W, Jerome R and Nazhat SN (2006), 'Poly(DL-lactide) (PDLLA) foams with TiO_2 nanoparticles and PDLLA/TiO_2-Bioglass foam composites for tissue engineering scaffolds', *J Mater Sci*, 41, 3999–4008.

Bognitzki M, Czad W, Frese T, Schaper A, Hellwig M, Steinhart M, Greiner A and Wendorff JH (2001), 'Nano-structured fibers via electrospinning', *Adv Mater*, 13, 70–72.

Bordes P, Pollet E and Averous L (2009), 'Nano-biocomposites: Biodegradable polyester/nanoclay systems', Prog Polym Sci, 34, 125–155.

Causa F, Netti PA and Ambrosio L (2007), 'A multi-functional scaffold for tissue regeneration: The need to engineer a tissue analogue', *Biomaterials*, 28, 5093–5099.

Chandra R and Rustgi R (1998), 'Biodegradable polymers', *Progr Polym Sci*, 23, 1273–1335.

Chen X, Tam UC, Czlapinski JL, Lee GS, Rabuka D, Zettle A and Bertozzi CR (2006), 'Interfacing carbon nanotubes with living cells', *J Am Chem Soc*, 128, 6292–6293.

Cui DX, Tian FR, Ozkan CS, Wang M and Gao HJ (2005), 'Effect of single wall carbon nanotubes on human HEK293 cells', *Toxicol Lett*, 155, 73–85.

Ding LH, Stilwell J, Zhang TT, Elboudwarej O, Jiang HJ, Selegue JP, Cooke PA, Gray JW and Chen FQF (2005), 'Molecular characterization of the cytotoxic mechanism of multiwall carbon nanotubes and nano-onions on human skin fibroblast', *Nano Lett*, 5, 2448–2464.

Discher DE, Janmey P and Wang YL (2005), 'Tissue cells feel and respond to the stiffness of their substrate', *Science*, 310, 1139–1143.

Dunn AS, Campbell PG and Marra KG (2001), 'The influence of polymer blend composition on the degradation of polymer/hydroxyapatite biomaterials', *J Mater Sci Mater Med*, 12, 673–677.

Dvir T, Timko BP, Brigham MD, Naik SR, Karajanagi SS, Levy O, Jin H, Parker KK, Langer R and Kohane DS (2011), 'Nanowired three-dimensional cardiac patches', *Nature Nanotech*, 6, 720–725.

El-Ghannam A, Ning CQ and Mehta J (2004), 'Cyclosilicate nanocomposite: a novel resorbable bioactive tissue engineering scaffold for BMP and bone-marrow cell delivery', *J Biomed Mater Res*, 71A, 377–390.

Engler AJ, Sen S, Sweeney HL and Disher DE (2006), 'Matrix elasticity directs stem cell lineage specification', *Cell*, 126, 677–689.

Gao F (2004), 'Clay/polymer composites: the story', *Mater Today*, 7, 50–55.

Gay S, Arostegui S and Lemaitre J (2009), 'Preparation and characterization of dense nanohydroxyapatite/PLLA composites', *Mater Sci Eng C*, 29, 172–177.

Gilding D and Reed AM (1979), 'Biodegradable polymers for use in surgery polyglycolic/poly(lactic acid) homo- and copolymers', *Polymer*, 20, 1459–1464

Goel SK and Beckman EJ (1994), 'Generation of microcellular polymeric foams using supercritical carbon dioxide. I: Effect of pressure and temperature on nucleation', *Polym Eng Sci*, 34, 1137–1147.

Greiner A and Wendorff JH (2007), 'Electrospinning: a fascinating method for the preparation of ultrathin fibers', *Angew Chem Int Ed*, 46, 5670–5703.

Gristina AG, Oga M, Webb LX and Hobgood CD (1985), 'Adherent bacterial colonization in the pathogenesis of osteomyelitis', *Science*, 228, 990–993.

Harrison BS and Atala A (2007), 'Review: carbon nanotube applications for tissue ngineering', *Biomaterials*, 28, 344–353.

He W, Yong T, Ma ZW, Inai R, Teo WE and Ramakrishna S (2006), 'Biodegradable polymer nanofiber mesh to maintain functions of endothelial cells', *Tissue Eng*, 12, 2457–2466.

Heisel J (1987), 'Animal experiment studies of callus formation by hydoxyapatite injection', *Unfallchirurgie*, 13, 179–186.

Hench LL (1998), 'Bioceramics', *J Am Ceram Soc*, 81, 1705–1728.

Hersel U, Dahmen C and Kessler H (2003), 'RGD modified polymers: biomaterials for stimulated cell adhesion and beyond', *Biomaterials*, 24, 4385–4415.

Hong Z, Zhang P, He C, Qiu X, Liu A, Chen L, Chen X and Jing X (2005), 'Nano-composite of poly(L-lactide) and surface grafted hydroxyapatite: mechanical properties and biocompatibility', *Biomaterials*, 26, 6296–6304.

Hong ZK, Qiu XY, Sun JR, Deng MX, Chen XS and Jing XB (2004), 'Grafting polymerization of L-Lactide on the surface of hydroxyapatite nano-crystals', *Polymer*, 45, 6705–6713.

Huang HJ, Pierstorff E, Osawa E and Ho D (2008), 'Protein-mediated assembly of nanodiamond hydrogels into a biocompatible and biofunctional multilayer nanofilm', *ACS Nano*, 2, 203–212.

Huczko A, Lange H, Bystrzejewski M, Baranowski P, Grubek-Jaworska H, Nejman P, Przybylowski T, Czuminska K, Glapinski J, Walton DRM and Kroto HW (2005), 'Pulmonary toxicity of 1-D nanocarbon materials', *Fuller Nanotub Carbon Nanostruct*, 13, 141–145.

Jagur-Grodzinski J (1999), 'Biomedical application of functional polymers', *React Funct Polym*, 39, 99–138.

Jia G, Wang HF, Yan L, Wang X, Pei RJ, Yan T, Zhao YL and Guo XB (2005), 'Cytotoxicity of carbon nanomaterials: Single-wall nanotube, multi-wall nanotube, and fullerene', *Environ Sci Technol*, 39, 1378–1383.

Johnson RM, Mwaikambo LY and Tucker N (2003), Biopolymers. Rapra Review Report No. 159, London, Rapra Technology Ltd.

Jose MV, Thomas V, Johnson KT, Dean DR and Nyairo E (2009), 'Aligned PLGA/HA nanofibrous nanocomposite scaffolds for bone tissue engineering', *Acta Biomater*, 5, 305–315.

Karageorgiou V and Kaplan D (2005), 'Porosity of 3D biomaterial scaffolds and osteogenesis', *Biomaterials*, 26, 5474–5491.

Karbushev VV, Konstantinov II, Parsamyan IL, Kulichikhin VG, Popov VA and George TF (2008), 'Preparation of polymer-nanodiamond composites with improved properties', *Adv Mater Res*, 59, 275–278.

Kim HW, Knowles JC and Kim HE (2004), 'Hydroxyapatite/poly(epsilon)-caprolactone) composite coating on hydroxyapatite porous bone scaffold for drug delivery', *Biomaterials*, 25, 1279–1287.

Kim SS, Ahn KM, Park MS, Lee JH, Choi CY and Kim BS (2007), 'A poly(lactide-co-glycolide)/hydroxyapatite composite scaffold with enhanced osteoconductivity', *J Biomed Mater Res Part A*, 80, 206–215.

Knowles JC (2003), 'A review article: phosphate glasses for biomedical applications', *J Mater Chem*, 13, 2395–2401.

Kohn J and Langer R (1996), 'Bioresorbable and bioerodible materials', in Ratner BD, Hoffman AS, Schoen FJ, *et al.*, editors, *Biomaterials Science: An Introduction to Materials in Medicine*, New York, Academic Press, pp. 64–72.

Kokubo T and Takadama H (2006), 'How useful is SBF in predicting in vivo bone bioactivity?', *Biomaterials*, 27, 2907–2915.

Kretlow JD and Mikos AG (2007), 'Review: mineralization of synthetic polymer scaffolds for bone tissue engineering', *Tissue Eng*, 13, 927–938.

Langer R and Vacanti JP (1993), 'Tissue engineering', *Science*, 260, 920–926

Lazzeri L, Cascone MG, Danti S, Serino LP, Moscato S and Bernardini N(2007), 'Geratine/PLLA sponge-like scaffolds: morphological and biological characterization', *J Mater Sci: Mater Med*, 18, 1399–1405.

Lee JH, Park TG, Park HS, Lee DS, Lee YK, Yoond SC, Nam JD and Jae-Do Nam

(2003), 'Thermal and mechanical characteristics of poly(l-lactic acid) nanocomposite scaffold', *Biomaterials*, 24, 2773–2778.

Lee JY, Nagahata JLR and Horiuchi S (2006), 'Effect of metal nanoparticles on thermal stabilization of polymer/metal nanocomposites prepared by a one-step dry process', *Polymer*, 47, 7970–7979.

Li J, Lu XL and Zheng YF (2008), 'Effect of surface modified hydroxyapatite on the tensile property improvement of HA/PLA composite', *Appl Surf Sci*, 255, 494–497.

Liu Q, Wu J, Tan T, Zhang L, Chen D and Tian W (2009), 'Preparation, properties and cytotoxicity evaluation of a biodegradable polyester elastomer composite', *Polym Degrad Stab*, 94, 1427–1435.

Lu L, Peter SJ, Lyman MD, Lai HL, Leite SM, Tamada JA, Vacanti JP, Langer R and Mikos AG (2000), 'In vitro degradation of porous poly(l-lactic acid) foams', *Biomaterials*, 21, 1595–1605.

Ma PX (2004), 'Scaffolds for tissue fabrication', *Mater Today*, 7, 30–40.

Margrez A, Kasas S, Salicio V, Pasquier N, Seo JW, Celio M, Catsicas S, Schwaller B and Forro L (2006), 'Cellular toxicity of carbon-based nanomaterials', *Nano Lett*, 6, 1121–1125.

Martin C, Winet H and Bao JY (1996), 'Acidity near eroding polylactidepolyglycolide in vitro and in vivo in rabbit tibial bone chambers', *Biomaterials*, 17, 2373–2380.

Mei F, Zhong J, Yang X, Ouyang X, Zhang S, Hu X, Ma Q, Lu J, Ryu S and Deng X (2007), 'Improved biological characteristics of poly(L-lactic acid) electrospun membrane by incorporation of multiwalled carbon nanotubes/hydroxyapatite nanoparticles', *Biomacromolecules*, 8, 3729–3735.

Mikos AG, Sarakinos G, Leite SM, Vacanti JP and Langer R (1993), 'Laminated three-dimensional biodegradable foams for use in tissue engineering', *Biomaterials*, 14, 323–330.

Mikos AG, Thorsen AJ, Czerwonka LA, Bao Y, Winslow DN and Vacanti JP (1994), 'Preparation and characterization of poly(l-lactic acid) foams', *Polymer*, 35, 1068–1077.

Mima Y, Fukumoto S, Koyama1 H, Okada M, Tanaka S, Shoji T, Emoto M, Furuzono T, Nishizawa Y and Inaba M (2012), 'Enhancement of cell-based therapeutic angiogenesis using a novel type of injectable scaffolds of hydroxyapatite-polymer nanocomposite microspheres', *Plos One*, 7, e35199.

Misra SK, Watts PCP, Valappil SP, Silva SRP, Roy I and Boccaccini AR (2007), 'Poly(3-hydroxybutyrate)/bioglass composite films containing carbon nanotubes', *Nanotechnology*, 18, 075701–075708.

Mistry AS and Mikos AG (2005), 'Tissue engineering strategies for bone regeneration', *Adv Biochem Eng Biotechnol*, 94, 1–22.

Mitragotri S and Lahann J (2009), 'Cell and biomolecular mechanics in silico', *Nat Mater*, 8, 15–23.

Mochalin VN, Shenderova O, Ho D and Gogotsi Y (2012), 'The properties and applications of nanodiamonds', *Nat Nano*, 7, 11–23.

Mohanty AK, Misra M and Hinrichsen G (2000), 'Biofibers. Biodegradable polymers and biocomposites: an overview', *Macromol Mater Eng*, 276, 1–24.

Monterio-Riviere NA, Nemanich RJ, Inman AO, Wang YYY and Riviere JE (2005),

'Multi-walled carbon nanotube interactions with human epidermal kerationcytes', *Toxicol Lett*, 155, 377–384.

Muller J, Huaux F, Moreau N, Misson P, Heilier JF, Delos M, Arras M, Fonseca A, Nagy JB and Lison D (2005), 'Respiratory toxicity of multi-wall carbon nanotubes', *Toxicol Appl Pharmacol*, 207, 221–231.

Mwenifumbo S, Shaffer MS and Stevens MM (2007), 'Exploring cellular behaviour with multi-walled carbon nanotube constructs', *J Mater Chem*, 17, 1894–1902.

Nam YS and Park TG (1999), 'Biodegradable polymeric microcellular foams by modified thermally induced phase separation method', *Biomaterials*, 20, 1783–1790.

Nejati E, Mirzadeh H and Zandi M (2008), 'Synthesis and characterization of nanohydroxyapatite rods/poly(L-lactide acid) composite scaffolds for bone tissue engineering', *Composites: Part A*, 39, 1589–1596.

Niiranen H, Pyhältö T, Rokkanen P, Kellomäki M and Törmälä P (2004), 'In vitro and in vivo behavior of self-reinforced bioabsorbable polymer and self-reinforced bioabsorbable polymer/bioactive glass composites', *J Biomed Mater Res A*, 69, 699–708.

Nimmagadda A, Thurston K, Nollert MU and McFetridge PSF (2006), 'Chemical modification of SWNT alters in vitro cell-SWNT interactions', *J Biomed Mater Res A*, 76, 614–625.

Okada A and Usuki A (2006), 'Twenty years of polymer-clay nanocomposites', *Macromol Mater Eng*, 291, 1449–1476.

Okamoto M (2006a), 'Recent advances in polymer/layered silicate nanocomposites: an overview from science to technology', *Mater Sci Tech*, 22, 756–779.

Okamoto M (2006b), 'Biodegradable polymer/layered silicate nanocomposites, A review', in Mallapragada S and Narasimhan B editors, *Handbook of Biodegradable Polymeric Materials and their Applications*, Los Angeles, CA, American Scientific Publishers, pp. 153–197.

Pektok E, Nottelet B, Tille JC, Gurny R, Kalangos A, Moeller M and Walpoth BH (2008), 'Degradation and healing characteristics of small-diameter poly(ε-caprolactone) vascular grafts in the rat systemic arterial', *Circulation*, 118, 2563–2570.

Pitt CG, Gratzel MM and Kimmel GL (1981), 'Aliphatic polyesters. 2. The degradation of poly(DL-lactide), poly(e-caprolactone) and their copolymers in vivo', *Biomaterials*, 2, 215–220.

Platt DK (2006), Biodegradable Polymers. Market Report, London, Rapra Technology Ltd.

Quirk RA, France RM, Shakesheff KM and Howdle SM (2004), 'Supercritical fluid technologies and tissue engineering scaffolds', *Curr Opin Solid State Mater Sci*, 8, 313–821.

Rahaman MN, Day DE, Bal BS, Fu Q, Jung SB, Bonewald LF and Tomsia AP (2011), 'Bioactive glass in tissue engineering', *Acta Biomater*, 7, 2355–2373.

Rai M, Yadav A and Gade A (2009), 'Silver nanoparticles as a new generation of antimicrobials', *Biotec Adv*, 27, 76–83.

Ramay Hassna RR and Zhang M (2004), 'Biphasic calcium phosphate nanocomposite porous scaffolds for load-bearing bone tissue engineering', *Biomaterials*, 25, 5171–5180.

Rezwan K, Chen QZ, Blaker JJ and Boccaccini AR (2006), 'Biodegradable and bioactive porous polymer/inorganic composite scaffolds for bone tissue engineering', *Biomaterials*, 27, 3413–3431.

Rich J, Jaakkola T, Tirri T, Narhi T, Yli-Urpo A and Seppala J (2002), 'In vitro evaluation of poly([var epsilon]-caprolactone-co-DL-lactide)/bioactive glass composites', *Biomaterials*, 23, 2143–2150.

Rohman G, Pettit JJ, Isaure F, Cameron NR and Southgate J (2007), 'Influence of the physical properties of two-dimensional polyester substrates on the growth of normal human urothelial and urinary smooth muscle cells in vitro', *Biomaterials*, 28, 2264–2274.

Sakai R, John B, Okamoto M, Seppälä JV, Vaithilingam J, Hussein H and Goodridge R (2012), 'Fabrication of polylactide based biodegradable thermoset scaffolds for tissue engineering applications', *Macromol Mater Eng*, doi: 10.1002/mame.201100436.

Sayes CM, Liang F, Hudson JL, Mendez J, Guo WH, Beach JM, Moore VC, Doyle CD, West JL, Billups WE, Ausman KD and Colvin VL (2006), 'Functionalization density dependence of single-walled carbon nanotubes cytotoxicity in vitro', *Toxicol Lett*, 161, 135–142.

Schneider OD, Loher S, Brunner TJ, Schmidlin P and Stark WJ (2008), 'Flexible, silver containing nanocomposites for the repair of bone defects: antimicrobial effect against E. coli infection and comparison to tetracycline containing scaffolds', *J Mater Chem*, 18, 2679–2684.

Seal BL, Otero TC and Panitch A (2001), 'Polymeric biomaterials for tissue and organ regeneration', *Mater Sci Eng: R: Rep*, 34, 147–230.

Shi X and Mikos AG (2006), 'Poly(propylene fumarate)', in Guelcher SA and Hollinger JO editors, *An Introduction to Biomaterials*, Boca Raton, FL, CRC Press, pp. 205–218.

Shi X, Sitharaman B, Pham QP, Spicer PP, Hudson JL, Wilson LJ, Tour JM, Raphael RM and Mikos AG (2008), 'In vitro cytotoxicity of single-walled carbon nanotube/biodegradable polymer nanocomposites', *J Biomed Mater Res Part A*, 86, 813–823.

Shvedova AA, Castranova V, Kisin ER, Schwegler-Berry D, Murray AR, Gandelsman VZ, Maynard A and Baronn P (2003), 'Exposure to carbon nanotube material: Assessment of nanotube cytotoxicity using human keratinocyte cells', *J Toxicol Environ Health A*, 66, 1909–1926.

Sinha Ray S and Bousmina M (2005), 'Biodegradable polymer and their layered silicate nanocomposits: In greeing the 21st century materials world', Prog Mater Sci, 50, 962–1079.

Sinha Ray S and Okamoto M (2003a), 'Biodegradable polylactide and its nanocomposites: Opening a new dimension for plastics and composites', *Macromol Rapid Commun*, 24, 815–840.

Sinha Ray S and Okamoto M (2003b), 'Polymer/layered silicate nanocomposites: a review from preparation to processing', Progr Polym Sci, 28, 1539–1641.

Smart SK, Cassady AI, Lu GQ and Martin DJ (2006), 'The biocompatibility of carbon nanotubes', *Carbon*, 44, 1034–1047.

Smith R (editor) (2005), *Biodegradable Polymers for Industrial Applications*, New York, CRC Press.

Socioa M, Lottia N, Finellia L, Gazzanob M and Munari A (2008), 'Influence of transesterification reactions on the miscibility and thermal properties of poly (butylene/diethylene succinate) copolymers', *Eur Polym J*, 44, 1722–1732.

Supronowicz PR, Ajayan PM, Ullmann KR, Arulanandam BP, Metzger DW and Bizios R (2002), 'Novel current-conducting composite substrates for exposing osteoblasts to alternating current stimulation', *J Biomed Mater Res*, 59, 499–506.

Tantini B, Fiumana E, Cetrullo S, Pignatti C, Bonavita F, Shantz LM, Giordano E, Muscari C, Flamigni F, Guarnieri C, Stefanelli C and Caldarera CM (2006), 'Involvement of polyamines in apoptosis of cardiac myoblasts in a model of simulated ischemia', *J Mol Cell Cardiol*, 40, 775–782.

Tsivintzelis I, Marras SI, Zuburtikudis I and Panayiotou C (2007), 'Porous poly(L-lactic acid) nanocomposite scaffolds prepared by phase inversion using supercritical CO_2 as antisolvent', *Polymer*, 48, 6311–6318.

Van de Witte P, Esselbrugge H, Dijkstra PJ, Van den Berg JWA and Feijen J (1996), 'Phase transitions during membrane formation of polylactides. I. A morphological study of membranes obtained from the system polylactide-chloroform-methanol', *J Membr Sci*, 113, 223–236.

Warheit DB, Laurence BR, Reed KL, Roach DH, Reynolds GAM and Webb TR (2004), 'Comparative pulmonary toxicity assessment of single-wall carbon nanotubes in rats', *Toxicol Sci*, 77, 117–125.

Webster TJ, Ergun C, Doremus RH, Siegel RW and Bizios R (2000), 'Enhanced functions of osteoblasts on nanophase ceramics', *Biomaterials*, 2, 1803–1810.

Wei G and Ma PX (2004), 'Structure and properties of nano-hydroxyapatite/polymer composite scaffolds for bone tissue engineering', *Biomaterials*, 25, 4749–4757.

Whang K, Thomas CH, Healy KE and Nuber GA (1995), 'A novel method to fabricate bioabsorbable scaffolds', *Polymer*, 36, 837–842.

White RA, Weber JN and White EW (1972), 'Replamineform: a new process for preparing porous ceramic, metal, and polymer prosthetic materials', *Science*, 176, 922–924.

WHO (2007), The International Decade for Action: Water for Life 2005–2015, http://www.who.int/water_sanitation_health/wwd7_water_scarcity_final_rev_1.pdf (accessed 27 October 2011).

Wu S, Liu X, Hu T, Chu PK, Ho JPY, Chan YL, Yeung KWK, Chu CL, Hung TF, Huo KF, Chung CY, Lu WW, Cheung KMC and Luk KDK (2008), 'A biomimetic hierarchical scaffold: natural growth of nanotitanates on three-dimensional microporous Ti-based metals', *Nano Lett*, 8, 3803–3808.

Xu X, Yang Q, Wang Y, Yu H, Chen X and Jing X (2006), 'Biodegradable electrospun poly (l-lactide) fibers containing antibacterial silver nanoparticles', *Eur Polym J*, 42, 2081–2087.

Xu X, Yang Q, Bai J, Lu T, Li Y and Jing X (2008), 'Fabrication of biodegradable electrospun poly(L-lactide-co-glycolide) fibers with antimicrobial nanosilver particles', *J Nanosci Nanotechnol*, 8, 5066–5070.

Zhang H, Lin CY and Hollister SJ (2009), 'The interaction between bone marrow stromal cells and RGD-modified three-dimensional porous polycaprolactone scaffolds', *Biomaterials*, 30, 4063–4069.

Zhang Q, Mochalin VN, Neitzel I, Hazeli K, Niu J, Kontsos A, Zhou JG, Lelkes PI and Gogotsi Y (2012), 'Mechanical properties and biomineralization of multifunctional nanodiamond-PLLA composites for bone tissue engineering', *Biomaterials*, doi:10.1016/j.biomaterials.2012.03.063.

Zhang QW, Mochalin VN, Neitzel I, Knoke IY, Han JJ, Klug CA, Zhou JG, Lelkes PI and Gogotsi Y (2011), 'Fluorescent PLLA-nanodiamond composites for bone tissue engineering', *Biomaterials*, 32, 87–94.

Zhang R and Ma PX (1999), 'Porous poly(l-lactic acid)/apatite composites created by biomimetic process', *J Biomed Mater Res*, 45, 285–293.

17.9   Appendix: abbreviations

Ag	silver
BMNC	bone marrow mononuclear cell
BMSC	bone marrow stromal cell
CNF	carbon nanofiber
CNT	carbon nanotube
CO_2	carbon dioxide
ECM	extracellular matrix
EGFP	enhanced green fluorescent protein
ES	embryonic stem
FDA	US Food and Drug Administration
GF/PL	carbon dioxide foaming and solid porogen leaching
g-HA	surface-grafted hydroxyapatite
HA	hydroxyapatite
HEK	human epidermal keratinocyte
hMSC	human mesenchymal stem cell
HRP	horseradish peroxidase
LA	bare poly(L-lactic acid) microsphere
mHA	micro-hydroxyapatite
MSC	mesenchymal stem cell
MWNT	multi-walled nanotube
ND	nanodiamond
ND-ODA	octadecylamine-functionalized nanodiamond
nHA	nano-sized hydroxyapatite
NS	nano-scaffold
$nTiO_2$	TiO_2 nanoparticles
PBS	poly(butylene succinate)
PBSPTDGS	poly(butylene/thiodiethylene succinate) block copolymer
PCL	poly(ε-caprolactone)
PDLLA	racemic mixture of D,L-poly(lactic acid)
PGA	poly(glycolic acid)
PHB	poly(hydroxy butyrate)
PLA	poly(lactic acid)
PLGA	poly(lactic acid-*co*-glycolic acid)
PLLA	poly(L-lactic acid)
PPF	poly(propylene fumarate)

PVA	polyvinyl alcohol
RGD	Arg–Gly–Asp
SBF	simulated body fluid
SEM	scanning electron microscope
SWNT	single-walled carbon nanotube
TIPS	thermally-induced phase separation
WHO	World Health Organization

Index

CPSIA information can be obtained at www.ICGtesting.com
Printed in the USA
BVOW02*0401030214

343683BV00006B/187/P

9 780857 093387